Lecture Notes in Computer Science 11628

Commenced Publication in 1973
Founding and Former Series Editors:
Gerhard Goos, Juris Hartmanis, and Jan van Leeuwen

More information about this series at http://www.springer.com/series/7407

Mikoláš Janota · Inês Lynce (Eds.)

Theory and Applications of Satisfiability Testing – SAT 2019

22nd International Conference, SAT 2019
Lisbon, Portugal, July 9–12, 2019
Proceedings

Springer

Editors
Mikoláš Janota ⓘ
University of Lisbon
Lisbon, Portugal

Inês Lynce ⓘ
University of Lisbon
Lisbon, Portugal

ISSN 0302-9743　　　　　　　ISSN 1611-3349　(electronic)
Lecture Notes in Computer Science
ISBN 978-3-030-24257-2　　　ISBN 978-3-030-24258-9　(eBook)
https://doi.org/10.1007/978-3-030-24258-9

LNCS Sublibrary: SL1 – Theoretical Computer Science and General Issues

This Springer imprint is published by the registered company Springer Nature Switzerland AG
The registered company address is: Gewerbestrasse 11, 6330 Cham, Switzerland

Preface

This volume contains the papers presented at SAT 2019: the 22nd International Conference on Theory and Applications of Satisfiability Testing held during July 9–12, 2019, in Lisbon.

The International Conference on Theory and Applications of Satisfiability Testing (SAT) is the premier annual meeting for researchers focusing on the theory and applications of the propositional satisfiability problem, broadly construed. Aside from plain propositional satisfiability, the scope of the meeting includes Boolean optimization, including MaxSAT and pseudo-Boolean (PB) constraints, quantified Boolean formulas (QBF), satisfiability modulo theories (SMT), and constraint programming (CP) for problems with clear connections to Boolean reasoning.

Many hard combinatorial problems can be tackled using SAT-based techniques, including problems that arise in formal verification, artificial intelligence, operations research, computational biology, cryptology, data mining, machine learning, mathematics, etc. Indeed, the theoretical and practical advances in SAT research over the past 25 years have contributed to making SAT technology an indispensable tool in a variety of domains.

SAT 2019 welcomed scientific contributions addressing different aspects of SAT interpreted in a broad sense, including (but not restricted to) theoretical advances (such as exact algorithms, proof complexity, and other complexity issues), practical search algorithms, knowledge compilation, implementation-level details of SAT solvers and SAT-based systems, problem encodings and reformulations, applications (including both novel application domains and improvements to existing approaches), as well as case studies and reports on findings based on rigorous experimentation.

SAT 2019 received 64 submissions, comprising 45 long papers, 15 short papers, and four tool papers. Each submission was reviewed by at least three Program Committee members. The reviewing process included an author response period, during which the authors of the submitted papers were given the opportunity to respond to the initial reviews for their submissions. To reach a final decision, a Program Committee discussion period followed the author response period. External reviewers supporting the Program Committee were also invited to participate directly in the discussion for the papers they reviewed. This year, most submissions received a meta-review, summarizing the discussion that occurred after the author response and an explanation of the final recommendation. In the end, the committee decided to accept a total of 28 papers; 19 long, seven short, and two tool papers.

The Program Committee singled out the following two submissions for the Best Paper Award and the Best Student Paper Award, respectively:

- Nikhil Vyas and Ryan Williams: "On Super Strong ETH"
- Katalin Fazekas, Armin Biere, and Christoph Scholl: "Incremental Inprocessing in SAT Solving"

In addition to presentations on the accepted papers, the scientific program of SAT included two invited talks:

- Carla Gomes (Cornell University, USA): "Computational Sustainability: Computing for a Better World"
- Josef Urban (Czech Technical University in Prague, Czech Republic): "Machine Learning in Automated and Interactive Theorem Proving"

SAT 2019 hosted various associated events.

In particular, the SAT/SMT/AR summer school was held July 3–6, organized by Alexey Ignatiev, António Morgado, Nina Narodytska, and Vasco Manquinho.

In addition, the following five workshops were held July 7–8, affiliated with SAT:

- Workshop on Satisfiability Modulo Theories, organized by Joe Hendrix and Natasha E. Sharygina
- QBF Workshop, organized by Hubie Chen, Florian Lonsing, Martina Seidl, and Friedrich Slivovsky
- Vampire Workshop, organized by Laura Kovacs and Andrei Voronkov
- Pragmatics of SAT Workshop, organized by Matti Järvisalo and Daniel Le Berre
- Workshop on Logic and Search, organized by David Mitchell

As in the previous years, the results of several competitive events were announced at SAT:

- SAT Race 2019, organized by Marijn Heule, Matti Järvisalo, and Martin Suda
- MaxSAT Evaluation 2019, organized by Fahiem Bacchus, Matti Järvisalo, and Ruben Martins
- QBFEVAL 2019, organized by Luca Pulina, Martina Seidl, and Ankit Shukla

We thank everyone who contributed to making SAT 2019 a success. We are indebted to the Program Committee members and the external reviewers, who dedicated their time to review and evaluate the submissions to the conference. We thank the authors of all submitted papers for their contributions, the SAT association for their guidance and support in organizing the conference, the EasyChair conference management system for facilitating the submission and selection of papers as well as the assembly of these proceedings. We wish to thank the workshop chair, Vasco Manquinho, and all the organizers of the SAT affiliated summer schools, workshops, and competitions. We gratefully thank the sponsors of SAT 2019: the SAT Association, for providing travel support for students attending the conference, Springer and SATALIA, for sponsoring the best paper awards for SAT 2019, the *Artificial Intelligence* journal, Instituto Superior Técnico, INESC-ID, Turismo de Lisboa, and Conserveira de Lisboa for financial and organizational support for SAT 2019. Thank you.

May 2019   Mikoláš Janota
                                                                                  Inês Lynce

Organization

Program Committee

Fahiem Bacchus	University of Toronto, Canada
Olaf Beyersdorff	Friedrich Schiller University Jena, Germany
Armin Biere	Johannes Kepler University Linz, Austria
Nikolaj Bjorner	Microsoft, USA
Maria Luisa Bonet	Universitat Politècnica de Catalunya, Spain
Sam Buss	University of California San Diego, USA
Pascal Fontaine	Université de Lorraine, CNRS, Inria, LORIA, France
Emmanuel Hebrard	LAAS, CNRS, France
Marijn Heule	The University of Texas at Austin, USA
Alexey Ignatiev	University of Lisbon, Portugal
Mikoláš Janota	INESC-ID/IST, University of Lisbon, Portugal
Jan Johannsen	Ludwig Maximilian University of Munich, Germany
Matti Järvisalo	University of Helsinki, Finland
Oliver Kullmann	Swansea University, UK
Daniel Le Berre	CNRS - Université d'Artois, France
Inês Lynce	INESC-ID/IST, University of Lisbon, Portugal
Vasco Manquinho	INESC-ID, IST/University of Lisbon, Portugal
Felip Manyà	IIIA-CSIC, Spain
Joao Marques-Silva	University of Lisbon, Portugal
Ruben Martins	Carnegie Mellon University, USA
Kuldeep S. Meel	National University of Singapore, Singapore
Alexander Nadel	Intel, Israel
Aina Niemetz	Stanford University, USA
Jakob Nordström	KTH Royal Institute of Technology, Sweden
Luca Pulina	University of Sassari, Italy
Markus Rabe	Google, USA
Roberto Sebastiani	University of Trento, Italy
Martina Seidl	Johannes Kepler University Linz, Austria
Natasha Sharygina	Università della Svizzera italiana, USI Lugano, Switzerland
Laurent Simon	Labri, Bordeaux Institute of Technology, France
Friedrich Slivovsky	Vienna University of Technology, Austria
Takehide Soh	Kobe University, Japan
Martin Suda	Czech Technical University, Czech Republic
Stefan Szeider	Vienna University of Technology, Austria
Ralf Wimmer	Concept Engineering GmbH and Albert-Ludwigs-Universität Freiburg, Germany
Christoph M. Wintersteiger	Microsoft, UK

Additional Reviewers

Asadi, Sepideh
Blicha, Martin
Bliem, Bernhard
Blinkhorn, Joshua
Chakraborty, Supratik
Chew, Leroy
Devriendt, Jo
Fichte, Johannes K.
Ge-Ernst, Aile
Giráldez-Cru, Jesus
Gocht, Stephan
Hecher, Markus
Hinde, Luke
Hjort, Håkan
Hyvarinen, Antti
Katsirelos, George
Kiesl, Benjamin
Kokkala, Janne
Koshimura, Miyuki
Levy, Jordi

Lonsing, Florian
Manthey, Norbert
Marescotti, Matteo
Morgado, Antonio
Möhle, Sibylle
Nabeshima, Hidetomo
Nötzli, Andres
Papadopoulos, Alexandre
Paxian, Tobias
Peitl, Tomáš
Razgon, Igor
Rebola Pardo, Adrian
Risse, Kilian
Ryvchin, Vadim
Scheder, Dominik
Siala, Mohamed
Sokolov, Dmitry
Soos, Mate
Trentin, Patrick

Abstracts

Computational Sustainability: Computing for a Better World

Carla P. Gomes

Department of Computer Science, Cornell University, Ithaca, USA

Computational sustainability is a new interdisciplinary research field with the overarching goal of developing computational models, methods, and tools to help manage the balance between environmental, economic, and societal needs for a sustainable future. I will provide an overview of computational sustainability, with examples ranging from wildlife conservation and biodiversity, to poverty mitigation and materials discovery for renewable energy materials. I will also highlight cross-cutting computational themes and challenges in Artificial Intelligence to address sustainability problems.

Machine Learning in Automated and Interactive Theorem Proving

Josef Urban

Czech Technical University in Prague

The talk will discuss the main methods that combine machine learning with automated and interactive theorem proving. This includes learning-based guidance of saturation-style and tableau-style automated theorem provers (ATPs), guiding tactical interactive theorem provers (ITPs) and using machine learning for selecting of relevant facts from large ITP libraries in the "hammer" linkups of ITPs with ATPs. I will also mention several feedback loops between proving and learning in this setting, and discuss some connections to SAT and related areas.

Contents

Circular (Yet Sound) Proofs

Albert Atserias[1] and Massimo Lauria[2(✉)]

[1] Universitat Politècnica de Catalunya, Barcelona, Spain
atserias@cs.upc.edu
[2] Sapienza - Università di Roma, Rome, Italy
massimo.lauria@uniroma1.it

Abstract. We introduce a new way of composing proofs in rule-based proof systems that generalizes tree-like and dag-like proofs. In the new definition, proofs are directed graphs of derived formulas, in which cycles are allowed as long as every formula is derived at least as many times as it is required as a premise. We call such proofs circular. We show that, for all sets of standard inference rules, circular proofs are sound. We first focus on the circular version of Resolution, and see that it is stronger than Resolution since, as we show, the pigeonhole principle has circular Resolution proofs of polynomial size. Surprisingly, as proof systems for deriving clauses from clauses, Circular Resolution turns out to be equivalent to Sherali-Adams, a proof system for reasoning through polynomial inequalities that has linear programming at its base. As corollaries we get: (1) polynomial-time (LP-based) algorithms that find circular Resolution proofs of constant width, (2) examples that separate circular from dag-like Resolution, such as the pigeonhole principle and its variants, and (3) exponentially hard cases for circular Resolution. Contrary to the case of circular resolution, for Frege we show that circular proofs can be converted into tree-like ones with at most polynomial overhead.

1 Introduction

In rule-based proof systems, proofs are traditionally presented as sequences of formulas, where each formula is either a hypothesis, or follows from some previous formulas in the sequence by one of the inference rules. Equivalently, such a proof can be represented by a directed acyclic graph, or *dag*, with one vertex for each formula in the sequence, and edges pointing forward from the premises to the conclusions. In this paper we introduce a new way of composing proofs: we allow cycles in this graph as long as every formula is derived at least as many times as it is required as a premise, and show that this structural condition is enough to guarantee soundness. Such proofs we call *circular*.

More formally, our definition is phrased in terms of *flow assignments*: each rule application must carry a positive integer, its *flow* or *multiplicity*, which intuitively means that in order to produce that many copies of the conclusion of the rule we must have produced at least that many copies of each of the premises first. Flow assignments induce a notion of *balance* of a formula in the

© Springer Nature Switzerland AG 2019
M. Janota and I. Lynce (Eds.): SAT 2019, LNCS 11628, pp. 1–18, 2019.
https://doi.org/10.1007/978-3-030-24258-9_1

proof, which is the difference between the number of times that the formula is produced as a conclusion and the number of times that it is required as a premise. Given these definitions, a proof-graph will be a valid circular proof if it admits a flow assignment that satisfies the following *flow-balance* condition: the only formulas of strictly negative balance are the hypotheses, and the formula that needs to be proved displays strictly positive balance. While proof-graphs with unrestricted cycles are, in general, unsound, we show that circular proofs *are* sound.

Proof Complexity of Circular Proofs. With all the definitions in place, we study the power of circular proofs from the perspective of propositional proof complexity.

For Resolution, we show that circularity *does* make a real difference. First we show that the standard propositional formulation of the pigeonhole principle has circular Resolution proofs of polynomial size. This is in sharp contrast with the well-known fact that Resolution *cannot count*, and that the pigeonhole principle is exponentially hard for (tree-like and dag-like) Resolution [16]. Second we observe that the LP-based proof of soundness of circular Resolution can be formalized in the Sherali-Adams proof system (with twin variables), which is a proof system for reasoning with polynomial inequalities that has linear programming at its base. Sherali-Adams was originally conceived as a hierarchy of linear programming relaxations for integer programs, but it has also been studied from the perspective of proof complexity in recent years.

Surprisingly, the converse simulation holds too! For deriving clauses from clauses, Sherali-Adams proofs translate efficiently into circular Resolution proofs. Moreover, both translations, the one from circular Resolution into Sherali-Adams and its converse, are efficient in terms of their natural parameters: length/size and width/degree. As corollaries we obtain for Circular Resolution all the proof complexity-theoretic properties that are known to hold for Sherali-Adams: (1) a polynomial-time (LP-based) proof search algorithm for proofs of bounded width, (2) length-width relationships, (3) separations from dag-like length and width, and (4) explicit exponentially hard examples.

Going beyond resolution we address the question of how circularity affects more powerful proof systems. For Frege systems, which operate with arbitrary propositional formulas through the standard textbook inference rules, we show that circularity adds no power: the circular, dag-like and tree-like variants of Frege polynomially simulate one another. The equivalence between the dag-like and tree-like variants of Frege is well-known [18]; here we add the circular variant to the list.

Earlier Work. While the idea of allowing cycles in proofs is not new, all the instances from the literature that we are aware of are designed for reasoning about inductive definitions, and not for propositional logic, nor for arbitrary inference-based proofs.

Shoesmith and Smiley [22] initiate the study of inference based proofs with multiple conclusions. In order to do so they introduce a graphical representation of proofs where nodes represents either formulas or inference steps, in a way similar to our definition in Sect. 2. While most of that book does not consider proof with cycles, in Sect. 10.5 they do mention briefly this possibility but they do not analyze it any further.

Niwińksi and Walukiewicz [19] introduced an infinitary tableau method for the modal μ-calculus. The proofs are regular infinite trees that are represented by finite graphs with cycles, along with a decidable *progress condition* on the cycles to guarantees their soundness. A sequent calculus version of this tableau method was proposed in [13], and explored further in [23]. In his PhD thesis, Brotherston [8] introduced a *cyclic* proof system for the extension of first-order logic with inductive definitions; see also [9] for a journal article presentation of the results. The proofs in [9] are ordinary proofs of the first-order sequent calculus extended with the rules that define the inductive predicates, along with a set of *backedges* that link equal formulas in the proof. The soundness is guaranteed by an additional *infinite descent condition* along the cycles that is very much inspired by the progress condition in Niwiński-Walukiewicz' tableau method. We refer the reader to Sect. 8 from [9] for a careful overview of the various flavours of proofs with cycles for logics with inductive definitions. From tracing the references in this body of the literature, and as far as we know, it seems that our flow-based definition of circular proofs had not been considered before.

The Sherali-Adams hierarchy of linear programming relaxations has received considerable attention in recent years for its relevance to combinatorial optimization and approximation algorithms; see the original [21], and [4] for a recent survey. In its original presentation, the Sherali-Adams hierarchy can already be thought of as a proof system for reasoning with polynomial inequalities, with the levels of the hierarchy corresponding to the degrees of the polynomials. For propositional logic, the system was studied in [11], and developed further in [3,20]. Those works consider the version of the proof system in which each propositional variable X comes with a formal *twin variable* $\bar{X}$, that is to be interpreted by the negation of X. This is the version of Sherali-Adams that we use. It was already known from [12] that this version of the Sherali-Adams proof system polynomially simulates standard Resolution, and has polynomial-size proofs of the pigeonhole principle.

2   Preliminaries

Formulas and Resolution Proofs. A *literal* is a variable X or the negation of a variable $\bar{X}$. A *clause* is a disjunction (or set) of literals, and a formula in conjunctive normal form, a *CNF formula*, is a conjunction (or set) of clauses. We use 0 to denote the empty clause. A *truth-assignment* is a mapping that assigns a truth-value *true* (1) or *false* (0) to each variable. A clause is true if one of its literal is true, and false otherwise. A CNF is true if all its clauses are true and false otherwise.

A *resolution* proof of a clause A from a CNF formula $C_1 \wedge \ldots \wedge C_m$ is a sequence of clauses $A_1, A_2, \ldots, A_r$ where $A_r = A$ and each A_i is either contained in $C_1, \ldots, C_m$ or is obtained by one of the following inference rules from earlier clauses in the sequence:

$$\frac{}{X \vee \overline{X}} \qquad \frac{C \vee X \qquad D \vee \overline{X}}{C \vee D} \qquad \frac{C}{C \vee D}. \tag{1}$$

Here C and D are clauses, and X must be some variable occurring in $C_1 \wedge \ldots \wedge C_m$. The inference rules in (1) are called *axiom*, *cut*, and *weakening*, respectively.

Resolution with Symmetric Rules. Most standard inference rules in the literature are defined to derive a single consequent formula from one or more antecedents. For standard, non-circular proofs, this is no big loss in generality. However, for the proof complexity of circular proofs a particular rule with two consequent formulas will play an important role. Consider the variant of Resolution defined through the axiom rule and the following two nicely symmetric-looking inference rules:

$$\frac{C \vee X \qquad C \vee \overline{X}}{C} \qquad \frac{C}{C \vee X \qquad C \vee \overline{X}}. \tag{2}$$

These rules are called *symmetric cut* and *symmetric weakening*, or *split*, respectively. Note the subtle difference between the symmetric cut rule and the standard cut rule in (1): in the symmetric cut rule, both antecedent formulas have the same *side formula* C. This difference is minor: an application of the non-symmetric cut rule that derives $C \vee D$ from $C \vee X$ and $D \vee \overline{X}$ may be efficiently simulated as follows: derive $C \vee X \vee D$ and $D \vee \overline{X} \vee C$ by sequences of $|D|$ and $|C|$ splits on $C \vee X$ and $D \vee \overline{X}$, respectively, and then derive $C \vee D$ by symmetric cut. Here $|C|$ and $|D|$ denote the widths of C and D. The standard weakening rule may be simulated also by a sequence of splits. Thus, Resolution may well be defined this way with little conceptual change. The choice of this form is not just due to elegance and symmetry: it makes it easier to deal with the concept of flow assignments that we introduce later, and to highlight the connection with Sherali-Adams proofs.

Resolution Complexity Measures. A *resolution refutation of $C_1 \wedge \ldots \wedge C_m$* is a resolution proof of the empty formula 0 from $C_1 \wedge \ldots \wedge C_m$. Its *length*, also called *size*, is the length of the sequence of clauses that constitutes it. The *width* of a resolution proof is the number of literals of its largest clause. If only the cut inference rule is allowed, resolution is sound and complete as a refutation system, meaning that a CNF is unsatisfiable if and only if it has a resolution refutation. Adding the axiom rule and weakening rules, resolution is sound and complete for deriving clauses from clauses.

Proofs are defined as sequences but are naturally represented through directed acyclic graphs, a.k.a. *dags*; see Fig. 1. The graph has one *formula-vertex* for each formula in the sequence (the boxes), and one *inference-vertex* for each inference step that produces a formula in the sequence (the circles). Each formula-vertex is labelled by the corresponding formula, and each inference-vertex is labelled by the corresponding instance of the corresponding inference rule. Each inference-vertex has an incoming edge from any formula-vertex that corresponds to one of its premises, and at least one outgoing edge towards the corresponding consequent formula-vertices. The *proof-graph* of a proof Π is its associated dag and is denoted $G(\Pi)$. A proof Π is *tree-like* if $G(\Pi)$ is a tree.

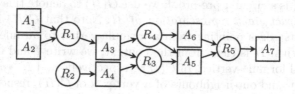

Fig. 1. A proof graph. All rules except R_4 have exactly one consequent formula; R_4 has two. All rules except R_2 have at least one antecedent formula; R_2 has none.

Sherali-Adams Proof System. Let $A_1, \ldots, A_m$ and A be polynomials on $X_1, \ldots, X_n$ and $\bar{X}_1, \ldots, \bar{X}_n$; variables X_i and $\bar{X}_i$ are twins with the intended meaning that $\bar{X}_i = 1 - X_i$. A *Sherali-Adams proof of* $A \geq 0$ *from* $A_1 \geq 0, \ldots, A_m \geq 0$ is an identity

$$\sum_{j=1}^{t} Q_j P_j = A, \tag{3}$$

where each Q_j is a non-negative linear combination of monomials on the variables $X_1, \ldots, X_n$ and $\bar{X}_1, \ldots, \bar{X}_n$, and each P_j is a polynomial among $A_1, \ldots, A_m$ or one among the following set of *basic* polynomials: $X_i - X_i^2$, $X_i^2 - X_i$, $1 - X_i - \bar{X}_i$, $X_i + \bar{X}_i - 1$, and 1. The *degree* of the proof is the maximum of the degrees of the polynomials $Q_j P_j$ in (3). The *monomial size* of the proof is the sum of the monomial sizes of the polynomials $Q_j P_j$ in (3), where the monomial size of a polynomial is the number of monomials with non-zero coefficient in its unique representation as a linear combination of monomials.

Simulation. A proof system P *polynomially simulates* another proof system P' if there is a polynomial-time algorithm that, given a proof Π' in P' as input, computes a proof Π in P, such that Π has the same goal formula and the same hypothesis formulas as Π'.

3 Circular Proofs

Circular Pre-Proofs. A *circular pre-proof* is just an ordinary proof with *backedges* that match equal formulas. More formally, a *circular pre-proof* from a set $\mathscr{H}$ of hypothesis formulas is a proof $A_1, \ldots, A_\ell$ from an augmented set of hypothesis formulas $\mathscr{H} \cup \mathscr{B}$, together with a set of *backedges* that is represented by a set $M \subseteq [\ell] \times [\ell]$ of pairs (i, j), with $j < i$, such that $A_j = A_i$ and $A_j \in \mathscr{B}$. The formulas in the set $\mathscr{B}$ of additional hypotheses are called *bud formulas*.

Just like ordinary proofs are natually represented by directed acyclic graphs, circular pre-proofs are natually represented by directed graphs; see Fig. 2. For each pair (i, j) in M there is a *backedge* from the formula-vertex of A_i to the formula-vertex of the bud formula A_j; note that $A_j = A_i$ by definition. By contracting the backedges of a circular pre-proof we get an ordinary directed graph with cycles. If Π is a circular pre-proof, we use $G(\Pi)$ to denote this graph, which we call the *compact graph representation of* Π. Note that $G(\Pi)$ is a bipartite graph with all its edges pointing from a formula-vertex to an inference-vertex, or vice-versa. When Π is clear from the context, we write I and J for the sets of inference- and formula-vertices of $G(\Pi)$, respectively, and $N^-(u)$ and $N^+(u)$ for the sets of in- and out-neighbours of a vertex u of $G(\Pi)$, respectively.

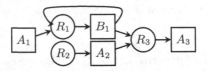

Fig. 2. The compact graph representation of a circular pre-proof.

In general, circular pre-proofs need not be sound; for an example we refer to the full version of the paper. In order to ensure soundness we need to require a global condition as defined next.

Circular Proofs. Let Π be a circular pre-proof. A *flow assignment* for Π is an assignment $F : I \to \mathbb{R}^+$ of positive real weights, or *flows*, where I is the set of inference-vertices of the compact graph representation $G(\Pi)$ of Π. The flow-extended graph that labels each inference-vertex w of $G(\Pi)$ by its flow $F(w)$ is denoted $G(\Pi, F)$. The *inflow* of a formula-vertex in $G(\Pi, F)$ is the sum of the flows of its in-neighbours. Similarly, the *outflow* of a formula-vertex in $G(\Pi, F)$ is the sum of the flows of its out-neighbours. The *balance* of a formula-vertex u of $G(\Pi, F)$ is the inflow minus the outflow of u, and is denoted $B(u)$. In symbols,

$$B(u) := \sum_{w \in N^-(u)} F(w) - \sum_{w \in N^+(u)} F(w). \tag{4}$$

The formula-vertices of strictly negative balance are the *sources* of $G(\Pi, F)$, and those of strictly positive balance are the *sinks* of $G(\Pi, F)$. We think of flow assignments as witnessing a proof of a formula that labels a sink, from the set of formulas that label the sources. Concretely, for a given set of hypothesis formulas $\mathscr{H}$ and a given goal formula A, we say that the flow assignment *witnesses a proof of A from $\mathscr{H}$* if every source of $G(\Pi, F)$ is labelled by a formula in $\mathscr{H}$, and some sink of $G(\Pi, F)$ is labelled by the formula A.

Finally, a *circular proof of A from $\mathscr{H}$* is a circular pre-proof for which there exists a flow assignment that witnesses a proof of A from $\mathscr{H}$. The *length* of a circular proof Π is the number of vertices of $G(\Pi)$, and the *size* of Π is the sum of the sizes of the formulas in the sequence. Note that this definition of size does not depend on the weights that witness the proof. The next lemma states that such weights can be found efficiently, may be assumed to be integral, and have small bit-complexity. For the proof of this lemma we refer to the full version of the paper.

Lemma 1. *There is a polynomial-time algorithm that, given a circular pre-proof Π, a finite set of hypothesis formulas $\mathcal{H}$, and a goal formula A as input, returns a flow assignment for Π that witnesses a proof of A from $\mathcal{H}$, if it exists. Moreover, if the length of the pre-proof is ℓ, then the flows returned by the algorithm are positive integers bounded by $\ell!$.*

Soundness of Circular Resolution Proofs. We still need to argue that the existence of a witnessing flow assignment guarantees soundness. In this section we develop the soundness proof for resolution as defined in (2). See Sect. 2 for a discussion on this choice of rules. The proof of the following theorem generalizes to circular proof systems based on more powerful inference rules with essentially no changes, but here we keep the discussion focused on resolution. In the full paper we develop the general case.

Theorem 1. *Let $\mathcal{H}$ be a set of hypothesis formulas and let A be a goal formula. If there is a circular resolution proof of A from $\mathcal{H}$ then every truth assignment that satisfies every formula in $\mathcal{H}$ also satisfies A.*

Proof. Fix a truth assignment α. We prove the stronger claim that, for every circular pre-proof Π from an unspecified set of hypothesis formulas, every integral flow assignment F for Π, and every sink s of $G(\Pi, F)$, if α falsifies the formula that labels s, then α also falsifies the formula that labels some source of $G(\Pi, F)$. The restriction to integral flow assignments is no loss of generality by Lemma 1, and allows a proof by induction on the sum of the flow among all inference-vertices, which we will call the total flow-sum of F.

If the total flow-sum is zero, then there are no inference-vertices, hence there are no sinks, and the statement holds vacuously. Assume then that the total flow-sum is positive, and let s be some sink of $G(\Pi, F)$ with balance $B(s) > 0$ so that the formula labels it is falsified by α. Since its balance is positive, s must have at least one in-neighbour r. Since the consequent formula of the rule at r is falsified by α, some antecedent formula of the rule at r must exist that is also falsified by α. Let u be the corresponding in-neighbour of r, and let $B(u)$ be its balance. If $B(u)$ is negative, then u is a source of $G(\Pi, F)$, and we are done. Assume then that $B(u)$ is non-negative.

Let $\delta := \min\{B(s), F(r)\}$ and note that $\delta > 0$ since $B(s) > 0$ and $F(r) > 0$. We define a new circular pre-proof Π' and an integral flow assignment F' for Π' to which we will apply the induction hypothesis. The construction will guarantee the following properties:

1. $G(\Pi')$ is a subgraph of $G(\Pi)$ with the same set of formula-vertices,
2. the total flow-sum of F' is smaller than the total flow-sum of F.

3. u is a sink of $G(\Pi', F')$ and s is not a source of $G(\Pi', F')$,
4. if t is a source of $G(\Pi', F')$, then t is a source of $G(\Pi, F)$ or an out-neighbour of r in $G(\Pi)$.

From this the claim will follow by applying the induction hypothesis to Π', F' and u. Indeed the induction hypothesis applies to them by Properties 1, 2 and the first half of 3, and it will give a source t of $G(\Pi', F')$ whose labelling formula is falsified by α. We argue that t must also be a source of $G(\Pi, F)$, in which case we are done. To argue for this, assume otherwise and apply Property 4 to conclude that t is an out-neighbour of r in $G(\Pi)$, which by the second half of Property 3 must be different from s because t is a source of $G(\Pi', F')$. Recall now that s is a second out-neighbour of r. This can be the case only if r is a split inference, in which case the formulas that label s and t must be of the form $C \vee X$ and $C \vee \overline{X}$, respectively, for appropriate formula C and variable X. But, by assumption, α falsifies the formula that labels s, let us say $C \vee X$, which means that α satisfies the formula $C \vee \overline{X}$ that labels t. This is the contradiction we were after.

It remains to construct Π' and F' that satisfy properties 1, 2, 3, and 4. We define them by cases according to whether $F(r) > B(s)$ or $F(r) \leq B(s)$, and then argue for the correctness of the construction. In case $F(r) > B(s)$, and hence $\delta = B(s)$, let Π' be defined as Π without change, and let F' be defined by $F'(r) := F(r) - \delta$ and $F'(w) := F(w)$ for every other $w \in I \setminus \{r\}$. Obviously Π' is still a valid pre-proof and F' is a valid flow assignment for Π' by the assumption that $F(r) > B(s) = \delta$. In case $F(r) \leq B(s)$, and hence $\delta = F(r)$, let Π' be defined as Π with the inference-step that labels r removed, and let F' be defined by $F'(w) := F(w)$ for every $w \in I \setminus \{r\}$. Note that in this case Π' is still a valid pre-proof but perhaps from a larger set of hypothesis formulas.

In both cases the proof of the claim that Π' and F' satisfy Properties 1, 2, 3, and 4 is the same. Property 1 is obvious in both cases. Property 2 follows from the fact that the total flow-sum of F' is the total flow-sum of F minus δ, and $\delta > 0$. The first half of Property 3 follows from the fact that the balance of u in $G(\Pi', F')$ is $B(u) + \delta$, while $B(u) \geq 0$ by assumption and $\delta > 0$. The second half of Property 3 follows from the fact that the balance of s in $G(\Pi', F')$ is $B(s) - \delta$, while $B(s) \geq \delta$ by choice of δ. Property 4 follows from the fact that the only formula-vertices of $G(\Pi', F')$ of balance smaller than that in $G(\Pi, F)$ are the out-neighbours of r. This completes the proof of the claim, and of the theorem. □

4   Circular Resolution

In this section we investigate the power of Circular Resolution. Recall from the discussion in Sect. 2 that Resolution is traditionally defined to have cut as its only rule, but that an essentially equivalent version of it is obtained if we define it through symmetric cut, split, and axiom, still all restricted to clauses. This more liberal definition of Resolution, while staying equivalent vis-a-vis the tree-like and dag-like versions of Resolution, will play an important role for the circular version of Resolution.

In this section we show that circular Resolution can be exponentially stronger than dag-like Resolution. Indeed, we show that Circular Resolution is polynomially equivalent with the Sherali-Adams proof system, which is already known to be stronger than dag-like Resolution:

Theorem 2. *Sherali-Adams and Circular Resolution polynomially simulate each other. Moreover, the simulation one way converts degree into width (exactly), and the simulation in the reverse way converts width into degree (also exactly).*

For the statement of Theorem 2 to even make sense, Sherali-Adams is to be understood as a proof system for deriving clauses from clauses, under an appropriate encoding.

Pigeonhole Principles. Let G be a bipartite graph with vertex bipartition (U, V), and set of edges $E \subseteq U \times V$. For a vertex $w \in U \cup V$, we write $N_G(w)$ to denote the set of neighbours of w in G, and $\deg_G(w)$ to denote its degree. The Graph Pigeonhole Principle of G, denoted G-PHP, is a CNF formula that has one variable $X_{u,v}$ for each edge (u,v) in E and the following set clauses:

$$X_{u,v_1} \vee \cdots \vee X_{u,v_d} \text{ for } u \in U \text{ with } N_G(u) = \{v_1, \ldots, v_d\},$$
$$\overline{X_{u_1,v}} \vee \overline{X_{u_2,v}} \quad \text{for } u_1, u_2 \in U, \ u_1 \neq u_2, \text{ and } v \in N_G(u_1) \cap N_G(u_2).$$

If $|U| > |V|$, and in particular if $|U| = n + 1$ and $|V| = n$, then G-PHP is unsatisfiable by the pigeonhole principle. For $G = K_{n+1,n}$, the complete bipartite graph with sides of sizes $n + 1$ and n, the formula G-PHP is the standard CNF encoding PHP_n^{n+1} of the pigeonhole principle.

Even for certain constant degree bipartite graphs with $|U| = n + 1$ and $|V| = n$, the formulas are hard for Resolution.

Theorem 3 [5,16]. *There are families of bipartite graphs $(G_n)_{n \geq 1}$, where G_n has maximum degree bounded by a constant and vertex bipartition (U, V) of G_n that satisfies $|U| = n + 1$ and $|V| = n$, such that every Resolution refutation of G_n-PHP has width $\Omega(n)$ and length $2^{\Omega(n)}$. Moreover, this implies that every Resolution refutation of PHP_n^{n+1} has length $2^{\Omega(n)}$.*

In contrast, we show that these formulas have Circular Resolution refutations of polynomial length and, simultaneously, constant width. This result already follows from Theorem 2 plus the fact that Sherali-Adams has short refutations for G-PHP, of degree proportional to the maximum degree of G. Here we show a self-contained separation of Resolution from Circular Resolution that does not rely of Sherali-Adams.

Theorem 4. *For every bipartite graph G of maximum degree d with bipartition (U, V) such that $|U| > |V|$, there is a Circular Resolution refutation of G-PHP of length polynomial in $|U| + |V|$ and width d. In particular, PHP_n^{n+1} has a Circular Resolution refutation of polynomial length.*

Proof. We are going to build the refutation in pieces. Concretely, for every $u \in U$ and $v \in V$, we describe a Circular Resolution proof $\Pi_{u \to}$ and $\Pi_{\to v}$, with their associated flow assignments. These proofs will have width bounded by $\deg_G(u)$ and $\deg_G(v)$, respectively, and size polynomial in $\deg_G(u)$ and $\deg_G(v)$, respectively. Moreover, the following properties will be ensured:

1. The proof-graph of $\Pi_{u \to}$ contains a formula-vertex labelled by the empty clause 0 with balance $+1$ and a formula-vertex labelled $\overline{X_{u,v}}$ with balance -1 for every $v \in N_G(u)$; any other formula-vertex that has negative balance is labelled by a clause of G-PHP.
2. The proof-graph $\Pi_{\to v}$ contains a formula-vertex labelled by the empty clause 0 with balance -1 and a formula-vertex labelled by $\overline{X_{u,v}}$ with balance $+1$ for every $u \in N_G(v)$; any other formula-vertex that has negative balance is labelled by a clause of G-PHP.

We join these pieces by identifying the two formula-vertices labeled $\overline{X_{u,v}}$, for each $\{u, v\} \in E(G)$. Each such vertex gets balance -1 from $\Pi_{u \to}$ and $+1$ from $\Pi_{\to v}$, thus its balance is zero. The empty clause occurs $|U|$ times on formula vertices with balance $+1$, and $|V|$ times on formula vertices with balance -1. Now we add back-edges from $|V|$ of the $|U|$ formula-vertices with balance $+1$ to all the formula-vertices with balance -1, forming a matching. The only remaining formula-vertices labeled by 0 have positive balance. Now in the proof all the formula-vertices that have negative balance are clauses of G-PHP, and the empty clause 0 has positive balance. This is indeed a Circular Resolution refutation of G-PHP. See Fig. 3 for a diagram of the proof for PHP_3^4.

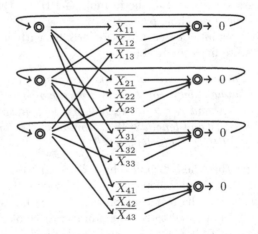

Fig. 3. The diagram of the circular proof of PHP_3^4. The double circles indicate multiple inferences. The empty clause 0 is derived four times and used only three times.

For the construction of $\Pi_{u \to}$, rename the neighbours of u as $1, 2, \ldots, \ell$. Let C_j denote the clause $X_{u,1} \vee \cdots \vee X_{u,j}$ and note that C_ℓ is a clause of G-PHP. Split

$\overline{X_{u,\ell}}$ on $X_{u,1}$, then on $X_{u,2}$, and so on up to $X_{u,\ell-1}$ until we produce $\overline{X_{u,\ell}} \vee C_{\ell-1}$. Then resolve this clause with C_ℓ to produce $C_{\ell-1}$. The same construction starting at $\overline{X_{u,\ell-1}}$ and $C_{\ell-1}$ produces $C_{\ell-2}$. Repeating ℓ times we get down to the empty clause.

For the construction of $\Pi_{\to v}$ we need some more work. Again rename the neighbours of v as $1, 2, \ldots, \ell$. We define a sequence of proofs $\Pi_1, \ldots, \Pi_\ell$ inductively. The base case Π_1 is just one application of the split rule to the empty clause to derive $X_{1,v}$ and $\overline{X_{1,v}}$, with flow 1. Proof Π_{i+1} is built using Π_i as a component. Let $X_{i+1,v} \vee \Pi_i$ denote the proof that is obtained from adding the literal $X_{i+1,v}$ to every clause in Π_i. First we observe that $X_{i+1,v} \vee \Pi_i$ has balance -1 on $X_{i+1,v}$ and balance $+1$ on, among other clauses, $X_{i+1,v} \vee \overline{X_{j,v}}$ for $j = 1, \ldots, i$. Each such clause can be resolved with clause $\overline{X_{i+1,v}} \vee \overline{X_{j,v}}$ to produce the desired clauses $\overline{X_{j,v}}$ with balance $+1$. Splitting the empty clause on variable $X_{i+1,v}$ would even out the balance of the formula-vertex labelled by $X_{i+1,v}$ and produce a vertex labelled by $\overline{X_{i+1,v}}$ of balance $+1$. Take $\Pi_\ell = \Pi_{\to v}$. $\qquad\square$

Equivalence with Sherali-Adams. In this section we prove each half of Theorem 2 in a separate lemma. We need some preparation. Fix a set of variables $X_1, \ldots, X_n$ and their twins $\bar{X}_1, \ldots, \bar{X}_n$. For a clause $C = \bigvee_{j \in Y} X_j \vee \bigvee_{j \in Z} \overline{X_j}$ with $Y \cap Z = \emptyset$, define

$$T(C) := - \prod_{j \in Y} \bar{X}_j \prod_{j \in Z} X_j, \tag{5}$$

Observe that a truth assignment satisfies C if and only if the corresponding 0–1 assignment for the variables of $T(C)$ makes the inequality $T(C) \geq 0$ true. There is an alternative encoding of clauses into inequalities that is sometimes used. Define $L(C) := \sum_{j \in Y} X_j + \sum_{j \in Z} \bar{X}_j - 1$, and observe that a truth assignment satisfies C if and only if the corresponding 0–1 assignment makes the inequality $L(C) \geq 0$ true. We state the results of this section for the T-encoding of clauses, but the same result would hold for the L-encoding because there are efficient SA proofs of $T(C) \geq 0$ from $L(C) \geq 0$, and vice-versa. We will use the following lemma, which is a variant of Lemma 4.4 in [3]:

Lemma 2. *Let $w \geq 2$ be an integer, let C be a clause with at most w literals, let D be a clause with at most $w - 1$ literals, and let X be a variable that does not appear in D. Then the following four inequalities have Sherali-Adams proofs (from nothing) of constant monomial size and degree w:*

1. $T(X \vee \overline{X}) \geq 0$,
2. $-T(D \vee \overline{X}) - T(D \vee X) + T(D) \geq 0$,
3. $-T(D) + T(D \vee \overline{X}) + T(D \vee X) \geq 0$,
4. $-T(C) \geq 0$.

Proof. Let $D = \bigvee_{i \in Y} X_i \vee \bigvee_{j \in Z} X_j$ and $C = \bigvee_{i \in Y'} X_i \vee \bigvee_{j \in Z'} X_j$. Then

1. $T(X \vee \overline{X}) = (1 - X - \bar{X}) \cdot X + (X^2 - X)$,

2. $-T(D \vee \overline{X}) - T(D \vee X) + T(D) = (X + \overline{X} - 1) \cdot \prod_{i \in Y} \overline{X}_i \prod_{j \in Z} X_j,$
3. $-T(D) + T(D \vee \overline{X}) + T(D \vee X) = (1 - X - \overline{X}) \cdot \prod_{i \in Y} \overline{X}_i \prod_{j \in Z} X_j,$
4. $-T(C) = 1 \cdot \prod_{i \in Y'} \overline{X}_i \prod_{j \in Z'} X_j.$

The claim on the monomial size and the degree follows. $\square$

Lemma 3. *Let $A_1, \ldots, A_m$ and A be non-tautological clauses. If there is a Circular Resolution proof of A from $A_1, \ldots, A_m$ of length s and width w, then there is a Sherali-Adams proof of $T(A) \geq 0$ from $T(A_1) \geq 0, \ldots, T(A_m) \geq 0$ of monomial size $O(s)$ and degree w.*

Proof. Let Π be a Circular Resolution proof of A from $A_1, \ldots, A_m$, and let F be the corresponding flow assignment. Let I and J be the sets of inference- and formula-vertices of $G(\Pi)$, and let $B(u)$ denote the balance of formula-vertex $u \in J$ in $G(\Pi, F)$. For each formula-vertex $u \in J$ labelled by formula A_u, define the polynomial $P_u := T(A_u)$. For each inference-vertex $w \in I$ labelled by rule R, with sets of in- and out-neighbours N^- and N^+, respectively, define the polynomial

$P_w := T(A_a)$ if R = axiom with $N^+ = \{a\}$,
$P_w := -T(A_a) - T(A_b) + T(A_c)$ if R = cut with $N^- = \{a, b\}$ and $N^+ = \{c\}$,
$P_w := -T(A_a) + T(A_b) + T(A_c)$ if R = split with $N^- = \{a\}$ and $N^+ = \{b, c\}$.

By double counting, the following polynomial identity holds:

$$\sum_{u \in J} B(u) P_u = \sum_{w \in I} F(w) P_w. \tag{6}$$

Let s be the sink of $G(\Pi, F)$ that is labelled by the derived clause A. Since $B(s) > 0$, Eq. (6) rewrites into $\sum_{w \in I} F(w)/B(s) P_w + \sum_{u \in J \setminus \{s\}} -B(u)/B(s) P_u = P_s$. We claim that this identity is a legitimate Sherali-Adams proof of $T(A) \geq 0$ from the inequalities $T(A_1) \geq 0, \ldots, T(A_m) \geq 0$. First, $P_s = T(A_s) = T(A)$, i.e. the right-hand side is correct. Second, each term $(F(w)/B(s))P_w$ for $w \in I$ is a sum of legitimate terms of a Sherali-Adams proof by the definition of P_w and Parts 1, 2 and 3 of Lemma 2. Third, since each source $u \in I$ of $G(\Pi, F)$ has $B(u) < 0$ and is labelled by a formula in $A_1, \ldots, A_m$, the term $(-B(u)/B(s))P_u$ of a source $u \in I$ is a positive multiple of $T(A_u)$ and hence also a legitimate term of a Sherali-Adams proof from $T(A_1) \geq 0, \ldots, T(A_m) \geq 0$. And forth, since each non-source $u \in I$ of $G(\Pi, F)$ has $B(u) \geq 0$, each term $(-B(u)/B(s))P_u$ of a non-source $u \in I$ is a sum of legitimate terms of a Sherali-Adams proof by the definition of P_u and Part 4 of Lemma 2. The monomial size and degree of this Sherali-Adams proof are as claimed. $\square$

Lemma 4. *Let $A_1, \ldots, A_m$ and A be non-tautological clauses. If there is a Sherali-Adams proof of $T(A) \geq 0$ from $T(A_1) \geq 0, \ldots, T(A_m) \geq 0$ of monomial size s and degree d, then there is a Circular Resolution proof of A from $A_1, \ldots, A_m$ of length $O(s)$ and width d.*

Proof. Fix a Sherali-Adams proof of $T(A) \geq 0$ from $T(A_1) \geq 0, \ldots, T(A_m) \geq 0$, say $\sum_{j=1}^{t} Q_j P_j = T(A)$, where each Q_j is a non-negative linear combination of monomials on the variables $X_1, \ldots, X_n$ and $\bar{X}_1, \ldots, \bar{X}_n$, and each P_j is a polynomial from among $T(A_1), \ldots, T(A_m)$ or from among the list of basic polynomials from the definition of Sherali-Adams in Sect. 2.

Our goal is to massage the proof until it becomes a Circular Resolution proof in disguise. Towards this, as a first step, we claim that the proof can be transformed into a *normalized proof* of the form $\sum_{j=1}^{t'} Q'_j P'_j = T(A)$ that has the following properties: (1) each Q'_j is a positive multiple of a multilinear monomial, and $Q'_j P'_j$ is multilinear, and (2) each P'_j is a polynomial among $T(A_1), \ldots, T(A_m)$, or among the polynomials in the set $\{-X_i \bar{X}_i, 1 - X_i - \bar{X}_i, X_i + \bar{X}_i - 1 : i \in [n]\} \cup \{1\}$. Comparing the list of Boolean axioms in (2) with the original list of basic polynomials in the definition of Sherali-Adams, note that we have replaced the polynomials $X_i - X_i^2$ and $X_i^2 - X_i$ by $-X_i \bar{X}_i$. Note also that, by splitting the Q_j's into their terms, we may assume without loss of generality that each Q_j is a positive multiple of a monomial on the variables $X_1, \ldots, X_n$ and $\bar{X}_1, \ldots, \bar{X}_n$.

In order to prove the claim we rely on the well-known fact that each real-valued function over Boolean domain has a unique representation as a multilinear polynomial. With this fact in hand, it suffices to convert each $Q_j P_j$ in the left-hand side of the proof into a $Q'_j P'_j$ of the required form (or 0), and check that $Q_j P_j$ and $Q'_j P'_j$ are equivalent over the 0–1 assignments to its variables (without relying on the constraint that $\bar{X}_i = 1 - X_i$). The claim will follow from the fact that $T(A)$ is multilinear since, by assumption, A is non-tautological.

We proceed to the conversion of each $Q_j P_j$ into a $Q'_j P'_j$ of the required form. Recall that we assumed already, without loss of generality, that each Q_j is a positive multiple of a monomial. The multilinearization of a monomial Q_j is the monomial $M(Q_j)$ that results from replacing every factor Y^k with $k \geq 2$ in Q_j by Y. Obviously Q_j and $M(Q_j)$ agree on 0–1 assignments, but replacing each Q_j by $M(Q_j)$ is not enough to guarantee the normal form that we are after. We need to proceed by cases on P_j.

If P_j is one of the polynomials among $T(A_1), \ldots, T(A_m)$, say $T(A_i)$, then we let Q'_j be $M(Q_j)$ with every variable that appears in A_i deleted, and let P'_j be $T(A_i)$ itself. It is obvious that this works. If P_j is $1 - X_i - \bar{X}_i$, then we proceed by cases on whether Q_j contains X_i or $\bar{X}_i$ or both. If Q_j contains neither X_i nor $\bar{X}_i$, then the choice $Q'_j = M(Q_j)$ and $P'_j = P_j$ works. If Q_j contains X_i or $\bar{X}_i$, call it Y, but not both, then the choice $Q'_j = M(Q_j)/Y$ and $P'_j = -X_i \bar{X}_i$ works. If Q_j contains both X_i and $\bar{X}_i$, then the choice $Q'_j = M(Q_j)/(X_i \bar{X}_i)$ and $P'_j = -X_i \bar{X}_i$ works. If P_j is $X_i + \bar{X}_i - 1$, then again we proceed by cases on whether Q_j contains X_i or $\bar{X}_i$ or both. If Q_j contains neither X_i nor $\bar{X}_i$, then the choice $Q'_j = M(Q_j)$ and $P'_j = P_j$ works. If Q_j contains X_i or $\bar{X}_i$, call it Y, but not both, then the choice $Q'_j = M(Q_j)\bar{Y}$ and $P'_j = 1$ works. If Q_j contains both X_i and $\bar{X}_i$, then the choice $Q'_j = M(Q_j)$ and $P'_j = 1$ works. If P_j is the polynomial 1, then the choice $Q'_j = M(Q_j)$ and $P'_j = 1$ works. Finally, if P_j is of the form $X_i - X_i^2$ or $X_i^2 - X_i$, then we replace $Q_j P_j$ by 0. Observe that in this case $Q_j P_j$ is always 0 over 0–1 assignments, and the conversion is correct. This completes the proof that the normalized proof exists.

It remains to be seen that the normalized proof is a Circular Resolution proof in disguise. For each $j \in [m]$, let a_j and M_j be the positive real and the multilinear monomial, respectively, such that $Q'_j = a_j \cdot M_j$. Let also C_j be the unique clause on the variables $X_1, \ldots, X_n$ such that $T(C_j) = -M_j$. Let $[t']$ be partitioned into five sets $I_0 \cup I_1 \cup I_2 \cup I_3 \cup I_4$ where

1. I_0 is the set of $j \in [t']$ such that $P'_j = T(A_{i_j})$ for some $i_j \in [m]$,
2. I_1 is the set of $j \in [t']$ such that $P'_j = -X_{i_j} \bar{X}_{i_j}$ for some $i_j \in [n]$,
3. I_2 is the set of $j \in [t']$ such that $P'_j = 1 - X_{i_j} - \bar{X}_{i_j}$ for some $i_j \in [n]$,
4. I_3 is the set of $j \in [t']$ such that $P'_j = X_{i_j} + \bar{X}_{i_j} - 1$ for some $i_j \in [n]$,
5. I_4 is the set of $j \in [t']$ such that $P'_j = 1$.

Define new polynomials P''_j as follows:

$P''_j := T(C_j \vee A_{i_j})$ for $j \in I_0$,
$P''_j := T(C_j \vee \overline{X_{i_j}} \vee X_{i_j})$ for $j \in I_1$,
$P''_j := -T(C_j) + T(C_j \vee \overline{X_{i_j}}) + T(C_j \vee X_{i_j})$ for $j \in I_2$,
$P''_j := -T(C_j \vee \overline{X_{i_j}}) - T(C_j \vee X_{i_j}) + T(C_j)$ for $j \in I_3$,
$P''_j := T(C_j)$ for $j \in I_4$.

With this notation, the normalized proof rewrites into

$$\sum_{j \in I_0} a_j P''_j + \sum_{j \in I_1} a_j P''_j + \sum_{j \in I_2} a_j P''_j + \sum_{j \in I_3} a_j P''_j = T(A) + \sum_{j \in I_4} a_j P''_j. \qquad (7)$$

Finally we are ready to construct the circular proof. We build it by listing the inference-vertices with their associated flows, and then we identify together all the formula-vertices that are labelled by the same clause.

Intuitively, I_0's are weakenings of hypothesis clauses, I_1's are weakenings of axioms, I_2's are cuts, and I_3's are splits. Formally, each $j \in I_0$ becomes a chain of $|C_j|$ many split vertices that starts at the hypothesis clause A_{i_j} and produces its weakening $C_j \vee A_{i_j}$; all split vertices in this chain have flow a_j. Each $j \in I_1$ becomes a sequence that starts at one axiom vertex that produces $X_{i_j} \vee \overline{X_{i_j}}$ with flow a_j, followed by a chain of $|C_j|$ many split vertices that produces its weakening $C_j \vee X_{i_j} \vee \overline{X_{i_j}}$; all split vertices in this chain also have flow a_j. Each $j \in I_2$ becomes one cut vertex that produces C_j from $C_j \vee X_{i_j}$ and $C_j \vee \overline{X_{i_j}}$ with flow a_j. And each $j \in I_3$ becomes one split vertex that produces $C_j \vee X_{i_j}$ and $C_j \vee \overline{X_{i_j}}$ from C_j with flow a_j.

This defines the inference-vertices of the proof graph. The construction is completed by introducing one formula-vertex for each different clause that is an antecedent or a consequent of these inference-vertices. The construction was designed in such a way that Eq. (7) is the proof that, in this proof graph and its associated flow assignment, the following hold (see the full version for details): (1) there is a sink with balance 1 and that is labelled by A, (2) all sources are labelled by formulas among $A_1, \ldots, A_m$, and (3) all other formula-vertices have non-negative balance. This proves that the construction is a correct Circular Resolution proof. The claim that the length of this proof is $O(s)$ and its width is d follows by inspection.   □

5   Further Remarks

One immediate consequence of Theorem 2 is that there is a polynomial-time algorithm that automates the search for Circular Resolution proofs of bounded width:

Corollary 1. *There is an algorithm that, given an integer parameter w and a set of clauses $A_1, \ldots, A_m$ and A with n variables, returns a width-w Circular Resolution proof of A from $A_1, \ldots, A_m$, if there is one, and the algorithm runs in time polynomial in m and n^w.*

The proof-search algorithm of Corollary 1 relies on linear programming because it relies on our translations to and from Sherali-Adams, whose automating algorithm does rely on linear programming [21]. Based on the fact that the number of clauses of width w is about n^w, a direct proof of Corollary 1 is also possible, but as far as we see it still relies on linear programming for finding the flow assignment. It remains as an open problem whether a more combinatorial algorithm exists for the same task.

Another consequence of the equivalence with Sherali-Adams is that Circular Resolution has a length-width relationship in the style of the one for Dag-like Resolution [5]. This follows from Theorem 2 in combination with the size-degree relationship that is known to hold for Sherali-Adams (see [1,20]). Combining this with the known lower bounds for Sherali-Adams (see [15,20]), we get the following:

Corollary 2. *There are families of 3-CNF formulas $(F_n)_{n\geq1}$, where F_n has $O(n)$ variables and $O(n)$ clauses, such that every Circular Resolution refutation of F_n has width $\Omega(n)$ and size $2^{\Omega(n)}$.*

It should be noticed that, unlike the well-known observation that tree-like and dag-like width are equivalent measures for Resolution, dag-like and circular width are *not* equivalent for Resolution. The sparse graph pigeonhole principle from Sect. 4 illustrates the point. This shows that bounded-width circular Resolution proofs cannot be *unfolded* into bounded-width tree-like Resolution proofs (except by going infinitary?).

For Resolution it makes a big difference whether the proof-graph has tree-like structure or not [7]. For Frege proof systems this is not the case, since Tree-like Frege polynomially simulates Dag-like Frege, and this holds true of any inference-based proof system with the set of all formulas as its set of allowed formulas, and a finite set of inference rules that is implicationally complete [18]. In the full version of the paper we show that, contrary to resolution, circular Frege proofs are no more powerful than standard ones. The main idea is to simulate in Dag-like Frege an alternative proof of the soundness of circular Frege that is based on linear programming. To do that we use a formalization of linear arithmetic, due to Buss [10] and Goerdt [14], which was originally designed to simulate counting arguments and Cutting Planes proofs in Dag-like Frege. Since Cutting Planes subsumes linear programming, the proof of soundness of circular Frege based on linear programming can be formalized in Dag-ike Frege.

Theorem 5 [2]. *Tree-like Frege and Circular Frege polynomially simulate each other.*

It is known that Tree-like Bounded-Depth Frege simulates Dag-like Bounded-Depth Frege, at the cost of increasing the depth by one. Could the simulation of Circular Frege by Tree-like Frege be made to preserve bounded depth? The (negative) answer is also provided by the pigeonhole principle which is known to be hard for Bounded-Depth Frege but is easy for Circular Resolution, and hence for Circular Depth-1 Frege.

One last aspect of the equivalence between Circular Resolution and Sherali-Adams concerns the theory of SAT-solving. As is well-known, state-of-the-art SAT-solvers produce Resolution proofs as certificates of unsatisfiability and, as a result, will not be able to handle counting arguments of pigeonhole type. This has motivated the study of so-called *pseudo-Boolean solvers* that handle counting constraints and reasoning through specialized syntax and inference rules. The equivalence between Circular Resolution and Sherali-Adams suggests a completely different approach to incorporate counting capabilities: instead of enhancing the syntax, keep it to clauses but enhance the *proof-shapes*. Whether circular proof-shapes can be handled in a sufficiently effective and efficient way is of course in doubt, but certainly a question worth studying.

It turns out that Circular Resolution has unexpected connections with Dual Rail MaxSAT Resolution [17]. MaxSAT Resolution is a variant of resolution where proofs give upper bounds on the number of clauses of the CNF that can be satisfied simultaneously. At the very least, when the upper bound is less than the number of clauses, MaxSAT resolution provides a refutation of the formula. The Dual Rail encoding is a special encoding of CNF formulas, and Dual Rail MaxSAT Resolution is defined to be MaxSAT resolution applied to the Dual Rail encoding of the input formula. In [6], the authors show that Dual Rail encoding gives strength to the proof system, providing a polynomial refutation of the pigeonhole principle formula. Recently [24] has argued that Circular Resolution polynomially simulates Dual Rail MaxSAT resolution, in the sense that when the Dual Rail encoding of a CNF formula F has a MaxSAT Resolution refutation of length ℓ and width w, then F has a Circular Resolution refutation of length $O(\ell w)$. This is interesting per se, and also provides yet another proof of Theorem 4.

Acknowledgments. Both authors were partially funded by European Research Council (ERC) under the European Union's Horizon 2020 research and innovation programme, grant agreement ERC-2014-CoG 648276 (AUTAR). First author partially funded by MICINN through TIN2016-76573-C2-1P (TASSAT3). We acknowledge the work of Jordi Coll who conducted experimental results for finding and visualizing actual circular resolution proofs of small instances of the sparse pigeonhole principle.

References

1. Atserias, A., Hakoniemi, T.: Size-degree trade-offs for sums-of-squares and Positivstellensatz proofs. To appear in Proceedings of 34th Annual Conference on Computational Complexity (CCC 2019) (2019). Long version in arXiv:1811.01351 [cs.CC] 2018
2. Atserias, A., Lauria, M.: Circular (yet sound) proofs. CoRR, abs/1802.05266 (2018)
3. Atserias, A., Lauria, M., Nordström, J.: Narrow proofs may be maximally long. ACM Trans. Comput. Log. **17**(3), 19:1–19:30 (2016)
4. Au, Y.H., Tunçel, L.: A comprehensive analysis of polyhedral lift-and-project methods. SIAM J. Discrete Math. **30**(1), 411–451 (2016)
5. Ben-Sasson, E., Wigderson, A.: Short proofs are narrow - resolution made simple. J. ACM **48**(2), 149–169 (2001)
6. Bonet, M.L., Buss, S., Ignatiev, A., Marques-Silva, J., Morgado, A.: MaxSAT resolution with the dual rail encoding. In: Proceedings of the 32nd AAAI Conference on Artificial Intelligence (2018)
7. Bonet, M.L., Esteban, J.L., Galesi, N., Johannsen, J.: On the relative complexity of resolution refinements and cutting planes proof systems. SIAM J. Comput. **30**(5), 1462–1484 (2000)
8. Brotherston, J.: Sequent calculus proof systems for inductive definitions. Ph.D. thesis, University of Edinburgh, November 2006
9. Brotherston, J., Simpson, A.: Sequent calculi for induction and infinite descent. J. Log. Comput. **21**(6), 1177–1216 (2011)
10. Buss, S.R.: Polynomial size proofs of the propositional pigeonhole principle. J. Symb. Log. **52**(4), 916–927 (1987)
11. Dantchev, S.S.: Rank complexity gap for Lovász-Schrijver and Sherali-Adams proof systems. In: Proceedings of the 39th Annual ACM Symposium on Theory of Computing, pp. 311–317 (2007)
12. Dantchev, S.S., Martin, B., Rhodes, M.N.C.: Tight rank lower bounds for the Sherali-Adams proof system. Theor. Comput. Sci. **410**(21–23), 2054–2063 (2009)
13. Dax, C., Hofmann, M., Lange, M.: A proof system for the linear time μ-calculus. In: Arun-Kumar, S., Garg, N. (eds.) FSTTCS 2006. LNCS, vol. 4337, pp. 273–284. Springer, Heidelberg (2006). https://doi.org/10.1007/11944836_26
14. Goerdt, A.: Cutting plane versus frege proof systems. In: Börger, E., Kleine Büning, H., Richter, M.M., Schönfeld, W. (eds.) CSL 1990. LNCS, vol. 533, pp. 174–194. Springer, Heidelberg (1991). https://doi.org/10.1007/3-540-54487-9_59
15. Grigoriev, D.: Linear lower bound on degrees of positivstellensatz calculus proofs for the parity. Theor. Comput. Sci. **259**(1), 613–622 (2001)
16. Haken, A.: The intractability of resolution. Theor. Comp. Sci. **39**, 297–308 (1985)
17. Ignatiev, A., Morgado, A., Marques-Silva, J.: On tackling the limits of resolution in SAT solving. In: Gaspers, S., Walsh, T. (eds.) SAT 2017. LNCS, vol. 10491, pp. 164–183. Springer, Cham (2017). https://doi.org/10.1007/978-3-319-66263-3_11
18. Krajíček, J.: Bounded Arithmetic, Propositional Logic and Complexity Theory. Cambridge University Press, Cambridge (1994)
19. Niwiński, D., Walukiewicz, I.: Games for the μ-calculus. Theor. Comp. Sci. **163**(1), 99–116 (1996)

20. Pitassi, T., Segerlind, N.: Exponential lower bounds and integrality gaps for tree-like Lovász-Schrijver procedures. SIAM J. Comput. **41**(1), 128–159 (2012)
21. Sherali, H.D., Adams, W.P.: A hierarchy of relaxations between the continuous and convex hull representations for zero-one programming problems. SIAM J. Disc. Math. **3**(3), 411–430 (1990)
22. Shoesmith, J., Smiley, T.J.: Multiple-Conclusion Logic. Cambridge University Press, Cambridge (1978)
23. Studer, T.: On the proof theory of the modal mu-calculus. Studia Logica **89**(3), 343–363 (2008)
24. Vinyals, M.: Personal communication (2018)

Short Proofs in QBF Expansion

Olaf Beyersdorff[1], Leroy Chew[2], Judith Clymo[2(✉)], and Meena Mahajan[3]

[1] Institute of Computer Science, Friedrich Schiller University Jena, Jena, Germany
[2] School of Computing, University of Leeds, Leeds, UK
scjc@leeds.ac.uk
[3] The Institute of Mathematical Sciences, HBNI, Chennai, India

Abstract. For quantified Boolean formulas (QBF) there are two main
different approaches to solving: conflict-driven clause learning (QCDCL)
and expansion solving. In this paper we compare the underlying proof
systems and show that expansion systems admit strictly shorter proofs
than QCDCL systems for formulas of bounded quantifier complexity,
thus pointing towards potential advantages of expansion solving techniques over QCDCL solving.

Our first result shows that *tree-like* expansion systems allow short
proofs of QBFs that are a source of hardness for QCDCL, i.e. tree-like
∀Exp+Res is strictly stronger than tree-like Q-Resolution.

In our second result we efficiently transform *dag-like* Q-Resolution
proofs of QBFs with bounded quantifier complexity into ∀Exp+Res
proofs. This is theoretical confirmation of experimental findings by Lonsing and Egly, who observed that expansion QBF solvers often outperform
QCDCL solvers on instances with few quantifier alternations.

Keywords: QBF · Proof complexity · Resolution · SC ·
Polynomial hierarchy

1 Introduction

Quantified Boolean formulas (QBFs) generalise propositional logic by adding
Boolean quantification. While not more expressive than propositional formulas,
QBFs allow more succinct encodings of many practical problems, including automated planning [12], verification [2], and ontologies [19]. From a complexity point
of view, they capture all problems from PSPACE.

Following the enormous success of SAT solving [25], there has been increased
attention on QBF solving in the past two decades. Currently, many QBF solvers
such as DepQBF [21], RAReQS [14], GhostQ [18], and CAQE [24], to name but a
few, compete on thousands of QBF instances. However, lifting the success of SAT
to QBF presents significant additional challenges stemming from quantification,
and solvers use quite different techniques to do this.

We consider two popular paradigms for QBF solving. In the first, QBF
solvers use *Conflict Driven Clause Learning (CDCL)* techniques from SAT solving, together with a 'reduction' rule to deal with universally quantified literals.

© Springer Nature Switzerland AG 2019
M. Janota and I. Lynce (Eds.): SAT 2019, LNCS 11628, pp. 19–35, 2019.
https://doi.org/10.1007/978-3-030-24258-9_2

In the second method, QBF solvers use quantifier expansion and remove the universally quantified literals in order to use SAT-based reasoning on the formulas.

These two paradigms can be modelled by different QBF proof systems. Modern SAT solvers correspond to the Resolution proof system [10,23]; similarly QBF solvers correspond to different QBF Resolution calculi. CDCL-style QBF (QCDCL) systems correspond to the QBF resolution system Q-resolution (Q-Res) [17], although algorithms implementing QCDCL, such as DepQBF, typically also implement additional reasoning techniques. In contrast, ∀Exp+Res was developed to model RAReQS [15], an expansion based QBF solver.

The proof systems ∀Exp+Res and Q-Res are known to be incomparable, i.e. there are families of QBFs that have polynomial-size refutations in one system, but require exponential-size refutations in the other [4,15]. As such we would not expect either QCDCL or expansion-based algorithms to be consistently stronger than the other, but would instead anticipate that solvers implementing the two systems would complement each other. An experimental comparison of this situation was recently conducted in the paper [22] where the authors conclude that QCDCL solvers perform better on instances with high quantifier complexity, whereas expansion-based solving appears superior on QBFs with few quantifier alternations.

In this paper we offer a theoretical explanation of these experimental findings. For this we closely re-examine the relations between the base systems Q-Res for QCDCL and ∀Exp+Res for expansion-based solving. As we know already that the two systems are incomparable in general, we consider two natural and practically important settings to obtain a more fine-grained analysis.

In the *first setting*, we require proofs to be *tree-like*, i.e. derived clauses may not be reused, a model corresponding to the classic (Q)DPLL algorithm [10]. While it is known that tree-like ∀Exp+Res p-simulates tree-like Q-Res [5,15], we show that this simulation is strict by providing an exponential separation. This separation uses the QPARITY formulas [4], which are known to be hard in (even dag-like) Q-Res [4]. Here we construct short tree-like proofs in ∀Exp+Res, using a Prover-Delayer game that characterises tree-like Resolution [7].

We generalise this technique for Q-*C* formulas based on descriptions of circuits *C*. For suitably chosen circuits, such QBFs yield lower bounds for quite strong QBF calculi, even including QBF Frege systems [3]. In contrast, we show that under certain width conditions on the circuit they have short proofs in tree-like ∀Exp+Res. In particular, for bounded-width circuits we obtain polynomial-size tree-like ∀Exp+Res proofs, and for circuits from the class SC we get quasi-polynomial-size proofs in tree-like ∀Exp+Res for the corresponding QBF family Q-*C*. Using this construction for the QINNERPRODUCT formulas [6], we show that tree-like ∀Exp+Res and the QBF extension of the Cutting Planes proof system CP+∀red are incomparable and exponentially separated.

In our *second setting* we consider families of QBFs of *bounded quantifier complexity*, which express exactly all problems from the polynomial hierarchy and thus cover most application scenarios. In this case, we show that (dag-like) Q-Res is p-simulated by ∀Exp+Res. The technically involved simulation increases in complexity as the number of quantification levels grows, and indeed there is an exponential separation between the two systems on a family of QBFs with an unbounded number of quantification levels [15]. For practitioners, our result offers a partial explanation for the observation that "solvers implementing orthogonal solving paradigms, such as variable expansion or backtracking search with clause learning, perform better on instances having either few or many alternations, respectively" [12].

2 Preliminaries

A literal is a propositional variable or its negation. The literals x, $\neg x$ are complementary to each other. A *clause* is a disjunction of literals. A *CNF* formula is a conjunction of clauses.

Quantified Boolean Formulas (QBFs). Propositional logic extended with the quantifiers $\forall, \exists$ gives rise to quantified Boolean logic [16]. Semantically, $\forall x. \Psi$ is satisfied by the same truth assignments as $\Psi[0/x] \wedge \Psi[1/x]$ and $\exists x. \Psi$ is satisfied by the same truth assignments as $\Psi[0/x] \vee \Psi[1/x]$.

A closed QBF is a QBF where all variables are quantified. We consider only closed QBFs in this paper. A closed QBF is either true or false, since if we semantically expand all the quantifiers we have a Boolean connective structure on $0, 1$. A prenex QBF is a QBF where all quantification is done outside of the propositional connectives. A prenex QBF Φ thus has a propositional part ϕ called the matrix and a prefix of quantifiers Π and can be written as $\Phi = \Pi \cdot \phi$. When the propositional matrix of a prenex QBF is a CNF, then we have a PCNF.

Let $\mathcal{Q}_1 X_1 \ldots \mathcal{Q}_k X_k . \phi$ be a prenex QBF, where for $1 \leq i \leq k$ $Q_i \in \{\exists, \forall\}$, $Q_i \neq Q_{i+1}$, X_i are pairwise disjoint sequences of variables. If $x \in X_i$, we say that x is at *level* i and write $\mathrm{lv}(x) = i$. We write $\mathrm{lv}(l)$ for $\mathrm{lv}(\mathrm{var}(l))$, where $\mathrm{var}(l)$ is the variable such that $l = \mathrm{var}(l)$ or $l = \neg \mathrm{var}(l)$. For fixed k, the class Σ_k^b contains all prenex QBFs of the above form with k quantifier blocks where the first block is existential.

Without loss of generality, we assume that the last block is existential in all QBFs we consider.

Proof Systems. Formally, a *proof system* [11] for a language L over alphabet Γ is a polynomial-time computable partial function $f : \Gamma^\star \to \Gamma^\star$ with $rng(f) = L$. For $x \in L$, a $y \in \Gamma^\star$ such that $f(y) = x$ is a proof of $x \in L$. If f and g are proof systems for the same language, then f *p-simulates* g if and only if there is a polynomially computable function mapping a proof in g to a proof of the same formula in f [11].

Q-Resolution. Introduced by Kleine Büning, Karpinski, and Flögel in [17], Q-Resolution (Q-Res) is a refutational system that operates on PCNFs. It uses the propositional resolution rule on existential variables with a side condition that prevents tautological clauses (cf. Fig. 1). Tautologies are also forbidden from being introduced in the proof as axioms. In addition, Q-Res has a universal reduction rule ($\forall$-Red) to remove universal variables. A Q-Res proof of false QBF

$$\frac{}{C}\ (\text{Ax}) \qquad\qquad \frac{C \vee x \qquad D \vee \neg x}{C \vee D}\ (\text{Res})$$

Ax: C is a non-tautological clause in the propositional matrix.
Res: variable x is existential, and if literal $z \in C$, then $\neg z \notin D$.

$$\frac{C \vee u}{C} \qquad\qquad \frac{C \vee \neg u}{C}\ (\forall\text{-Red})$$

u is universal and all other variables $x \in C$ are left of u in the prefix.

Fig. 1. The rules of Q-Res [17]

Φ is a sequence of clauses $C_1 \ldots C_m$ such that C_m is the empty clause (denoted $\bot$) and each C_i is either introduced by rule Ax, derived by rule Res from C_j and C_k on pivot x ($j, k < i$), or derived by $\forall$-Red from C_j ($j < i$). Clauses need not be unique in the proof. The parent(s) of a clause are the clause(s) used to derive it; the derived clause is called the child.

The size of a Q-Res proof π is the number of clauses it contains, denoted $|\pi|$. The size of refuting QBF Φ in Q-Res is minimum over the size of all possible Q-Res proofs.

A proof $C_1 \ldots C_m$ induces a directed acyclic graph (dag) with nodes corresponding to clauses, and edges to derivation steps. Clauses introduced by the rule Ax are source nodes and edges point from parent(s) of a proof step to the clause derived. If the induced dag is a tree then the proof is called tree-like. In tree-like Q-Res, only tree-like proofs are allowed.

QBF Expansion. In addition to Q-Res we consider an alternative way of extending the resolution calculus based on *instantiation* of universal variables that was introduced to model *expansion-based QBF solving*. We operate on clauses that comprise only existential variables from the original QBF, which are additionally *annotated* by a substitution to universal variables, e.g. $\neg x^{0/u,1/v}$. For any annotated literal l^σ, the substitution σ must not make assignments to variables right of l, i.e. if $u \in \text{dom}(\sigma)$, then u is universal and $\text{lv}(u) < \text{lv}(l)$. To preserve this invariant, we use the *auxiliary notation* $l^{[\sigma]}$, which for an existential literal l and an assignment σ to the universal variables filters out all assignments that are not permitted, i.e. $l^{[\sigma]} = l^{\{c/u \in \sigma\ |\ \text{lv}(u) < \text{lv}(l)\}}$. We say that an assignment is complete if its domain is all universal variables. Likewise, we say that a literal x^τ is fully annotated if all universal variables u with $\text{lv}(u) < \text{lv}(x)$ in the QBF are in $\text{dom}(\tau)$, and a clause is fully annotated if all its literals are fully annotated.

The calculus $\forall$Exp+Res from [15] works with fully annotated clauses on which resolution is performed. For each clause C from the matrix and an assignment τ to all universal variables, falsifying all universal literals in C, $\forall$Exp+Res can use the axiom $\{l^{[\tau]}\ |\ l \in C, l\ \text{existential}\}$.

As its only rule it uses the resolution rule on annotated clauses, the pivot literals must have matching annotations.

$$\frac{C \vee x^{\tau} \qquad D \vee \neg x^{\tau}}{C \cup D} \text{ (Res)}.$$

An ∀Exp+Res proof is a sequence of clauses where every clause is either a valid expansion of a clause in the input formula, or is derived by resolution from previous clauses. The size of a proof is the number of clauses it contains, and the size of refuting a formula is the smallest valid proof. A proof induces a dag, and we may restrict our attention to tree-like proofs.

The Prover-Delayer Game. The Prover-Delayer game from [7] gives a characterisation of the size of tree-like resolution proofs of unsatisfiable formulas. The game has two players, who construct a partial assignment. The game terminates when the (partial) assignment constructed falsifies some clause. The Delayer tries to score as many points as possible before the inevitable termination, while the Prover tries to limit the Delayer's score. Starting with the empty assignment, the game proceeds in rounds:

- Prover chooses an unassigned variable x.
- Delayer assigns weights $0 \leq w_0, w_1 \leq 1$, such that $w_0 + w_1 = 1$.
- Prover now chooses to set the variable x either to 0 or to 1.
- If the variable is set to $i \in \{0, 1\}$ then Delayer scores $\log(\frac{1}{w_i})$ points.

Note that if the Delayer chooses $w_b = 1$ for some b, then the Prover can respond in this round by setting x to b, and the Delayer scores nothing in this round. For such a Delayer response ($w_0 = 1$ or $w_1 = 1$), we say that the Delayer chooses the value of the variable. Otherwise the actual choice is made by the Prover.

Theorem 1 (Beyersdorff, Galesi and Lauria; [7]). *Let ϕ be an unsatisfiable false CNF with shortest tree-like Resolution refutation of size S. Then, there is a Delayer strategy such that the Delayer scores at least $\log(\lceil \frac{S}{2} \rceil)$ points in any game on ϕ against any Prover. Furthermore, no Delayer strategy can guarantee more; there is a Prover strategy that limits the Delayer score to $\log(\lceil \frac{S}{2} \rceil)$ points.*

3 Short Tree-Like Expansion Proofs for QBFs Based on Parity and Thin Circuits

The false QBFs QPARITY, introduced in [4], express the contradiction that for some binary string x, for every z, the parity of the number of 1s in x is different from z. The propositional matrix is falsified whenever $z = \text{PARITY}(x_1, \ldots, x_n)$. The QPARITY$_n$ QBFs are given as

$$\exists x_1 \ldots \exists x_n \forall z \exists t_0 \ldots \exists t_n (\neg t_0) \wedge (z \vee t_n) \wedge (\neg z \vee \neg t_n) \wedge \bigwedge_{i=1}^{n} (t_i = t_{i-1} \oplus x_i),$$

where the formulas $(t_i \equiv t_{i-1} \oplus x_i)$ are expressed by the four clauses $(\neg t_{i-1} \vee x_i \vee t_i)$, $(t_{i-1} \vee \neg x_i \vee t_i)$, $(t_{i-1} \vee x_i \vee \neg t_i)$, $(\neg t_{i-1} \vee \neg x_i \vee \neg t_i)$.

These formulas are known to be hard for both tree-like and dag-like Q-Res [4]. Here we show that they have short tree-like proofs in ∀Exp+Res.

Theorem 2. QPARITY$_n$ *have polynomial-size tree-like* $\forall$*Exp+Res proofs.*

Proof. In order to analyse $\forall$Exp+Res proofs, it suffices to simply expand all clauses in every universal assignment and look at all propositional resolution refutations from tree-like resolution. We expand out all the clauses of QPARITY$_n$ based on the two settings to the single universal variable z. The formula now contains twice as many clauses. We now need to construct a tree-like resolution proof from these clauses. We use the asymmetric Prover-Delayer game. The Prover strategy is given in Algorithm 1.

Algorithm 1. Prover Strategy

$i = 0$, $j = n$.
The Prover queries the variables of the unit clauses $\neg t_0^{0/z}, \neg t_0^{1/z}, t_n^{0/z}, \neg t_n^{1/z}$. The Delayer is forced to satisfy these clauses or get a constant score.
while $j - i > 1$ **do**
    $k \leftarrow \lfloor \frac{i+j}{2} \rfloor$.
    The Prover queries the variable $t_k^{1/z}$.
    The Prover chooses the value that gives the Delayer the least score.
    The Prover queries the variable $t_k^{0/z}$.
    The Prover chooses the value that gives the Delayer the least score.
    if $t_k^{0/z} = t_k^{1/z}$ **then** $i \leftarrow k$ **else** $j \leftarrow k$
The Prover now queries x_j, and chooses the value that gives the Delayer the least score.
  $\triangleright$ The game ends here because $t_i^{0/z} = t_i^{1/z}$, $t_j^{0/z} \neq t_j^{1/z}$, and for $c \in \{0,1\}$, there are clauses expressing $t_j^{c/z} \equiv x_j \oplus t_i^{c/z}$.

The Delayer can score at most 2 points per loop iteration, hence a total score of at most $2\lceil \log(n) \rceil$. By Theorem 1, there are tree-like resolution refutations with $O(n^2)$ leaves.

One nice feature of the Prover-Delayer technique is that we can construct a tree-like Resolution proof from a Prover strategy. To simplify, let us suppose $n = 2^r$. We proceed in rounds. The idea is that inductively, in the kth round, for each $0 \leq i < 2^{r-k}$, we produce 2^{r-k} copies of the clauses

$$t_{2^k i}^{0/z} \vee t_{2^k i}^{1/z} \vee \neg t_{2^k(i+1)}^{0/z} \vee t_{2^k(i+1)}^{1/z}, \qquad \neg t_{2^k i}^{0/z} \vee \neg t_{2^k i}^{1/z} \vee \neg t_{2^k(i+1)}^{0/z} \vee t_{2^k(i+1)}^{1/z},$$

$$t_{2^k i}^{0/z} \vee t_{2^k i}^{1/z} \vee t_{2^k(i+1)}^{0/z} \vee \neg t_{2^k(i+1)}^{1/z}, \qquad \neg t_{2^k i}^{0/z} \vee \neg t_{2^k i}^{1/z} \vee t_{2^k(i+1)}^{0/z} \vee \neg t_{2^k(i+1)}^{1/z}.$$

For the base case $k = 0$, these clauses are produced using the axioms. For instance, the clause $t_{i-1}^{0/z} \vee t_{i-1}^{1/z} \vee \neg t_i^{0/z} \vee t_i^{1/z}$ is produced by resolving $t_{i-1}^{0/z} \vee x_i \vee \neg t_i^{0/z}$ and $t_{i-1}^{1/z} \vee \neg x_i \vee t_i^{1/z}$.

Once we reach round r, we have the clause $t_0^{0/z} \vee t_0^{1/z} \vee \neg t_{2^r}^{0/z} \vee t_{2^r}^{1/z}$ which we resolve with $\neg t_0^{0/z}, \neg t_0^{1/z}, t_{2^r}^{0/z}$ and $\neg t_{2^r}^{1/z}$ to get a contradiction.

Corollary 1. *Tree-like* ∀*Exp+Res p-simulates tree-like Q-Res, but tree-like Q-Res does not simulate* ∀*Exp+Res, and there are QBFs providing an exponential separation.*

Proof. From [15] we know that tree-like ∀Exp+Res p-simulates tree-like Q-Res. The QBF QPARITY requires exponential size proofs for tree-like Q-Res (shown in [4]). Consequently, by Theorem 2, we conclude that tree-like ∀Exp+Res is strictly stronger than tree-like Q-Res.

We extend the method demonstrated above for other QBFs based on Boolean circuits.

Definition 1. *Let C be a Boolean circuit over variables $x_1, \ldots, x_n$, with gates computing binary Boolean functions. Let the gates of C be $g_1, \ldots, g_m$ in topological order, with g_m being the output gate. The QBF Q-C expresses the contradiction $\exists x_1 \ldots x_n \forall z \, [z \neq C(x_1, \ldots, x_n)]$ and is constructed as follows.*

The variable set $X = \{x_1, \ldots, x_n\}$ contains all the input variables of C. The variable set T has one variable t_j corresponding to each gate g_j of C; $T = \{t_1, \ldots, t_m\}$. The quantifier prefix is $\exists X \forall z \exists T$. The QBF is of the form

$$(z \vee t_m) \wedge (\neg z \vee \neg t_m) \wedge \bigwedge_{j=1}^{S} [t_j \text{ is consistent with the inputs to gate } g_j].$$

The predicate $[t_j$ is consistent$]$ depends on t_j and at most two other variables in $X \cup T$, depending on the connections in C. Hence it can be written as a short CNF formula; we include these clauses. (E.g. if $g_2 = g_1 \wedge x_1$ then we add clauses $(t_2 \vee \neg t_1 \vee \neg x_1), (\neg t_2 \vee t_1), (\neg t_2 \vee x_1)$.)

Note that the QBF Q-C is of size $O(m)$, where m is the number of gates in the circuit C. In particular, if C is of size poly(n), then so is Q-C.

The following result appears in [4].

Proposition 1 (Proposition 28 in [4]). *For every family of polynomial-size circuits $\{C_n\}$, the QBF family $\{Q\text{-}C_n\}$ has poly(n)-size proofs in dag-like* ∀*Exp+Res.*

The proof of this result reuses derivations of clauses expressing $t_j^{0/z} = t_j^{1/z}$ at all stages i where g_j is an input to g_i. For tree-like ∀Exp+Res, we cannot reuse them. Instead we generalise our idea from Theorem 2 of a Prover strategy using binary search. Note that this technique works precisely because the circuits underlying QPARITY formulas have very small "width".

Definition 2 (Layered Circuits, and Circuit Width). *A circuit is said to be* layered *if its gates can be partitioned into disjoint sets S_i for $1 \leq i \leq \ell$, such that for each i, for each gate in layer S_i, all its inputs are either input variables or the outputs of gates in S_{i-1}. The output gate is in the final layer S_ℓ. The* width *of a layered circuit is the maximum number of gates in any layer; width(C) = $\max\{|S_i| \mid i \in \ell\}$.*

Theorem 3. *Let C be a layered circuit of size m and width w, and let Q-C be the corresponding QBF (as in Definition 1). Then Q-C has a proof, in tree-like $\forall Exp+Res$, of size $m^{O(w)}$.*

Proof (Proof Sketch). Consider the expanded CNF obtained from Q-C. Let X be the set of x variables, U be the set of $t^{0/z}$ variables and V be the set of $t^{1/z}$ variables. Let the clauses expressing that the t variables correspond to the computation of C on x be denoted $F(X,T)$. A proof in (tree-like) $\forall Exp+Res$ that Q-C is false is a proof in (tree-like) Resolution that the CNF $G(X,U,V)$, defined below, is unsatisfiable.

$$G(X,U,V) = F(X,U) \wedge F(X,V) \wedge t_m^{0/z} \wedge \neg t_m^{1/z}$$

Let the number of layers be ℓ; note that $\ell \leq m$. We will describe a Prover strategy that limits the Delayer's score to $O(w \log \ell)$ and then invoke Theorem 1.

Let W_ℓ be the variables corresponding to the output gate; $W_\ell = \{t_m^{0/z}, t_m^{1/z}\}$. For $i \in [\ell]$, let $W_{i-1} \subseteq X \cup U \cup V$ be the variables feeding into gates at layer i of C.

The Prover uses binary search to identify a layer j where all corresponding variables of W_{j-1} from U and V are in agreement, whereas some corresponding variables of W_j from U and V disagree. There are variables $t_b^{0/z} \neq t_b^{1/z}$ in W_j, but for all t_a variables in W_{j-1}, $t_a^{0/z} = t_a^{1/z}$. Furthermore all X variables in W_{j-1} are set. So $t_b^{0/z} \neq t_b^{1/z}$ must cause a contradiction with the copies of the clauses expressing consistency of gate g_b with its inputs.

For every layer i where the Prover queries variables from W_i, there are at most $4w$ variables to query, and this is the maximum score for the Delayer on W_i. The Prover looks at no more than $\lceil \log \ell \rceil$ sets W_i. Hence the score for the Delayer is at most $4w \log \ell + 1$ and by Theorem 1, there are tree-like Resolution proofs of size $2l^{4w}$.

Corollary 2. *Suppose $\{C_n\}$ is a family of layered circuits with width bounded by a constant. Then the family of QBFs Q-C_n has polynomial-size proofs in tree-like $\forall Exp+Res$.*

If we relax our desire for polynomial-size proofs to just quasi-polynomial size we can allow circuits with non-constant width: QBFs constructed from circuits with poly-logarithmic width have quasi-polynomial size proofs.

Let C be a circuit that is layered, where every gate has fan-in (number of incoming wires) at most two, where the number of variables is n, the total size (number of gates) is s, the width of any layer is at most w, and the depth is d. Let C compute an n-variate Boolean function $f : \{0,1\}^n \rightarrow \{0,1\}$.

A language $L \subseteq \{0,1\}^*$ is in P/poly if there is a family of circuits $\{C_n\}_{n \geq 1}$ where the size of each C_n is $n^{O(1)}$, such that C_n computes the characteristic function of $L \cap \{0,1\}^n$. If, furthermore, the depth of C_n is $(\log n)^{O(1)}$, then the language is in NC. If, instead, the width of C_n is $(\log n)^{O(1)}$, then the language is in SC. Note that P/poly and SC so defined are the non-uniform analogues of the complexity classes P and TIME, SPACE$(n^{O(1)}, (\log n)^{O(1)})$.

Corollary 3. *Suppose $\{C_n\}$ is an* SC *family of circuits. Then the family of QBFs Q-C_n has quasi-polynomial size proofs in tree-like* ∀Exp+Res.

In essence, we show that these Q-C formulas, which can give lower bounds in QCDCL style systems [3], are easy even for tree-like ∀Exp+Res. Notice that the false Q-C formulas are all Σ_3^b; this is the minimum quantifier complexity needed to separate Q-Res and ∀Exp+Res.

The QINNERPRODUCT formulas, introduced in [6], extend the QPARITY formulas in a simple way: each variable x_i in QPARITY is replaced by the logical AND of two new variables y_i and z_i. As with parity, the Inner Product function also has constant-width circuits. Hence, by Theorem 3, the QINNERPRODUCT formulas have short tree-like proofs in ∀Exp+Res. However, it is shown in [6] that these formulas require $\exp(\Omega(n))$ size in the proof system CP+∀red, that augments the propositional Cutting Planes proof system with the universal reduction rule. Thus CP+∀red cannot even p-simulate tree-like ∀Exp+Res. On the other hand, it is also shown in [6] that CP+∀red is incomparable with ∀Exp+Res, with exponential separation. Thus we now obtain the following strengthening of Theorem 7 from [6].

Corollary 4. *Tree-like* ∀Exp+Res *and* CP+∀red *are incomparable, and there are QBFs providing exponential separation in both directions.*

4 ∀Exp+Res Simulates Dag-Like Q-Res for Bounded Alternations

We now move from the tree-like systems to the stronger dag-like model. While it is known that ∀Exp+Res and Q-Res are in general incomparable [4], we will show here a simulation of dag-like Q-Res by ∀Exp+Res for QBFs of *bounded* quantifier complexity.

A single clause in a Q-Res refutation of Φ can be naturally turned into a clause in a ∀Exp+Res refutation in the same way that axioms are instantiated in ∀Exp+Res. We define some complete assignment α to the universal variables of Φ that does not satisfy the clause. All universal literals are removed (since they are falsified under α), and each existential literal x is replaced by $x^{[\alpha]}$ (recall that $[\alpha]$ indicates that the assignment α is restricted to those variables that appear before the annotated literal in the quantifier prefix). In a ∀Exp+Res proof the pivot literal of a resolution step must have the same annotation in both parent clauses. In general it is not possible to annotate the clauses in a Q-Res proof so that this requirement is respected.

Consider the simple example in Fig. 2. $(y \vee x)$ is resolved once with a clause containing universal literal u, and separately with another clause containing $\neg u$. It is not possible to define a single annotation for x in $(y \vee x)$ so that both these steps are valid in ∀Exp+Res. As shown, $(y \vee x)$ could be duplicated to accommodate both annotations. Overall a clause could be repeated once for each path between that clause and the root to ensure that it can be annotated consistently with its neighbours. In the worst case this means making the proof

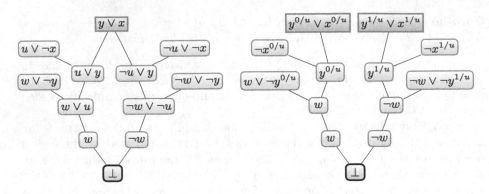

Fig. 2. Duplicating clauses to create an expansion refutation of QBF with prefix $\exists w \forall u \exists x y$

fully tree-like and incurring an exponential increase in the proof size. We show how to modify this simple idea to efficiently transform a Q-Res proof of QBF with bounded quantifier complexity into a ∀Exp+Res proof.

Recall that a Q-Res proof is a sequence of clauses, not necessarily unique, and induces a dag by the inference relationship. Our aim is to construct from a Q-Res proof of PCNF Φ a sequence of new Q-Res proofs of Φ, the last of which can be readily turned into a valid ∀Exp+Res proof.

4.1   Expanding a Q-Resolution Proof

We start with a few useful definitions and observations regarding Q-Res proofs.

In a Q-Res refutation there is no benefit to delaying a ∀-Red step since it cannot cause a later step to be blocked, and the result of ∀-Red is stronger than its parent, so it is safe to assume that ∀-Red is carried out as soon as possible. If a clause is used in a ∀-Red step then it is not used in any other proof step, since this would delay a possible ∀-Red on that path, so all possible ∀-Red steps for a clause are carried out consecutively, without branching.

Where the proof contains consecutive ∀-Red steps, they may be re-ordered so that innermost literals are removed first. This allows consecutive ∀-Red which remove literals from the same level in the quantifier prefix to be treated as a single step in the construction below. Literals removed in consecutive ∀-Red steps cannot be complementary, since they must all appear together in the clause prior to the sequence and clauses in a Q-Res proof are never tautologous.

For the remainder of this subsection let $\pi = C_1 \ldots C_m$ be a Q-Res proof of PCNF $\Phi = Q_1 X_1 \ldots Q_k X_k. \phi$. We assume that $Q_1 = \forall$, since X_1 may be empty.

Definition 3. *A ∀-Red step in π is at level i if the universal literal(s) removed by the step belong to X_i.*

Definition 4. *Let A and B be two clauses in π. Then A is i-connected to B if there is a subsequence $C_{a_1} \ldots C_{a_n}$ of π such that $C_{a_1} = A$, $C_{a_n} = B$, $\forall l \in$*

$\{1 \ldots n-1\}$ C_{a_l} is a parent of $C_{a_{l+1}}$ in π, and no member of the sequence is derived by $\forall$-Red at any level $j \leq i$ in π.

Definition 5. The level i derivation of a clause C in π, denoted $\pi(C, i)$, is the subsequence of π ending at C and containing exactly those clauses which are i-connected to C.

Definition 6. $\mathcal{A}_\pi^i$ is the set of clauses that are parents of a $\forall$-Red step at level i in π.

The main idea in the following construction is to use the level i derivation of $\forall$-Red steps at level i to find and isolate sections of a proof that contain no complementary universal literals at level i and which therefore could be given the same (level i) annotation in a $\forall$Exp+Res proof.

Definition 7 (Level i expansion of π). For π a Q-Res refutation of PCNF $\Phi = \mathcal{Q}_1 X_1 \ldots \mathcal{Q}_k X_k. \phi$ with $\mathcal{Q}_i = \forall$, the level i expansion of π is defined by the following construction.

Let $P \in \mathcal{A}_\pi^i$ and C the unique child of P. We have assumed that consecutive $\forall$-Red steps are carried out in reverse level order, and with steps at the same level collapsed together so P was not derived by $\forall$-Red at level i or at any level $j < i$ in π.

Find $\pi(P, i)$, the level i derivation of P, and copy this section of the proof. The original is not discarded until later. Each identified clause $D \in \pi(P, i)$ generates a new (identical) clause D'. Suppose $C_a \in \pi(P, i)$ and C_b is a parent of C_a in π. Then C_a' has parent C_b' if it exists, and otherwise C_b. Update C to be derived from the copy P' of its parent P.

Repeat this process for each member of $\mathcal{A}_\pi^i$. Then the clauses are ordered so that clause C_a comes before any copies of C_a, and if $a > b$ then every copy of C_b comes before C_a or any of its copies. This ensures that the parent(s) of a proof step always appear before the clause they derive. Among copies of the same clause, assume an ordering based on the order in which $\forall$-Red steps appear in π.

Clauses from the original refutation which no longer have any children (i.e. if copies of that clause are used for every derivation it was involved in) are removed. The result is the level i expansion of π, written $e_i(\pi)$.

Lemma 1. Let π be a valid Q-Res refutation of Φ. For every universal level i in Φ, $e_i(\pi)$ is a valid Q-Res refutation of Φ.

Proof. Every derivation is an exact copy of a derivation in π, and the imposed ordering respects the order of derivations.

Lemma 2. Let π be a Q-Res proof of Φ and i a universal level in Φ, then $e_i(\pi)$ and π have the same number of $\forall$-Red steps at levels $j \leq i$, for j a universal level.

Proof. A clause which is the result of a $\forall$-Red step at level $j \leq i$ does not belong to any level i derivation of a member of $\mathcal{A}_{e_i(\pi)}^i$, by definition, so will never be copied.

Lemma 3. *Let π be a Q-Res proof of Φ and i a universal level in Φ. The level i derivations of clauses in $\mathcal{A}^i_{e_i(\pi)}$ are disjoint.*

Proof. The clauses copied for $P \in \mathcal{A}^i_\pi$ are exactly the level i derivation of $P' \in \mathcal{A}^i_{e_i(\pi)}$.

Lemma 4. *Any clause in $e_i(\pi)$ that is not in a level i derivation of some $P' \in \mathcal{A}^i_{e_i(\pi)}$ does not contain any literal at level i.*

Proof. Let clause C contain level i literal u. For every path in $e_i(\pi)$ between C and the empty clause there is sub-path beginning at C and ending at the first clause not containing u, which must immediately follow $\forall$-Red since there is no other way to remove universal literals from a clause. The parent of that $\forall$-Red is P' and by definition C belongs to the level i derivation of P'.

Lemma 5. *Let π be a Q-Res proof of Φ and i a universal level in Φ. The parent and child of any proof step in $e_i(\pi)$ cannot belong to level i derivations of different members of $\mathcal{A}^i_{e_i(\pi)}$. Similarly, the two parents of a resolution step in $e_i(\pi)$ cannot belong to level i derivations of different members of $\mathcal{A}^i_{e_i(\pi)}$.*

Proof. Let B be a parent of A in $e_i(\pi)$. By construction, if A is in the level i derivation of $P \in \mathcal{A}^i_{e_i(\pi)}$ and B is not then B is the result of a $\forall$-Red step at some level $j \leq i$ and so cannot be included in the level i derivation of any clause. If A is derived by resolution and does not belong to any level i derivation of $P \in \mathcal{A}^i_{e_i(\pi)}$ then neither can its parents.

Lemma 6. *Let π be a Q-Res proof of Φ and i a universal level in Φ. The size of the level i expansion of Q-Res refutation π is at most $|\pi|^2$.*

Proof. If there are S distinct $\forall$-Red steps at level i, then each clause in π may be copied up to S times, so the size of the level i expansion of π is at most $|\pi| \cdot (S + 1)$. Clearly $S < |\pi|$.

Since the level i expansion of a Q-Res refutation is itself a Q-Res refutation, we can apply the process iteratively for different values of i. We will expand the proof for each universal level, starting from the innermost.

Definition 8 (Complete expansion of π). *Let π be a Q-Res proof of PCNF Φ. The complete expansion of π, denoted $E(\pi)$ is $E(\pi) = e_1(e_3 \dots (e_{k-1}(\pi)))$. Intermediate stages are labelled π_i (where $\mathcal{Q}_i = \forall$), so that $\pi_i = e_i(\pi_{i+2}) = e_i(e_{i+2} \dots (e_{k-1}(\pi)))$.*

Repeated applications of Lemmas 1 and 2 respectively give the following Lemmas.

Lemma 7. *Let π be a Q-Res proof of PCNF Φ. Then $E(\pi)$ is a Q-Res refutation of Φ.*

Lemma 8. *Let π be a Q-Res proof of PCNF Φ and $\mathcal{Q}_i = \mathcal{Q}_{i+2} = \forall$ in Φ. Then the number of $\forall$-Red steps at level i in π_{i+2} equals the number of $\forall$-Red steps at level i in π.*

Lemma 9. *Given Q-Res proof π of PCNF $\Phi = \mathcal{Q}_1 X_1 \ldots \mathcal{Q}_k X_k. \phi$, $|E(\pi)| \leq |\pi|^k$.*

Proof. The argument proceeds by a simple induction on the number of universal levels that have been expanded, showing that for every level i with $\mathcal{Q}_i = \forall$, $\pi_i \leq |\pi|^{k-i+1}$.

Let S be the maximum number of $\forall$-Red steps on any single level in π. Then, following Lemma 6, $|\pi_{k-1}| \leq |\pi| \cdot (S + 1) \leq |\pi|^2 \leq |\pi|^{k-(k-1)+1}$. Assume the hypothesis for $i = k - 1, \ldots, i + 2$. Since π_{i+2} has the same number of level i $\forall$-Red steps as π (Lemma 8), then applying Lemma 6 to π_{i+2}, $|\pi_i| \leq |\pi_{i+2}| \cdot (S + 1) \leq |\pi|^{k-(i+2)+1} \cdot |\pi| \leq |\pi|^{k-i+1}$.
    $E(\pi) = \pi_1$ and $|E(\pi)| \leq |\pi|^k$.

$E(\pi)$ can now be made into a $\forall$Exp+Res refutation of Φ. We introduce a system of labelling clauses in the proofs π_i with partial assignments to the universal variables in the formula being refuted. Each clause in $E(\pi)$ will be associated with a complete assignment to universal variables which is then used to define annotations for the existential literals in that clause.

4.2   Annotating the Expanded Proof

Definition 9. *Let π be a Q-Res of PCNF $\Phi = \mathcal{Q}_1 X_1 \ldots \mathcal{Q}_k X_k. \phi$, $\mathcal{Q}_i = \forall$. For a clause $D \in \mathcal{A}^i_{\pi_i}$ let α^D_i be the assignment to X_i that sets variables appearing in D so that D is not satisfied, and all other variables in X_i to 0.*

Lemma 10. *Let $D \in \mathcal{A}^i_{\pi_i}$. Then α^D_i does not satisfy any $C \in \pi_i(D, i)$.*

Proof. $C \in \pi_i(D, i)$, so by construction there is a path between C and D that does not include any $\forall$-Red step at level i. Therefore any level i literal $u \in C$ also appears in D, and any assignment to variables at level i that does not satisfy D also will not satisfy C.

Immediately after generating π_i from π_{i+2} add the following labels: for each $D \in \mathcal{A}^i_{\pi_i}$, label all clauses in $\pi_i(D, i)$ with α^D_i. Any clause in π_i that is not in a level i derivation of some $D \in \mathcal{A}^i_{\pi_i}$ is not satisfied by any assignment to level i variables (Lemma 4). Label such clauses with the assignment setting all level i variables to 0. In subsequent expansions, clauses are copied with their labels. This means that all clauses in π_i will be labelled with complete assignments to all levels $\geq i$, and that $E(\pi)$ will have all clauses labelled with a complete assignment to all universal variables in Φ. No clause is labelled twice (Lemma 3).

Lemma 11. *In π_i the parent and child of any proof step are labelled with the same assignment to universals from all levels $j \geq i$ for which both clauses contain some existential literal at a level greater than j. Similarly, if the proof step* *is resolution then the two parents are labelled with the same assignment to universals from all levels $j \geq i$ for which they both contain an existential literal at a level greater than j.*

Proof. For universal level $j > i$ assume that the result holds for refutation π_{i+2} and recall that every derivation in π_i is an exact copy of a derivation in π_{i+2}. Labels are copied with clauses, so the result also holds for π_i. In the base case where $i = k - 1$ there are no universal levels $j > i$.

For level i we consider whether the parent and child of a proof step belong to some $\pi_i(P, i)$ for $P \in \mathcal{A}^i_{\pi_i}$. If neither parent nor child belong to any such $\pi_i(P, i)$ then both have been labelled with the level i assignment setting all variables to 0. If the parent is in $\pi_i(P, i)$ for some $P \in \mathcal{A}^i_{\pi_i}$ but the child is not, then the child must be the result of $\forall$-Red at level i and so contains no existential literals at a level $> i$. If the child is in $\pi_i(P, i)$ for some $P \in \mathcal{A}^i_{\pi_i}$ but the parent is not, then the parent is the result of $\forall$-Red at some level $\leq i$ and so contains no existential literals at a level $> i$. If both parent and child are in $\pi_i(P, i)$ for some $P \in \mathcal{A}^i_{\pi_i}$ then they are both labelled with α^P_i. It is not possible for the parent and child of a proof step to belong to level i derivation of different clauses in $\mathcal{A}^i_{\pi_i}$ (Lemma 5).

The second statement follows by exactly the same argument. If neither parent belongs to any such $\pi_i(P, i)$, they are labelled identically at level i. If both parents belong to the same $\pi_i(P, i)$ then they are in the same section and are labelled identically at level i. If only one parent belongs to some $\pi_i(P, i)$, then the other is the result of $\forall$-Red at some level $\leq i$ and so contains no existential literals at a level $> i$. They cannot belong to level i derivation of different clauses in $\mathcal{A}^i_{\pi_i}$ (Lemma 5).

4.3   Putting Everything Together for the Simulation

To create a $\forall$Exp+Res proof from $E(\pi)$, we simply use the clause labels to generate annotations for the existential literals in each clause and our main result now follows easily:

Theorem 4. *For any constant k, $\forall$Exp+Res p-simulates Q-Res on Σ^b_k formulas.*

Proof. For any Σ^b_k formula Φ and its Q-Res proof π we can generate a Q-Res refutation $E(\pi)$ of Φ and label the clauses of $E(\pi)$ as described. Remove all universal literals from clauses of $E(\pi)$. Any existential literal x in a clause C with label α is replaced by the annotated literal $x^{[\alpha]}$. $\forall$-Red steps are now meaningless and can be removed. Resolution steps remain, acting on annotated literals with matching annotations (Lemma 11). The leaves of $E(\pi)$ were all copies of leaves in π, i.e. they are clauses from Φ, so the leaves of the constructed $\forall$Exp+Res refutation are annotated versions of those same clauses from Φ. Therefore we have a valid $\forall$Exp+Res proof of Φ constructed from the Q-Res proof, and by Lemma 9 the size of the new refutation is bounded above by $|\pi|^k$.

As for the tree-like model, we obtain that $\forall$Exp+Res is strictly stronger than Q-Res also for dag-like proofs, when restricted to QBFs of bounded quantifier complexity.

Corollary 5. *For each* $k \geq 3$, $\forall Exp+Res$ *p-simulates* Q-Res *on* Σ_k^b *formulas, but the reverse simulation does not hold, and there are* Σ_3^b *formulas providing an exponential separation.*

Proof. The simulation is stated as Theorem 4 and the exponential separation is given by QPARITY [4], which is a family of Σ_3^b formulas.

This is tight since we know that relaxing the requirement for bounded quantifier complexity allows formulas with polynomial sized dag-like Q-Res proofs but only exponential sized $\forall Exp+Res$ proofs [15], and also that for QBFs with only one or two quantifier levels, Q-Res and $\forall Exp+Res$ are p-equivalent.

5   Conclusion

Our results demonstrate proof-theoretic advantages of $\forall Exp+Res$ over Q-Res, both for tree-like proofs and for QBFs with bounded quantifier complexity.

These advantages are not meant to suggest that QCDCL systems are inferior on all accounts. The simulation on bounded quantifier levels becomes less efficient as the number of alternations increase, and the existence of short proofs does not guarantee that proof search will find them. The models of $\forall Exp+Res$ and Q-Res simplify the QBF solving algorithms: QCDCL solvers can introduce proof steps that are better represented in the LD-Q-Res proof system from [1,26]; also QBF solvers regularly use dependency schemes [20] which we do not take into account here. Both long-distance steps and dependency schemes are known to shorten proofs in comparison to Q-Res [9,13].

Nor should it be inferred that $\forall$-Red is an inherently weaker way of dealing with universally quantified variables than expansion. For example the systems Frege+$\forall$-Red and eFrege+$\forall$-Red [3] are very strong, and finding lower bounds for them is equivalent to solving major open problems in circuit complexity or propositional proof complexity [8].

Acknowledgements. Some of this work was done at Dagstuhl Seminar 18051, Proof Complexity. Research supported by the John Templeton Foundation and the Carl Zeiss Foundation (1st author) and EPSRC (2nd author).

References

1. Balabanov, V., Jiang, J.H.R.: Unified QBF certification and its applications. Formal Methods Syst. Des. **41**(1), 45–65 (2012)
2. Benedetti, M., Mangassarian, H.: QBF-based formal verification: experience and perspectives. J. Satisfiability Boolean Model. Comput. (JSAT) **5**(1–4), 133–191 (2008)
3. Beyersdorff, O., Bonacina, I., Chew, L.: Lower bounds: from circuits to QBF proof systems. In: Proceedings of the ACM Conference on Innovations in Theoretical Computer Science (ITCS 2016), pp. 249–260. ACM (2016)

4. Beyersdorff, O., Chew, L., Janota, M.: Proof complexity of resolution-based QBF calculi. In: Proceedings of the Symposium on Theoretical Aspects of Computer Science, pp. 76–89. LIPIcs Series (2015)
5. Beyersdorff, O., Chew, L., Mahajan, M., Shukla, A.: Are short proofs narrow? QBF resolution is not so simple. ACM Trans. Comput. Logic **19**(1), 1:1–1:26 (2018). (preliminary version in STACS 2016)
6. Beyersdorff, O., Chew, L., Mahajan, M., Shukla, A.: Understanding cutting planes for QBFs. Inf. Comput. **262**, 141–161 (2018)
7. Beyersdorff, O., Galesi, N., Lauria, M.: A characterization of tree-like resolution size. Inf. Process. Lett. **113**(18), 666–671 (2013)
8. Beyersdorff, O., Pich, J.: Understanding Gentzen and Frege systems for QBF. In: Proceedings of the ACM/IEEE Symposium on Logic in Computer Science (LICS 2016) (2016)
9. Blinkhorn, J., Beyersdorff, O.: Shortening QBF proofs with dependency schemes. In: Gaspers, S., Walsh, T. (eds.) SAT 2017. LNCS, vol. 10491, pp. 263–280. Springer, Cham (2017). https://doi.org/10.1007/978-3-319-66263-3_17
10. Buss, S.R.: Towards NP-P via proof complexity and search. Ann. Pure Appl. Logic **163**(7), 906–917 (2012)
11. Cook, S.A., Reckhow, R.A.: The relative efficiency of propositional proof systems. J. Symb. Logic **44**(1), 36–50 (1979)
12. Egly, U., Kronegger, M., Lonsing, F., Pfandler, A.: Conformant planning as a case study of incremental QBF solving. Ann. Math. Artif. Intell. **80**(1), 21–45 (2017)
13. Egly, U., Lonsing, F., Widl, M.: Long-distance resolution: proof generation and strategy extraction in search-based QBF solving. In: McMillan, K., Middeldorp, A., Voronkov, A. (eds.) LPAR 2013. LNCS, vol. 8312, pp. 291–308. Springer, Heidelberg (2013). https://doi.org/10.1007/978-3-642-45221-5_21
14. Janota, M., Klieber, W., Marques-Silva, J., Clarke, E.: Solving QBF with counterexample guided refinement. In: Cimatti, A., Sebastiani, R. (eds.) SAT 2012. LNCS, vol. 7317, pp. 114–128. Springer, Heidelberg (2012). https://doi.org/10.1007/978-3-642-31612-8_10
15. Janota, M., Marques-Silva, J.: Expansion-based QBF solving versus Q-resolution. Theor. Comput. Sci. **577**, 25–42 (2015)
16. Kleine Büning, H., Bubeck, U.: Theory of quantified Boolean formulas. In: Biere, A., Heule, M., van Maaren, H., Walsh, T. (eds.) Handbook of Satisfiability, Frontiers in Artificial Intelligence and Applications, vol. 185, pp. 735–760. IOS Press (2009)
17. Kleine Büning, H., Karpinski, M., Flögel, A.: Resolution for quantified Boolean formulas. Inf. Comput. **117**(1), 12–18 (1995)
18. Klieber, W., Sapra, S., Gao, S., Clarke, E.: A non-prenex, non-clausal QBF solver with game-state learning. In: Strichman, O., Szeider, S. (eds.) SAT 2010. LNCS, vol. 6175, pp. 128–142. Springer, Heidelberg (2010). https://doi.org/10.1007/978-3-642-14186-7_12
19. Kontchakov, R., et al.: Minimal module extraction from DL-lite ontologies using QBF solvers. In: Proceedings of the International Joint Conference on Artificial Intelligence (IJCAI), pp. 836–841. AAAI Press (2009)
20. Lonsing, F.: Dependency schemes and search-based QBF solving: theory and practice. Ph.D. thesis, Johannes Kepler University (2012)
21. Lonsing, F., Biere, A.: DepQBF: a dependency-aware QBF solver. JSAT **7**(2–3), 71–76 (2010)
22. Lonsing, F., Egly, U.: Evaluating QBF solvers: quantifier alternations matter. In: Hooker, J. (ed.) CP 2018. LNCS, vol. 11008, pp. 276–294. Springer, Cham (2018). https://doi.org/10.1007/978-3-319-98334-9_19

23. Nordström, J.: On the interplay between proof complexity and SAT solving. SIGLOG News **2**(3), 19–44 (2015)
24. Rabe, M.N., Tentrup, L.: CAQE: a certifying QBF solver. In: Proceedings of the 15th Conference on Formal Methods in Computer-Aided Design, pp. 136–143. FMCAD Inc. (2015)
25. Vardi, M.Y.: Boolean satisfiability: theory and engineering. Commun. ACM **57**(3), 5 (2014)
26. Zhang, L., Malik, S.: Conflict driven learning in a quantified Boolean satisfiability solver. In: ICCAD, pp. 442–449 (2002)

Proof Complexity of QBF Symmetry Recomputation

Joshua Blinkhorn and Olaf Beyersdorff[(✉)]

Institut für Informatik, Friedrich-Schiller-Universität Jena, Jena, Germany
olaf.beyersdorff@uni-jena.de

Abstract. For quantified Boolean formulas (QBF), a resolution system
with a symmetry rule was recently introduced by Kauers and Seidl (Inf.
Process. Lett. 2018). In this system, many formulas hard for QBF resolution admit short proofs.

Kauers and Seidl apply the symmetry rule on symmetries of the original formula. Here we propose a new formalism where symmetries are
dynamically recomputed during the proof on restrictions of the original
QBF. This presents a theoretical model for the potential use of symmetry recomputation as an inprocessing rule in QCDCL solving.

We demonstrate the power of symmetry recomputation by proving
an exponential separation between Q-resolution with the symmetry rule
and Q-resolution with our new symmetry recomputation rule. In fact, we
show that bounding the number of symmetry recomputations gives rise
to a hierarchy of QBF resolution systems of strictly increasing strength.

1 Introduction

The last decade has seen tremendous advances in our understanding and algorithmic handling of quantified Boolean formulas (QBF), both theoretically and
practically. QBF solving has emerged as a powerful technique to apply to hard
problems from many application domains (e.g. [6,14,15]). To theoretically model
and analyse the success of QBF solving, a number of QBF proof systems have
been developed and analysed, yielding a surge in QBF proof complexity research
(e.g. [4,7,8]). Ideally, this interaction works in both directions: QBF resolution
systems aim to model central solving features and lower bounds to proof size
in these systems translate to lower bounds for solver running time. Conversely,
new theoretical models can also stimulate practical improvements.

This paper explores the power of symmetry recomputation for QBF from a
theoretical proof-complexity perspective. It is well known that many combinatorial principles exhibit symmetry properties [22], both propositionally and in
QBF. Breaking these symmetries is an effective technique for SAT and QBF
that can significantly reduce the search space and speed up search.

In SAT solving, symmetry breaking is done both statically [12] – as a preprocessing technique – as well as dynamically during the search [1,13,23]. Part
of the work on static symmetry breaking was lifted to the more complex setting

M. Janota and I. Lynce (Eds.): SAT 2019, LNCS 11628, pp. 36–52, 2019.
https://doi.org/10.1007/978-3-030-24258-9_3

of QBF [2,3], while dynamic symmetry breaking has not yet been realised in the QBF domain (cf. [17] for a recent overview of symmetry breaking in QBF).

On the proof-theoretic side, the propositional resolution system – underlying CDCL solving [5,21] – has been augmented with symmetry rules of different strengths [19,24,25]. In the most basic setting of [19], a symmetry rule is added that from a clause C allows to derive $\sigma(C)$ for each symmetry σ of the input CNF. Already this yields a quite powerful proof system, which e.g. admits short proofs of the pigeonhole principle [25].

Recently, the system of [19] was generalised to QBF by Kauers and Seidl [16]. This proof system Q-Res+S builds on Q-resolution (Q-Res[18]) and again augments it with a symmetry rule for symmetries of the original formula. In [16] the power of this proof system was demonstrated by the construction of short proofs for the formulas of Kleine Büning et al. [18] and of the parity formulas [8], two notorious examples of QBFs hard for Q-resolution.

In this paper we continue the proof-theoretic study of proof systems exploiting QBF symmetries. Our contributions can be summarised as follows.

1. QBF Resolution Systems for Symmetry Recomputation. We introduce a framework for symmetry recomputation during proof search. While the system Q-Res+S only allows to use symmetries of the input CNF, our new system Q-Res+SR additionally exploits symmetries of restrictions of the input formula. These restrictions correspond to certain partial assignments (called linear assignments here, Definition 5) that arise during runs of QCDCL algorithms. During such a run of QCDCL, we allow to recompute symmetries for the restricted formula currently considered by the solver. These 'newly recomputed' symmetries then replace the existing set of symmetries. When the QCDCL solver backtracks and unassigns variables, symmetries based on the corresponding restrictions of these variables are discarded as well.

We proof-theoretically model this algorithmic approach using the framework of Lonsing, Egly, and Seidl [20], who propose a general approach with additional axioms (corresponding to learned clauses) for a proof-theoretic modelling of inprocessing techniques in QCDCL. This framework has also been previously used for dependency recomputation in QBF [10]. This gives rise to our new QBF resolution system Q-Res+SR, where at each point in the proof, the current symmetries are made available to the symmetry rule. Using the approach of [10,20] we show soundness of the system (Theorem 9); completeness follows as the calculus extends Q-Res.

We can parameterise the system Q-Res+SR by keeping track of the maximal number d of symmetry recomputations on a QCDCL branch without any backtracking. We call this number d the degree of the Q-Res+SR proof. Restricting proofs in Q-Res+SR to degree 0 yields the system Q-Res+S.

2. Exponential Separations. Our main results assess the proof complexity of the new calculi for symmetry recomputation. We show that Q-Res+SR is exponentially stronger than Q-Res+S, and in fact, the subsystem of Q-Res+SR restricted to degree d proofs is exponentially stronger than Q-Res+SR restricted to degree $d-1$ proofs (Theorem 18). Thus allowing to successively recompute symmetries corresponds to a hierarchy of proof systems of strictly increasing strength.

To show this result we start by noticing that also the equality formulas – known to be hard in Q-Res and even stronger systems such as QBF cutting planes and polynomial calculus [7] – admit short proofs in Q-Res+S (Theorem 4).

We then devise a *symmetry blocker* (Definition 11): a simple gadget that when applied to an arbitrary QBF yields a 'blocked' QBF without any symmetries (Proposition 12). Such blocking can also be repeated, and applying it d times results in the d-blocked version of the formula. However, the symmetries can be unlocked again with symmetry recomputation, and in particular the d-blocked versions of the equality formulas have short proofs in Q-Res+SR of degree d (Lemma 17).

The main technical difficulty lies in showing that lower bounds for QBFs Φ_n for Q-Res lift to lower bounds for the d-blocked versions of Φ_n in the subsystem of Q-Res+SR restricted to degree $d-1$ proofs (Lemma 13). In combination, these upper and lower bounds yield the strictness of the hierarchy for the symmetry recomputation proof system (Theorem 18).

To ease the technical presentation, our model of QCDCL assignments used to define the system Q-Res+SR does neither incorporate unit propagation nor pure literal elimination (in contrast to the model of [20]). However, we argue (Sect. 6) that all our hardness results can be lifted to the practically more realistic setting where unit propagation and pure literal elimination is by default built into the proof system.

2   Preliminaries

Conjunctive Normal Form. A *literal* is a Boolean variable z or its negation $\bar{z}$. The *complement* $\bar{a}$ of a literal a is z if $a = \bar{z}$, and $\bar{z}$ if $a = z$. A *clause* is a finite disjunction of literals, and a *conjunctive normal form* formula (CNF) is a finite conjunction of clauses. We denote clauses as sets of literals and CNFs as sets of clauses. A clause is *tautological* if it contains some literal and its complement, otherwise it is *non-tautological*.

The variable of a literal $a = z$ or $a = \bar{z}$ is $\mathrm{var}(a) := z$. The variable set of a clause is $\mathrm{vars}(C) := \{\mathrm{var}(a) : a \in C\}$, and the variable set of a CNF is $\mathrm{vars}(f)$, the union of the variable sets of the clauses in f.

An assignment to a set $Z \subseteq \mathrm{vars}(f)$ is a mapping $\alpha : Z \to \{0, 1\}$, typically represented as a set of literals $\{a_1, \ldots, a_k\}$, where literals $\bar{z}$ and z represent the respective assignments $z \mapsto 0$ and $z \mapsto 1$. The *restriction* of f by α, denoted $f[\alpha]$, is obtained from f by removing any clause containing a literal in α, and removing the complementary literals $\bar{a}_1, \ldots, \bar{a}_k$ from the remaining clauses.

Quantified Boolean Formulas. A *prenex quantified Boolean formula* (QBF) $F := Q \cdot f$ consists of a *quantifier prefix* $Q = Q_1 z_1 \cdots Q_n z_n$, in which each Q_i

is a quantifier in $\{\exists, \forall\}$, and a CNF f called the *matrix*, for which vars$(f) = \{z_1, \ldots, z_n\}$. The variable set of a QBF is vars$(F) := $ vars(f). The prefix Q defines a total order $<_Q$ on vars(F) such that $z_i <_Q z_j$ holds whenever $i < j$, in which case we say that z_i *is left of* z_j and z_j *is right of* z_i. Variables in the first and last blocks are termed *leftmost* and *rightmost*, respectively.

The *restriction* of F by an assignment α is $F[\alpha] := Q[\alpha] \cdot f[\alpha]$, where $Q[\alpha]$ is obtained from Q by removing the variables vars$(f) \setminus$ vars$(f[\alpha])$ along with their associated quantifiers.

A model g for a QBF $F := Q \cdot f$ is a set $\{G_x : x \in \text{vars}_\exists(F)\}$ for which (a) each G_x is a Boolean circuit over the universal variables left of x, and (b) the simultaneous substitution of each G_x for x in f is a tautology, i.e. a circuit computing the constant function 1. A QBF is true iff it has a model, otherwise it is false.

Proof Systems. Given any literal p, a clause $C = C_1 \cup C_2$ is called a *resolvent* of $C_1 \cup \{p\}$ and $C_2 \cup \{\bar{p}\}$. Given any prefix Q, a clause C is a *Q-reduction* of a clause $C \cup R$ if every variable in R is universally quantified and right of every variable in C (with respect to Q).

A Q-resolution (Q-Res) [18] refutation of a QBF $Q \cdot f$ is a sequence $C_1, \ldots, C_k$ of non-tautological clauses in which (a) C_k is the empty clause, and (b) each clause either belongs to f, or is a resolvent or Q-reduction of preceding clauses.

A proof system P *p-simulates* a proof system Q if there exists a polynomial-time computable function that takes a Q-proof to a P-proof of the same formula [11].

3   Static Symmetries in Q-resolution

This section provides some background on Q-Res+S and its proof complexity.

Definition 1 (symmetry). *A* symmetry σ *for a QBF* $Q \cdot f$ *is a bijection on its literals for which* (a) *applying* σ *to every literal in every clause preserves* f, *and* (b) *for each literal* $a \in \text{dom}(\sigma)$, $\sigma(a), \sigma(\bar{a})$ *are complementary literals and* var(a), var$(\sigma(a))$ *belong to the same block of* Q.

The set of all symmetries of a QBF forms a group under composition, and is denoted $\mathcal{S}(F)$.

Symmetries are incorporated into Q-Res with the addition of a single inference rule: any symmetry of the input QBF can be applied to a derived clause. This rule is labelled 'S' in the following definition.

Definition 2 (Kauers and Seidl [17]). *A* Q-Res+S *refutation of a QBF* $Q \cdot f$ *is a sequence of non-tautological clauses* $\pi := C_1, \ldots, C_k$ *in which* C_k *is the empty clause and one of the following holds for each* $i \in [k]$:

A **axiom:** *C_i is a clause in the matrix f;*
R **resolution:** *C_i is the resolvent of two preceding clauses;*

U **universal reduction:** C_i is a Q-reduction of a preceding clause;
W **weakening:** C_i is subsumed by a preceding clause;
S **symmetry:** C_i is the image of a preceding clause under a symmetry of $Q \cdot f$.

The size of π is $|\pi| = k$.

It was shown in [16] that Q-Res+S is exponentially stronger than Q-Res, the separation was demonstrated by two different QBF families. Here, we briefly point out that the separation is also performed by another QBF family, namely the equality formulas [7]. The particulars of the linear-size Q-Res+S refutations of equality are used later on in Sect. 5.

Definition 3 (equality family [7]). The equality family is the QBF family whose n^{th} instance is $\mathrm{EQ}_n := \exists x_1 \cdots x_n \forall u_1 \cdots u_n \exists z_1 \cdots z_n \cdot \mathrm{eq}_n$, where

$$\mathrm{eq}_n := \left(\bigcup_{i \in [n]} \{\{\bar{x}_i \bar{u}_i \bar{z}_i\}, \{x_i u_i \bar{z}_i\}\} \right) \cup \{\{z_1 \cdots z_n\}\}.$$

Theorem 4. The equality family has linear-size refutations in Q-Res+S.

Proof. For $i \in [n]$, let σ_i be the bijection on the literals of EQ_n that sends literals in x_i and u_i to their complements, and is the identity everywhere else. It is easy to see that each σ_i is a symmetry of EQ_n. In n resolution steps, resolving $\{z_1 \cdots z_n\}$ against each $\{\bar{x}_i \bar{u}_i \bar{z}_i\}$ over z_i, we obtain the clause $\{\bar{x}_1 \cdots \bar{x}_n \bar{u}_1 \cdots \bar{u}_n\}$. By universal reduction, we then obtain $\{\bar{x}_1 \cdots \bar{x}_n\}$. From any $\{\bar{x}_1 \cdots \bar{x}_i\}$, we obtain $\{\bar{x}_1 \cdots \bar{x}_{i-1}\}$ by application of σ_i and resolution over x_i.   □

4 A Theoretical Model of Symmetry Recomputation

In this section we introduce the proof system Q-Res+SR, a theoretical model for QBF symmetry recomputation.

Our model is built on the foundation of 'Q-Resolution with Generalised Axioms' due to Lonsing et al. [20], whose primary motivation was to model the integration of preprocessing techniques into the QCDCL search process itself; in other words, to reformulate preprocessing as inprocessing. The authors, however, noted that their setup offers a more general 'interface to Q-resolution' capable of modelling 'the direct combination of orthogonal solving techniques' [20].

This direction was subsequently taken up in the paper [10]. The setup was reformulated to interface with QBF dependency schemes, thereby modelling the integration of dependency-awareness into QCDCL search. An important development there was the introduction of *proof referencing*. A refutation becomes a nested structure of 'subrefutations' of restrictions of the instance, allowing the work behind the interface to contribute to the overall proof size. This addition is key for proof complexity, since without it, every QBF has a trivial refutation (cf. [20]).

Here, in much the same spirit, we propose to interface with the computation of symmetries, thereby modelling the integration of symmetry techniques into

QCDCL search. This offers a new possibility: otherwise unidentified symmetries of the current formula can be found by recomputation at arbitrary search nodes. New symmetries can be applied to learned clauses, thereby strengthening the knowledge base.

The Trail. Central to the model of Lonsing et al. is a particular kind of QBF assignment (QCDCL assignment [20, p. 437]), intended to represent the current assignment, or *trail*, of the solver at an arbitrary search node. Here, we have chosen to omit constraint propagation, and work instead with *linear assignments*. Propagation can be safely detached from the proof complexity discussion, with considerable simplification of technical content – this is explained in greater detail in Sect. 6.

Definition 5 (linear assignment). *A partial assignment $\{a_1, \ldots, a_n\}$ to a QBF F is* linear *if, for each i in $[n]$, $\mathrm{var}(a_i)$ is in the first block of $F[\{a_1, \ldots, a_{i-1}\}]$.*

The proof system Q-Res+SR. Derivations in Q-Res+SR are defined recursively by *degree*, using the proof referencing method of [10].

Definition 6 (Q-Res+SR). *Given a QBF F, Q-Res+SR refutations of F are defined inductively by degree:*

- *a degree-0 refutation is a Q-Res+S refutation of F;*
- *for $d \in \mathbb{N}$, a degree-d refutation is a sequence $\pi_0 \circ \rho_1 \circ \cdots \circ \rho_k$ satisfying*
  - *π_0 is a Q-Res+S refutation of F with extra axioms $A_1, \ldots, A_k$;*
  - *each A_i is the negation of a linear assignment α_i to F;*
  - *each ρ_i is a Q-Res+SR refutation of $F[\alpha_i]$;*
  - *the maximum degree of the ρ_i is $d - 1$.*

The size of a Q-Res+SR *refutation is the number of clauses in the sequence.*

The extra axioms $A_1, \ldots, A_k$ are said to *reference* the lower-degree refutations $\rho_1, \ldots, \rho_k$. We illustrate the system, and the use of proof referencing, with the following example.

Example 7. The QBF $F := \exists a x_1 \forall u_1 \exists z_1 \cdot \{\{a x_1 u_1 z_1\}\{a \bar{x}_1 \bar{u}_1 z_1\}\{a \bar{z}_1\}\{\bar{a}\}\}$ has a degree-1 Q-Res+SR refutation $\pi \circ \rho$, where π and ρ are the sequences:

$\pi :=$	1 $\{a\}$	extra axiom	$\rho :=$	1 $\{x_1 u_1 z_1\}$	axiom
	2 $\{\bar{a}\}$	axiom		2 $\{\bar{z}_1\}$	axiom
	3 $\square$	resolution		3 $\{x_1 u_1\}$	resolution
				4 $\{x_1\}$	universal reduction
				5 $\{\bar{x}_1\}$	symmetry
				6 $\square$	resolution

Consider the linear assignment $\alpha := \{\bar{a}\}$ to F. Notice that $F[\alpha] = \mathrm{EQ}_1$, and that the sequence ρ is the Q-Res+S refutation of EQ_1 described in the proof of Theorem 4. In line 1 of π, the clause $\{a\}$, being the negation of α, can be introduced as an extra axiom, referencing the degree-0 refutation ρ. The refutation is concluded in π by resolution against the unit clause $\{\bar{a}\}$.

Notice that the application of the symmetry rule in line 5 of ρ would not be allowed in π, since σ_1 (the symmetry of the restricted formula EQ_1 that sends x_1 and u_1 to their complements, and is the identity everywhere else) is not a symmetry of F itself. $\qquad\square$

Modelling Symmetry Recomputation. We provide some high-level intuition on how symmetry recomputation would work in practice, and its connection to proof referencing in Q-Res+SR.

In our model, we consider the symmetry groups of the input formula and its restrictions; derived clauses and free axioms do not contribute to these symmetry groups. This is analogous to the solving approach in which learned clauses are the target of the new recomputed symmetries, and not the source. While there are other possibilities, the current approach is perhaps the most straightforward for a first theoretical model.

The degree of a refutation can be understood as the maximum depth of nested symmetry recomputations. When recomputation takes place, a pointer to the new set of symmetries is placed on the trail. The current symmetry set can be applied to any newly learned clauses until it is either replaced by recomputation, or it is removed from the trail by backtracking step. In the latter case, the solver reverts to an earlier set of symmetries, following the highest level pointer that remains on the trail. In this way, higher degree refutations are associated with symmetries recomputed deeper into the search, under increasingly larger restrictions of the input formula. Degree-0 refutations, which coincide with Q-Res+S, represent the traditional setting in which no recomputation takes place.

One might ask whether the symmetries of the parent formula should be available to referenced refutations. However, a simple example demonstrates that this is not sound, even at the propositional level. Consider the true QBF

$$F := \exists ax \cdot \{\{ax\}\{\bar{a}\bar{x}\}\},$$

and the symmetry σ that sends both literals a and x to their complements. Applying σ to either of

$$F[\{\bar{a}\}] = \exists x \cdot \{\{x\}\}, \quad F[\{a\}] = \exists x \cdot \{\{\bar{x}\}\}$$

permits a refutation, but $F[\{\bar{a}\}]$ and $F[\{a\}]$ are both true QBFs. Moreover, the unit clauses $\{a\}, \{\bar{a}\}$ can be introduced as extra axioms, refuting F itself.

There is a subtle point here: application of symmetries and restriction of clauses do not associate. In practice, symmetries must be applied to the whole learned clause, not merely to its restriction under the current trail assignment. The restriction, if performed first, may remove literals which would have been satisfied under the symmetry – this is the issue with the foregoing example.

Soundness. It is already known from [10, 20] that the method of proof referencing admits a soundness proof by induction on degree. We follow this method to prove the soundness of Q-Res+SR. The following lemma constitutes the chief observation. It is the analogue of [20, Theorem 2] and [10, Lemma 14].

Lemma 8. *Let A be the negation of a linear assignment α to a QBF $Q \cdot f$. If $Q \cdot f$ is false under α, then $Q \cdot f \vDash Q \cdot \{A\}$.*

It is known that the rules of Q-Res+S preserve the models of the input QBF [16], so degree-0 refutations are sound. Moreover, if we can show that models of the input QBF also satisfy the extra axioms, then we prove the soundness of refutations at the next degree. This is merely the contrapositive statement of Lemma 8, since extra axioms reference refutations of lower degree, which refute false formulas by the inductive hypothesis.

Theorem 9. *If a QBF has a Q-Res+SR refutation, then it is false.*

Since our setting differs at the technical level from both [20] and [10], we provide a full proof of the core lemma.

Proof (of Lemma 8). The lemma is trivially true if $F := Q \cdot f$ is false, so we assume otherwise.

Let $A := \{a_1 \cdots a_k\}$ be the negation of a linear assignment α to F. For each $0 \le i \le k$, let A_i be the first i literals of A, let α_i be its negation, and define

$$E_i := \mathrm{vars}_\exists(F[\alpha_{i-1}]) \setminus \mathrm{vars}_\exists(F[\alpha_i]),$$
$$U_i := \mathrm{vars}_\forall(F[\alpha_{i-1}]) \setminus \mathrm{vars}_\forall(F[\alpha_i]).$$

Now, let g be a model for F. Set $g_0 := g$, and for each i in $[k]$, obtain g_i from g_{i-1} by discarding the circuits for variables in E_i and restricting the rest by the assignment

$$\beta_i := \{\bar{a}_i\} \cup \{\bar{z} : z \in U_i \setminus \{\mathrm{var}(a_k)\}\}.$$

Lastly, let T_i be the circuit obtained by substituting the circuits in g_i for the existential variables in the CNF $f[\alpha_i]$.

By induction on k, we prove the following: if g models F but not $Q \cdot A_k$, then g_k models $F[\alpha_k]$. The base case $k = 0$ is trivial. For the inductive step, let $k \ge 1$. Suppose that g models F but not $Q \cdot \{A_k\}$. Then g does not model $Q \cdot \{A_{k-1}\}$, so g_{k-1} models $F[\alpha_{k-1}]$ by the inductive hypothesis, as the negation of A_{k-1} is a linear assignment. Hence T_{k-1} is a tautology.

On the other hand, $T_k = T_{k-1}[\beta_i]$, so T_k is a tautology, and g_k models $F[\alpha_k]$. Indeed, given $f[\alpha_{k-1}]$, applying the substitution based on g_{k-1} followed the restriction β_k has the same effect as applying the restriction β_k first, followed by substitution of the restricted circuits g_k. To see this, one must note the following: if $\{\bar{a}_k\}$ represents an existential assignment, say $x \mapsto b$, then the circuit for x in g_{k-1} computes the constant b, so substituting that circuit for x is the same as applying the assignment $\{\bar{a}_k\}$. $\qquad\square$

5   Proof Complexity of Symmetry Recomputation

In this section we show that degree-d Q-Res+SR refutations may be exponentially shorter than refutations of degree $d - 1$, for each natural number d.

We make use of the product operation [9] on CNFs and QBFs. The product of two CNFs f and g is $f \otimes g := \{C \cup D : C \in f, D \in g\}$, and the product of two QBFs $F := Q \cdot f, G := R \cdot g$ is $F \otimes G := QR \cdot f \otimes g$.

Provided that the concatenation of prefixes does not create a longer block in the middle, taking a product has the natural effect on the symmetry groups.

Proposition 10. *Let F and G be variable-disjoint QBFs. If the rightmost block of F and the leftmost block of G are oppositely quantified, then*

$$\mathcal{S}(F \otimes G) = \{\sigma \cup \tau : \sigma \in \mathcal{S}(F), \tau \in \mathcal{S}(G)\}.$$

Proof. It is clear that $\sigma \cup \tau$ is a symmetry of $F \otimes G$ whenever σ, τ are respective symmetries of F, G. For the reverse direction, suppose that $\sigma \cup \tau$ is a symmetry of $F \otimes G$, where the respective domains of σ, τ are the literals of F, G. Let C be a clause in $f \otimes g$, and let C_f, C_g be the respective intersections of C with the literals of F, G. Since each block of $F \otimes G$ is either a block of F or of G, we have

$$\begin{aligned}
C_f \in f \text{ and } C_g \in g &\Leftrightarrow C \in f \otimes g \\
&\Leftrightarrow (\sigma \cup \tau)(C) \in f \otimes g \\
&\Leftrightarrow \sigma(C_f) \in f \text{ and } \tau(C_g) \in g,
\end{aligned}$$

where the last equivalence is due to the fact that both σ, τ are bijections. $\square$

Symmetry Blocking. In [10], a technique for obfuscating the independencies of a particular QBF family was introduced. Here, we devise a similar method for blocking QBF symmetries. The main idea is to add literals in fresh variables, such that only the identity symmetry survives; meanwhile, an assignment to the fresh variables returns the original instance with the symmetries intact.

Definition 11 (blocker). *Given any QBF $F := Q \cdot f$ over variables $Z = \{z_1, \ldots, z_k\}$, the symmetry blocker for F is the QBF $\mathcal{A}(F)$ whose prefix is $\exists a_0 a_1 \cdots a_k \forall b Q \forall c$ and matrix is*

$$\{\{a_0 bc\}\} \cup (\{\{a_0 a_1 \cdots a_i z_i\} : i \in [k]\} \otimes \{\{bc\}\}) \cup (\{\{\bar{a}_0 c\}\} \otimes f),$$

where $a_0, a_1, \ldots, a_k, b, c$ are fresh variables not in Z. For each natural number d, the d-blocker of F is

$$\mathcal{D}_d(F) := \mathcal{A}^d(F) \otimes \cdots \otimes \mathcal{A}^1(F),$$

where $\mathcal{A}^i(F)$ is obtained from $\mathcal{A}(F)$ by adding the superscript i to each occurrence of a variable.

Proposition 12. *For any natural number d and QBF F whose left- and right-most variables are existential, the only symmetry of $\mathcal{D}_d(F)$ is the identity.*

Proof. By Proposition 10, every symmetry of $\mathcal{D}_d(F)$ is of the form $\sigma_1 \cup \cdots \cup \sigma_d$, where each σ_i is a symmetry of $\mathcal{A}^i(F)$. Hence, by syntactic equivalence, it suffices to show that the identity is the only symmetry of $\mathcal{A}(F)$.

Let σ be a symmetry of $\mathcal{A}(F)$. Since $\{b\}$ and $\{c\}$ are both singleton universal blocks of $\mathcal{A}(F)$, and both variables b, c occur only in positive polarity, we must have $\sigma(b) = b$ and $\sigma(c) = c$. The only literal ℓ for which $\{\ell bc\}$ is a clause in the matrix is $\ell = a_0$, hence we must have $\sigma(a_0) = a_0$. For each $i \in k$, the positive literal a_i occurs in a clause of size $i + 4$ containing both literals b and c; there is only one such clause in the matrix, hence we must have $\sigma(a_i) = a_i$. For each $i \in k$, the only literal ℓ for which $\{a_0 a_1 \cdots a_i b \ell c\}$ is a clause in the matrix of $\mathcal{A}(F)$ is $\ell = z_i$, hence we must have $\sigma(z_i) = z_i$. The remainder of σ is defined by the property $\overline{\sigma(a)} = \sigma(\bar{a})$; hence σ is the identity.   □

Lower Bound. We use symmetry blocking to prove the following lemma.

Lemma 13. *If a QBF family requires $T(n)$-size Q-Res refutations, and the left- and rightmost variable of every instance is existential, then its d-blocker requires $T(n)$-size Q-Res+SR refutations of degree $d - 1$.*

Our argument uses three low-level propositions. Each of them details a situation in which a refutation-size lower bound can be inferred from that of a simpler QBF. The first is well known, and states the closure of Q-Res under existential restrictions.

Proposition 14. *Given a Q-Res refutation π of a QBF F and an assignment ε to its existentials, $\pi[\varepsilon]$ is a Q-Res refutation of $F[\varepsilon]$ whose size is at most $|\pi|$.*

For the second proposition, we say that a QBF $F := Q \cdot f$ subsumes another $G := R \cdot g$ if the following two conditions hold:

- each clause in f is a subset of some clause in g;
- $\mathrm{vars}(F) \subseteq \mathrm{vars}(G)$, and, for each $x, y \in \mathrm{vars}(F)$, $x <_Q y \Rightarrow x <_R y$.

It is easy to see that QBFs cannot have smaller Q-Res refutations than those which subsume them.

Proposition 15. *Let F and G be false QBFs. If F subsumes G, then the shortest Q-Res refutation of G is no smaller than that of F.*

The third proposition is rather more specific to Q-Res+SR and symmetry blockers; therefore we include a proof.

Proposition 16. *Let F and G be variable-disjoint QBFs satisfying:*

(a) *the rightmost block of F and the leftmost block of G are oppositely quantified;*
(b) *a rightmost variable appears in every clause of F;*
(c) *a leftmost variable appears in every clause of G;*
(d) *the only symmetry of G is the identity.*

Then the shortest degree-d Q-Res+SR *refutation of* $F \otimes G$ *is no smaller than that of* G.

Proof. Let $\pi := \pi_0 \circ \rho_1 \circ \cdots \circ \rho_k$ be a degree d Q-Res+SR derivation from $F \otimes G$, and suppose that the identity is the only symmetry of G. Let $\pi' := \pi'_0 \circ \rho'_1 \circ \cdots \circ \rho'_k$ be the sequence obtained from π by removing all literals in variables of F; in particular, let $\pi_0 = C_1, \ldots, C_n$ and $\pi'_0 = C'_1, \ldots, C'_n$. By induction on n and d, we show that π' is a valid degree d Q-Res+SR derivation from G.

The base case $n = 0$ is trivial, since the derivation is empty. For the inductive step, we branch on the inference rule with which C_n was derived.

(a) Suppose that C_n was derived as a standard axiom. Then C'_n belongs to the matrix of G, and can be derived as an axiom.

(b) Suppose that C_n was derived as the resolvent of C_a and C_b over a pivot variable p. If p is in vars(G), then C'_n can be derived as a resolvent of C'_a and C'_b. On the other hand, if p is in vars(F), then C'_n can be derived from one of C'_a and C'_b by weakening.

(c) Suppose that C_n was derived by universal reduction from C_a. Then C'_n may be derived by universal reduction from C'_a.

(d) Suppose that C_n was derived by weakening from C_a. Then C'_n may be derived by weakening from C'_a.

(e) Suppose that C_n was derived by application of the symmetry σ to C_a. Then, by Proposition 10, σ is the identity on G, and $C'_n = C'_a$ can be derived by application of the identity symmetry.

(f) Suppose that C_n was introduced as an extra axiom, let α_i be its negation and let ρ_i be the corresponding referenced refutation.

If vars$(F) \subseteq$ vars(C_n), then ρ'_i is identical to ρ_i; that is, it is a valid refutation of $F \otimes G[\alpha_i]$ of degree at most $d - 1$. Let α_i^F and α_i^G be the subassignments of α_i on the variables of F and G, respectively. It is easy to see that, since $F \otimes G[\alpha_i]$ is false by the soundness of Q-Res+SR, α_i^F falsifies every clause in the matrix of F. It follows that $G[\alpha_i^G] = F \otimes G[\alpha_i]$. Therefore C'_n can be introduced as an extra axiom, referencing ρ'_i.

On the other hand, if vars$(F) \not\subseteq$ vars(C_n), then we must have vars$(C_n) \subset$ vars(F), by conditions (a), (b) and (c), and the fact that the negation of C_n is a linear assignment to $F \otimes G$. Hence ρ_i is a refutation of $F[\alpha_i] \otimes G$ of degree at most $d - 1$. By induction on degree, ρ'_i is a valid degree $d - 1$ refutation of G. Thus C'_n, which is the empty clause, may be introduced as an extra axiom, referencing ρ'_i. □

With these three propositions, we are ready to prove Lemma 13.

Proof (of Lemma 13). Let $\{F_n\}_{n \in \mathbb{N}}$ be QBFs requiring $T(n)$-size Q-Res refutations, and let $\{\pi_n\}_{n \in \mathbb{N}}$ be degree $d - 1$ Q-Res+SR refutations of $\{\mathcal{D}_d(F_n)\}_{n \in \mathbb{N}}$. We prove that $|\pi_n| \geq T(n)$ by induction on the degree d.

For the base case $d = 1$, recall that the only symmetry of $\mathcal{D}_1(F)$ is the identity, by Proposition 12. It follows that each π_n is a valid Q-Res refutation of $\mathcal{D}_1(F_n)$. Now, let α be the assignment $\{a_0 a_1 \bar{c}\}$. It is easy to see that $\mathcal{D}_1(F_n)[\alpha]$

is syntactically equivalent to F_n. Moreover, by the monotonic existential closure of Q-Res (Proposition 14), restriction of π_n by α yields a valid Q-Res refutation of $\mathcal{D}_1(F_n)[\alpha]$, whose size is no larger than $|\pi_n|$. Thus $|\pi_n| \geq T(n)$.

For the inductive step, let $d \geq 2$. We call an extra axiom A in π_n *short* if $c^d \notin \text{vars}(A)$. We consider two cases.

(a) Suppose that π_n uses a short extra axiom A, and let β be the negation of A. Note that c^d appears in every clause of the false QBF $\mathcal{A}^d(F_n)$ and forms a singleton universal block. Hence β, which is a linear assignment to $\mathcal{D}_d(F_n)$, contains only variables from $\mathcal{A}^d(F_n)$. Hence A in π_n references some refutation ρ, whose degree is at most $d - 1$, of the QBF

$$\mathcal{A}^d(F_n)[\beta] \otimes \mathcal{A}^{d-1}(F_n) \otimes \cdots \otimes \mathcal{A}^1(F_n),$$

which is syntactically equivalent to $\mathcal{A}^d(F_n)[\beta] \otimes \mathcal{D}_{d-1}(F_n)$. Note that the matrix of $\mathcal{A}^d(F_n)[\beta]$ contains the rightmost variable c^d in every clause. Thus, by Proposition 16 and the inductive hypothesis, $|\pi_n| \geq |\rho| \geq T(n)$.

(b) On the other hand, suppose that π_n uses no short extra axiom. By definition, π_n is of the form $\pi' \circ \rho_1 \circ \cdots \circ \rho_k$, where π is a Q-Res+S refutation of $\mathcal{D}_d(F_n)$ using $k \geq 0$ extra axioms $A_1, \ldots, A_k$, referencing the refutations $\rho_1, \ldots, \rho_k$. By Proposition 12, the only symmetry of $\mathcal{D}_d(F_n)$ is the identity; therefore π' is in fact a Q-Res refutation of $\mathcal{D}_d(F_n)$ with the same extra axioms. We claim that every extra axiom in π_n is subsumed by some clause in $\mathcal{A}^d(F_n)$, in which case π' is a Q-Res refutation of a QBF that is subsumed by $\mathcal{A}^d(F_n)$. Since $\mathcal{A}^d(F_n)$ and $\mathcal{D}_1(F_n)$ are syntactically equivalent, and the latter requires $T(n)$-size Q-Res refutations, we have $|\pi_n| \geq T(n)$ by Proposition 15.

It remains to show that every extra axiom in π' is subsumed by some clause in $\mathcal{A}^d(F_n)$. To that end, let A_i be an extra axiom in π_n, and let α_i be its negation. Now, the rightmost variable c^d of $\mathcal{A}^d(F_n)$ appears in every clause, and forms a single universal block in $\mathcal{D}_d(F_n)$. Hence, since α_i is a linear assignment, $\mathcal{A}^d(F_n)[\alpha_i]$ has no variables. Moreover, its matrix cannot be empty, for otherwise $\mathcal{D}_d(F_n)[\alpha_i]$, which is refuted by ρ_i, would be true, contradicting the soundness of Q-Res+SR. It follows that the matrix of $\mathcal{A}^d(F_n)[\alpha_i]$ contains the empty clause. Thus some clause of $\mathcal{A}^d(F_n)$ subsumes the negation of α_i, and the claim follows.   □

Upper Bound. For the corresponding upper bound, we construct short degree-d refutations of the d-blocker for the equality family. A general construction, matching the upper-bound argument, is not possible here; to prove the correctness of our construction, we need to use the specifics of the formulas inside the d-blocker.

We give a brief description of the construction. The central idea is that restriction of $\mathcal{A}(\text{EQ}_n)$ by the assignment $a_0 \mapsto 1$ yields $\text{EQ}_n \otimes \forall c \cdot \{\{c\}\}$, whose symmetries are (essentially) those of equality itself. This establishes a short refutation for $d = 1$.

For larger d, the construction is iterated. Restriction of $\mathcal{A}^1(\text{EQ}_n)$ by $a_0^1 \mapsto 1$ unlocks new symmetries (again, essentially those of EQ_n), which are used in conjunction with extra axioms that reference short proofs in a repeated, nested fashion.

Lemma 17. *For each d in $\mathbb{N}$, the d-blocker of the equality family has $O(n)$-size Q-Res+SR refutations of degree d.*

Proof. Let d be a fixed natural number. For each i in $[d]$, we let EQ_n^i be the QBF obtained from EQ_n by adding the superscript i to each variable occurrence, and let eq_n^i be its matrix; further, we define a set of clauses

$$f_n^i := \{\{x_1^i \cdots x_n^i u_1^i \cdots u_n^i\}\} \otimes \{\{\bar{z}_1^i\} \cdots \{\bar{z}_n^i\}\{z_1^i \cdots z_n^i\}\} \otimes \{\{c^i\}\},$$

and a sequence π_n^i, read as two columns:

$$
\begin{array}{ll}
\{x_1^i \cdots x_n^i u_1^i \cdots u_n^i z_1^i \cdots z_n^i\} & \{x_1^i \cdots x_n^i\} \\
\{x_1^i \cdots x_n^i u_1^i \cdots u_n^i \bar{z}_n^i\} & \{x_1^i \cdots \bar{x}_n^i\} \\
\{x_1^i \cdots x_n^i u_1^i \cdots u_n^i z_1^i \cdots z_{n-1}^i\} & \{x_1^i \cdots x_{n-1}^i\} \\
\{x_1^i \cdots x_n^i u_1^i \cdots u_n^i \bar{z}_{n-1}^i\} & \{x_1^i \cdots \bar{x}_{n-1}^i\} \\
\vdots & \vdots \\
\{x_1^i \cdots x_n^i u_1^i \cdots u_n^i z_1^1\} & \{x_1^i\} \\
\{x_1^i \cdots x_n^i u_1^i \cdots u_n^i \bar{z}_1^1\} & \{\bar{x}_1^i\} \\
\{x_1^i \cdots x_n^i u_1^i \cdots u_n^i\} & \square
\end{array}
$$

Based on the proof of Theorem 4, it is easy to see that $\mathrm{seq}(f_n^i) \circ \pi_n^i$ is a Q-Res+S refutation of $\mathrm{EQ}_n^i \otimes F_c^i$ using extra axioms f_n^i, where $F_c^i := \forall c^i \cdot \{\{c^i\}\}$ and $\mathrm{seq}()$ denotes the clauses of a CNF written in an arbitrary fixed sequence.

Now we build short degree-d refutations ζ_n^d of $\mathcal{D}_d(\mathrm{EQ}_n)$. We define

$$\zeta_n^1 := \{a_0^1\}, \{\bar{a}_0^1\}, \square \circ \rho_n^1 \quad \text{and} \quad \rho_n^1 := \mathrm{seq}(\mathrm{eq}_n^1) \circ \pi_n^1;$$

further, for each $2 \le i \le d$, we define $\zeta_n^i := \{a_0^i\}, \{\bar{a}_0^i\}, \square \circ \zeta_n^{i-1} \circ \rho_n^i$, where

$$\rho_n^i := \mathrm{seq}(f_n^i \otimes \{\{a_0^{i-1}\}\}) \circ \mathrm{seq}(f_n^i \otimes \{\{\bar{a}_0^{i-1}\}\}) \circ \pi_n^i \circ \zeta_n^{i-2} \circ \rho_n^{i-1}.$$

Here, we take $\zeta_n^0 := \emptyset$.

We observe that $|\zeta_n^d| = O(n)$. To see this, observe that there exists constant c_1, c_2 such that $|\zeta_n^1| \le c_1 \cdot n$ and, for each i in $[d]$, $|\zeta_n^i| \le c_2 |\zeta_n^{i-1}|$. Hence $|\zeta_n^d| \le c_1 c_2^{d-1} \cdot n$.

To finish the proof, we show two invariants by induction on d:

(1) ρ_n^{d-1} is a refutation of $\mathcal{D}_d(\mathrm{EQ}_n)[\alpha_d]$ of degree $d-1$;
(2) ζ_n^d is a refutation of $\mathcal{D}_d(\mathrm{EQ}_n)$ of degree d.

We make use of $\alpha_d : a_0^d \mapsto 1$ and $\bar{\alpha}_d : a_0^d \mapsto 0$, which are both linear assignments to $\mathcal{D}_d(\mathrm{EQ}_n)$.

For the base case $d = 1$, observe that ζ_n^1 is a degree-0 refutation of $\mathcal{D}_1(\mathrm{EQ}_n)$ using a single extra axiom $\{\bar{a}_0^d\}$ whose negation is α_1. Since $\mathrm{eq}_n^1 \otimes \{\{c^1\}\}$ subsumes f_n^1, ρ_n^1 is a degree 0 refutation of $\mathrm{EQ}_n^1 \otimes F_c^1$. This establishes invariant (1), since $\mathcal{D}_1(\mathrm{EQ}_n)[\alpha_1] = \mathrm{EQ}_n^1 \otimes F_c^1$. It follows that ζ_n^1 is indeed a degree-1 refutation of $\mathcal{D}_1(\mathrm{EQ}_n)$, establishing invariant (2).

For the inductive step, let $d \geq 2$. Every symmetry of $\mathrm{EQ}_n^d \otimes F_c^d$, when extended by the identity on the variables of $\mathcal{A}^{d-1}(\mathrm{EQ}_n) \otimes \cdots \otimes \mathcal{A}^1(\mathrm{EQ}_n)$, is a symmetry of

$$\mathcal{D}_d(\mathrm{EQ}_n)[\alpha_d] = \mathrm{EQ}_n^d \otimes F_c^d \otimes \mathcal{A}^{d-1}(\mathrm{EQ}_n) \otimes \cdots \otimes \mathcal{A}^1(\mathrm{EQ}_n)$$

by Proposition 10. It follows that

$$\mathrm{seq}(f_n^d \otimes \{\{a_0^{d-1}\}\}) \circ \mathrm{seq}(f_n^d \otimes \{\{\bar{a}_0^{d-1}\}\}) \circ \pi_n^d$$

is a degree-0 refutation of $\mathcal{D}_d(\mathrm{EQ}_n)[\alpha_d]$, given the first two terms as extra axioms. Let β_1 and β_2 be the negations of any clause in $\mathrm{seq}(f_n^d \otimes \{\{a_0^{d-1}\}\})$ and $\mathrm{seq}(f_n^d \otimes \{\{\bar{a}_0^{d-1}\}\})$, respectively. It is readily verified that both β_1 and β_2 are a linear assignments to $(\mathcal{D}_d(\mathrm{EQ}_n) \otimes F_c^d)[\alpha_d]$. By the inductive hypothesis, ζ_n^{d-2} is a refutation of $\mathcal{D}_d(\mathrm{EQ}_n)[\alpha_d][\beta_1] = \mathcal{D}_{d-2}(\mathrm{EQ}_n)$ of degree $d-2$, where $\mathcal{D}_0(\mathrm{EQ}_n)$ is the QBF on the empty set of variables whose matrix contains only the empty clause. Also by the inductive hypothesis, ρ_n^{d-1} is a refutation of $\mathcal{D}_d(\mathrm{EQ}_n)[\alpha_d][\beta_2] = \mathcal{D}_{d-1}(\mathrm{EQ}_n)[\alpha_{d-1}]$ of degree $d-1$. This establishes invariant (1).

Now, observe that ζ_n^d is a degree-0 refutation of $\mathcal{D}_d(\mathrm{EQ}_n)$ using two extra axioms $\{a_0^d\}$ and $\{\bar{a}_0^d\}$, whose respective negations are $\bar{\alpha}_d$ and α_d. By the inductive hypothesis, ζ_n^{d-1} is a refutation of $\mathcal{D}_{d-1}(\mathrm{EQ}_n)$ of degree $d-1$, and that QBF is equal to $\mathcal{D}_d(\mathrm{EQ}_n)[\bar{\alpha}_d]$. Hence, by invariant (1), ζ_n^d is indeed a degree-d refutation of $\mathcal{D}_d(\mathrm{EQ}_n)$, which establishes invariant (2). □

Our main result is an immediate consequence of Lemmata 13 and 17, and the fact that the equality family requires 2^n-size Q-Res refutations [7].

Theorem 18. *For each d in $\mathbb{N}$, there exists a QBF family that has $O(n)$-size Q-Res+SR refutations of degree d and requires $2^{\Omega(n)}$-size refutations of degree $d-1$.*

6 Constraint Propagation

From a practical point of view, the following observation could be made: linear assignments do not cover constraint propagation (i.e. unit propagation and pure literal elimination, cf. [20]), whereas our lower bound formulas contain both unit clauses and pure literals.

However, as we show below, any lower bounds are easily adapted to hold in the presence of constraint propagation; in fact, one can easily modify a formula in such a way that propagation is rendered ineffective. Moreover, the current setup shows tighter upper bounds, since the system without constraint propagation is certainly no stronger. Thus, as far as proof complexity is concerned, including propagation in the system merely introduces unnecessary complications.

Indeed, with a simple addition, one can block all propagation, while (essentially) preserving the symmetry group. Given a QBF $F := Q \cdot f$ whose last block is universal, add a fresh block of existential variables p_1, p_2, q_1, q_2 to the end of the prefix Q, and replace the matrix f with

$$(f \otimes \{\{\bar{p}_1, \bar{p}_2\}, \{\bar{p}_1, p_2\}, \{p_1, \bar{p}_2\}, \{p_1, p_2\}\}) \cup (g \otimes \{\{q_1, \bar{q}_2\}, \{\bar{q}_1, q_2\}\}),$$

where g consists of the full set of unit clauses for F:

$$g := \{\{\bar{z}\} : z \in \mathrm{vars}(F)\} \cup \{\{z\} : z \in \mathrm{vars}(F)\}.$$

Let us call the result of this modification F'.

It is easy to see that no unit propagation or pure literal elimination can take place until all the variables of F are assigned, leaving only the fresh variables in the final block. Hence, propagation cannot enlarge the set of linear assignments. Notice also that $\{\{q_1, \bar{q}_2\}, \{\bar{q}_1, q_2\}\}$ is a satisfiable CNF. As a result, the clauses in $g \otimes \{\{q_1, \bar{q}_2\}, \{\bar{q}_1, q_2\}\}$ never contribute positively to a refutation, and one can assume without loss of generality that they never appear. Thus, Q-Res+SR hardness for F lifts straightforwardly to F', even if propagation is built into the proof system (viz. [20]).

On the other hand, the modified formulas are certainly no harder to refute; the symmetries of F are preserved when extended by the identity to the fresh variables, and every clause of the original matrix can be recovered in three resolution steps.

If the final block of F is existential, the same effect is achieved by inserting a fresh universal variable v directly before p_1, and taking a further matrix product with $\{\{v\}, \{\bar{v}\}\}$.

7   Conclusions

We introduced a theoretical model of QBF symmetry recomputation, which forms a hierarchy of proof systems of strictly increasing strength when the degree is bounded by a constant. Bounding the degree of a refutation corresponds to bounding the number of recomputations allowed on any single search branch of runs of QCDCL algorithms.

Since the number of paths explored correlates approximately with total running time of solvers, a sensible bound on degree – some fraction of the number of variables, for example – limits the total number of recomputations in terms of the length of the search. Our strict hierarchy shows the best possible separations for bounds of this type.

Our investigation of QBF here also applies to SAT as a special case. Considering purely existentially quantified formulas, the system Q-Res+S coincides with resolution with the symmetry rule as introduced by Krishnamurthy [19]. Similarly, when only allowing existential formulas, we obtain a propositional version of Q-Res+SR for symmetry recomputation in SAT. The methods we employed for QBF are sufficient to show that propositional symmetry recomputation is exponentially stronger than Krishnamurthy's system. We need only apply our symmetry blocking technique to formulas separating the latter from resolution (e.g. the pigeonhole formulas [25]).

Acknowledgments. Research was supported by grants from the John Templeton Foundation (grant no. 60842) and the Carl-Zeiss Foundation.

References

1. Aloul, F.A., Ramani, A., Markov, I.L., Sakallah, K.A.: Dynamic symmetry-breaking for boolean satisfiability. Ann. Math. Artif. Intell. **57**(1), 59–73 (2009)
2. Audemard, G., Jabbour, S., Sais, L.: Symmetry breaking in quantified boolean formulae. In: Veloso, M.M. (ed.) Proceedings of the 20th International Joint Conference on Artificial Intelligence, IJCAI 2007, pp. 2262–2267 (2007)
3. Audemard, G., Mazure, B., Sais, L.: Dealing with symmetries in quantified boolean formulas. In: SAT 2004 - The Seventh International Conference on Theory and Applications of Satisfiability Testing (2004)
4. Balabanov, V., Widl, M., Jiang, J.-H.R.: QBF resolution systems and their proof complexities. In: Sinz, C., Egly, U. (eds.) SAT 2014. LNCS, vol. 8561, pp. 154–169. Springer, Cham (2014). https://doi.org/10.1007/978-3-319-09284-3_12
5. Beame, P., Kautz, H.A., Sabharwal, A.: Towards understanding and harnessing the potential of clause learning. J. Artif. Intell. Res. (JAIR) **22**, 319–351 (2004)
6. Benedetti, M., Mangassarian, H.: QBF-based formal verification: experience and perspectives. J. Satisfiability Boolean Model. Comput. **5**(1–4), 133–191 (2008)
7. Beyersdorff, O., Blinkhorn, J., Hinde, L.: Size, cost and capacity: a semantic technique for hard random QBFs. In: Karlin, A.R. (ed.) ACM Conference on Innovations in Theoretical Computer Science (ITCS). Leibniz International Proceedings in Informatics (LIPIcs), vol. 94, pp. 9:1–9:18. Schloss Dagstuhl - Leibniz-Zentrum für Informatik (2018)
8. Beyersdorff, O., Chew, L., Janota, M.: Proof complexity of resolution-based QBF calculi. In: Mayr, E.W., Ollinger, N. (eds.) International Symposium on Theoretical Aspects of Computer Science (STACS). Leibniz International Proceedings in Informatics (LIPIcs), vol. 30, pp. 76–89. Schloss Dagstuhl - Leibniz-Zentrum für Informatik (2015)
9. Beyersdorff, O., Hinde, L., Pich, J.: Reasons for hardness in QBF proof systems. In: Lokam, S.V., Ramanujam, R. (eds.) Conference on Foundations of Software Technology and Theoretical Computer Science (FSTTCS). LIPIcs, vol. 93, pp. 14:1–14:15. Schloss Dagstuhl - Leibniz-Zentrum für Informatik (2017)
10. Blinkhorn, J., Beyersdorff, O.: Shortening QBF proofs with dependency schemes. In: Gaspers, S., Walsh, T. (eds.) SAT 2017. LNCS, vol. 10491, pp. 263–280. Springer, Cham (2017). https://doi.org/10.1007/978-3-319-66263-3_17
11. Cook, S.A., Reckhow, R.A.: The relative efficiency of propositional proof systems. J. Symb. Logic **44**(1), 36–50 (1979)
12. Devriendt, J., Bogaerts, B., Bruynooghe, M., Denecker, M.: Improved static symmetry breaking for SAT. In: Creignou, N., Le Berre, D. (eds.) SAT 2016. LNCS, vol. 9710, pp. 104–122. Springer, Cham (2016). https://doi.org/10.1007/978-3-319-40970-2_8

13. Devriendt, J., Bogaerts, B., Cat, B.D., Denecker, M., Mears, C.: Symmetry propagation: improved dynamic symmetry breaking in SAT. In: IEEE 24th International Conference on Tools with Artificial Intelligence, ICTAI 2012, pp. 49–56. IEEE Computer Society (2012)
14. Egly, U., Kronegger, M., Lonsing, F., Pfandler, A.: Conformant planning as a case study of incremental QBF solving. Ann. Math. Artif. Intell. **80**(1), 21–45 (2017)
15. Faymonville, P., Finkbeiner, B., Rabe, M.N., Tentrup, L.: Encodings of bounded synthesis. In: Legay, A., Margaria, T. (eds.) TACAS 2017. LNCS, vol. 10205, pp. 354–370. Springer, Heidelberg (2017). https://doi.org/10.1007/978-3-662-54577-5_20
16. Kauers, M., Seidl, M.: Short proofs for some symmetric quantified boolean formulas. Inf. Process. Lett. **140**, 4–7 (2018)
17. Kauers, M., Seidl, M.: Symmetries of quantified boolean formulas. In: Beyersdorff, O., Wintersteiger, C.M. (eds.) SAT 2018. LNCS, vol. 10929, pp. 199–216. Springer, Cham (2018). https://doi.org/10.1007/978-3-319-94144-8_13
18. Kleine Büning, H., Karpinski, M., Flögel, A.: Resolution for quantified Boolean formulas. Inf. Comput. **117**(1), 12–18 (1995)
19. Krishnamurthy, B.: Short proofs for tricky formulas. Acta Inf. **22**(3), 253–275 (1985)
20. Lonsing, F., Egly, U., Seidl, M.: Q-resolution with generalized axioms. In: Creignou, N., Le Berre, D. (eds.) SAT 2016. LNCS, vol. 9710, pp. 435–452. Springer, Cham (2016). https://doi.org/10.1007/978-3-319-40970-2_27
21. Pipatsrisawat, K., Darwiche, A.: On the power of clause-learning SAT solvers as resolution engines. Artif. Intell. **175**(2), 512–525 (2011)
22. Sakallah, K.A.: Symmetry and satisfiability. In: Biere, A., Heule, M., van Maaren, H., Walsh, T. (eds.) Handbook of Satisfiability, Frontiers in Artificial Intelligence and Applications, vol. 185, pp. 289–338. IOS Press (2009)
23. Schaafsma, B., Heule, M.J.H., van Maaren, H.: Dynamic symmetry breaking by simulating Zykov contraction. In: Kullmann, O. (ed.) SAT 2009. LNCS, vol. 5584, pp. 223–236. Springer, Heidelberg (2009). https://doi.org/10.1007/978-3-642-02777-2_22
24. Szeider, S.: The complexity of resolution with generalized symmetry rules. Theory Comput. Syst. **38**(2), 171–188 (2005)
25. Urquhart, A.: The symmetry rule in propositional logic. Discrete Appl. Math. **96–97**, 177–193 (1999)

Satisfiability Threshold for Power Law Random 2-SAT in Configuration Model

Oleksii Omelchenko and Andrei A. Bulatov$^{(\boxtimes)}$

Simon Fraser University, Burnaby, Canada
{oomelche,abulatov}@sfu.ca

Abstract. The Random Satisfiability problem has been intensively studied for decades. For a number of reasons the focus of this study has mostly been on the model, in which instances are sampled uniformly at random from a set of formulas satisfying some clear conditions, such as fixed density or the probability of a clause to occur. However, some non-uniform distributions are also of considerable interest. In this paper we consider Random 2-SAT problems, in which instances are sampled from a wide range of non-uniform distributions.

The model of random SAT we choose is the so-called configuration model, given by a distribution ξ for the degree (or the number of occurrences) of each variable. Then to generate a formula the degree of each variable is sampled from ξ, generating several *clones* of the variable. Then 2-clauses are created by choosing a random paritioning into 2-element sets on the set of clones and assigning the polarity of literals at random.

Here we consider the random 2-SAT problem in the configuration model for power-law-like distributions ξ. More precisely, we assume that ξ is such that its right tail $F_\xi(x)$ satisfies the conditions $W\ell^{-\alpha} \le F_\xi(\ell) \le V\ell^{-\alpha}$ for some constants V, W. The main goal is to study the satisfiability threshold phenomenon depending on the parameters α, V, W. We show that a satisfiability threshold exists and is determined by a simple relation between the first and second moments of ξ.

Keywords: Satisfiability · Power law · Phase transition

1 Introduction

The Random Satisfiability problem (Random SAT) and its special cases Random k-SAT as a model of 'typical case' instances of SAT has been intensively studied for decades. Apart from algorithmic questions related to the Random SAT, much attention has been paid to such problems as satisfiability thresholds and the structure of the solution space. The most widely studied model of the Random k-SAT is the uniform one parametrized by the (expected) density or clause-to-variable ratio ϱ of input formulas. Friedgut in [26] proved that depending on the parameter ϱ (and possibly the number of variables) Random k-SAT

This work was supported by an NSERC Discovery grant.

M. Janota and I. Lynce (Eds.): SAT 2019, LNCS 11628, pp. 53–70, 2019.
https://doi.org/10.1007/978-3-030-24258-9_4

exhibits a sharp satisfiability threshold: a formula of density less than a certain value ϱ_0 (or possibly $\varrho_0(n)$) is satisfiable with high probability, and if the density is greater than ϱ_0, it is unsatisfiable with high probability. Moreover, a recent work of Friedrich and Rothenberger [28], which may be regarded as an extension of Fridgut's result to non-uniform random SAT instances, shows that if a distribution of variable's occurence in random formulas satisfies some criteria, then such formulas must undergo a sharp satisfiability threshold.

Hence, an impressive line of research aims at locating the satisfiability threshold for each random generating model. This includes more and more sophisticated methods of algorithms analysis [1,17,18,22] and applications of the second moment method [2] to find lower bounds, and a variety of probabilistic and proof complexity tools to obtain upper bounds [23,24,33]. In the case of sufficiently large k the exact location of the satisfiability threshold was identified by Ding, Sly and Sun [22]. The satisfiability threshold and the structure of random k-CNFs received special attention for small values of k, see [15,30,40] for $k = 2$, and [21,31,33] for $k = 3$.

The satisfiability threshold phenomenon turned out to be closely connected with algorithmic properties of the Random SAT, as well as with the structure of its solution space. Experimental and theoretical results [19,39] demonstrate that finding a solution or proving unsatisfiability is hardest around the satisfiability threshold. The geometry of the solution space also exhibits phase transitions not far from the satisfiability threshold, related to various clustering properties [35]. This phenomenon has been exploited by applications of methods from statistical physics that resulted in some of the most efficient algorithms for Random SAT with densities around the satisfiability threshold [14,36].

Random k-SAT can be formulated using one of the three models whose statistical properties are very similar. In the model with fixed density ϱ, one fixes n distinct propositional variables $v_1, \ldots, v_n$ and then chooses ϱn k-clauses uniformly at random [25,39]. Alternatively, for selected variables every possible k-clause is included with probability tuned up so that the expected number of clauses equals ϱn. Finally, Kim [32] showed that one can also use the configuration model, which he called Poisson Cloning model. In this model for each variable v_i we first select a positive integer d_i accordingly to the Poisson distribution with expectation $k\varrho$, the degree of the variable. Then we create d_i *clones* of variable v_i, and choose $(d_1 + \cdots + d_n)/k$ random k-element subsets of the set of clones, then converting them into clauses randomly. The three models are largely equivalent and can be used whichever suits better to the task at hand.

The configuration model opens up a possibility for a wide range of different distributions of k-CNFs arising from different degree distributions. Starting with any random variable ξ that takes positive integer values one obtains a distribution $\Phi(\xi)$ on k-CNFs as above using ξ in place of the Poisson distribution. Note that ξ may depend on n, the number of variables, and even be different for different variables. One 'extreme' case of such a distribution is Poisson Cloning described above. Another case is studied by Cooper, Frieze, and Sorkin [20]. In their case each variable of a 2-SAT instance has a prescribed degree, which can be viewed as assigning a degree to every variable according to a random variable that only takes one value. We will be often returning to that paper, as our criterion for a satisfiability threshold is a generalization of that in [20]. Boufkhad et al. [13] considered another case of this kind—regular Random k-SAT.

In this paper we consider Random 2-SAT in the configuration model given by distribution $\Phi(\xi)$, where ξ is distributed according to the power law distribution in the following sense. Let $F_\xi(\ell) = \Pr[\xi \geq \ell]$ denote the tail function of a positive integer valued random variable ξ. We say that ξ is distributed according to the power law with parameter α if there exist constants V, W such that

$$W\ell^{-\alpha} \leq F_\xi(\ell) \leq V\ell^{-\alpha}. \tag{1}$$

Power law type distributions have received much attention. They have been widely observed in natural phenomena [16,37], as well as in more artificial structures such as networks of various kinds [10]. Apart from the configuration model, graphs (and therefore 2-CNFs) whose degree sequences are distributed accordingly to a power law of some kind can also be generated in a number of ways. These include preferential attachment [3,10–12], hyperbolic geometry [34], and others [5,6]. Although the graphs resulting from all such processes satisfy the power law distributions of their degrees, other properties can be very different. We will encounter the same phenomenon in this paper.

The approach most closely related to this paper was suggested by Ansótegui et al. [5,6]. Given the number of variables n, the number of clauses m, and a parameter β, the first step in their construction is to create m k-clauses without naming the variables. Then for every variable-place X in every clause, X is assigned to be one of the variables $v_1, \ldots, v_n$ according to the distribution

$$\Pr[X = v_i, \beta, n] = \frac{i^{-\beta}}{\sum_{j=1}^n j^{-\beta}}.$$

Ansótegui et al. argue that this model often well matches the experimental results on industrial instances, see also [4,8,29]. Interesting to note that although the model studied in these papers differ from the configuration model, it exhibits the same criterion of unsatisfiability $\mathbb{E}K^2 > 3\mathbb{E}K$, where K is the r.v. that governs the number of times a variable appears in 2-SAT formula ϕ [7].

The satisfiability threshold of this model has been studied by Friedrich et al. in [27]. Since the model has two parameters, β and $r = m/n$, the resulting picture is complicated. Friedrich et al. proved that a random CNF is unsatisfiable with high probability if r is large enough (although constant), and if $\beta < \frac{2k-1}{k-1}$. If $\beta \geq \frac{2k-1}{k-1}$, the formula is satisfiable with high probability provided r is smaller than a certain constant. The unsatisfiability results in [27] are mostly proved using the local structure of a formula.

In this paper we aim at a similar result for Random 2-SAT in the configuration model. Although the configuration model has only one parameter, the overall picture is somewhat more intricate, because there are more reasons for unsatisfiability than just the local structure of a formula. We show that for 2-SAT the parameter α from the tail condition (1) is what decides the satisfiability of such CNF. The main result of this paper is a satisfiability threshold given by the following

Theorem 1. *Let ϕ be a random 2-CNF in the configuration model, such that the number of occurrences of each variable in ϕ is an independent copy of the random variable ξ, satisfying the tail condition (1) for some α. Then for $n \to \infty$*

$$\Pr[\phi \text{ is satisfiable}] = \begin{cases} 0, & \text{when} & 0 < \alpha < 2 \\ 0, & \text{when} & \alpha = 2 \text{ or } \mathbb{E}\xi^2 > 3\mathbb{E}\xi, \\ 1, & \text{when} & \mathbb{E}\xi^2 < 3\mathbb{E}\xi. \end{cases}$$

In the first case of Theorem 1 we show that ϕ is unsatisfiable with high probability due to very local structure of the formula, such as the existence of variables of sufficiently high degree. Moreover, same structures persist with high probability in k-CNF formulas for *any* $k \geq 2$ obtained from the configuration model, when $\alpha < \frac{k}{k-1}$.

In the remaining cases we apply the approach of Cooper, Frieze, and Sorkin [20]. It makes use of the structural characterization of unsatisfiable 2-CNFs: a 2-CNF is unsatisfiable if and only if it contains so-called *contradictory paths*. If $\mathbb{E}\xi^2 < 3\mathbb{E}\xi$ we prove that w.h.p. formula ϕ does not have long paths, and contradictory paths are unlikely to form. If $\mathbb{E}\xi^2 > 3\mathbb{E}\xi$, we use the analysis of the dynamics of the growth of ϕ to show that contradictory paths appear w.h.p. However, the original method by Cooper et al. only works with strong restrictions on the maximal degree of variables that are not affordable in our case, and so it requires substantial modifications.

2   Notation and Preliminaries

We use the standard terminology and notation of variables, positive and negative literals, clauses and 2-CNFs, and degrees of variables. The degree of variable v will be denoted by $deg(v)$, or when our CNF contains only variables $v_1, \ldots, v_n$, we use $d_i = deg(v_i)$. By $C(\phi)$ we denote the set of clauses in ϕ.

2.1   Configuration Model

We describe the configuration model for k-CNFs, but will only use it for $k = 2$, see also [32]. In the configuration model of k-CNFs with n variables $v_1, \ldots, v_n$ we are given a positive integer-valued random variable (r.v.) ξ from which we sample independently n integers $\{d_i\}_{i=1}^n$. Then d_i is the degree of v_i, that is, the number of occurrences of v_i in the resulting formula ϕ. Each occurrence of v_i in ϕ we call a *clone* of v_i. Hence, d_i is the number of clones of v_i. Then we sample k-element sets of clones from the set of all clones without replacement. Finally, every such subset is converted into a clause by choosing the polarity of every clone in it uniformly at random. If the total number of clones is not a multiple of k, we discard the set and repeat the procedure. Algorithm 1 gives a

---

Algorithm 1. Configuration Model $\mathbb{C}_n^k(\xi)$

1: **procedure** SAMPLECNF(n, k, ξ)
2:     Form a sequence of n numbers $\{d_i\}_{i=1}^n$ each sampled independently from ξ
3:     **if** $S_n := \sum_{i=1}^n d_i$ is not a multiple of k **then**
4:         discard the sequence, and go to step 2
5:     **end if**
6:     Otherwise, introduce multi-set $S \leftarrow \bigcup_{i=1}^n \underbrace{\{v_i, v_i, \ldots, v_i\}}_{d_i \text{ times}}$
7:     Let $\phi \leftarrow \emptyset$
8:     **while** $S \neq \emptyset$ **do**
9:         Pick u.a.r. k elements $\{v_1, v_2, \ldots, v_k\}$ from S without replacement
10:        Let $C \leftarrow \{v_1, v_2, \ldots, v_k\}$
11:        $S \leftarrow S - C$
12:        Negate each element in C u.a.r with probability $1/2$
13:        $\phi \leftarrow \phi \cup C$
14:    **end while**
15:    **return** ϕ
16: **end procedure**

---

more precise description of the process. We will sometimes say that a clone p is associated with variable v if p is a clone of v. In a similar sense we will say a clone associated with a literal if we need to emphasize the polarity of the clone.

We will denote a random formula ϕ obtained from $\mathbb{C}_n^k(R)$ by $\phi \sim \mathbb{C}_n^k(R)$. Clearly, formulas $\phi \sim \mathbb{C}_n^k(R)$ are defined over a set of n Boolean variables, which we denote by $V(\phi)$. By $L(\phi)$ we denote the set of all literals in ϕ. Let d_i^+ denote the number of occurrences of v_i as a positive literal (or the number of positive clones of v_i), and let d_i^- denote the number of negative clones of v_i.

2.2   Power Law Distributions

We focus our attention on the configuration model $\mathbb{C}_n^k(\xi)$, in which every variable is an i.i.d. copy of the random variable ξ having power-law distribution. In this paper we define such distributions through the properties of their tail functions. If ξ is an integer-valued r.v., its tail function is defined to be $F_\xi(\ell) = \Pr[\xi \geq \ell]$, where $\ell \geq 1$.

Definition 1. *An integer-valued positive r.v. ξ has power-law probability distribution, if $F_\xi(\ell) = \Theta\left(\ell^{-\alpha}\right)$, where $\alpha > 0$. We denote this fact as $\xi \sim \mathcal{P}(\alpha)$.*

Clearly, if $\xi \sim \mathcal{P}(\alpha)$, then there exist constants $V, W > 0$, such that $W\,\ell^{-\alpha} \leq F_\xi(\ell) \leq V\,\ell^{-\alpha}$, for every $\ell \geq 1$.

The existence of the moments of $\xi \sim \mathcal{P}(\alpha)$ depend only on α.

Lemma 1. *Let $\xi \sim \mathcal{P}(\alpha)$. Then $\mathbb{E}\xi^m < \infty$ iff $0 < m < \alpha$.*

We will write $\mathbb{E}\xi^m = \infty$ when the m-th moment of some r.v. ξ is not finite or does not exist. We will have to deal with cases when the second or even first moment of ξ does not exist.

Nevertheless, we can obtain good bounds on useful quantities formed from such variables with a good level of confidence, despite the absence of expectation or variance. One such quantity is the sum of *independent* variables drawn from $\mathcal{P}(\alpha)$: $S_n = \sum_{i=1}^{n} \xi_i$, where $\xi_i \sim \mathcal{P}(\alpha)$. Note that ξ_i's are not required to be identically distributed. They can come from different distributions, as long as their right tail can be bounded with some power-law functions with exponent α. But we do require their independence.

The next two theorems provide bounds on the values of S_n, depending on α in a slightly more general case of r.vs. admitting negative values.

Theorem 2 (Corollary 1 from [38]). *Let $S_n = \sum_{i=1}^{n} \xi_i$, where ξ_i's are independent integer-valued random variables, with*

$$\Pr[\xi_i \geq \ell] \leq V \ell^{-\alpha}, \quad and \quad \Pr[\xi_i \leq -\ell] \leq V \ell^{-\alpha},$$

where $V > 0$ and $0 < \alpha \leq 1$ are constants. Then w.h.p. $S_n \leq C n^{\frac{1}{\alpha}}$, where $C > 0$ is some constant.

As for the second theorem, we deal with a similar sum of random variables, but each variable's tail can be majorized with a power-law function with exponent $\alpha > 1$. Then, as it follows from Lemma 1, such variables have finite expectation, and due to the linearity of expectation, the sum itself has well defined mean value.

Theorem 3 (Corollary 5 from [38]). *Let $S_n = \sum_{i=1}^{n} \xi_i$, where ξ_i's are independent integer-valued random variables, with*

$$\Pr[\xi_i \geq \ell] \leq V \ell^{-\alpha}, \quad and \quad \Pr[\xi_i \leq -\ell] \leq V \ell^{-\alpha},$$

where $V > 0$ and $\alpha > 1$ are constants. Then w.h.p. $S_n = \sum_{i=1}^{n} \mathbb{E}\xi_i + o(n)$.

Hence, as the theorem states, when ξ_i's are independent r.vs. with power-law boundable tails with tail exponent $\alpha > 1$, then the sum of such variables does not deviate much from its expected value.

Note that from now on we will deal with strictly positive power-law r.vs. ξ_i's, hence, their expectation (given that it exists) is a positive constant. Then when $\alpha > 1$, we have $S_n = \sum_{i=1}^{n} \mathbb{E}\xi_i + o(n) = (1 + o(1)) \sum_{i=1}^{n} \mathbb{E}\xi_i$.

Another important quantity we need is the maximum, Δ, of the sequence of n independent random variables (or the maximum degree of a CNF in our case).

Lemma 2. *Let $\Delta = \max(\xi_1, \xi_2, \cdots, \xi_n)$, where ξ_i's are independent copies of an r.v. $\xi \sim \mathcal{P}(\alpha)$ with $\alpha > 0$. Then w.h.p. $\Delta \leq C n^{1/\alpha}$, where $C > 0$ is some constant.*

We will also need some bounds on the number of pairs of complementary clones of a variable v_i, that is, the value $d_i^+ d_i^-$. By the definition of the configuration model

$$d_i^+ \sim Bin\left(\deg(v_i), 1/2 \right) \quad \text{and} \quad d_i^- = \deg(v_i) - d_i^+,$$

where $Bin(n, p)$ is the binomial distribution with n trials and the success probability p.

Lemma 3. *Let ξ be some positive integer-valued r.v., and let $d^+ \sim Bin\left(\xi, 1/2\right)$, while $d^- = \xi - d^+$. Then*

$$F_{d^+}(\ell) = F_{d^-}(\ell) \leq F_\xi(\ell), \quad \text{and} \quad F_{d^+ d^-}(\ell) \leq 2 F_\xi\left(\ell^{1/2}\right).$$

Hence, for $\xi \sim \mathcal{P}(\alpha)$, we have

Corollary 1. *Let $\xi \sim \mathcal{P}(\alpha)$, where $\alpha > 0$, be some positive integer-valued r.v., and let $d^+ \sim Bin\left(\xi, 1/2\right)$, while $d^- = \xi - d^+$. Then*

$$F_{d^+}(\ell) = F_{d^-}(\ell) \leq V \ell^{-\alpha}, \tag{2}$$

$$F_{d^+ d^-}(\ell) \leq 2V \ell^{-\alpha/2}. \tag{3}$$

The expectations of d^+ and d^- are easy to find: $\mathbb{E}d^+ = \mathbb{E}d^- = \frac{\mathbb{E}\xi}{2}$. However, the expected value of $d^+ d^-$ requires a little more effort.

Lemma 4. *Let ξ be some positive integer-valued r.v., and let $d^+ \sim Bin\left(\xi, 1/2\right)$, while $d^- = \xi - d^+$. Then $\mathbb{E}\left[d^+ d^-\right] = \frac{\mathbb{E}\xi^2 - \mathbb{E}\xi}{4}$.*

We use $T_n = \sum_{i=1}^n d_i^+ d_i^-$ to denote the total number of pairs of complementary clones, i.e. the sum of unordered pairs of complementary clones over all n variables.

Note, that when $\alpha > 2$, the r.v. $d_i^+ d_i^-$ has finite expectation due to Lemma 1. Then by Theorem 3 w.h.p.

$$T_n = (1 + o(1)) \sum_{i=1}^n \mathbb{E}\left[d_i^+ d_i^-\right].$$

We finish this subsection with *Azuma-like inequality* first appeared in [20], which will be used in the proofs. Informally, the inequality states that a discrete-time random walk $X = \sum_{i=1}^n X_i$ with positive drift, consisting of not necessary independent steps, each having a right tail, which can be bounded by a power function with exponent at least 1, is very unlikely to drop much below the expected level, given n is large enough. Although, the original proof was relying on the rather artificial step of introducing a sequence of uniformly distributed random numbers, we figured out that the same result can be obtained by exploiting the tower property of expectation.

Lemma 5 (Azuma-like inequality). *Let $X = X_0 + \sum_{i=1}^{t} X_i$ be some random walk, such that $X_0 \geq 0$ is constant initial value of the process, $X_i \geq -a$, where $a > 0$ is constant, are bounded from below random variables, not necessary independent, and such that $\mathbb{E}[X_i \mid X_1, \ldots, X_{i-1}] \geq \mu > 0$ (μ is constant) and $\Pr[X_i \geq \ell \mid X_1, \ldots, X_{i-1}] \leq V \ell^{-\alpha}$ for every $\ell \geq 1$ and constants $V > 0$, $\alpha > 1$. Then for any $0 < \varepsilon < \frac{1}{2}$, the following inequality holds*

$$\Pr[X \leq \varepsilon \mu t] \leq \exp\left(-\frac{t + X_0}{4 \log^2 t} \mu^2 \left(\frac{1}{2} - \varepsilon\right)^2\right).$$

2.3   Contradictory Paths and Bicycles

Unlike k-CNFs for larger values of k, 2-CNFs have a clear structural feature that indicates whether or not the formula is satisfiable. Let ϕ be a 2-CNF on variables $v_1, \ldots, v_n$. A sequence of clauses $(l_1, l_2), (\bar{l}_2, l_3), \ldots, (\bar{l}_{s_1}, l_s)$ is said to be a *path* from literal l_1 to literal l_s. As is easily seen, if there are variables u, v, w in ϕ such that there are paths from u to v and $\bar{v}$, and from $\bar{u}$ to w and $\bar{w}$, then ϕ is unsatisfiable, see also [9]. Such a collection of paths is sometimes called *contradictory paths*.

On the other hand, if ϕ is unsatisfiable, it has to contain a *bicycle*, see [15]. A bicycle of length s is a path $(u, l_1), (\bar{l}_1, l_2), \ldots, (\bar{l}_s, v)$, where the variables associated with literals $l_1, l_2, \ldots, l_s$ are distinct, and $u, v \in \{l_1, \bar{l}_1, l_2, \bar{l}_2, \ldots, l_s, \bar{l}_s\}$.

2.4   The Main Result

Now we are ready to state our main result:

Theorem 4. *Let $\phi \sim \mathbb{C}_n^2(\xi)$, where $\xi \sim \mathcal{P}(\alpha)$. Then for $n \to \infty$*

$$\Pr[\phi \text{ is } SAT] = \begin{cases} 0, & \text{when} \quad 0 < \alpha < 2, \\ 0, & \text{when} \quad \alpha = 2 \text{ or } \mathbb{E}\xi^2 > 3\mathbb{E}\xi, \\ 1, & \text{when} \quad \mathbb{E}\xi^2 < 3\mathbb{E}\xi. \end{cases}$$

If the r.v. ξ is distributed according to the zeta distribution, that is, $\Pr[\xi = \ell] = \frac{\ell^{-\beta}}{\zeta(\beta)}$ for some $\beta > 1$ and where $\zeta(\beta) = \sum_{d \geq 1} d^{-\beta}$ is the Riemann zeta function (note that in this case $\xi \sim \mathcal{P}(\beta - 1)$), then the satisfiabitliy threshold is given by a certain value of β.

Corollary 2. *Let $\phi \sim \mathbb{C}_n^2(\xi)$, where the pdf of ξ is $\Pr[\xi = \ell] = \frac{\ell^{-\beta}}{\zeta(\beta)}$ for some $\beta > 1$ and all $\ell \geq 1$. Then there exists β_0 such that for $n \to \infty$*

$$\Pr[\phi \text{ is } SAT] = \begin{cases} 0, & \text{when} \quad 1 < \beta < \beta_0, \\ 1, & \text{when} \quad \beta > \beta_0. \end{cases}$$

The value β_0 is the positive solution of the equation $\mathbb{E}\xi^2 = 3\mathbb{E}\xi$, and $\beta_0 \approx 3.26$.

A proof of this theorem constitutes the rest of the paper. We consider each case separately, and the first case is proved in Proposition 1, while the other two cases are examined in Propositions 2 and 3.

3   Satisfiability of $\mathbb{C}_n^2(\xi)$, When $\xi \sim \mathcal{P}(\alpha)$ and $0 < \alpha < 2$

This case is the easiest to analyze. Moreover, we show that the same result holds for any $\phi \sim \mathbb{C}_n^k(\xi)$, where $k \geq 2$, when $\alpha < \frac{k}{k-1}$. Hence, the case $0 < \alpha < 2$ for unsatisfiable 2-CNFs follows. In other words, if $\alpha < \frac{k}{k-1}$, then *any* k-CNF formulas from $\mathbb{C}_n^k(\xi)$ will be unsatisfiable w.h.p.

What happens here, is that we expect many variables to have degree $\gg S_n^{(k-1)/k}$. Let us fix k such variables. Then, as it is shown in the proof, the formula ϕ contains at least $(k-1)! \log^k n$ clauses that are formed only from literals of these k variables. However, one of the possible subformulas, which is formed from only k variables, that renders the whole k-CNF formula unsatisfiable consists of only 2^k clauses.

The next proposition establishes a lower bound of satisfiability threshold for any power-law distributed k-CNF from configuration model.

Proposition 1. *Let* $\phi \sim \mathbb{C}_n^k(\xi)$, *where* $\xi \sim \mathcal{P}(\alpha)$, $k \geq 2$ *and* $0 < \alpha < \frac{k}{k-1}$. *Then w.h.p.* ϕ *is unsatisfiable.*

After proving the above proposition, result for 2-CNF from $\mathbb{C}_n^2(\xi)$ naturally follows.

Corollary 3. *Let* $\phi \sim \mathbb{C}_n^2(\xi)$, *where* $\xi \sim \mathcal{P}(\alpha)$, *such that* $0 < \alpha < 2$. *Then w.h.p.* ϕ *is unsatisfiable.*

4   Satisfiability of $\mathbb{C}_n^2(\xi)$, When $\xi \sim \mathcal{P}(\alpha)$ and $\alpha = 2$ or $\mathbb{E}\xi^2 > 3\mathbb{E}\xi$

4.1   The Inequality $\mathbb{E}\xi^2 > 3\mathbb{E}\xi$

Analysis of this and subsequent cases mainly follows the approach suggested in [20], where they deal with random 2-SAT instances having prescribed literal degrees. In other words, the assumption in [20] is that the degree sequences $d_1^+, \ldots, d_n^+$ and $d_1^-, \ldots, d_n^-$ are fixed, and a random 2-CNF is generated as in the configuration model. Then two quantities play a very important role. The first one is the sum of all degrees $S_n = \sum_{i=1}^n (d_i^+ + d_i^-)$ (we use our notation) and the second one is the number of pairs of complementary clones $T_n = \sum_{i=1}^n d_i^+ d_i^-$. It is then proved that a 2-CNF with a given degree sequence is satisfiable w.h.p. if and only if $2T_n < (1 - \varepsilon)S_n$ for some $\varepsilon > 0$. We will quickly show that the conditions $\alpha = 2$ and $\mathbb{E}\xi^2 > 3\mathbb{E}\xi$ imply the inequality $2T_n > (1 + \varepsilon)S_n$ w.h.p., see Lemma 6, and therefore a random 2-CNF in this case should be unsatisfiable w.h.p. The problem however is that Cooper et al. only prove their result under a significant restrictions on the maximal degree of literals, $\Delta < n^{1/11}$. By Lemma 2 the maximal degree of literals in our case tends to be much higher, and we cannot directly utilize the result from [20]. Therefore we follow the main steps of the argument in [20] changing parameters, calculations, and in a number of cases giving a completely new proofs.

Lemma 6. *Let* $\phi \sim \mathbb{C}_n^2(\xi)$, *where* $\xi \sim \mathcal{P}(\alpha)$ *and* $\alpha = 2$ *or* $\mathbb{E}\xi^2 > 3\mathbb{E}\xi$. *Let also* $S_n = \sum_{i=1}^n d_i$ *and* $T_n = \sum_{i=1}^n d_i^+ d_i^-$. *Then w.h.p.* $2T_n > (1+\varepsilon)S_n$.

Proof. Let us first consider the case, when $\alpha > 2$ and $\mathbb{E}\xi^2 > 3\mathbb{E}\xi$. Then by Lemma 1 and Theorem 3, we have that w.h.p.

$$S_n = \sum_{i=1}^n d_i = (1+o(1)) \sum_{i=1}^n \mathbb{E}d_i = (1+o(1))\, n\mathbb{E}\xi,$$

since $d_i \overset{d}{=} \xi$. Likewise, since $\alpha > 2$, we also have that w.h.p.

$$T_n = \sum_{i=1}^n d_i^+ d_i^- = (1+o(1)) \sum_{i=1}^n \mathbb{E}\left[d_i^+ d_i^- \right] = (1+o(1))\, n \frac{\mathbb{E}\xi^2 - \mathbb{E}\xi}{4},$$

where the last equality follows from Lemma 4.

Hence, when $\mathbb{E}\xi^2 > 3\mathbb{E}\xi$, we have that w.h.p.

$$\frac{2T_n}{S_n} = (1 \pm o(1)) \frac{\mathbb{E}\xi^2 - \mathbb{E}\xi}{2\mathbb{E}\xi} = (1 \pm o(1)) \left(\frac{\mathbb{E}\xi^2}{2\mathbb{E}\xi} - \frac{1}{2} \right) > 1.$$

Now we consider the case $\alpha = 2$. Unfortunately, then $\mathbb{E}\left[d_i^+ d_i^- \right] = \infty$ for any $i \in [1 \ldots n]$, and so we cannot claim that T_n is concentrated around its mean. Nevertheless, the quantity $\frac{2T_n}{S_n}$ is still greater than 1 in this case.

Since $\xi \sim \mathcal{P}(2)$, there are constants V, W such that $W\,\ell^{-2} \leq F_\xi(\ell) \leq V\,\ell^{-2}$. We construct auxiliary random variables $\xi_\varepsilon \sim \mathcal{P}(2+\varepsilon)$ for $\varepsilon > 0$. Later we will argue that ξ_ε can be chosen such that $\mathbb{E}\xi_\varepsilon^2 > 3\mathbb{E}\xi_\varepsilon$. Specifically, let ξ_ε be such that $F_{\xi_\varepsilon}(1) = 1$ and $F_{\xi_\varepsilon}(\ell) = W\,\ell^{-2-\varepsilon}$ for $\ell > 1$.

Let T_n^ε be the number of pairs of complementary clones in formula $\phi_0 \sim \mathbb{C}_n^2(\xi_\varepsilon)$. Since $\Pr[\xi_\varepsilon \geq \ell] \leq \Pr[\xi \geq \ell]$ for any $\ell \geq 1$, we have that

$$\Pr\left[2T_n > S_n\right] \geq \Pr\left[2T_n^\varepsilon > S_n\right], \tag{4}$$

due to the stochastic dominance of the r.v. T_n over T_n^ε. As is easily seen, for sufficiently small ε we have $\mathbb{E}\xi_\varepsilon^2 > 3\mathbb{E}\xi_\varepsilon$. Therefore, by the first part of the proof $2T_n^\varepsilon > S_n$ w.h.p. The result follows.

Thus, in either case we obtain that for some $\mu > 0$ w.h.p. $\frac{2T_n}{S_n} = 1 + \mu$.

In what follows, we will always assume that $\alpha > 2$.

4.2 TSPAN

The process of generating a random 2-CNF in the configuration model can be viewed as follows. After creating a pool of clones, we assign each clone a polarity, making it a clone of a positive or negative literal. Then we choose a random paritioning of the set of clones into 2-element sets. The important point here is that in the process of selection of a random matching we pair clones up one after another, and it does not matter in which order a clone to match is selected, as long as it is paired with a random unpaired clone.

Our goal is to show that our random 2-CNF ϕ contains contradictory paths. In order to achieve this we exploit the property above as follows. Starting from a random literal p we will grow a set $span(p)$ of literals reachable from p in the sense of paths introduced in Sect. 2.3. This is done by trying to iteratively extend $span(p)$ by pairing one the unpaired clones of the negation of a literal from $span(p)$. The details of the process will be described later. The hope is that at some point $span(p)$ contains a pair of literals of the form $v, \bar{v}$, and therefore ϕ contains a part of the required contradictory paths. To obtain the remaining part we run the same process starting from $\bar{p}$.

To show that this approach works we need to prove three key facts:

- that $span(p)$ grows to a certain size with reasonable probability (Lemma 10),
- that if $span(p)$ has grown to the required size, it contains a pair $v, \bar{v}$ w.h.p. (Lemma 11), and
- that the processes initiated at p and $\bar{p}$ do not interact too much w.h.p. (Lemma 8).

Since the probability that $span(p)$ grows to the required size is not very high, most likely this process will have to be repeated multiple times. It is therefore important that the probabilities above are estimated when some clones are already paired up, and that all the quantities involved are carefully chosen.

We now fill in some details. The basic "growing" algorithm is TSPAN (short for *truncated span*), see Algorithm 2. Take a literal and pick a clone p associated with it. Then partition the set $\mathcal{S}$ of all clones into 3 subsets: the set $\mathcal{L}(p)$ of "live" clones from which we can grow the span, the set $\mathcal{C}$ of paired (or "connected") clones, and the set $\mathcal{U}$ of "untouched" yet clones. We start with $\mathcal{L}(p) = \{p\}$, $\mathcal{U} = \mathcal{S} - \{p\}$, and empty set $\mathcal{C}$.

TSPAN works as follows: while the set of live clones is not empty, pick u.a.r. clone c_1 from the live set, and pair it u.a.r. with any non-paired clone $c_2 \in \mathcal{U} \cup \mathcal{L}(p) \setminus \{c_1\}$. Since clones c_1 and c_2 are paired now, we move them into the set of paired clones $\mathcal{C}$, while removing them from both sets $\mathcal{L}(p)$ and $\mathcal{U}$ to preserve the property that the sets $\mathcal{C}, \mathcal{U}$, and $\mathcal{L}(p)$ form a partition of $\mathcal{S}$.

Next, we identify the literal l which clone c_2 is associated with, and we move all the clones of complementary literal $\bar{l}$ from the set of untouched clones $\mathcal{U}$ into $\mathcal{L}(p)$. The idea of this step is, when we add an edge (c_1, c_2), where c_2 is one of the l's clones, to grow the span further we will need to add another directed edge $(c_3, \cdot)$, where c_3 is one of the clones belonging to $\bar{l}$. Hence, we make all clones of $\bar{l}$ live, making them available to pick as a starting point during next iterations of TSPAN. This way we can grow a span, starting from the clone p, and then the set

$$span(p) = \{c \in \mathcal{S} \mid c \text{ is reachable from } p\},$$

contains all the clones, which are reachable from the clone p (or literal that is associated with p) at a certain iteration of TSPAN. We call this set a *p-span*.

The version of TSPAN given in Algorithm 2 takes as input sets $\mathcal{C}, \mathcal{L}, \mathcal{U}$ (which therefore do not have to be empty in the beginning of execution of the procedure), a maximal number of iterations τ, and a maximal size of the set of live clones. It starts by using the given sets, $\mathcal{C}, \mathcal{L}, \mathcal{U}$, stops after at most τ iterations or when $\mathcal{L}$ reaches size σ.

Algorithm 2. Procedure TSPAN

1: **procedure** TSPAN$(\mathcal{C}, \mathcal{L}, \mathcal{U}, \sigma, \tau)$
2:      **while** $0 < |\mathcal{L}| \leq \sigma$ **and** less than τ pairings performed **do**
3:          Pick u.a.r. a live clone $c_1 \in \mathcal{L}$
4:          Pick u.a.r. an unpaired clone $c_2 \in \mathcal{U} \cup \mathcal{L} \setminus \{c_1\}$
5:          Pair clones c_1 and c_2, i.e.
6:              $\mathcal{C} \leftarrow \mathcal{C} \cup \{c_1, c_2\}$
7:              $\mathcal{L} \leftarrow \mathcal{L} \setminus \{c_1, c_2\}$
8:              $\mathcal{U} \leftarrow \mathcal{U} \setminus \{c_1, c_2\}$
9:          Let w be the literal associated with c_2
10:         Make live the clones associated with $\bar{w}$, i.e
11:         Let
12:             $\kappa(\bar{w}) = \{c \in \mathcal{S} \,|\, c \text{ is associated with } \bar{w}\}$
13:             $L \leftarrow L \cup (U \cap \kappa(\bar{w}))$
14:             $U \leftarrow U \setminus \kappa(\bar{w})$
15:     **end while**
16: **end procedure**

4.3   Searching for Contradictory Paths

The procedure TSPAN is used to find contradictory paths as follows:

STEP 1. Pick a variable and a pair of its complementary clones p, q.
STEP 2. Run TSPAN starting from p for at most $s_1 = n^{\frac{\alpha+4}{6(\alpha+1)}}$ steps. If $\mathcal{L}(p)$ becomes empty during the process, or if q gets included into $span(p)$, or if in the end $|\mathcal{L}(p)| < \sigma = s_1 \mu/6$ (μ is determined by the value $2T_n/S_n$, see Lemma 7), declare failure.
STEP 3. Run TSPAN starting from q and the current set $\mathcal{C}$ of paired clones for at most $s_1 = n^{\frac{\alpha+4}{6(\alpha+1)}}$ steps. If $\mathcal{L}(q)$ becomes empty during the process, or if $|\mathcal{L}(q) \cap \mathcal{L}(q)| = \Theta(s_1)$, or if in the end $|\mathcal{L}(q)| < \sigma$, declare failure.
STEP 4. Run TSPAN starting from $\mathcal{L}(p)$ and the current set $\mathcal{C}$ of paired clones for at most $s_2 = n^{\frac{11\alpha^2+3\alpha-2}{12\alpha(\alpha+1)}}$ steps. If $\mathcal{L}(p)$ becomes empty during the process, declare failure.
STEP 5. Similarly, run TSPAN starting from q and the current set $\mathcal{C}$ of paired clones for at most $s_2 = n^{\frac{11\alpha^2+3\alpha-2}{12\alpha(\alpha+1)}}$ steps. If $\mathcal{L}(q)$ becomes empty during the process, declare failure.

If a failure is declared at any step, we abandon the current pair p, q and pick another variable and a pair of clones keeping the current set $\mathcal{C}$ of paired clones that will be used in the next round. Also, even if all the Steps are successful, but the constructed span does not contain contradictory paths, we also declare a failure. It is important that the set $\mathcal{C}$ never grows too large, that is, it remains of size $|\mathcal{C}| = o(n)$. This implies that the number of restarts does not exceed $K = n^{\frac{7\alpha+10}{12(\alpha+1)}}$.

The next lemma shows how we choose the value of μ. It also shows that we expect a positive change in the size of the live set when $\frac{2T_n}{S_n} > 1$. However, first, we need to introduce several variables. Let $\mathcal{L}_i, \mathcal{U}_i$, and $\mathcal{C}_i$ are the live, untouched, and connected sets respectively after the i-th iteration of some execution of TSPAN. Additionally we have $L_i = |\mathcal{L}_i|$, $C_i = |\mathcal{C}_i|$, $U_i = |\mathcal{U}_i|$. Also let X_i indicate the change in the number of live clones after performing the ith iteration, i.e. $X_i = L_i - L_{i-1}$.

Lemma 7. *Let $\frac{2T_n}{S_n} = 1 + \mu$, where $\mu > 0$. Then for any $t \leq |\mathcal{C}| = o(n)$, we have*

$$\mathbb{E}\left[X_t \mid X_1, \ldots, X_{t-1}\right] \geq \mu/2.$$

Next, we bound the probability of failure in each of Steps 2–5. We start with STEP 2 assuming that the number of paired clones is $o(n)$.

Lemma 8 (STEP 2). *(1) Let $s_1 = n^{\frac{\alpha+4}{6(\alpha+1)}}$. If TSPAN starts with a live set containing only a single point $L_0 = 1$, time bound $\tau = s_1$, the live set size bound $\sigma = s_1\mu/6$, and the number of already paired clones $|\mathcal{C}| = o(n)$, then with probability at least $\frac{1}{2s_1}$ TSPAN terminates with the live set of size at least σ.*

(2) For any fixed clone q, the probability it will be paired in $s_1 = n^{\frac{\alpha+4}{6(\alpha+1)}} \leq t = o(n)$ steps of the algorithm, is at most $o\left(\frac{1}{s_1}\right)$.

Note that in Lemma 8(1) the size of $\mathcal{L}(p)$ can be slightly greater than σ, as it may increase by more than 1 in the last iteration. Also, in Lemma 8(2) the bound on the probability is only useful when s_1 is sufficiently large.

Proof. We prove item (1) here. The TSPAN procedure may terminate at the moment $i < \tau$ due to one of two reasons: first, when L_i hits 0, and second, when $L_i = \sigma$. To simplify analysis of the lemma, instead of dealing with conditional probabilities that the live set hasn't paired all its clones, we suggest to use a slightly modified version of TSPAN, which *always* runs for τ steps.

The modified version works exactly as the original TSPAN procedure when the live set has at least one clone. But if at some moment, the live set has no clones to pick, we perform a "restart": we restore the sets $\mathcal{L}, \mathcal{C}$, and $\mathcal{D}$ to the states they'd been before the first iteration of TSPAN procedure occurred. After that we continue the normal pairing process. Although during restarts we reset the values of the sets, the counter that tracks the number of iterations the TSPAN has performed is never reset, and keeps increasing with every iteration until the procedure has performed pairings τ times, or the live set was able to grow up to size σ, and only then the TSPAN terminates.

Now, let $r_i = 1$ represents a "successfull" restart that started at i iteration, meaning during this restart the live set accumulated σ clones, while $r_i = 0$ means there was no restart or the live set became empty. What we are looking for $\Pr[r_1 = 1]$, since this probability is identical to the probability that the original TSPAN was able to grow the live set to the desired size. Next, we can have at most τ restarts, and, since the very first restart has the most time and we expect the live set to grow in the long run, it follows that it stochastically dominates over other r_i's. Thus,

$$\Pr\left[L_{s_1} \geq s_1\mu/6\right] \leq \Pr\left[\sum_{i=1}^{s_1} r_i \geq 1\right] \leq \mathbb{E}\sum_{i=1}^{s_1} r_i \leq s_1\mathbb{E}r_1 = s_1\Pr[r_1 = 1]$$

from which we obtain the probability that the TSPAN terminates with large enough live set from the very first try:

$$P := \Pr[r_1 = 1] \geq \frac{\Pr[L_{s_1} \geq s_1\mu/6]}{s_1}. \tag{5}$$

Now what is left is to obtain bounds on the right-hand side probability. We have a random process

$$L_{s_1} = \sum_{i=1}^{s_1} (L_i - L_{i-1}) = \sum_{i=1}^{s_1} X_i,$$

which consists of steps X_i, each of which can be proved to have the right tail bounded by $\Pr[X_i \geq \ell \mid X_1, \ldots, X_{i-1}] \leq V\ell^{-\alpha}$, and positive expectation (Lemma 7) $\mathbb{E}[X_i \mid X_1, \ldots, X_{i-1}] \geq \frac{\mu}{2}$.

Therefore, according to *Azuma-like inequality* (Lemma 5), we obtain that

$$\Pr\left[L_{s_1} \leq s_1\mu/6\right] = \Pr\left[L_{s_1} \leq \left(s_1\frac{\mu}{2}\right)\frac{1}{3}\right] \leq \exp\left(-\frac{s_1}{4\log^2 s_1}\frac{\mu^2}{576}\right)$$

Fixing $s_1 = n^{\frac{\alpha+4}{6(\alpha+1)}}$, we have for some constant $C > 0$

$$\Pr\left[L_{s_1} \leq s_1\mu/6\right] \leq \exp\left(-C\frac{n^{\frac{\alpha+4}{6(\alpha+1)}}}{\log^2 n}\right) = o(1) \leq 1/2.$$

Thus, from (5) follows $P \geq \frac{\Pr[L_{s_1} \geq s_1\mu/6]}{s_1} = \frac{1}{2s_1}$, which proves item (1) of the lemma.   □

The probability that both runs of TSPAN for p and q are successful is given by the following

Lemma 9 (STEP 3). *The probability that two specific clones p and q accumulate $s_1\mu/7$ clones in their corresponding live sets $\mathcal{L}$ during the execution of* STEPS 2, 3, *such that the span from clone p doesn't include q nor make it live, is at least $\frac{1}{5s_1^2}$.*

Next, we show that we can grow the spans for another s_2 steps, while keeping the size of the respective live sets of order at least $s_1\mu/8$.

Lemma 10 (STEPS 4, 5). *Assume that p- and q-spans were both able to accumulate at least $s_1\mu/8$ live clones after $s_1 = n^{\frac{\alpha+4}{6(\alpha+1)}}$ steps, and q is not in the p-span. Then with probability $1 - o(1)$, TSPAN will be able to perform another $s_2 = n^{\frac{11\alpha^2+3\alpha-2}{12\alpha(\alpha+1)}}$ iterations, and $L_{s_1+j} \geq s_1\mu/8$ for every $0 \leq j \leq s_2$ for each clone p and q.*

Finally, we show that w.h.p. the spans produced in STEPS 1–5, provided no failure occurred, contain contradictory paths. In other words, we are looking for the probability that spans do not contain complement clones after growing them for $s_1 + s_2$ steps, i.e. at each step TSPAN was choosing only untouched clones from the set $\mathcal{U}$.

Lemma 11 (Contradiction paths). *If for a pair of complementary clones p, q Steps 1–5 are completed successfully, the probability that span(p) or span(q) contain no 2 complementary clones is less than $\exp\left(-n^{\frac{\alpha^2-\alpha-2}{12\alpha(\alpha+1)}}\right)$.*

This completes the proof in the case $\alpha = 2$ or $\mathbb{E}\xi^2 > 3\mathbb{E}\xi$, and the next proposition summarizes the result.

Proposition 2. *Let $\phi \sim \mathbb{C}_n^2(\xi)$, where $\xi \sim \mathcal{P}(\alpha)$ and $\alpha = 2$ or $\mathbb{E}\xi^2 > 3\mathbb{E}\xi$. Then w.h.p. ϕ is unsatisfiable.*

5 Satisfiability of $\mathbb{C}_n^2(\xi)$, When $\xi \sim \mathcal{P}(\alpha)$ and $\mathbb{E}\xi^2 < 3\mathbb{E}\xi$

Chvátal and Reed [15] argue that if 2-SAT formula ϕ is unsatisfiable, then it contains a *bicycle*, see Sect. 2.3. Thus, the absence of *bicycles* may serve as a convenient witness of formula's satisfiability. The general idea of this section is to show that w.h.p. there are no bicycles in $\phi \sim \mathbb{C}_n^2(\xi)$, when $\xi \sim \mathcal{P}(\alpha)$ and $\mathbb{E}\xi^2 < 3\mathbb{E}\xi$.

Intuitively, when $\mathbb{E}\xi^2 < 3\mathbb{E}\xi$, then we expect $\frac{2T_n}{S_n} = 1 - \mu' > 0$, where $\mu' > 0$ is some small number. As it was showed in Lemma 7, the latter quantity approximates the number of newly added live clones, when running the TSPAN procedure. Since TSPAN always performs at least one iteration of growing the span, it may add at most Δ clones into the live set after constructing the very first span from the root. After that each subsequent iteration adds *on average* $\approx \frac{2T_n}{S_n}$ new live clones. So after running the TSPAN for j iterations, where $j \to \infty$, when $n \to \infty$, then we expect the live set to contain around

$$L_{t^*} = \Delta\left(\frac{2T_n}{S_n}\right)^j = \Delta(1-\mu')^j \leq \Delta e^{-j\mu'}$$

clones. Therefore, after $O(\log n)$ iterations, the live set becomes empty, and TSPAN terminates. Thus, we expect paths of length at most $O(\log n)$, which is not enough for bicycles to occur.

More formally we first show that in the case $\frac{2T_n}{S_n} = 1 - \mu'$ a random formula is unlikely to contain long paths.

Lemma 12. *If $\frac{2T_n}{S_n} = 1 - \mu' < 1$, then paths in ϕ are of length $O(\log n)$, w.h.p.*

Then we give a straightforward estimation of the number of 'short' bicycles

Lemma 13. *If $\frac{2T_n}{S_n} = 1 - \mu' < 1$, then for any k the probability that formula ϕ contains a bicycle of length r is at most $(1 + o(1))^r (1 - \mu')^r$.*

These two lemmas imply that ϕ contains no bicycles.

Corollary 4. *If $\frac{2T_n}{S_n} = 1 - \mu' < 1$, then ϕ contains no bicycles, w.h.p.*

It remains to argue that the inequality $\frac{2T_n}{S_n} = 1 - \mu' < 1$ holds w.h.p.

Proposition 3. *Let $\phi \sim \mathbb{C}_n^2(\xi)$, where $\xi \sim \mathcal{P}(\alpha)$ and $\mathbb{E}\xi^2 < 3\mathbb{E}\xi$. Then w.h.p. ϕ is satisfiable.*

References

1. Achlioptas, D.: Lower bounds for random 3-SAT via differential equations. Theor. Comput. Sci. **265**(1–2), 159–185 (2001)
2. Achlioptas, D., Moore, C.: Random k-SAT: two moments suffice to cross a sharp threshold. SIAM J. Comput. **36**(3), 740–762 (2006)
3. Aiello, W., Graham, F.C., Lu, L.: A random graph model for power law graphs. Exp. Math. **10**(1), 53–66 (2001)
4. Ansótegui, C., Bonet, M.L., GirÁldez-Cru, J., Levy, J.: Community structure in industrial SAT instances (2016)
5. Ansótegui, C., Bonet, M.L., Levy, J.: On the structure of industrial SAT instances. In: Gent, I.P. (ed.) CP 2009. LNCS, vol. 5732, pp. 127–141. Springer, Heidelberg (2009). https://doi.org/10.1007/978-3-642-04244-7_13
6. Ansótegui, C., Bonet, M.L., Levy, J.: Towards industrial-like random SAT instances. In: IJCAI 2009, Proceedings of the 21st International Joint Conference on Artificial Intelligence, Pasadena, California, USA, 11–17 July 2009, pp. 387–392 (2009)
7. Ansótegui, C., Bonet, M.L., Levy, J.: Scale-free random SAT instances (2017). https://arxiv.org/abs/1708.06805v2
8. Ansótegui, C., Bonet, M.L., Levy, J., Manyà, F.: Measuring the hardness of SAT instances. In: Proceedings of the 23rd National Conference on Artificial Intelligence, AAAI 2008, vol. 1, pp. 222–228 (2008)
9. Aspvall, B., Plass, M.F., Tarjan, R.E.: A linear-time algorithm for testing the truth of certain quantified Boolean formulas. Inf. Process. Lett. **8**(3), 121–123 (1979)
10. Barabási, A.L., Albert, R.: Emergence of scaling in random networks. Science **286**, 509–512 (1999)
11. Bollobás, B., Riordan, O.: Mathematical results on scale-free random graphs. In: Handbook of Graphs and Networks, pp. 1–34. Wiley-VCH (2002)
12. Bollobás, B., Riordan, O., Spencer, J., Tusnády, G.E.: The degree sequence of a scale-free random graph process. Random Struct. Algorithms **18**(3), 279–290 (2001)

13. Boufkhad, Y., Dubois, O., Interian, Y., Selman, B.: Regular random k-SAT: properties of balanced formulas. J. Autom. Reason. **35**(1–3), 181–200 (2005)
14. Braunstein, A., Mézard, M., Zecchina, R.: Survey propagation: an algorithm for satisfiability. Random Struct. Algorithms **27**(2), 201–226 (2005)
15. Chvátal, V., Reed, B.A.: Mick gets some (the odds are on his side). In: 33rd Annual Symposium on Foundations of Computer Science, Pittsburgh, Pennsylvania, USA, 24–27 October 1992, pp. 620–627 (1992)
16. Clauset, A., Shalizi, C., Newman, M.: Power-law distributions in empirical data. SIAM Rev. **51**(4), 661–703 (2009)
17. Coja-Oghlan, A.: A better algorithm for random k-SAT. SIAM J. Comput. **39**(7), 2823–2864 (2010)
18. Coja-Oghlan, A., Panagiotou, K.: Going after the k-SAT threshold. In: Symposium on Theory of Computing Conference, STOC 2013, Palo Alto, CA, USA, 1–4 June 2013, pp. 705–714 (2013)
19. Cook, S.A., Mitchell, D.G.: Finding hard instances of the satisfiability problem: a survey. In: Proceedings of a DIMACS Workshop on Satisfiability Problem: Theory and Applications, Piscataway, New Jersey, USA, 11–13 March 1996, pp. 1–18 (1996)
20. Cooper, C., Frieze, A., Sorkin, G.B.: Random 2-SAT with prescribed literal degrees. Algorithmica **48**(3), 249–265 (2007)
21. Díaz, J., Kirousis, L.M., Mitsche, D., Pérez-Giménez, X.: On the satisfiability threshold of formulas with three literals per clause. Theor. Comput. Sci. **410**(30–32), 2920–2934 (2009)
22. Ding, J., Sly, A., Sun, N.: Proof of the satisfiability conjecture for large k. In: Proceedings of the Forty-Seventh Annual ACM on Symposium on Theory of Computing, STOC 2015, Portland, OR, USA, 14–17 June 2015, pp. 59–68 (2015)
23. Dubios, O., Boufkhad, Y.: A general upper bound for the satisfiability threshold of random r-SAT formulae. J. Algorithms **24**(2), 395–420 (1997)
24. Dubois, O., Boufkhad, Y., Mandler, J.: Typical random 3-SAT formulae and the satisfiability threshold. Electron. Colloq. Comput. Complex. (ECCC) **10**(007) (2003)
25. Franco, J., Paull, M.C.: Probabilistic analysis of the Davis Putnam procedure for solving the satisfiability problem. Discrete Appl. Math. **5**(1), 77–87 (1983)
26. Friedgut, E.: Sharp thresholds of graph properties, and the k-SAT problem. J. ACM **12**(4), 1017–1054 (1999)
27. Friedrich, T., Krohmer, A., Rothenberger, R., Sauerwald, T., Sutton, A.M.: Bounds on the satisfiability threshold for power law distributed random SAT. In: 25th Annual European Symposium on Algorithms, ESA 2017, Vienna, Austria, 4–6 September 2017, pp. 37:1–37:15 (2017)
28. Friedrich, T., Rothenberger, R.: Sharpness of the satisfiability threshold for nonuniform random k-SAT. In: Beyersdorff, O., Wintersteiger, C.M. (eds.) SAT 2018. LNCS, vol. 10929, pp. 273–291. Springer, Cham (2018). https://doi.org/10.1007/978-3-319-94144-8_17
29. Giráldez-Cru, J., Levy, J.: Generating SAT instances with community structure. Artif. Intell. **238**(C), 119–134 (2016)
30. Goerdt, A.: A threshold for unsatisfiability. J. Comput. Syst. Sci. **53**(3), 469–486 (1996)
31. Kaporis, A.C., Kirousis, L.M., Lalas, E.G.: The probabilistic analysis of a greedy satisfiability algorithm. Random Struct. Algorithms **28**(4), 444–480 (2006)

32. Kim, J.H.: The Poisson cloning model for random graphs, random directed graphs and random k-SAT problems. In: Chwa, K.-Y., Munro, J.I.J. (eds.) COCOON 2004. LNCS, vol. 3106, p. 2. Springer, Heidelberg (2004). https://doi.org/10.1007/978-3-540-27798-9_2

33. Kirousis, L.M., Kranakis, E., Krizanc, D., Stamatiou, Y.C.: Approximating the unsatisfiability threshold of random formulas. Random Struct. Algorithms **12**(3), 253–269 (1998)

34. Krioukov, D.V., Papadopoulos, F., Kitsak, M., Vahdat, A., Boguñá, M.: Hyperbolic geometry of complex networks. CoRR abs/1006.5169 (2010)

35. Krzakała, F., Montanari, A., Ricci-Tersenghi, F., Semerjian, G., Zdeborová, L.: Gibbs states and the set of solutions of random constraint satisfaction problems. PNAS **104**(25), 10318–10323 (2007)

36. Mézard, M., Parisi, G., Zecchina, R.: Analytic and algorithmic solution of random satisfiability problems. Science **297**(5582), 812–815 (2002)

37. Newman, M.: Power laws, Pareto distributions and Zipf's law. Contemp. Phys. **46**(5), 323–351 (2005)

38. Omelchenko, O., Bulatov, A.: Concentration inequalities for sums of random variables, each having power bounded tails (2018). https://arxiv.org/abs/1903.02529

39. Selman, B., Mitchell, D.G., Levesque, H.J.: Generating hard satisfiability problems. Artif. Intell. **81**(1–2), 17–29 (1996)

40. de la Vega, W.F.: Random 2-SAT: results and problems. Theor. Comput. Sci. **265**(1–2), 131–146 (2001)

DRAT Proofs, Propagation Redundancy, and Extended Resolution

Sam Buss[1]($\boxtimes$) and Neil Thapen[2]

[1] Department of Mathematics, U.C. San Diego, La Jolla, CA, USA
sbuss@ucsd.edu
[2] Institute of Mathematics, Czech Academy of Sciences, Prague, Czech Republic
thapen@math.cas.cz

Abstract. We study the proof complexity of RAT proofs and related systems including BC, SPR, and PR which use blocked clauses and (subset) propagation redundancy. These systems arise in satisfiability (SAT) solving, and allow inferences which preserve satisfiability but not logical implication. We introduce a new inference SR using substitution redundancy. We consider systems both with and without deletion. With new variables allowed, the systems are known to have the same proof theoretic strength as extended resolution. We focus on the systems that do not allow new variables to be introduced.

Our first main result is that the systems DRAT⁻, DSPR⁻ and DPR⁻, which allow deletion but not new variables, are polynomially equivalent. By earlier work of Kiesl, Rebola-Pardo and Heule, they are also equivalent to DBC⁻. Without deletion and without new variables, we show that SPR⁻ can polynomially simulate PR⁻ provided only short clauses are inferred by SPR inferences. Our next main results are that many of the well-known "hard" principles have polynomial size SPR⁻ refutations (without deletions or new variables). These include the pigeonhole principle, bit pigeonhole principle, parity principle, Tseitin tautologies, and clique-coloring tautologies; SPR⁻ can also handle or-fication and xor-ification. Our final result is an exponential size lower bound for RAT⁻ refutations, giving exponential separations between RAT⁻ and both DRAT⁻ and SPR⁻.

1 Introduction

SAT solvers are routinely used for many large-scale instances of satisfiability. It is widely realized that when a solver reports a SAT instance Γ is unsatisfiable, it should also produce a *proof* of unsatisfiability. This is of particular importance as SAT solvers become increasingly complex, and thus more subject to software bugs or even design problems.

S. Buss—This work was initiated on a visit of the first author to the Czech Academy of Sciences in July 2018, supported by ERC advanced grant 339691 (FEALORA). Also supported by Simons Foundation grant 578919.
N. Thapen—Partially supported by GA ČR project 19-05497S and by ERC advanced grant 339691 (FEALORA) and RVO:67985840.

© Springer Nature Switzerland AG 2019
M. Janota and I. Lynce (Eds.): SAT 2019, LNCS 11628, pp. 71–89, 2019.
https://doi.org/10.1007/978-3-030-24258-9_5

The first proof systems proposed for SAT solvers were based on reverse unit propagation (RUP) inferences [8,27] as this is sufficient to handle resolution and the usual CDCL clause learning schemes. However, RUP inferences only support logical implication, and do not accommodate many "inprocessing" rules. Inprocessing rules may not respect logical implication; instead they only guarantee *equisatisfiability* [15]. Inprocessing inferences have been formalized with sophisticated inference rules including DRAT (*deletion, reverse asymmetric tautology*), PR (*propagation redundancy*) and SPR (*subset* PR) in a series of papers including [10,11,15,28]. These systems can be used both as proof systems to verify unsatisfiability, and as inference systems to facilitate searching for either a satisfying assignment or a proof of unsatisfiability.

The ability to introduce new variables makes DRAT very powerful, and it can simulate extended resolution [16]. Moreover there are recent results [9,12–14] indicating that DRAT and PR are still powerful when restricted to use few new variables, or even no new variables. In particular, [12–14] showed that the pigeonhole principle clauses have short (polynomial size) refutations in the PR proof system; in fact, these refutations can be found *automatically* by an appropriately configured SAT solver [14]. There are at present no good proof search heuristics for how to introduce new variables with the extension rule. It is possible however that there are good heuristics for searching for proofs that do not use new variables in DRAT and PR and related systems. Thus, these systems could potentially lead to dramatic improvements in the power of SAT solvers.

This paper studies these proof systems viewed as refutation systems, paying particular attention to proof systems that do not allow new variables. The proof systems BC (*blocked clauses*), RAT, SPR, PR, and SR are defined below. (Only SR is new to this paper.) These systems have variants which allow deletion, called DBC, DRAT, DSPR, DPR and DSR. There are also variants of all these systems restricted to not allow new variables: we denote these with a superscript "−" as BC$^-$, DBC$^-$, RAT$^-$, DRAT$^-$ etc.

Section 2 studies the relation between these systems and extended resolution. We show that any proof system containing BC$^-$ and closed under restrictions simulates extended resolution. A proof system $\mathcal{P}$ *simulates* a proof system $\mathcal{Q}$ if any $\mathcal{Q}$-proof can be converted, in polynomial time, into a $\mathcal{P}$-proof of the same result. It is known that DBC$^-$ simulates DRAT$^-$ [16], and that DRAT simulates DPR with the use of only one extra variable [9]. Section 3.1 shows DRAT$^-$ simulates DPR$^-$. As a consequence, DBC$^-$ can also simulate DPR$^-$. Section 3.2 gives a method to convert SPR$^-$ refutations into PR$^-$ refutations with a size increase that is exponential in the "discrepancy" of the PR inferences. However, in many cases, the discrepancy will be logarithmic or even smaller.

Section 4 proves new polynomial upper bounds on the size of SPR$^-$ proofs for many of the "hard" tautologies from proof complexity. This includes the pigeonhole principle, the bit pigeonhole principle, the parity principle, the clique-coloring principle, and the Tseitin tautologies. We also show that obfuscation by or-fication and xor-ification does not work against SPR$^-$. Note that SPR$^-$

allows neither deletion nor the use of new variables. Prior results gave SPR^- proofs for the pigeonhole principle (PHP) [12,13], and PR^- proofs for the Tseitin tautologies and the 2-1 PHP [9]. These results raise the question of whether SPR^- can simulate, for instance, Frege systems.

Section 5 shows RAT^- cannot simulate either DRAT^- or SPR^-, by proving size and width lower bounds for RAT^- proofs of the bit pigeonhole principle (BPHP).

Most of the known inclusions for these systems, including our new results, are summarized in (1)–(3). Allowing new variables (and with or without deletion), we have

$$\text{Res} < \text{BC} \equiv \text{RAT} \equiv \text{SPR} \equiv \text{PR} \equiv \text{SR} \equiv \text{ER}. \tag{1}$$

With deletion and no new variables (except ER may use new variables):

$$\text{Res} < \text{DBC}^- \equiv \text{DRAT}^- \equiv \text{DSPR}^- \equiv \text{DPR}^- \le \text{DSR}^- \le \text{ER}. \tag{2}$$

With no deletion and no new variables (except ER may use new variables):

$$\text{Res} < \text{BC}^- \le \text{RAT}^- < \text{SPR}^- \le^* \text{PR}^- \le \text{SR}^- \le \text{ER}. \tag{3}$$

In these equations, equivalence ($\equiv$) indicates the systems simulate each other. Inequality ($\le$) indicates only one direction is known for the simulation. Strict inequality ($<$) means that it is known there is no simulation in the other direction. The symbol $\le^*$ in (3) means PR^- simulates SPR^-, and there is a simulation in the other direction under the additional assumption that the discrepancies of PR inferences are logarithmically bounded.

There are still a number of open questions about the systems with no new variables. Of particular importance is the question of the relative strengths of DPR^-, DSR^- and ER. The system DPR^- is a promising system for effective proof search algorithms, and ER is known to be strong. The results of [9,12,13] and the present paper show that DPR^- is also strong. Indeed, Sect. 4 shows even (the possibly weaker) SPR^- is strong.

Another important question is to understand the strength of deletion for these systems. Deletion is well-known to help the performance of SAT solvers in practice, and for systems such as RAT, it is known that deletion can allow new inferences. Our results in Sects. 4 and 5 show that in fact RAT^- does not simulate DRAT^-. This strengthens the case for the importance of deletion.

We thank the reviewers for suggestions and comments that improved the paper.

1.1 Preliminaries

We use the usual conventions for clauses, variables, literals, truth assignments, etc. *Var* and *Lit* are the sets of all variables and all literals. A set of literals is called *tautological* if it contains a pair of complementary literals p and $\bar{p}$. A *clause* is a non-tautological set of literals; we use $C, D, \ldots$ to denote clauses. The empty clause is denoted $\bot$, and is always false. 0 and 1 denote respectively

False and *True*; and $\overline{0}$ and $\overline{1}$ are respectively 1 and 0. We use both $C \cup D$ or $C \vee D$ to denote unions of clauses, but usually write $C \vee D$ when the union is a clause. The notation $C = D \dot{\vee} E$ indicates that $C = D \vee E$ is a clause and D and E have no variables in common. If Γ is a set of clauses, $C \vee \Gamma$ is the set $\{C \vee D : D \in \Gamma \text{ and } C \vee D \text{ is a clause}\}$.

A *partial assignment* τ is a mapping from a set of variables to $\{0,1\}$. It acts on literals by letting $\tau(\overline{p}) = \overline{\tau(p)}$. We sometimes identify a partial assignment τ with the set of unit clauses asserting that τ holds. For C a clause, $\overline{C}$ denotes the partial assignment whose domain is the variables of C and which asserts that C is false. For example, if $C = x \vee \overline{y} \vee z$ then, depending on context, $\overline{C}$ will denote either the set containing the three unit clauses $\overline{x}$ and y and $\overline{z}$, or the partial assignment α with domain $dom(\alpha) = \{x,y,z\}$ such that $\alpha(x) = \alpha(z) = 1$ and $\alpha(y) = 0$.

A *substitution* generalizes the notion of a partial assignment by allowing variables to be mapped also to literals. Formally, a substitution σ is a map from $Var \cup \{0,1\}$ to $Lit \cup \{0,1\}$ which is constant on $\{0,1\}$. Note that a substitution may cause different literals to become identified. A partial assignment τ can be viewed as a substitution, by defining $\tau(x) = x$ for all variables x outside the domain of τ.

Suppose C is a clause and σ is a substitution. Let $\sigma(C) = \{\sigma(p) : p \in C\}$. We say σ *satisfies* C, written $\sigma \models C$, if $1 \in \sigma(C)$ or $\sigma(C)$ is tautological. When $\sigma \not\models C$, the *restriction* $C_{|\sigma}$ is defined by letting $C_{|\sigma}$ equal $\sigma(C) \setminus \{0\}$. Thus $C_{|\sigma}$ is a clause expressing the meaning of C under σ. For Γ a set of clauses, the restriction of Γ under σ is denoted $\Gamma_{|\sigma}$ and equals $\{C_{|\sigma} : C \in \Gamma \text{ and } \sigma \not\models C\}$. The composition of substitutions τ and π is defined by $(\tau \circ \pi)(x) = \tau(\pi(x))$, and in particular $(\tau \circ \pi)(x) = \pi(x)$ if $\pi(x) \in \{0,1\}$. For partial assignments τ and π, this means that $dom(\tau \circ \pi) = dom(\tau) \cup dom(\pi)$ and that $(\tau \circ \pi)(x)$ equals $\pi(x)$ for $x \in dom(\pi)$ and $\tau(x)$ for $x \in dom(\tau) \setminus dom(\pi)$.

Lemma 1. *For a set of clauses Γ and substitutions τ and π, $\Gamma_{|\tau \circ \pi} = (\Gamma_{|\pi})_{|\tau}$. In particular, $\tau \models \Gamma_{|\pi}$ if and only if $\tau \circ \pi \models \Gamma$.*

We write $\Gamma \models C$, if every total assignment satisfying Γ also satisfies C. Recall that a *unit propagation* refutation of Γ is a resolution refutation in which, in every resolution step, one of the clauses resolved is a unit clause.

Definition 2. *We write $\Gamma \vdash_1 \bot$ to denote that there is a unit propagation refutation of Γ. We define $\Gamma \vdash_1 C$ to mean $\Gamma \cup \overline{C} \vdash_1 \bot$. For a set of clauses Δ, we write $\Gamma \vdash_1 \Delta$ to mean $\Gamma \vdash_1 C$ for every $C \in \Delta$.*

Fact 3. *If $\Gamma \vdash_1 \bot$ and α is any partial assignment or substitution, then $\Gamma_{|\alpha} \vdash_1 \bot$.*

When $\Gamma \vdash_1 C$, then C is said to be derivable from Γ by *reverse unit propagation* (RUP), or is called an *asymmetric tautology* (AT) with respect to Γ [11,15,27]. Of course, $\Gamma \vdash_1 C$ implies that $\Gamma \models C$. The advantage of working with $\vdash_1$ is that there is a simple polynomial time algorithm to determine whether $\Gamma \vdash_1 C$. We have:

Lemma 4. *If C is derivable from Γ by a single resolution inference, then $\Gamma \vdash_1 C$. Conversely, if $\Gamma \vdash_1 C$, then some $C' \subseteq C$ has a resolution derivation from Γ of length at most $n + 1$, where n is the number of variables appearing in Γ.*

Lemma 5. *Let $C \vee D$ be a clause (so $C \cup D$ is not tautological), and set $\alpha = \overline{C}$. Then*

$$\Gamma_{|\alpha} \vdash_1 D \setminus C \quad \Longleftrightarrow \quad \Gamma_{|\alpha} \vdash_1 D \quad \Longleftrightarrow \quad \Gamma \vdash_1 C \vee D.$$

1.2   RAT and Propagation Redundancy

We next describe inference rules which can be used to add a clause C to a set of clauses Γ, maintaining satisfiability. In non-strictly increasing order of strength, they are

$$\mathrm{BC} \to \mathrm{RAT} \to \mathrm{SPR} \to \mathrm{PR} \to \mathrm{SR}.$$

The definitions follow [11,13,15], except for the new SR ("substitution redundancy"). All of these rules can be viewed as allowing the introduction of clauses that hold "without loss of generality" [22].

Let Γ be a set of clauses and C a clause with a distinguished literal p, so that C has the form $p \mathbin{\dot\vee} C'$.

Definition 6. *The clause C is a* blocked clause *(BC) with respect to p and Γ if, for every clause D of the form $\overline{p} \mathbin{\dot\vee} D'$ in Γ, the set $C' \cup D'$ is tautological.*

Definition 7. *A clause C is a* resolution asymmetric tautology *(RAT) with respect to p and Γ if, for every clause D of the form $\overline{p} \mathbin{\dot\vee} D'$ in Γ, either $C' \cup D'$ is tautological or $\Gamma \vdash_1 p \vee C' \vee D'$.*

We write $p \vee C'$ instead of C to emphasize that we include the literal p (some definitions of RAT omit it). Clearly, being BC implies being RAT.

Two sets Γ and Π of clauses are *equisatisfiable* if they are both satisfiable or both unsatisfiable. A well-known, important property of BC and RAT is:

Theorem 8 ([15]). *If C is BC or RAT w.r.t. Γ, then Γ and $\Gamma \cup \{C\}$ are equisatisfiable.*

For the rest of this section, let α be the partial assignment $\overline{C}$.

Definition 9 ([13]). *A clause C is* propagation redundant *(PR) with respect to Γ if there is a partial assignment τ such that $\tau \vDash C$ and $\Gamma_{|\alpha} \vdash_1 \Gamma_{|\tau}$.*

Theorem 10 ([13]). *If C is PR with respect to Γ, then Γ and $\Gamma \cup \{C\}$ are equisatisfiable.*

Of the remaining rules, SPR is a restriction of PR, but is more general than RAT [13]. The new substitution redundancy rule (SR) generalizes PR, allowing τ to be a substitution rather than a partial assignment. The condition is still polynomial-time checkable.

Definition 11 ([13]). *A clause C is* subset propagation redundant *(SPR) w.r.t. Γ if there is a partial assignment τ with $dom(\tau) = dom(\alpha)$ such that $\tau \vDash C$ and $\Gamma_{|\alpha} \vdash_1 \Gamma_{|\tau}$.*

Definition 12. *A clause C is* substitution redundant *(SR) with respect to Γ if there is a substitution τ such that $\tau \vDash C$ and $\Gamma_{|\alpha} \vdash_1 \Gamma_{|\tau}$.*

Theorem 13. *If C is SPR or SR w.r.t. Γ, then Γ and $\Gamma \cup \{C\}$ are equisatisfiable.*

Proof. This follows by Theorem 10 or, in the case of SR, by an identical proof. $\square$

We next give a technical lemma giving a kind of normal form for propagation redundancy. It implies that if C is PR with respect to Γ, then without loss of generality $dom(\tau)$ includes $dom(\alpha)$.

Lemma 14. *If C is PR w.r.t. Γ, witnessed by partial assignment τ, then $\Gamma_{|\alpha} \vdash_1 \Gamma_{|\alpha \circ \tau}$.*

Proof. Let $\pi = \alpha \circ \tau$. Suppose $E \in \Gamma$ is such that $\pi \nVdash E$. We must show that $\Gamma_{|\alpha} \vdash_1 E_{|\pi}$. We can decompose E as $E_1 \vee E_2 \vee E_3$ where E_1 contains the literals in $dom(\tau)$, E_2 the literals in $dom(\alpha) \setminus dom(\tau)$ and E_3 the remaining literals. Then $E_{|\tau} = E_2 \vee E_3$ and by the PR assumption $\Gamma_{|\alpha} \vdash_1 E_{|\tau}$, so there is a derivation $\Gamma_{|\alpha} \cup \overline{E_2} \cup \overline{E_3} \vdash_1 \bot$. But neither $\Gamma_{|\alpha}$ nor $\overline{E_3}$ contain any variables from $dom(\alpha)$, so the literals in $\overline{E_2}$ are not used in this derivation. Hence $\Gamma_{|\alpha} \cup \overline{E_3} \vdash_1 \bot$, which completes the proof since $E_3 = E_{|\pi}$. $\square$

1.3 Proof Systems

This section introduces proof systems based on the BC, RAT, SPR, PR and SR inferences. Some of the systems also allow the use of the deletion rule: these systems are denoted DBC, DRAT, etc. All the proof systems are *refutation systems*. They start with a set of clauses Γ, and successively derive sets Γ_i of clauses, first $\Gamma_0 = \Gamma$, then $\Gamma_1, \Gamma_2, \ldots, \Gamma_m$ until reaching a set Γ_m containing the empty clause. It will always be the case that if Γ_i is satisfiable, then Γ_{i+1} is satisfiable. Since the empty clause $\bot$ is in Γ_m, this last set is not satisfiable. This suffices to show that Γ is not satisfiable.

Definition 15. *A BC, RAT, SPR, PR, or SR proof (or refutation) of Γ is a sequence $\Gamma_0, \ldots, \Gamma_m$ such that $\Gamma_0 = \Gamma$, $\bot \in \Gamma_m$ and each $\Gamma_{i+1} = \Gamma_i \cup \{C\}$, where either*

- *$\Gamma_i \vdash_1 C$ (that is, "C is RUP with respect to Γ_i"), or*
- *C is BC, RAT, SPR, PR, or SR (respectively) with respect to Γ_i.*

For BC or RAT steps, the proof must specify some p, and for SPR, PR or SR, it must specify some τ.

There is no constraint on the variables that appear in clauses C introduced in BC, RAT etc. steps. They are free to include new variables that did not occur in $\Gamma_0, \ldots, \Gamma_i$.

Definition 16. *A DBC, DRAT, DSPR, DPR, or DSR proof allows the same rules of inference (respectively) as Definition 15, plus the* deletion *inference rule:*

- $\Gamma_{i+1} = \Gamma_i \setminus \{C\}$ *for some* $C \in \Gamma_i$.

Since RUP inferences simulate resolution, these systems all simulate resolution. By Theorems 8, 10 and 13, they are sound. Since the inferences are defined using $\vdash_1$ they are polynomial time verifiable, as the description of τ is included with every SPR, PR or SR inference. Hence they are all proof systems in the sense of Cook-Reckhow [6, 7].

The deletion rule allows *any* clause to be deleted, even initial clauses. So it can happen that Γ_i is unsatisfiable but Γ_{i+1} is satisfiable. This is okay for us since we focus on refuting sets of unsatisfiable clauses. Surprisingly, deletion is important because the property of being BC, RAT etc. involves a universal quantification over the current set of clauses Γ_i. Thus deletion can make the systems more powerful, by making more inferences possible. For this, see Corollary 47. (Also, an early paper on this by Kullmann [19] exploited deletions to generalize the power of BC inferences.)

All the systems defined so far are equivalent to extended resolution (ER), because of their ability to freely introduce new variables. The main topic of the paper is the systems in the next definition, which lack this ability.

Definition 17. *A BC refutation of* Γ *without new variables, or, for short, a* BC^- *refutation of* Γ, *is a BC refutation of* Γ *in which only variables from* Γ *appear. The systems* RAT$^-$, SPR$^-$, PR$^-$, SR$^-$ *and* DBC$^-$, DRAT$^-$, DSPR$^-$, DPR$^-$, DSR$^-$ *are defined similarly.*

2   Relations with Extended Resolution

2.1   With New Variables

It is well-known that RAT, and even BC, can simulate extended resolution (ER) if new variables are allowed. To see this, consider an extended resolution inference which uses the extension rule to introduce a new variable x to stand for the conjunction $p \wedge q$ of two literals. This means that the three extension clauses

$$x \vee \overline{p} \vee \overline{q} \qquad \overline{x} \vee p \qquad \overline{x} \vee q \qquad (4)$$

are introduced. We can instead add these clauses using the BC rule. Let Γ be the original set of clauses, and let $\Gamma_1, \Gamma_2, \Gamma_3$ be Γ with the three clauses above successively added. Then $x \vee \overline{p} \vee \overline{q}$ is BC with respect to Γ and x because no clause in Γ contains $\overline{x}$. The clause $\overline{x} \vee p$ is BC with respect to Γ_1 and $\overline{x}$ because the only clause in Γ_1 containing x is $x \vee \overline{p} \vee \overline{q}$, and resolving this with $\overline{x} \vee p$ gives a tautological conclusion. The clause $\overline{x} \vee q$ is BC with respect to Γ_2 and $\overline{x}$ in a similar way. Thus BC, and hence all the other systems which allow new variables, simulate ER. The converse holds as well:

Theorem 18. *The system* ER *simulates* DSR, *and hence every other system above.*

For space reasons, we omit the proof here, but it is known already by [9,16] that ER simulates DPR. A similar proof works for DSR.

2.2   Without New Variables

In the systems without the ability to freely add new variables, we can still imitate extended resolution by adding dummy variables to the formula we want to refute.

For $m \geq 1$, define X^m to be the set consisting of only the two clauses

$$y \vee x_1 \vee \cdots \vee x_m \quad \text{and} \quad y.$$

Lemma 19. *Suppose* Γ *has an* ER *refutation* Π *of size* m, *and that* Γ *and* X^m *have no variables in common. Then* $\Gamma \cup X^m$ *has a* BC$^-$-*refutation* Π^* *of size* $O(m)$, *which can furthermore be constructed from* Π *in polynomial time.*

Proof. We describe how to change Π into Π^*. We first rename all extension variables to use names from $\{x_1, \ldots, x_m\}$ and replace all resolution steps with $\vdash_1$ inferences. Now consider an extension rule in Π which introduces the three extension clauses (4) expressing $x_i \leftrightarrow (p \wedge q)$, where we may assume that p and q are either variables of Γ or from $\{x_1, \ldots, x_{i-1}\}$. We simulate this by introducing successively the three clauses

$$x_i \vee \overline{p} \vee \overline{q} \qquad \overline{x}_i \vee p \vee \overline{y} \qquad \overline{x}_i \vee q \vee \overline{y}$$

using the BC rule. The first clause, $x_i \vee \overline{p} \vee \overline{q}$, is BC with respect to x_i, because $\overline{x}_i$ has not appeared yet. The second clause is BC with respect to $\overline{x}_i$, because x_i appears only in two earlier clauses, namely $y \vee x_1 \vee \cdots \vee x_m$, which contains y, and $x_i \vee \overline{p} \vee \overline{q}$, which contains $\overline{p}$. In both cases the resolvent with $\overline{x}_i \vee p \vee \overline{y}$ is tautological. The third clause is similar. The unit clause y is in X^m, so we can then derive the remaining two needed extension clauses $\overline{x}_i \vee p$ and $\overline{x}_i \vee q$ by two $\vdash_1$ inferences.   □

As the next corollary shows, this lemma can be used to construct examples of usually-hard formulas which have short proofs in BC$^-$. (We will give less artificial examples of short SPR$^-$ proofs in Sect. 4.) Let $m(n)$ be the polynomial size upper bound on ER refutations of the pigeonhole principle PHP$_n$ which follows from [7].

Corollary 20. *The set of clauses* PHP$_n \cup X^{m(n)}$ *has polynomial size proofs in* BC$^-$, *but requires exponential size proofs in constant depth Frege.*

Proof. The upper bound is by Lemma 19. For the lower bound, let Π be a refutation in depth-d Frege. Then we can restrict Π by setting $y = 1$ to obtain a depth-d refutation of PHP$_n$ which, by [18,20], must have exponential size.   □

The same argument can give a more general result. A propositional proof system $\mathcal{P}$ is *closed under restrictions* if given any $\mathcal{P}$-refutation of Γ and any partial assignment ρ, we can construct a $\mathcal{P}$-refutation of $\Gamma_{|\rho}$ in polynomial time.

Theorem 21. *Let $\mathcal{P}$ be any propositional proof system which is closed under restrictions. If $\mathcal{P}$ simulates* BC^-, *then $\mathcal{P}$ simulates ER.*

Proof. Suppose Γ has a refutation Π in ER of length m. Take a copy of X^m in disjoint variables from Γ. By Lemma 19 we can construct a BC^--refutation of $\Gamma \cup X^m$. Since $\mathcal{P}$ simulates BC^-, we can then construct a $\mathcal{P}$-refutation of $\Gamma \cup X^m$. Let ρ be the restriction which just sets $y = 1$, so that $(\Gamma \cup X^m)_{|\rho} = \Gamma$. Since $\mathcal{P}$ is closed under restrictions, we can construct a $\mathcal{P}$-refutation of Γ. All constructions are polynomial time. $\qquad\qquad\qquad\qquad\qquad\qquad\qquad\qquad\square$

Corollary 22. *If, as is expected, the Frege proof system is strictly weaker than ER, then Frege does not simulate* BC^-.

3   Simulations

3.1   DRAT$^-$ Simulates DPR$^-$

The following relations were known between DBC^-, $DRAT^-$ and DPR^-.

Theorem 23 ([16]). DBC^- *simulates* $DRAT^-$.

Theorem 24 ([9]). *Suppose Γ has a DPR refutation Π. Then it has a DRAT refutation constructible in polynomial time from Π, using at most one variable not used in Π.*

We will show, in Theorem 30 below, that DRAT$^-$ simulates DPR$^-$. Thus the systems DBC^-, $DRAT^-$, $DSPR^-$ and DPR^- are all equivalent. Our proof relies on the main step in the proof of Theorem 24:

Lemma 25 ([9]). *Suppose C is PR w.r.t. Γ. Then there is a polynomial size DRAT derivation of $\Gamma \cup \{C\}$ from Γ, using at most one variable not appearing in Γ or C.*

Definition 26. *Let Γ be a set of clauses and x any variable. Then $\Gamma^{(x)}$ consists of every clause in Γ which does not mention x, together with every clause of the form $E \vee F$ where both $x \dot{\vee} E$ and $\overline{x} \dot{\vee} F$ are in Γ.*

In other words, $\Gamma^{(x)}$ is formed from Γ by doing all possible resolutions with respect to x and then deleting all clauses containing either x or $\overline{x}$. (This is exactly like the first step of the Davis-Putnam procedure.)

Lemma 27. *There is a polynomial size DRAT derivation of Γ from $\Gamma^{(x)}$, using only variables from Γ.*

Proof. We first derive every clause of the form $E \dot\vee x$ in Γ, by RAT on x. As $\overline{x}$ has not appeared yet the RAT condition is satisfied. Then we derive each clause of the form $F \dot\vee \overline{x}$ in Γ, by RAT on $\overline{x}$. The only possible resolutions are with clauses of the form $E \dot\vee x$ which we have just introduced, but in this case either $E \cup F$ is tautological or $E \vee F$ is in $\Gamma^{(x)}$ so $\Gamma^{(x)} \vdash_1 \overline{x} \vee F \vee E$. Finally we delete all clauses not in Γ. □

The next two lemmas show that, under suitable conditions, if we can derive C from Γ in DPR$^-$, then we can derive it from $\Gamma^{(x)}$. We will use a kind of normal form for PR inferences. Say that a clause C is PR$_0$ with respect to Γ if there is a partial assignment τ such that $\tau \vDash C$, all variables in C are in $dom(\tau)$, and

$$C \vee \Gamma_{|\tau} \subseteq \Gamma. \tag{5}$$

The PR$_0$ inference rule lets us derive $\Gamma \cup \{C\}$ from Γ when (5) holds. It is not hard to see that (5) implies $\Gamma_{|\alpha} \vdash_1 \Gamma_{|\tau}$, where $\alpha = \overline{C}$, so this is a special case of the PR rule.

Lemma 28. *Any* PR *inference can be replaced with a* PR$_0$ *inference together with polynomially many* $\vdash_1$ *and deletion steps, using no new variables.*

Proof. Suppose $\Gamma_{|\alpha} \vdash_1 \Gamma_{|\tau}$, where $\alpha = \overline{C}$ and $\tau \vDash C$. By Lemma 14 we may assume $dom(\alpha) \subseteq dom(\tau)$ so $dom(\tau)$ contains all variables in C. Let $\Delta = C \vee \Gamma_{|\tau}$ and $\Gamma^* = \Gamma \cup \Delta$. Note that $\Delta_{|\tau}$ is empty, as τ satisfies C. This implies that $C \vee \Gamma^*_{|\tau} = C \vee \Gamma_{|\tau} \subseteq \Gamma^*$, so C is PR$_0$ w.r.t. Γ^*. Furthermore the condition $\Gamma_{|\alpha} \vdash_1 \Gamma_{|\tau}$ implies that every clause in Δ is derivable from Γ by a $\vdash_1$ step, by Lemma 5. Thus we can derive Γ^* from Γ by $\vdash_1$ steps, then introduce C by the PR$_0$ rule, and recover $\Gamma \cup \{C\}$ by deleting everything else. □

Lemma 29. *Suppose C is* PR$_0$ *with respect to Γ, witnessed by τ with $x \notin dom(\tau)$. Then C is* PR$_0$ *with respect to $\Gamma^{(x)}$.*

Proof. The PR$_0$ condition implies the variable x does not occur in C. We are given that $C \vee \Gamma_{|\tau} \subseteq \Gamma$ and want to show $C \vee (\Gamma^{(x)})_{|\tau} \subseteq \Gamma^{(x)}$. Let $D \in \Gamma^{(x)}$ with $\tau \nvDash D$. First suppose D is in Γ and x does not occur in D. Then $C \vee D_{|\tau} \in \Gamma$ by assumption, so $C \vee D_{|\tau} \in \Gamma^{(x)}$. Otherwise, $D = E \vee F$ where both $E \dot\vee x$ and $F \dot\vee \overline{x}$ are in Γ. Then by assumption both $C \vee E_{|\tau} \vee x$ and $C \vee F_{|\tau} \vee \overline{x}$ are in Γ. Hence $C \vee D_{|\tau} = C \vee E_{|\tau} \vee F_{|\tau} \in \Gamma^{(x)}$. □

Theorem 30. DRAT$^-$ *simulates* DPR$^-$.

Proof. We are given a DPR$^-$ refutation of some set Δ, using only the variables in Δ. By Lemma 28 we may assume without loss of generality that the refutation uses only $\vdash_1$, deletion and PR$_0$ steps. Consider a PR$_0$ inference in this refutation, which derives $\Gamma \cup \{C\}$ from a set of clauses Γ, witnessed by a partial assignment τ. We want to derive $\Gamma \cup \{C\}$ from Γ in DRAT using only variables in Δ.

Suppose τ is a total assignment to all variables in Γ. The set Γ is necessarily unsatisfiable, or it could not occur as a line in a refutation. Therefore $\Gamma_{|\tau}$ is simply $\bot$, so the PR_0 condition tells us that $C \in \Gamma$ and we do not need to do anything.

Otherwise, there is some variable x which occurs in Γ but is outside the domain of τ, and thus in particular does not occur in C. We first use $\vdash_1$ and deletion steps to replace Γ with $\Gamma^{(x)}$. By Lemma 29, C is PR_0, and thus PR, with respect to $\Gamma^{(x)}$. By Lemma 25 there is a short DRAT derivation of $\Gamma^{(x)} \cup \{C\}$ from $\Gamma^{(x)}$, using one new variable which does not occur in $\Gamma^{(x)}$ or C. We choose x for this variable. Finally, observing that here $\Gamma^{(x)} \cup \{C\} = (\Gamma \cup \{C\})^{(x)}$, we recover $\Gamma \cup \{C\}$ using Lemma 27. $\square$

3.2   Towards a Simulation of PR^- by SPR^-

Our next result shows how to replace a PR inference with SPR inferences, without additional variables. It is not a polynomial simulation of PR^- by SPR^- however, as it depends exponentially on the "discrepancy" as defined next. Recall that C is PR w.r.t. Γ if $\Gamma_{|\alpha} \vdash_1 \Gamma_{|\tau}$, where $\alpha = \overline{C}$ and τ is a partial assignment satisfying C. We will keep this notation throughout this section. C is SPR w.r.t. Γ if additionally $dom(\tau) = dom(\alpha)$.

Definition 31. *The* discrepancy *of a PR inference is* $|dom(\tau) \setminus dom(\alpha)|$. *That is, it is the number of variables which are assigned by τ but not by α.*

Theorem 32. *Suppose that Γ has a PR refutation Π of size S in which every PR inference has discrepancy bounded by δ. Then Γ has a SPR refutation of size $O(2^\delta S)$ which does not use any variables not present in Π.*

If the discrepancy is logarithmically bounded, Theorem 32 gives polynomial size SPR refutations automatically. We need a couple of lemmas before proving the theorem.

Lemma 33. *Suppose $\Gamma_{|\alpha} \vdash_1 \Gamma_{|\tau}$ and α^+ is a partial assignment extending α, such that $dom(\alpha^+) \subseteq dom(\tau)$. Then $\Gamma_{|\alpha^+} \vdash_1 \Gamma_{|\tau}$*

Proof. Suppose $E \in \Gamma_{|\tau}$. Then E contains no variables from α^+ and by assumption there is a refutation $\Gamma_{|\alpha}, \overline{E} \vdash_1 \bot$. Thus $\Gamma_{|\alpha^+}, \overline{E} \vdash_1 \bot$, by Fact 3. $\square$

Definition 34. *A clause C* subsumes *a clause D if $C \subseteq D$. A set Γ of clauses* subsumes *a set Γ' if each clause of Γ' is subsumed by some clause of Γ.*

Lemma 35. *Suppose $\Gamma \subseteq \Gamma'$ and Γ subsumes Γ'. Suppose α and τ are substitutions and $\Gamma_{|\alpha} \vdash_1 \Gamma_{|\tau}$ holds. Then $\Gamma'_{|\alpha} \vdash_1 \Gamma'_{|\tau}$. Consequently, if C can be inferred from Γ by an SPR, PR or SR rule, then C can also be inferred from Γ' by the same rule.*

Proof. Suppose $D \in \Gamma'$ and $\tau \not\vdash D$. We must show $\Gamma'_{|\alpha} \vdash_1 D_{|\tau}$. Let $E \in \Gamma$ with $E \subseteq D$. Then $\tau \not\vdash E$, so by assumption $\Gamma_{|\alpha} \vdash_1 E_{|\tau}$. Also $E_{|\tau} \subseteq D_{|\tau}$, so $\Gamma_{|\alpha} \vdash_1 D_{|\tau}$. It follows that $\Gamma'_{|\alpha} \vdash_1 D_{|\tau}$. $\square$

Proof (of Theorem 32). Our main task is to show that a PR inference with discrepancy at most δ can be simulated by multiple SPR inferences, while bounding the increase in proof size in terms of δ. Suppose C is derivable from Γ by a PR inference. That is, $\Gamma_{|\alpha} \vdash_1 \Gamma_{|\tau}$ where $\alpha = \overline{C}$ and $\tau \vDash C$; by Lemma 14 we may assume that $dom(\tau) \supseteq dom(\alpha)$. List the variables in $dom(\tau) \setminus dom(\alpha)$ as $p_1, \ldots, p_s$, where $s \leq \delta$.

Enumerate as $D_1, \ldots, D_{2^s}$ all clauses containing exactly the variables $p_1, \ldots, p_s$ with some pattern of negations. Let $\sigma_i = \overline{C \vee D_i}$, so that $\sigma_i \supseteq \alpha$ and $dom(\sigma_i) = dom(\tau)$. By Lemma 33, $\Gamma_{|\sigma_i} \vdash_1 \Gamma_{|\tau}$. Since $\tau \vDash C \vee D_j$ for every j, in fact $\Gamma_{|\sigma_i} \vdash_1 (\Gamma \cup \{C \vee D_1, \ldots, C \vee D_{i-1}\})_{|\tau}$. Thus we may introduce all clauses $C \vee D_1, \ldots, C \vee D_{2^s}$ one after another by SPR inferences. We can then use $2^s - 1$ resolution steps to derive C.

The result is a set $\Gamma' \supseteq \Gamma$ which contains C plus extra clauses subsumed by C. By Lemma 35 these extra clauses do not affect the validity of later PR inferences. $\square$

4   Upper Bounds for Some Hard Tautologies

This section proves that SPR⁻—without new variables—can give polynomial size refutations for essentially all the usual "hard" propositional principles. Heule, Kiesl and Biere [12,13] showed that the tautologies based on the pigeonhole principle (PHP) and the 2-1 pigeonhole principle have polynomial size SPR⁻ proofs, and Heule and Biere discuss polynomial size PR⁻ proofs of the Tseitin tautologies in [9]. The SPR⁻ proof of the PHP tautologies can be viewed as a version of the original extended resolution proof of PHP given by Cook and Reckhow [7]. Here we describe polynomial size SPR⁻ proofs for several well-known principles. We also show that orification and xorification can be handled in SPR⁻. This is surprising since the proofs contain only clauses in the original literals, and it is well-known that such clauses are limited in what they can express.

It is open whether extended resolution, or the Frege proof system, can be simulated by PR⁻ or DPR⁻, or more generally by DSR⁻. The examples below show that any separation of these systems must involve a new technique.

For space reasons, we omit the proofs and the descriptions of the clauses for Theorems 38, 41, 42 and 43. They can be found in the full version of the paper. We include the proof of Theorem 39 for the bit pigeonhole principle as an example, as it is relatively easy, and is used in Sect. 5.

Definition 36. *A Γ-symmetry is an invertible substitution π such that $\Gamma_{|\pi} = \Gamma$.*

If π is a Γ-symmetry and $\alpha = \overline{C}$ is a partial assignment, then by Lemma 1 we have $\Gamma_{|\alpha} = (\Gamma_{|\pi})_{|\alpha} = \Gamma_{|\alpha \circ \pi}$. Hence, if $\alpha \circ \pi \vDash C$, we can infer C from Γ by an SR inference with $\tau = \alpha \circ \pi$. If furthermore π is the identity outside $dom(\alpha)$, then $\alpha \circ \pi$ behaves as a partial assignment and $dom(\alpha \circ \pi) = dom(\alpha)$, so this becomes an SPR inference.

Below we write $\overline{\alpha}$ for the clause expressing that the partial assignment α does not hold (so $C = \overline{\alpha}$ if and only if $\alpha = \overline{C}$).

Lemma 37. *Suppose* $(\alpha_0, \tau_0), \ldots, (\alpha_m, \tau_m)$ *is a sequence of pairs of partial assignments such that for each* i,

1. $\Gamma_{|\alpha_i} = \Gamma_{|\tau_i}$
2. α_i *and* τ_i *are contradictory and have the same domain*
3. for all $j < i$, *either* α_j *and* τ_i *are disjoint or they are contradictory.*

Then we can derive $\Gamma \cup \{\overline{\alpha_i} : i = 0, \ldots, m\}$ *from* Γ *by a sequence of* SPR *inferences.*

Proof. We write C_i for $\overline{\alpha_i}$. By item 2, $\tau_i \vDash C_i$. Thus it is enough to show that for each i,

$$\left(\Gamma \cup \{C_0, \ldots, C_{i-1}\}\right)_{|\alpha_i} \supseteq \left(\Gamma \cup \{C_0, \ldots, C_{i-1}\}\right)_{|\tau_i}.$$

We have $\Gamma_{|\alpha_i} = \Gamma_{|\tau_i}$. For $j < i$, either α_j and τ_i are disjoint, and so $(C_j)_{|\alpha_i} = (C_j)_{|\tau_i} = C_j$, or they are contradictory and so $\tau_i \vDash C_j$ and C_j vanishes from the right hand side. $\qquad\qquad\square$

4.1   Pigeonhole Principles

Let $n \geq 1$. The *pigeonhole principle* PHP_n asserts that $n + 1$ pigeons can be mapped to n holes with no collisions.

Theorem 38 ([13]). PHP_n *has polynomial size* SPR^- *refutations.*

Let $n = 2^k$. The *bit pigeonhole principle* contradiction, BPHP_n, asserts that each of $n + 1$ pigeons can be assigned a distinct k-bit binary string. For each pigeon x, $0 \leq x < n + 1$, it has variables $p_1^x, \ldots, p_k^x$ for the bits of the string assigned to x. We think of strings $y \in \{0, 1\}^k$ as holes. When convenient we will identify holes with numbers $y < n$. We write $(x \rightarrow y)$ for the conjunction $\bigwedge_i (p_i^x = y_i)$ asserting that pigeon x goes to hole y. We write $(x \nrightarrow y)$ for its negation $\bigvee_i (p_i^x \neq y_i)$. The axioms of BPHP_n are then

$$(x \nrightarrow y) \vee (x' \nrightarrow y) \quad \text{for all holes } y \text{ and all distinct pigeons } x, x'.$$

The set $\{(x \nrightarrow y) : y < n\}$ consists of the 2^k clauses containing the variables $p_1^x, \ldots, p_k^x$ with all patterns of negations. We can derive $\bot$ from them in $2^k - 1$ resolution steps.

Theorem 39. *The* BPHP_n *clauses have polynomial size* SPR^- *refutations.*

The theorem is proved below. It is essentially the same as the proof of PHP in [13] (or Theorem 38 above). For each $m < n - 1$ and each pair $x, y > m$, we define a clause

$$C_{m,x,y} := (m \nrightarrow y) \vee (x \nrightarrow m).$$

Let Γ be the set of all such clauses $C_{m,x,y}$. We will show these clauses can be introduced by SPR inferences, but first we show they suffice to derive BPHP_n.

Lemma 40. $\mathrm{BPHP}_n \cup \Gamma$ *has a polynomial size resolution refutation.*

Proof. Using induction on $m = 0, 1, 2, \ldots, n-1$ we derive all clauses $\{(x \not\rightarrow m) : x > m\}$. So suppose $m < n$ and $x > m$. For each $y > m$, we have the clause $(m \not\rightarrow y) \vee (x \not\rightarrow m)$, as this is $C_{m,x,y}$. We also have the clause $(m \not\rightarrow m) \vee (x \not\rightarrow m)$, as this is an axiom of BPHP_n. Finally, for each $m' < m$, we have $(m \not\rightarrow m')$ by the inductive hypothesis (or, in the base case $m = 0$, there are no such clauses). Resolving all these together gives $(x \not\rightarrow m)$.

This derives all clauses in $\{(n \not\rightarrow m) : m < n\}$. Resolving these yields $\perp$. $\qquad\square$

Thus it is enough to show that we can introduce all clauses in Γ using SPR inferences. We use Lemma 37. For $m < n - 1$ and each pair $x, y > m$, define partial assignments

$$\alpha_{m,x,y} := (m \rightarrow y) \wedge (x \rightarrow m) \qquad \text{and} \qquad \tau_{m,x,y} := (m \rightarrow m) \wedge (x \rightarrow y)$$

so that $C_{m,x,y} = \overline{\alpha_{m,x,y}}$ and $\tau_{m,x,y} = \alpha_{m,x,y} \circ \pi$ where π swaps all variables for pigeon m and x. Hence $(\mathrm{BPHP}_n)_{|\alpha_{m,x,y}} = (\mathrm{BPHP}_n)_{|\tau_{m,x,y}}$ as required.

For the other conditions for Lemma 37, first observe that assignments $\alpha_{m,x,y}$ and $\tau_{m,x',y'}$ are always inconsistent, since they map m to different places. Now suppose that $m' < m$ and $\alpha_{m,x,y}$ and $\tau_{m',x',y'}$ are not disjoint. Then they must have some pigeon in common, so either $m' = x$ or $x' = x$. In both cases $\tau_{m',x',y'}$ contradicts $(x \rightarrow m)$, in the first case because it maps x to m' and in the second because it maps x to y' with $y' > m'$.

4.2   Other Tautologies

The *parity principle* states that there is no (undirected) graph on an odd number of vertices in which each vertex has degree exactly one (see [1,2]). For n odd, let PAR_n be a set of clauses expressing (a violation of) the parity principle on n vertices.

Theorem 41. *The* PAR_n *clauses have polynomial size* SPR^- *refutations.*

The *clique-coloring principle* $\mathrm{CC}_{n,m}$ states, informally, that a graph with n vertices cannot have both a clique of size m and a coloring of size $m - 1$ (see [17, 21]).

Theorem 42. *The* $\mathrm{CC}_{n,m}$ *clauses have polynomial size* SPR^- *refutations.*

The *Tseitin tautologies* $\mathrm{TS}_{G,\gamma}$ are hard examples for many proof systems (see [24, 25]). Let G be an undirected graph with n vertices, with each vertex i labelled with a charge $\gamma(i) \in \{0, 1\}$ such that the total charge on G is odd. For each edge e of G there is a variable x_e. Then $\mathrm{TS}_{G,\gamma}$ is the CNF consisting of clauses expressing that, for each vertex i, the parity of the values x_e over the edges e touching i is equal to the charge $\gamma(i)$. For a vertex i of degree d, this requires 2^{d-1} clauses, using one clause to rule out each assignment to the edges touching i with the wrong parity. If G has constant degree then this has size polynomial in n. It is well-known to be unsatisfiable.

Theorem 43. *The* $\mathrm{TS}_{G,\gamma}$ *clauses have polynomial size* SPR^- *refutations.*

Orification and xorification have also been used to make hard instances of propositional tautologies. (See [3,4,26].) Nonetheless, SPR-inferences can be used to "undo" the effects of orification and xorification, without using any new variables. (This is argued in the full version of the paper.) As a consequence, these techniques are not likely to be helpful in establishing lower bounds for the size of PR refutations.

The principles considered above exhaust most of the known "hard" tautologies that have been shown to require exponential size, constant depth Frege proofs. It is open whether SPR^- or SR^- simulates Frege; and by the above results, any separation of SPR^- and Frege systems will likely require new techniques.

Paul Beame [private comm., 2018] suggested the graph PHP principles (see [5]) may separate systems such as SPR^-, or even SR^-, from Frege systems. However, it is plausible that the graph PHP principles also have short SPR^- proofs. Namely, SPR inferences can infer a lot of clauses from the graph PHP clauses. If an instance of graph PHP has every pigeon with outdegree ≥ 2, then there must be an alternating cycle of pigeons $i_1, \ldots i_{\ell+1}$ and holes $j_1, \ldots j_\ell$ such that $i_\ell = i_1$, the edges (i_s, j_s) and (i_{s+1}, j_s) are all in the graph, and $\ell = O(\log n)$. An SPR inference can be used to learn the clause $\overline{x_{i_1,j_1}} \vee \overline{x_{i_2,j_2}} \vee \cdots \vee \overline{x_{i_\ell,j_\ell}}$, by using the fact that a satisfying assignment that falsifies this clause can be replaced by the assignment that maps instead each pigeon i_{s+1} to hole j_s.

This allows SPR inferences to infer many clauses from the graph PHP clauses. However, it remains open whether a polynomial size SPR^- refutation exists.

5   Lower Bounds

This section gives an exponential separation between DRAT^- and RAT^-, by showing that the bit pigeonhole principle BPHP_n requires exponential size refutations in RAT^-. This lower bound still holds if we allow some deletions, as long as no initial clause of BPHP_n is deleted. On the other hand, with unrestricted deletions, Theorems 23, 30 and 39 imply that BPHP_n has polynomial size refutations in DRAT^- and even DBC^-, and Theorem 39 shows that it has polynomial size SPR^- refutations.

We define the *pigeon-width* of a clause or assignment to be the number of distinct pigeons that it mentions. Our lower bound proof uses a conventional strategy: we first show a width lower bound (on pigeon-width), and then use a random restriction argument to show that a proof of subexponential size can be made into one of small pigeon-width. However, RAT^- refutation size may not behave well under restrictions (see Sect. 2.2). So, rather than using restrictions directly to reduce width, we will define a partial random matching ρ of pigeons to holes and show that if BPHP_n has a RAT^- refutation of small size, then $\mathrm{BPHP}_n \cup \rho$ has one of small pigeon-width.

We will sometimes identify resolution refutations of Γ with winning strategies for the Prover in the Prover-Adversary game on Γ, in which the Adversary claims to know a satisfying assignment and the Prover tries to force her into a contradiction by querying variables; the Prover can also forget variables to save memory and simplify his strategy.

Lemma 44. *Let β be a partial assignment corresponding to a partial matching of m pigeons to holes. Then $\mathrm{BPHP}_n \cup \beta$ requires pigeon-width $n+1-m$ to refute in resolution.*

Proof. A refutation of pigeon-width less than $n + 1 - m$ would give a Prover-strategy in which the Prover never has information about more than $n - m$ pigeons; namely, traverse Π upwards from $\perp$ to an initial clause, remembering only the values of variables mentioned in the current clause. This strategy is easy for the Adversary to defeat, as $\mathrm{BPHP}_n \cup \beta$ is essentially the pigeonhole principle with $n - m$ holes. $\square$

Theorem 45. *Let ρ be a partial matching of size at most $n/4$. Let Π be a DRAT refutation of $\mathrm{BPHP}_n \cup \rho$ in which no new variables are introduced and no clause of BPHP_n is ever deleted. Then some clause in Π has pigeon-width more than $n/3$.*

Proof. Suppose for a contradiction there is a such a refutation Π in pigeon-width $n/3$. We consider each RAT inference in Π in turn, and show that it can be eliminated and replaced with standard resolution reasoning, without increasing the pigeon-width.

Inductively suppose Γ is a set of clauses derivable from $\mathrm{BPHP}_n \cup \rho$ in pigeon-width $n/3$, using only resolution and weakening. Suppose a clause C in Π of the form $p \mathbin{\dot\vee} C'$ is RAT w.r.t. Γ and p. Let $\alpha = \overline{C}$, so $\alpha(p) = 0$ and α mentions at most $n/3$ pigeons. We consider three cases.

Case 1: the assignment α is inconsistent with ρ. This means that ρ satisfies a literal which appears in C, so C can be derived from ρ by a single weakening step.

Case 2: the assignment $\alpha \cup \rho$ can be extended to a partial matching β of the pigeons it mentions. We will show that this cannot happen. Let x be the pigeon associated with the literal p. Let $y = \beta(x)$ and let y' be the hole β would map x to if the bit p were flipped to 1. If $y' = \beta(x')$ for some pigeon x' in the domain of β, let $\beta' = \beta$. Otherwise let $\beta' = \beta \cup \{(x', y')\}$ for some pigeon x' outside the domain of β.

Let H be the hole axiom $(x \nrightarrow y') \vee (x' \nrightarrow y')$ in Γ. The clause $(x \nrightarrow y')$ contains the literal $\overline{p}$, since $(x \rightarrow y')$ contains p. So $H = \overline{p} \mathbin{\dot\vee} H'$ for some clause H'. By the RAT condition, either $C' \cup H'$ is a tautology or $\Gamma \vdash_1 C \vee H'$. Either way, $\Gamma \cup \overline{C} \cup \overline{H'} \vdash_1 \perp$. Since $\beta' \supseteq \alpha$, β' falsifies C. It also falsifies H', since it satisfies $(x \rightarrow y') \wedge (x' \rightarrow y')$ except at p. It follows that $\Gamma \cup \beta' \vdash_1 \perp$. By assumption, Γ is derivable from $\mathrm{BPHP}_n \cup \rho$ in pigeon-width $n/3$, and $\beta' \supseteq \rho$. As unit propagation does not increase pigeon-width, this implies that $\mathrm{BPHP}_n \cup \beta'$ is refutable in resolution in pigeon-width $n/3$, by first deriving Γ and then using unit propagation. This contradicts Lemma 44 as β' is a matching of at most $n/3 + n/4 + 1$ pigeons.

Case 3: the assignment $\alpha \cup \rho$ cannot be extended to a partial matching of the pigeons it mentions. Consider a position in the Prover-Adversary game on $\text{BPHP}_n \cup \rho$ in which the Prover knows α. The Prover can ask all remaining bits of the pigeons mentioned in α, and since there is no suitable partial matching this forces the Adversary to reveal a collision and lose the game. This strategy has pigeon-width $n/3$; it follows that C is derivable from $\text{BPHP}_n \cup \rho$ in resolution in this pigeon-width. $\qquad\square$

Theorem 46. *Let Π be a DRAT^- refutation of BPHP_n in which no clause of BPHP_n is ever deleted. Then Π has size at least $2^{n/80}$.*

Proof. Construct a random restriction ρ by selecting each pigeon independently with probability $1/5$ and then randomly matching them with distinct holes. Let $m = n/4$. Let C be a clause mentioning at least m distinct pigeons $x_1, \ldots, x_m$ and choose literals $p_1, \ldots, p_m$ in C such that p_i belongs to pigeon x_i. The probability p_i is satisfied by ρ is $1/10$. These events are not quite independent for different p_i, as the holes used by other pigeons are blocked for pigeon x_i. But since $m = n/4$, fewer than half of the holes that would satisfy p_i are blocked. That is, the probability that p_i is satisfied by ρ, on the worst-case condition that all other literals p_j are not satisfied by ρ, is at least $1/20$. Therefore the probability that C is not satisfied by ρ is at most $(1 - 1/20)^m < e^{-m/20} = e^{-n/80}$.

Now suppose Π contains no more than $2^{n/80}$ clauses. By the union bound, there is some restriction ρ which satisfies all clauses in Π of pigeon-width at least $n/4$, and by the Chernoff bound we may assume that ρ sets no more than $n/4$ pigeons.

We now observe inductively that for each clause C in Π, some subclause of C is derivable from $\text{BPHP}_n \cup \rho$ in resolution in pigeon-width $n/3$, ultimately contradicting Lemma 44. If C has pigeon-width more than $n/3$, this follows because C is subsumed by ρ. Otherwise, if C is derived by a RAT inference, we repeat the proof of Theorem 45; in case 2 we additionally use the observation that if $\Gamma \vdash_1 C \vee H'$ and Γ' subsumes Γ, then $\Gamma' \vdash_1 C \vee H'$. $\qquad\square$

Corollary 47. RAT^- *does not simulate* DRAT^-. RAT^- *does not simulate* SPR^-.

References

1. Ajtai, M.: Parity and the pigeonhole principle. In: Buss, S.R., Scott, P.J. (eds.) Feasible Mathematics, pp. 1–24. Birkhäuser, Boston (1990)
2. Beame, P., Impagliazzo, R., Krajíček, J., Pitassi, T., Pudlák, P.: Lower bounds on Hilbert's Nullstellensatz and propositional proofs. Proc. Lond. Math. Soc. **73**(3), 1–26 (1996)
3. Ben-Sasson, E.: Size space tradeoffs for resolution. SIAM J. Comput. **38**(6), 2511–2525 (2009)
4. Ben-Sasson, E., Impagliazzo, R., Wigderson, A.: Near optimal separation of tree-like and general resolution. Combinatorica **24**(4), 585–603 (2004)

5. Ben-Sasson, E., Wigderson, A.: Short proofs are narrow—resolution made simple. J. ACM **48**, 149–169 (2001)
6. Cook, S.A., Reckhow, R.A.: On the lengths of proofs in the propositional calculus, preliminary version. In: Proceedings of the Sixth Annual ACM Symposium on the Theory of Computing, pp. 135–148 (1974)
7. Cook, S.A., Reckhow, R.A.: The relative efficiency of propositional proof systems. J. Symb. Log. **44**, 36–50 (1979)
8. Goldberg, E.I., Novikov, Y.: Verification of proofs of unsatisfiability for CNF formulas. In: Design, Automation and Test in Europe Conference (DATE), pp. 10886–10891. IEEE Computer Society (2003)
9. Heule, M.J.H., Biere, A.: What a difference a variable makes. In: Beyer, D., Huisman, M. (eds.) TACAS 2018. LNCS, vol. 10806, pp. 75–92. Springer, Cham (2018). https://doi.org/10.1007/978-3-319-89963-3_5
10. Heule, M.J.H., Hunt Jr., W.A., Wetzler, N.: Trimming while checking clausal proofs. In: Formal Methods in Computer-Aided Design (FMCAD), pp. 181–188. IEEE (2013)
11. Heule, M.J.H., Hunt Jr., W.A., Wetzler, N.: Verifying refutations with extended resolution. In: Bonacina, M.P. (ed.) CADE 2013. LNCS (LNAI), vol. 7898, pp. 345–359. Springer, Heidelberg (2013). https://doi.org/10.1007/978-3-642-38574-2_24
12. Heule, M.J.H., Kiesl, B., Biere, A.: Short proofs without new variables. In: de Moura, L. (ed.) CADE 2017. LNCS (LNAI), vol. 10395, pp. 130–147. Springer, Cham (2017). https://doi.org/10.1007/978-3-319-63046-5_9
13. Heule, M.J.H., Kiesl, B., Biere, A.: Strong extension-free proof systems. J. Autom. Reason. 1–22 (2019). https://doi.org/10.1007/s10817-019-09516-0. Extended version of [12]
14. Heule, M.J.H., Kiesl, B., Seidl, M., Biere, A.: PRuning through satisfaction. Hardware and Software: Verification and Testing. LNCS, vol. 10629, pp. 179–194. Springer, Cham (2017). https://doi.org/10.1007/978-3-319-70389-3_12
15. Järvisalo, M., Heule, M.J.H., Biere, A.: Inprocessing rules. In: Gramlich, B., Miller, D., Sattler, U. (eds.) IJCAR 2012. LNCS (LNAI), vol. 7364, pp. 355–370. Springer, Heidelberg (2012). https://doi.org/10.1007/978-3-642-31365-3_28
16. Kiesl, B., Rebola-Pardo, A., Heule, M.J.H.: Extended resolution simulates DRAT. In: Galmiche, D., Schulz, S., Sebastiani, R. (eds.) IJCAR 2018. LNCS (LNAI), vol. 10900, pp. 516–531. Springer, Cham (2018). https://doi.org/10.1007/978-3-319-94205-6_34
17. Krajíček, J.: Interpolation theorems, lower bounds for proof systems, and independence results for bounded arithmetic. J. Symb. Log. **62**, 457–486 (1997)
18. Krajíček, J., Pudlák, P., Woods, A.: Exponential lower bound to the size of bounded depth Frege proofs of the pigeonhole principle. Random Struct. Algorithms **7**, 15–39 (1995)
19. Kullmann, O.: On a generalizaton of extended resolution. Discrete Appl. Math. **96–97**, 149–176 (1999)
20. Pitassi, T., Beame, P., Impagliazzo, R.: Exponential lower bounds for the pigeonhole principle. Comput. Complex. **3**, 97–140 (1993)
21. Pudlák, P.: Lower bounds for resolution and cutting planes proofs and monotone computations. J. Symb. Log. **62**, 981–998 (1997)
22. Rebola-Pardo, A., Suda, M.: A theory of satisfiability-preserving proofs in SAT solving. In: Proceedings 22nd International Conference on Logic for Programming, Artificial Intelligence and Reasoning (LPAR-22). EPiC Series in Computing, vol. 57, pp. 583–603. EasyChair (2018)

23. Siekmann, J., Wrightson, G.: Automation of Reasoning, vol. 1&2. Springer, Berlin (1983)
24. Tsejtin, G.S.: On the complexity of derivation in propositional logic. In: Studies in Constructive Mathematics and Mathematical Logic, part 2, pp. 115–125 (1968). Reprinted in: [23, vol. 2], pp. 466–483
25. Urquhart, A.: Hard examples for resolution. J. ACM **34**, 209–219 (1987)
26. Urquhart, A.: A near-optimal separation of regular and general resolution. SIAM J. Comput. **40**(1), 107–121 (2011)
27. Van Gelder, A.: Verifying RUP proofs of propositional unsatisfiability. In: 10th International Symposium on Artificial Intelligence and Mathematics (ISAIM) (2008). http://isaim2008.unl.edu/index.php?page=proceedings
28. Wetzler, N., Heule, M.J.H., Hunt, W.A.: DRAT-trim: efficient checking and trimming using expressive clausal proofs. In: Sinz, C., Egly, U. (eds.) SAT 2014. LNCS, vol. 8561, pp. 422–429. Springer, Cham (2014). https://doi.org/10.1007/978-3-319-09284-3_31

Knowledge Compilation Languages
as Proof Systems

Florent Capelli[(✉)] [iD]

Université de Lille, Inria, UMR 9189 - CRIStAL - Centre de Recherche en
Informatique Signal et Automatique de Lille, 59000 Lille, France
`florent.capelli@univ-lille.fr`

Abstract. In this paper, we study proof systems in the sense of Cook-
Reckhow for problems that are higher in the Polynomial Hierarchy than
coNP, in particular, #SAT and maxSAT. We start by explaining how the
notion of Cook-Reckhow proof systems can be apply to these problems
and show how one can twist existing languages in knowledge compilation
such as decision DNNF so that they can be seen as proof systems for
problems such as #SAT and maxSAT.

Keywords: Knowledge compilation ·
Propositional Proof Complexity · Propositional Model Counting ·
maxSAT

1 Introduction

Propositional Proof Complexity studies the hardness of finding a certificate that
a CNF formula is not satisfiable. A minimal requirement for such a certificate
is that it should be checkable in polynomial time in its size, so that it is easier
for an independent checker to assess the correctness of the proof than to redo
the computation made by a solver. While proof systems have been implicitly
used for a long time starting with resolution [10,11], their systematic study has
been initiated by Cook and Reckhow [7] who showed that unless NP = coNP, one
cannot design a proof system where all unsatisfiable CNF have short certificates.
Nevertheless, many unsatisfiable CNF may have short certificates if the proof
system is powerful enough, motivating the study of how such systems, such
as resolution [10] or polynomial calculus [5], compare in terms of succinctness
(see [18] for a survey). More recently, proof sytems found practical applications
as SAT solvers are expected – since 2013 – to output a proof of unsatisfiability
in SAT competitions to avoid implementation bugs.

While the proof systems implicitly defined by the execution trace of modern CDCL SAT solvers are fairly well understood [20], it is not the case for
tools solving harder problems on CNF formulas such as #SAT and maxSAT. For
maxSAT, a resolution-like system for maxSAT has been proposed by Bonet et
al. [2] for which a compressed version has been used in a solver by Bacchus and
Narodytska [17]. To the best of our knowledge, it is the only proof system for
maxSAT and no proof system has been proposed for #SAT.

© Springer Nature Switzerland AG 2019
M. Janota and I. Lynce (Eds.): SAT 2019, LNCS 11628, pp. 90–99, 2019.
https://doi.org/10.1007/978-3-030-24258-9_6

In this short paper, we introduce new proof systems for #SAT and maxSAT. Contrary to the majority of proof systems for SAT, our proof systems are not based on the iterative application of inference rules on the original CNF formula. In our proof systems, our certificates are restricted Boolean circuits representing the Boolean function computed by the input CNF formula. These restricted circuits originate from the field of knowledge compilation [9], whose primary focus is to study the succinctness and tractability of representations such as Read Once Branching Programs [23] or deterministic DNNF [8] and how CNF formula can be transformed into such representations. To use them as certificates for #SAT, we first have to add some extra information in the circuit so that one can check in polynomial time that they are equivalent to the original CNF. The syntactic properties of the input circuits then allow to efficiently count the number of satisfying assignments, resulting in the desired proof system. Moreover, we observe that most tools doing exact model counting are already implicitly generating such proofs. Our result generalizes known connections between regular resolution and Read Once Branching Programs (see [13, Sect. 18.2]).

The paper is organized as follows. Section 2 introduces all the notions that will be used in the paper. Section 3 contains the definition of certified dec-DNNF that allows us to define our proof systems for #SAT and maxSAT.

2 Preliminaries

Assignments and Boolean Functions. Let X be a finite set of variables and D a finite domain. We denote the set of functions from X to D as D^X. An *assignment on variables* X is an element of $\{0,1\}^X$. A *Boolean function* f *on variables* X is an element of $\{0,1\}^{\{0,1\}^X}$, that is, a function that maps an assignment to a value in $\{0,1\}$. An assignment $\tau \in \{0,1\}^X$ such that $f(\tau) = 1$ is called a *satisfying assignment* of f, denoted by $\tau \models f$. We denote by $\perp_X$ the Boolean function on variables X whose value is always 0. Given two Boolean functions f and g on variables X, we write $f \Rightarrow g$ if for every τ, $f(\tau) \leq g(\tau)$.

CNF. Let X be a set of variables. A *literal* on variables X is either a variable $x \in X$ or its negation $\neg x$. A *clause* is a disjunction of literals. A *conjunctive normal form formula*, CNF for short, is a conjunction of clauses. A CNF naturally defines a Boolean function on variables X: a *satisfying assignment* for a CNF F on variable X is an assignment $\tau \in \{0,1\}^X$ such that for every clause C of F, there exists a literal ℓ of C such that $\tau(\ell) = 1$ (where we define $\tau(\neg x) := 1 - \tau(x)$). We often identify a CNF with the Boolean function it defines. The problem SAT is the problem of deciding, given a CNF formula F, whether F has a satisfying assignment. It is the generic NP-complete problem [6]. The problem UNSAT is the problem of deciding, given a CNF formula F, whether F does not have a satisfying assignment. It is the generic coNP-complete problem. Given a CNF F, we denote by $\#F = |\{\tau \mid \tau \models F\}|$ the number of solutions of F and by $M(F) = \max_\tau |\{C \in F \mid \tau \models C\}|$ the maximum number of clauses of F that can be simultaneously satisfied. The problem #SAT is the problem of computing $\#F$

given a CNF F as input and the problem maxSAT is the problem of computing $M(F)$ given a CNF F as input.

Cook-Reckhow Proof Systems. Let Σ, Σ' be finite alphabets. A *(Cook-Reckhow) proof system* [7] for a language $L \subseteq \Sigma^*$ is a surjective polynomial time computable function $f : \Sigma' \to L$. Given $a \in L$, there exists, by definition, $b \in \Sigma'$ such that $f(b) = a$. We will refer to b as being *a certificate of a*.

In this paper, we will mainly be interested in proof systems for the problems #SAT and maxSAT, that is, we would like to design polynomial time verifiable proofs that a CNF formula has k solutions or that at most k clauses in the formula can be simultaneously satisfied. For the definition of Cook-Reckhow, this could translate to finding a proof system for the languages $\{(F, \#F) \mid F \text{ is a CNF}\}$ and $\{(F, M(F)) \mid F \text{ is a CNF}\}$. For example, a naive proof system for #SAT could be the following: a certificate that F has k solutions is the list of the k solutions together with a resolution proof that F' is not satisfiable where F' is the CNF F plus clauses encoding the fact that the k solutions found are not accepted. One could then check in polynomial time that each of the k assignments satisfies F and that the given proof of unsatisfiability of F' is correct and then output (F, k). This proof system is however not very interesting as one can construct very simple CNF with exponentially many solutions: for example the tautological CNF $\emptyset$ on n variables only has certificates of size at least 2^n.

dec-DNNF. A *decision Decomposable Negation Normal Form circuit D* on variables X, dec-DNNF for short, is a directed acyclic graph (DAG) with exactly one node of indegree 0 called the *source*. Nodes of outdegree 0, called the *sinks*, are labeled by 0 or 1. The other nodes have outdegree 2 and can be of two types:

- The *decision nodes* are labeled with a variable $x \in X$. One outgoing edge is labeled with 1 and the other by 0, represented respectively as a solid and a dashed edge in our figures.
- The $\land$-*nodes* are labeled with $\land$.

We introduce a few notations before explaining two other syntactic properties. If there is a decision node in D labeled with variable x, we say that x is *tested* in D. We denote by $\mathsf{var}(D)$ the set of variables tested in D. Given a node α of D, we denote by $D(\alpha)$ the dec-DNNF whose source is α and nodes are the nodes that can be reached in D starting from α. We also assume the following:

- Every $x \in X$ is tested at most once on every source-sink path of D.
- Every $\land$-gate of D is *decomposable*, that is, for every $\land$-node α with successors β, γ in D, it holds that $\mathsf{var}(D(\beta)) \cap \mathsf{var}(D(\gamma)) = \emptyset$.

Let $\tau \in \{0,1\}^X$. A source-sink path P in D is *compatible* with τ if and only if when x is tested on P, the outgoing edge labeled with $\tau(x)$ is in P. We say that τ *satisfies* D if only 1-sinks are reached by paths compatible with τ. A dec-DNNF and the paths compatible with the assignment $\tau(x) = \tau(y) = 0, \tau(z) = 1$ are

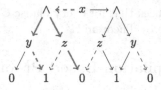

Fig. 1. A dec-DNNF computing $x = y = z$.

depicted in bold red on Fig. 1. Observe that a 0-sink is reached so τ does not satisfy D. We often identify a dec-DNNF with the Boolean function it computes.

Observation 1. *Given a* dec-DNNF *D on variables X and a source-sink path P in D, there exists $\tau \in \{0,1\}^X$ such that P is compatible with τ. Indeed, by definition, every variable $x \in X$ is tested at most once in P: if x is tested on P in a decision node α and P contains the outgoing edge labeled with v_x, we set $\tau(x) := v_x$. If x is not tested on P, we choose arbitrarily $\tau(x) = 0$.*

Tractable Queries. The main advantage of representing a Boolean function with a dec-DNNF is that it makes the analysis of the function easier. The size of a dec-DNNF D, denoted by $\mathsf{size}(D)$ is the number of edges of the underlying graph of D. Given a dec-DNNF D, one can find a satisfying assignment in linear time in the size of D by only following paths backward from 1-sinks. Similarly, one can also count the number of satisfying assignments or find one satisfying assignment with the least number of variables set to 1 etc. The relation between the queries that can be solved efficiently and the representation of the Boolean function has been one focus of Knowledge Compilation. See [9] for an exhaustive study of tractable queries depending on the representation. Let $f : 2^X \rightarrow \{0,1\}$ be a Boolean function. In this paper, we will mainly be interested in solving the following problems:

- Model Counting Problem (MC): return the number of satisfying assignment of f.
- Clause entailment (CE): given a clause C on variables X, does $f \Rightarrow C$?
- Maximal Hamming Weight (HW): given $Y \subseteq X$, compute

$$\max_{\tau \models f} |\{y \in Y \mid \tau(y) = 1\}|.$$

All these problems are tractable when the Boolean function is given as a dec-DNNF:

Theorem 1 [8,14]. *Given a* dec-DNNF *D, one can solve problems* MC, CE, HW *on the Boolean function computed by D in linear time in* $\mathsf{size}(D)$.

The tractability of CE on dec-DNNF has the following useful consequence:

Corollary 1. *Given a* dec-DNNF *D and a CNF formula F, one can check in time $O(\mathsf{size}(F) \times \mathsf{size}(D))$ whether $D \Rightarrow F$ (by slightly abusing notations to identify D and F with the Boolean functions they compute).*

Proof. One simply has to check that for every clause C of F, $D \Rightarrow C$, which can be done in polynomial time by Theorem 1.   □

3   Knowledge Compilation Based Proof Systems

Theorem 1 suggests that given a CNF F, one could use a dec-DNNF D computing F as a certificate for #SAT. We can check a certificate as follows:

1. Compute the number k of satisfying assignments of D.
2. Check whether F is equivalent to D.
3. If so, return (F, k).

While Step 3 can be done in polynomial time by Theorem 1, it turns out that Step 3 is not tractable:

Theorem 2. *The problem of checking, given a CNF F and a dec-DNNF D as input, whether $F \Rightarrow D$ is coNP-complete.*

Proof. The problem is clearly in coNP. For completeness, there is a straightforward reduction to UNSAT. Indeed, observe that a CNF F on variables X is not satisfiable if and only if $F \Rightarrow \bot_X$. Moreover, $\bot_X$ is easily computed by a dec-DNNF having only one node: a 0-labeled sink.   □

3.1   Certified dec-DNNF

The reduction used in the proof of Theorem 2 suggests that the coNP-completeness of checking whether $F \Rightarrow D$ comes from the fact that dec-DNNF can succinctly represent $\bot$. In this section, we introduce restrictions of dec-DNNF called certified dec-DNNF for which one can check whether a CNF formula entails the certified dec-DNNF. The idea is to add information on 0-sink to explain which clause would be violated by an assignment leading to this sink.

Our inspiration comes from a known connection between regular resolution and read once branching programs (*i.e.* a dec-DNNF without ∧-gates [1]) that appears to be folklore but we refer the reader to the book by Jukna [13, Sect. 18.2] for a thorough and complete presentation. It turns out that a regular resolution[1] proof of unsatisfiability of a CNF F can be computed by a read once branching program D whose sinks are labeled with clauses of F. Moreover, for every τ, if a sink labeled by a clause C is reached by a path compatible with τ, then $C(\tau) = 0$. We generalize this idea so that the function computed by a dec-DNNF is not only an unsatisfiable CNF:

Definition 1. *A certified dec-DNNF D on variables X is a dec-DNNF on variables X such that every 0-sink α of D is labeled with a clause C_α. D is said to be correct if for every $\tau \in \{0,1\}^X$ such that there is a path from the source of D to a 0-sink α compatible with τ, $C_\alpha(\tau) = 0$.*

---

[1] A regular resolution proof is a resolution proof where, on each path, a variable is resolved at most once.

Given a certified dec-DNNF, we denote by $Z(D)$ the set of 0-sinks of D and by $F(D) = \bigwedge_{\alpha \in Z(D)} C_\alpha$.

Intuitively, the clause labeling a 0-sink is an explanation on why one assignment does not satisfy the circuit. The degenerated case where there are only 0-sinks and no $\wedge$-gates corresponds to the characterization of regular resolution.

A crucial property of certified dec-DNNF is that their correctness can be tested in polynomial time:

Theorem 3. *Given a certified* dec-DNNF *D, one can check in polynomial time whether D is correct.*

Proof. By definition, D is not correct if and only if there exists a 0-sink α, a literal ℓ in C_α, an assignment τ such that $\tau(\ell) = 1$ and a path in D from the source to α compatible with τ. By Observation 1, it is equivalent to the fact that there exists a path from the source to α that: either does not test the underlying variable of ℓ or contains the outgoing edge corresponding to $\tau(\ell) = 1$ when the underlying variable of ℓ is tested.

In other words, D is correct if and only if for every 0-sink α and for every literal ℓ of C_α with variable x, every path from the source to α tests variable x and contains the outgoing edge corresponding to an assignment τ such that $\tau(\ell) = 0$.

This can be checked in polynomial time. Indeed, fix a 0-sink α and a literal ℓ of C_α. For simplicity, we assume that $\ell = x$ (the case $\ell = \neg x$ is completely symmetric). We have to check that every path from the source to α contains a decision node β on variable x and contains the outgoing edge of β labeled with 0. To check this, it is sufficient to remove all the edges labeled with 0, going out of a decision node on variable x and test that the source and α are now in two different connected components of D, which can obviously be done in polynomial time. Running this for every 0-sink α and every literal ℓ of C_α gives the expected algorithm. □

The clauses labeling the 0-sinks of a correct certified dec-DNNF naturally connect to the function computed by D:

Theorem 4. *Let D be a correct certified* dec-DNNF *on variables X. We have $F(D) \Rightarrow D$.*

Proof. Observe that $F(D) \Rightarrow D$ if and only if for every $\tau \in \{0,1\}^X$, if τ does not satisfy D then τ does not satisfy $F(D)$. Now let τ be an assignment that does not satisfy D. By definition, there exists a path compatible with τ from the source of D to a 0-sink α of D. Since D is correct, $C_\alpha(\tau) = 0$. Thus, τ does not satisfy $F(D)$ as C_α is by definition a clause of $F(D)$. □

Corollary 2. *Let F be CNF formula and D be a correct certified* dec-DNNF *such that every clause of $F(D)$ is also in F. Then $F \Rightarrow D$.*

3.2   Proof Systems

Proof system for #SAT. One can use certified dec-DNNF to define a proof system for #SAT. The *Knowledge Compilation based Proof System for* #SAT, kcps(#SAT) for short, is defined as follows: given a CNF F, a certificate that F has k satisfying assignments is a correct certified dec-DNNF D such that every clause of $F(D)$ is a clause of F, D computes F and D has k satisfying assignments.

To check a certificate D, one has to check that $D \Leftrightarrow F$ and that D has indeed k satisfying assignments, which can be done in polynomial time as follows:

– Check that D is correct, which is tractable by Theorem 3.
– Check that $D \Rightarrow F$, which is tractable by Corollary 1 and that every clause of $F(D)$ is indeed a clause of F. By Corollary 2, it means that $D \Leftrightarrow F$.
– Compute the number k of solutions of D, which is tractable by Theorem 1.
– Return (F, k).

This proof system for #SAT is particularly well-suited for the existing tools solving #SAT in practice. Many of them such as sharpSAT [22] or cachet [21] are based on a generalization of DPLL for counting which is sometimes refered as exhaustive DPLL in the literature. It has been observed by Huang and Darwiche [12] that these tools were implicitly constructing a dec-DNNF equivalent to the input formula. Tools such as c2d [19], D4 [15] or DMC [16] already exploit this connection and have the option to directly output an equivalent dec-DNNF. These solvers explore the set of satisfying assignments by branching on variables of the formula which correspond to a decision node and, when two variable independent components of the formula are detected, compute the number of satisfying assignments of both components and take the product, which corresponds to a decomposable ∧-gate. When a satisfying assignment is reached, it corresponds to a 1-sink. If a clause is violated by the current assignment, then it corresponds to a 0-sink. At this point, the solvers could also label the 0-sink by the violated clause which would give a correct certified dec-DNNF.

Proof System for maxSAT. As for #SAT, one can exploit the tractability of many problems on dec-DNNF to define a proof system for maxSAT. Given a CNF formula F, let $\tilde{F} = \bigwedge_{C \in F} C \vee \neg s_C$ be the formula where each clause is augmented with a fresh *selector* variable. The *Knowledge Compilation based Proof System for* maxSAT, kcps(maxSAT) for short, is defined as follows: given a CNF F, a certificate that at most k clauses of F can be simultaneously satisfied is a correct certified dec-DNNF D such that every clause of $F(D)$ is a clause of $\tilde{F}$ and D computes $\tilde{F}$. Moreover,

– there exists a satisfying assignment of D setting k selector variables to 1,
– every satisfying assignment of D sets at most k selector variables to 1.

To check a certificate D, one has to check that D is correct and equivalent to $\tilde{F}$ which is tractable for the same reasons as before. To check the other two conditions, one simply has to compute a satisfying assignment of D that maximize the number of selector variables set to 1 which can be done in polynomial time since the problem HW is tractable on dec-DNNF by Theorem 1.

Contrary to #SAT, we are not aware of any tool solving maxSAT based on this technique and thus the implementation of such a proof system in existing tools may not be realistic. The question on how kcps(maxSAT) compares with the resolution for maxSAT of Bonet et al. [2] is open.

In general, we observe that we can use this idea to build a proof system kcps(Q) for any tractable problem Q on dec-DNNF. This could for example be applied to weighted versions of #SAT and maxSAT.

Combining Proof Systems. An interesting feature of kcps-like proof systems is that they can be combined with other proof systems for UNSAT to be made more powerful. Indeed, one could label the 0-sink of the dec-DNNF with a clause C that are not originally in the initial CNF F but that is entailed by F, that is, $F \Rightarrow C$. In this case, Corollary 2 would still hold. The only thing that is needed to obtain a real proof system is that a proof that $F \Rightarrow C$ has to be given along the correct certified dec-DNNF, that is, a proof of unsatisfiability of $F \wedge \neg C$. Any proof system for UNSAT may be used here.

Lower Bounds. Lower bounds on the size of dec-DNNF representing CNF formulas may be directly lifted to lower bounds for kcps(#SAT) or kcps(maxSAT). There exist families of monotone 2-CNF that cannot be computed by polynomial size dec-DNNF [1,3,4]. It directly gives the following corollary:

Corollary 3. *There exists a family* $(F_n)_{n \in \mathbb{N}}$ *of monotone 2-CNF such that* F_n *is of size* $O(n)$ *and any certified* dec-DNNF *for* F_n *is of size at least* $2^{\Omega(n)}$, *that is,* #F_n *does not have polynomial size proofs in* kcps(#SAT).

An interesting open question is to find CNF formulas having polynomial size dec-DNNF but no small proof in kcps(#SAT).

4 Future Work

In this paper, we have developed techniques based on circuits used in knowledge compilation to extend existing proof systems for tautology to harder problems. It seems possible to implement these systems into existing tools for #SAT based on exhaustive DPLL, which would allow these tools to provide an independently checkable certificate that their output is correct, the same way SAT-solvers returns a proof on unsatisfiable instances. It would be interesting to see how adding the computation of this certificate to existing solver impacts their performances. Another interesting direction would be to compare the power of kcps(maxSAT) with the resolution for maxSAT of Bonet et al. [2] and to see how such proof systems could be implemented in existing tools for maxSAT. Finally, we think that a systematic study of other languages used in knowledge compilation such as deterministic DNNF should be done to see if they can be used as proof systems, by trying to add explanations on why an assignment does not satisfy the circuit.

References

1. Beame, P., Li, J., Roy, S., Suciu, D.: Lower bounds for exact model counting and applications in probabilistic databases. In: Proceedings of the Twenty-Ninth Conference on Uncertainty in Artificial Intelligence (2013)
2. Bonet, M.L., Levy, J., Manyà, F.: Resolution for Max-SAT. Artif. Intell. **171** (8—-9), 606–618 (2007)
3. Bova, S., Capelli, F., Mengel, S., Slivovsky, F.: Knowledge compilation meets communication complexity. In: Proceedings of the Twenty-Fifth International Joint Conference on Artificial Intelligence, IJCAI 2016, New York, NY, USA, 9–15 July 2016, pp. 1008–1014 (2016)
4. Capelli, F.: Structural restrictions of CNF formulas: application to model counting and knowledge compilation. PhD thesis, Université Paris Diderot (2016)
5. Clegg, M., Edmonds, J., Impagliazzo, R.: Using the groebner basis algorithm to find proofs of unsatisfiability. In Proceedings of the Twenty-Eighth Annual ACM Symposium on Theory of Computing, STOC 1996 (1996)
6. Cook, S.A.: The complexity of theorem-proving procedures. In: Proceedings of the Third Annual ACM Symposium on Theory of Computing, pp. 151–158. ACM (1971)
7. Cook, S.A., Reckhow, R.A.: The relative efficiency of propositional proof systems. J. Symb. Log. **44**(1), 36–50 (1979)
8. Darwiche, A.: On the tractable counting of theory models and its application to truth maintenance and belief revision. J. Appl. Non-Classical Log. **11**(1–2), 11–34 (2001)
9. Darwiche, A., Marquis, P.: A knowledge compilation map. J. Artif. Intell. Res. **17**, 229–264 (2002)
10. Davis, M., Logemann, G., Loveland, D.: A machine program for theorem-proving. Commun. ACM **5**(7), 394–397 (1962)
11. Davis, M., Putnam, H.: A computing procedure for quantification theory. J. ACM **7**(3), 201–215 (1960)
12. Huang, J., Darwiche, A.: DPLL with a trace: from SAT to knowledge compilation. In: Proceedings of the Nineteenth International Joint Conference on Artificial Intelligence, pp. 156–162 (2005)
13. Jukna, S.: Boolean Function Complexity - Advances and Frontiers. Algorithms and combinatorics, vol. 27. Springer, Heidelberg (2012)
14. Koriche, F., Le Berre, D., Lonca, E., Marquis, P.: Fixed-parameter tractable optimization under DNNF constraints. In: ECAI 2016–22nd European Conference on Artificial Intelligence, 29 August-2 September 2016, The Hague, The Netherlands - Including Prestigious Applications of Artificial Intelligence (PAIS 2016), pp. 1194–1202 (2016)
15. Lagniez, J.-M., Marquis, P.: An improved decision-DNNF compiler. In: Proceedings of the Twenty-Sixth International Joint Conference on Artificial Intelligence, IJCAI 2017 (2017)
16. Lagniez, J.-M., Marquis, P., Szczepanski, N.: DMC: a distributed model counter. In: IJCAI, pp. 1331–1338 (2018)
17. Narodytska, N., Bacchus, F.: Maximum satisfiability using core-guided MAXSAT resolution. In: Twenty-Eighth AAAI Conference on Artificial Intelligence (2014)
18. Nordström, J.: Pebble games, proof complexity, and time-space trade-offs. Logical Methods Comput. Sci. (LMCS) **9**(3) (2013)

19. Oztok, U., Darwiche, A.: A top-down compiler for sentential decision diagrams. In: Proceedings of the Twenty-Fourth International Joint Conference on Artificial Intelligence, IJCAI 2015, pp. 3141–3148 (2015)
20. Pipatsrisawat, K., Darwiche, A.: On the power of clause-learning sat solvers as resolution engines. Artif. Intell. **175**(2), 512–525 (2011)
21. Sang, T., Bacchus, F., Beame, P., Kautz, H.A., Pitassi, T.: Combining component caching and clause learning for effective model counting. Theory Appl. Satisf. Test. **4**, 7th (2004)
22. Thurley, M.: sharpSAT – counting models with advanced component caching and implicit BCP. In: Biere, A., Gomes, C.P. (eds.) SAT 2006. LNCS, vol. 4121, pp. 424–429. Springer, Heidelberg (2006). https://doi.org/10.1007/11814948_38
23. Wegener, I.: Branching Programs and Binary Decision Diagrams. SIAM, Philadelphia (2000)

The Equivalences of Refutational QRAT

Leroy Chew[(✉)] and Judith Clymo

School of Computing, University of Leeds, Leeds, UK
scslnc@leeds.ac.uk

Abstract. The solving of Quantified Boolean Formulas (QBF) has been advanced considerably in the last two decades. In response to this, several proof systems have been put forward to universally verify QBF solvers. QRAT by Heule et al. is one such example of this and builds on technology from DRAT, a checking format used in propositional logic. Recent advances have shown conditional optimality results for QBF systems that use extension variables. Since QRAT can simulate Extended Q-Resolution, we know it is strong, but we do not know if QRAT has the strategy extraction property as Extended Q-Resolution does. In this paper, we partially answer this question by showing that QRAT with a restricted reduction rule has strategy extraction (and consequentially is equivalent to Extended Q-Resolution modulo NP). We also extend equivalence to another system, as we show an augmented version of QRAT known as QRAT+, developed by Lonsing and Egly, is in fact equivalent to the basic QRAT. We achieve this by constructing a line-wise simulation of QRAT+ using only steps valid in QRAT.

Keywords: QBF · QRAT · Proof complexity · Herbrand functions · Certificate

1 Introduction

Quantified Boolean Formulas (QBFs) extend propositional logic with Boolean quantifiers. The languages of true and, symmetrically, of false QBFs are PSPACE-complete, meaning they can capture any problem within the class PSPACE, and that they may allow for more succinct problem representations than propositional logic.

Modern algorithms for solving propositional satisfiability (SAT) problems and deciding the truth of QBFs are able to solve large industrial problems, and may also provide verification by outputting a proof. It is desirable that this "certificate" should be in a standard format, which can be achieved by designing a proof system that is powerful enough to simulate all of the proof systems used in various solving algorithms. A proof can then be easily converted from a solver output into the universal format.

In propositional logic, there is a successful proof checking format known as DRAT [15] that has since been used as a basis for a QBF proof checking format

© Springer Nature Switzerland AG 2019
M. Janota and I. Lynce (Eds.): SAT 2019, LNCS 11628, pp. 100–116, 2019.
https://doi.org/10.1007/978-3-030-24258-9_7

known as QRAT [7]. A further extension of QRAT, QRAT+ [13] has also recently been developed.

These proof systems are designed to capture practical QBF solving and involve some complex sub-procedures between the lines. In contrast, proof systems developed for theory tend to be based on simple ideas. Some theoretical results about QRAT and QRAT+ are known: firstly, they were shown to simulate known QBF techniques such as pre-processing [7] and long distance resolution [9] in QBFs; secondly, Skolem functions that certify true QBFs are known to be extractable from QRAT [6] and QRAT+ [13]. However Herbrand functions that certify false QBFs have not been shown to have efficient extraction.

One possible way to get a better handle on these systems from a theoretical point of view is to show that they are equivalent to some simpler system. For propositional logic, DRAT was recently shown to be equivalent to Extended Resolution [10] so there is some hope that QRAT is equivalent to some QBF analogue of Extended Resolution.

In Sect. 4 we demonstrate such an equivalence partially. If we simplify one of the rules in QRAT in a natural way we get that Herbrand functions can be extracted efficiently. Using a relaxed framework of simulation by Beyersdorff et al. [2] we can say that Extended Q-Resolution (a QBF analogue of extended resolution [8]) simulates, and thus is equivalent in this model to, this restricted version of QRAT.

We also prove an unconditional equivalence between QRAT and QRAT+ for false QBFs which we show in Sect. 5.

2   Preliminaries

2.1   Proof Complexity

Formally, a *proof system* [4] for a language $\mathcal{L}$ over alphabet Γ is a polynomial-time computable partial function $f : \Gamma^* \to \Gamma^*$ with $rng(f) = \mathcal{L}$, where rng denotes the range. A proof system maps *proofs* to *theorems*. A *refutation* system is a proof system where the language $\mathcal{L}$ is of contradictions.

The partial function f actually gives a proof checking function. Soundness is given by $rng(f) \subseteq \mathcal{L}$ and completeness is given by $rng(f) \supseteq \mathcal{L}$. The polynomial-time computability is an indication of feasibility, relying on the complexity notion that an algorithm that runs in polynomial-time is considered feasible.

Proof size is given by the number of characters appearing in a proof. A proof system f is *polynomially bounded* if there is a polynomial p such that every theorem of size n has a proof in f of size at most $p(n)$.

Proof systems are compared by simulations. We say that a proof system f *simulates* a proof system g $(g \le f)$ if there exists a polynomial p such that for every g-proof π_g there is an f-proof π_f with $f(\pi_f) = g(\pi_g)$ and $|\pi_f| \le p(|\pi_g|)$. If in addition π_f can be constructed from π_g in polynomial-time, then we say that f *p-simulates* g $(g \le_p f)$. Two proof systems f and g are *(p-)equivalent* $(g \equiv_{(p)} f)$ if they mutually (p-)simulate each other.

In propositional logic a literal is a variable (x) or its negation ($\neg x$), a clause is a disjunction of literals and a formula in conjunctive normal form (CNF) is a conjunction of clauses. For a literal l, if $l = x$ then $\bar{l} = \neg x$, and if $l = \neg x$ then $\bar{l} = x$. An assignment τ for formula A over n variables is a partial function from the variables of A to $\{0,1\}^n$. For clause C, $\tau(C)$ is the result of evaluating C under assignment τ. For formula (or circuit) A, we define $A[b/x]$ so that all instances of variable x in A are replaced with $b \in \{0,1\}$.

Several kinds of inferences can be made on formulas in conjunctive normal form. *Unit propagation* simplifies a formula Φ in conjunctive normal form by building a partial assignment and applying it to Φ. It builds the assignment by satisfying any literal that appears in a singleton (unit) clause. Doing so may negate opposite literals in other clauses and result in them effectively being removed from that clause. In this way, unit propagation can create more unit clauses and can keep on propagating until no more unit clauses are left.

The advantage of unit propagation is that it reaches fix-point in time that is polynomial in the number of clauses, however unit propagation is not a complete procedure and so is not a refutational proof system for propositional logic.

An example of a proof system, and also in fact a refutation system, is *Resolution* (Res). It works on formulas in CNF. Two clauses $(C \vee x)$ and $(D \vee \neg x)$ can be merged removing the pivot variable x to get $(C \vee D)$. An enhanced version of Resolution known as *Extended Resolution* allows new variables known as extension variables to be introduced. Extension variables represent functions of existing variables, these variables are introduced alongside clauses, known as extension clauses, that define this function. In Fig. 1 we define these systems in the more general logic of QBF (to understand Fig. 1 in the context of SAT treat all quantifiers as existential).

2.2 Quantified Boolean Formulas

Quantified Boolean Formulas (QBF) extend propositional logic with quantifiers $\forall, \exists$ that work on propositional atoms [11]. The standard QBF semantics is that $\forall x\, \Psi$ is satisfied by the same truth assignments as $\Psi[0/x] \wedge \Psi[1/x]$ and $\exists x\, \Psi$ is satisfied by the same truth assignments as $\Psi[0/x] \vee \Psi[1/x]$.

A prenex QBF is a QBF where all quantification is done outside of the propositional connectives. A prenex QBF Ψ therefore consists of a propositional part Φ called the matrix and a prefix of quantifiers Π and can be written as $\Psi = \Pi\Phi$. Starting from left to right we give each bound variable x a numerical level, denoted $\mathrm{lv}(x)$, starting from 1 and increasing by one each time the quantifier changes (it stays the same whenever the quantifier is not changed). For literal l, $\mathrm{var}(l) = x$ if $l = x$ or $l = \neg x$ and $\mathrm{lv}(l) = \mathrm{lv}(x)$. We write $\mathrm{lv}(x) <_\Pi \mathrm{lv}(y)$ to indicate that x appears in an earlier quantifier level than y in Π (though the subscript can be omitted if the context is clear). When the propositional matrix of a prenex QBF is in conjunctive normal form then we have a PCNF. A QBF of arbitrary structure can be transformed in polynomial time in to a PCNF. If no variables appear in the QBF without quantification then the QBF is closed.

It is natural to understand a PCNF as a set of clauses, and a clause as a set of literals. As such we will use notation $C \in \Phi$ to indicate that QBF $\Psi = \Pi\Phi$ has the clause C in its matrix. Similarly $l \in C$ indicates that clause C contains literal l.

Set notation is also used to define sub-clauses and sub-formulas, for example we can define the sub-clause of C containing only literals bound at level i or earlier by $\{l \in C \mid \mathrm{lv}(l) \leq i\}$.

A closed prenex QBF may be thought of as a game between two players. One player is responsible for assigning values to the existentially quantified variables, and the other responsible for the universally quantified variables. The existential player wins the game if the formula evaluates to true once all assignments have been made, the universal player wins if the formula evaluates to false. The players take turns to make assignments according to the quantifier prefix, so the levels of the prefix correspond to turns in the game. A strategy for the universal player on QBF $\Pi\Phi$ is a method for choosing assignments for each universal u that depends only on variables earlier than u in Π. If this strategy ensures the universal player always wins games on $\Pi\Phi$ (however the existential player makes assignments), then we say the universal player has a winning strategy. A QBF is false if and only if the universal player has a winning strategy. Strategies for the existential player are defined analogously and the QBF is true if and only if the existential player has a winning strategy.

For a universal variable u in QBF $\Pi\Phi$, a function σ_u that acts on assignments to the existential variables prior to u in Π and has Boolean output, is called a Herbrand function. The collection of σ_u for all universal variables u is a strategy for $\Pi\Phi$ and is denoted σ. Given $\tau_\exists$, an assignment to the existentially quantified variables prior to u, and Herbrand function σ_u, we can extend this assignment to u by evaluating $\sigma_u(\tau_\exists)$. We say this is an extension of the partial assignment $\tau_\exists$ that is consistent with σ_u.

A proof system is said to admit strategy extraction if and only if it is possible to efficiently (i.e. in polynomial time in the size of the proof) construct a circuit representing a winning strategy for the universal player from a refutation of a QBF.

An example of a QBF proof system which admits strategy extraction is the refutation system Q-Resolution (Q-Res) by Kleine-Büning et al. [12]. It combines the Resolution rule with a QBF rule known as universal reduction ($\forall$red). Universal reduction allows you to locally set the value of a universal literal within a clause, but only under the condition that no other literal in that clause appears to the right of it in the prefix. In a clause, only one choice of value does not satisfy the clause. By assumption universal reduction set the value so that it does not satisfy the clause.

Like Resolution, Q-Resolution can be augmented with extension variables to get Extended Q-Resolution. In Fig. 1 we detail each proof system based on which rules we have. Note that for propositional proof systems Resolution and Extended Resolution that we assume the prefix is purely existential as this is equivalent to the propositional setting.

$$\frac{}{C} \text{ (Ax)} \qquad\qquad \frac{}{(\neg x \vee \neg y \vee \neg v), (x \vee v), (y \vee v)} \text{ (Ext)}$$

Ax: C is a clause in the propositional matrix.
Ext: x, y are variables already in the formula, v is a fresh variable, v is inserted into prefix as existentially quantified, after x and y in the prefix.

$$\frac{C \vee x \qquad D \vee \neg x}{C \vee D} \text{ (Res)} \qquad\qquad \frac{C \vee l}{C} \text{ (}\forall\text{red)}$$

Res: variable x is existential.
$\forall$*red*: literal l has variable u, which is universal, and all other existential variables $x \in C$ are left of u in the quantifier prefix. Literal $\bar{l}$ does not appear in C.

Fig. 1. Rules of our resolution systems in the language of QBF, resolution is given by (Ax)+(Res), extended resolution is given by (Ax)+(Ext)+(Res), Q-resolution is given by (Ax)+(Res)+($\forall$red) and extended Q-resolution is given by (Ax)+(Ext)+(Res)+($\forall$red) [12].

This propositional case is important for QBFs. Any logical propositional implication is also valid in QBF (provided it is model preserving) and we can use this to make important steps in QBF inference. For a QBF $\Pi\Phi$, a full abstraction $\mathsf{Abs}(\Pi)$ returns an identical prefix with all variables of Π, except that every variable is quantified existentially. We will use a more general version of abstraction in Sect. 5.

Definition 1 (NP Oracle derivations [2]). *For QBF refutational system g, a g^{NP} proof of a QBF Ψ is a derivation of the empty clause by any of the g rules or an* NP-*derivation rule:*

$$\frac{C_1, \ldots, C_l}{D} \text{ (NP-derivation)}$$

For any l, where $\mathsf{Abs}(\Pi) \bigwedge_{i=1}^{l} C_i \vDash \mathsf{Abs}(\Pi)(\bigwedge_{i=1}^{l} C_i) \wedge D$. $C_1, \ldots, C_l$ and D must be permitted in the system g.

An NP-derivation rule can infer ΠD from $\Pi \bigwedge_{i=1}^{l} C_i$ whenever $\bigwedge_{i=1}^{l} C_i \vDash D$. When we add D we do not change the prefix Π. Hence g^{NP} augments QBF proof system g with all propositional inferences.

Notice that g^{NP} is not a proof system unless we can check the NP-derivation in polynomial-time. This cannot be done unless $\mathsf{P} = \mathsf{NP}$.

Definition 2. *Let P, Q be QBF proof systems, then we write $f \equiv^{\mathsf{NP}} g$ whenever f^{NP} and g^{NP} mutually p-simulate each other.*

In [3] Extended Q-Res with an NP derivation rule (Extended Q-Res^{NP}) was investigated. In the NP-derivation rule the input arguments $C_1, \ldots, C_l$ and the output argument D are all clauses as Extended Q-Res works on clauses. D is added to the CNF. QRAT^{NP} is defined in the same way, adding clauses logically implied by existing clauses.

We can combine our understanding of NP-oracle derivations and strategy extraction to show a weak optimality result for Extended Q-Res.

Theorem 1 (Chew 18 [3]). *If a refutational QBF proof system has strategy extraction it can be simulated by Extended Q-Res^{NP}.*

As we can see, Extended Q-Res is very strong and so it simulates most of the known QBF proof systems. It is therefore important to understand its relation to other strong systems such as the systems QRAT and QRAT+ which we will study in this paper.

3   QRAT

QRAT was introduced as a universal proof checking format for QBF. It simulates many QBF preprocessing techniques and proof systems. QRAT is based on the propositional DRAT format which is an advancement of blocked clause addition. Blocked literals generalise extension variables and so DRAT simulates extended resolution, likewise QRAT p-simulates Extended Q-Res.

QRAT works by having a PCNF QBF $\Pi\Phi$ that is edited throughout the proof by a number of satisfiability preserving rules. In contrast to line-based systems like Extended Resolution, QRAT does not just accumulate lines based on other lines. New conjuncts can be added and clauses can be altered or even deleted and rules are usually based globally around the current status of $\Pi\Phi$.

In [7] six rules were listed for QRAT. These were named ATA, ATE, QRATA, QRATE, QRATU and EUR. However QRAT has two modes that make it a proof system: refutation and satisfaction. In this work we focus only on refutation, which uses only ATA, QRATA, QRATU, EUR, and a general deletion rule. We will state each of these rules. Their correctness is proved in [7], but the strategy extraction arguments we make in Sect. 4 double as arguments for correctness.

If C is a clause, then $\bar{C}$ is the conjunction of the negation of the literals in C. We denote that the clause D is derived by unit propagation applied to $\Pi\Phi$ by $\Pi\Phi \vdash_1 D$. Unit propagation is used because it is a polynomial-time procedure.

Definition 3 (Asymmetric Tautology Addition (ATA)). *Let $\Pi\Phi$ be a closed PCNF with prefix Π and CNF matrix Φ. Let C be a clause not in Φ. Let Π' be a prefix including the variables of C and Φ, Π is a sub-prefix of Π' containing the variables of Φ only.*
Suppose $\Pi'\Phi \wedge \bar{C} \vdash_1 \bot$. Then we can make the following inference

$$\frac{\Pi\Phi}{\Pi'\Phi \wedge C} \; (ATA)$$

Notice that in the way that we define QRAT, $\Pi'\Phi \wedge C$ replaces $\Pi\Phi$.

Definition 4 (Outer Clause, Outer Resolvent). *Let $\Pi\Phi$ be a PCNF with closed prefix Π and CNF matrix Φ. Let C be a clause not in Φ. Let Π' be a prefix including the variables of C and Φ, Π is a sub-prefix of Π' containing the variables of Φ only.*

Suppose C contains a literal l. Consider all clauses D in Φ with $\bar{l} \in D$. The outer clause O_D of D is $\{k \in D \mid \mathrm{lv}(k) \leq_\Pi \mathrm{lv}(l), k \neq \bar{l}\}$. The outer resolvent $\mathcal{R}(C, D, \Pi, l)$ is defined as $C \vee O_D$.[1]

Definition 5 (Quantified Resolution Asymmetric Tautology Addition (QRATA)). *Let $\Pi\Phi$ be a PCNF with closed prefix Π and CNF matrix Φ. Let C be a clause not in Φ. Let Π' be a prefix including the variables of C and Φ, Π is a sub-prefix of Π' containing the variables of Φ only.*

If C contains an existential literal l such that for every $D \in \Phi$ with $\bar{l} \in D$, $\Pi\Phi \wedge \bar{C} \wedge \bar{O}_D \vdash_1 \bot$ (or equivalently $\Pi\Phi \wedge \bar{\mathcal{R}}(C, D, \Pi, l) \vdash_1 \bot$) then we can derive

$$\frac{\Pi\Phi}{\Pi'\Phi \wedge C} \ (QRATA\ w.r.t.\ l)$$

Definition 6 (Quantified Resolution Asymmetric Tautology Universal (QRATU)). *Let $\Pi\Phi$ be a PCNF with closed prefix Π and CNF matrix Φ. Let $C \vee l$ be a clause with universal literal l.*

If for every $D \in \Phi$ with $\bar{l} \in D$, $\Pi\Phi \wedge \bar{C} \wedge \bar{O}_D \vdash_1 \bot$ then we can derive

$$\frac{\Pi\Phi \wedge (C \vee l)}{\Pi\Phi \wedge C} \ (QRATU\ w.r.t.\ l)$$

Definition 7 (Extended Universal Reduction (EUR)). *Given a clause $C \vee u$ with universal literal u, consider extending C by*
$$C := C \cup \{k \in D \mid \mathrm{lv}(k) >_\Pi \mathrm{lv}(u) \ or \ k = \bar{u}\} \,,$$
where $D \in \Phi$ is any clause with some $p : \mathrm{lv}(p) >_\Pi \mathrm{lv}(u)$, $p \in C$ and $\bar{p} \in D$,
until we reach a fixed point denoted ε. If $\bar{u} \notin \varepsilon$ then we can perform the following rule.

$$\frac{\Pi\Phi \wedge (C \vee u)}{\Pi'\Phi \wedge C} \ (EUR)$$

EUR encompasses the important reduction rule used in Q-Resolution. Along with the other rules, this allows us to simulate Extended Q-Res and thus we have refutational completeness. EUR is strictly stronger than the standard universal reduction (UR), because it uses a dependency scheme from [14]. While refutational QRAT works perfectly fine with a standard universal reduction rule, extended universal reduction was used because it allows QRAT to easily simulate expansion steps used in QBF preprocessing. Note that this was the *only* preprocessing rule that required the extended universal reduction rule. In theory QRAT could be augmented with any sound dependency scheme and that is used as the basis of its reduction rule.

Definition 8 (Clause Deletion). *In refutational QRAT clauses may be arbitrarily deleted without checking if they conform to a rule, note this is not the case in satisfaction QRAT.*

---

[1] Other authors have made two separate definitions of outer resolvents based on existential or universal l. In order to simplify, we only use the existential definition here and factor in the necessary changes in the description of our rules.

$$\frac{\Pi\Phi \wedge C}{\Pi\Phi}$$

It is important to note that because some of the rules work by looking at every clause contained in Φ, clause deletion may not be superficial. A clause may need to be deleted in order for EUR to be performed. The situation is true also for QRATA and QRATU with respect to l where our property needs to check all other clauses that have the complimentary literal $\bar{l}$.

4 Strategy Extraction and Simulations

In [6] it was shown that satisfaction QRAT has strategy extraction. The equivalent, however, was not proved for refutational QRAT. The problem is the asymmetry of the two systems as EUR is not needed in satisfaction QRAT. EUR causes particular problems which we were not able to get around. It reduces a universal variable u in a clause C even when there are variables in C to the right of u in the prefix. Now this is also true for QRATU however in that case we have the QRAT framework to work with, which is not similar to the dependency framework of EUR.

We have avoided this issue by removing EUR altogether. EUR is not essential to the underlying techniques of QRAT and QRAT can still simulate the main preprocessing techniques except ∀-Expansion. In addition EUR is not required to simulate Extended Q-Resolution as this can be done with a simpler reduction rule, as extension clauses can be added with QRATA wrt the extension variable, Resolution uses ATA and reduction can be performed with simple universal reduction.

Let QRAT(X) be QRAT with the EUR rule replaced with reduction rule X. This means that the standard QRAT is given by QRAT(EUR). An alternative would be to use the ∀-reduction rule from Q-Res, which allows

$$\frac{\Pi\Phi \wedge (C \vee u)}{\Pi\Phi \wedge C}$$

whenever $\mathrm{lv}(u)$ is greater than $\mathrm{lv}(x)$ for all existentially quantified x in C.

We call this simplest version QRAT(UR).

In order to show that QRAT(UR) has polynomial-time strategy extraction on false QBFs we will inductively compute a winning universal strategy for formulas at each step of the QRAT proof. For the inductive step we need to construct a winning strategy σ for the formula prior to some proof step, from a known winning strategy σ' for the formula after that proof step. We prove that this is possible for each derivation rule in refutational QRAT. The strategy σ is composed of Herbrand functions σ_u for each universal variable u.

Lemma 1. *If $\Pi'\Phi \wedge C$ is derived from $\Pi\Phi$ by ATA, and σ' is a winning universal strategy for $\Pi'\Phi \wedge C$, then we can can construct a winning universal strategy for $\Pi\Phi$.*

Proof. If clause C is added by ATA then $\Phi \wedge \bar{C} \vdash_1 \bot$

Because $\Phi \wedge \bar{C} \vdash_1 \bot$, any assignment that falsifies C also falsifies Φ. Therefore if σ' is a strategy that falsifies $\Pi'\Phi \wedge C$, σ' must falsify Φ. Let $\sigma = \sigma'$ except that if $\Pi \neq \Pi'$ then any existential input variable that does not appear in Φ can be restricted in σ to 0 or 1 arbitrarily (there is a winning strategy for either assignment, so we just pick one). If a universal variable is in Π' but not Π we do not need a strategy for it as it will have no effect on the outcome of the game for $\Pi\Phi$. □

Lemma 2. *If $\Pi'\Phi \wedge C$ is derived from $\Pi\Phi$ by QRATA, and σ' is a winning universal strategy for $\Pi'\Phi \wedge C$, then we can can construct a winning universal strategy for $\Pi\Phi$.*

Proof. C contains some existential literal l such that for every $D \in \Phi$ with $\bar{l} \in D$ and outer clause O_D, $\Pi'\Phi \wedge \bar{C} \wedge \bar{O}_D \vdash_1 \bot$. For notational convenience, let $A = \{k \in C \mid k \neq l, \text{lv}(k) \leq_\Pi \text{lv}(l)\}$, and $B = \{k \in C \mid \text{lv}(k) >_\Pi \text{lv}(l)\}$ so that $C = (A \vee l \vee B)$ and $\Pi'\Phi \wedge \bar{A} \wedge \bar{l} \wedge \bar{B} \wedge \bar{O}_D \vdash_1 \bot$. We derive:

$$\frac{\Pi\Phi}{\Pi'\Phi \wedge (A \vee l \vee B)} \; (\text{QRATA wrt. } l)$$

Initially we assume $\Pi = \Pi'$. Let u be a universal variable in Π'. If $\text{lv}(u) <_{\Pi'} \text{lv}(l)$ then $\sigma'_u = \sigma_u$. If $\text{lv}(u) >_{\Pi'} \text{lv}(l)$, we proceed by a case distinction. Let $\tau_\exists$ be an assignment to the existential variables x with $\text{lv}(x) <_{\Pi'} \text{lv}(u)$, and τ its extension consistent with the Herbrand functions of universal variables y with $\text{lv}(y) <_{\Pi'} \text{lv}(u)$.

$$\sigma_u(\tau_\exists) = \begin{cases} \sigma'_u(\tau'_\exists) & \tau(A \vee l) = \bot, \text{ but for every clause } D \text{ with } \bar{l} \in D \\ & \text{if } O \text{ is the outer clause of } D \text{ then } \tau(O) = \top \\ & \text{where } \tau'_\exists \text{ differs from } \tau_\exists \text{ only on variable } l \\ & \text{such that } l \text{ is satisfied in } \tau'_\exists, \\ \sigma'_u(\tau_\exists) & \text{otherwise.} \end{cases}$$

We have to show that σ actually falsifies $\Pi\Phi$. Assume we reach an assignment τ by playing according to σ.

Suppose τ satisfies A then $\tau(A \vee l) = \top$ so for all u, $\sigma_u(\tau_\exists) = \sigma'_u(\tau_\exists)$. τ is consistent with σ' so falsifies $\Pi'\Phi \wedge (A \vee l \vee B)$. It cannot falsify $(A \vee l \vee B)$ so τ falsifies $\Pi\Phi$.

Suppose τ falsifies A but satisfies the outer clauses of every D with $\bar{l} \in D$. If $\tau_\exists$ were modified so l is true then σ' yields τ' which satisfies $(A \vee l \vee B)$ so must falsify Φ. Changing l to be false in τ' cannot satisfy any additional clauses since the outer clauses of all D with $\bar{l} \in D$ are already satisfied (by construction τ and τ' are identical prior to l). Under σ all universal variables right of l are played according to σ' as if l were made true, this will falsify some clause in Φ regardless of how l is actually set.

If τ falsifies A but also falsifies the outer clause O of some clause D with $\bar{l} \in D$ then we know that the responses from σ here are defined to be consistent with the original σ'. This means that either τ falsifies Φ or τ falsifies $A \vee l \vee B$.

We know that $\Pi'\Phi \wedge \bar{A} \wedge \bar{l} \wedge \bar{B} \wedge \bar{O}_D \vdash_1 \bot$. If τ falsifies $A \vee l \vee B$ then also $\tau(O_D \vee A \vee l \vee B) = \bot$, and so Φ is also falsified by τ.

If in fact $\Pi \neq \Pi'$ then Π' contains more variables than Π. First construct the strategies as above, assuming prefix Π', then fix these as in Lemma 1 to not include the variables missing from Π. Universal variables not in Π simply have their strategies removed from σ. Existential variables not in Π are restricted in σ to 0 or 1 arbitrarily. □

Example 1. Consider clause $(a \vee l \vee x \vee \neg y)$ and QBF $\exists ablx \forall y\phi$ where
$$\phi = (a \vee b \vee \neg y) \wedge (\neg b \vee y) \wedge (b \vee l \vee y) \wedge (b \vee \neg l \vee y).$$

The only clause that contains $\neg l$ in ϕ is $(b \vee \neg l \vee y)$, so the only outer clause we need to consider is (b). $\exists ablx \forall y\phi \wedge \neg b \wedge \neg a \wedge \neg l \wedge \neg x \wedge y \vdash_1 \bot$ so QRATA is possible with respect to l.

A winning strategy for the universal player after the new clause is added is to play y to 1 if and only if a, l and x are all 0. In our strategy extraction we derive the strategy prior to when QRATA is used.

If any of a, l, x are 1, we can continue to set y to 0 and falsify some clause in ϕ not containing $\neg y$.

If a, b, l, x are all 0, we falsify the only outer clause (b), thus we know via unit propagation some other clause (here $(a \vee b \vee \neg y)$) will be falsified if we continue to falsify $(a \vee l \vee x \vee \neg y)$ only. We keep playing the old strategy for this reason.

If a, l, x are all 0 and b is 1, setting y to 1 no longer works as it only falsifies the added clause, we instead see what happens when l is flipped to 1, and set y to 0 according to the strategy, falsifying clause $(\neg b \vee y)$.

Lemma 3 (Balabanov, Jiang [1]). *If $\Pi'\Phi \wedge C$ is derived from $\Pi\Phi \wedge (C \vee l)$ by UR, and σ' is a winning universal strategy for $\Pi'\Phi \wedge C$, then we can can construct a winning universal strategy for $\Pi\Phi \wedge (C \vee l)$.*

Lemma 4. *If $\Pi'\Phi \wedge C$ is derived from $\Pi\Phi \wedge (C \vee l)$ by QRATU, and σ' is a winning universal strategy for $\Pi'\Phi \wedge C$, then we can construct a winning universal strategy for $\Pi\Phi \wedge (C \vee l)$.*

Proof (Sketch). As before, it is useful to have notation for subclauses of C having variables to the left (A) or right (B) of l in the prefix.

For every $D \in \Phi$ with $\bar{l} \in D$ and outer clause O_D, $\Pi'\Phi \wedge \bar{A} \wedge \bar{B} \wedge \bar{O}_D \vdash_1 \bot$.

Let u be a universal variable in Π. If $\text{var}(l) \neq u$ then $\sigma_u = \sigma'_u$. If $u = \text{var}(l)$ does not appear in Π' then $\sigma_u = c$ where $c \in \{0, 1\}$ falsifies l.

Otherwise, if $\text{var}(l) = u$, we define σ_u by

$$\sigma_u(\tau_\exists) = \begin{cases} c & \tau(A) = \bot, \text{ but for every clause } D \text{ with } \bar{l} \in D \\ & \text{if } O \text{ is the outer clause of } D \text{ then } \tau(O) = \top \\ & \text{where } c \in \{0, 1\} \text{ is such that setting } u \text{ to } c \text{ falsifies } l, \\ \sigma'_u(\tau_\exists) & \text{otherwise.} \end{cases}$$

Lemma 5. *If $\Pi'\Phi$ is derived from $\Pi\Phi \wedge C$ by clause deletion, and σ' is a winning universal strategy for $\Pi'\Phi$, then we can can construct a winning universal strategy for $\Pi\Phi \wedge C$.*

Theorem 2. *QRAT(UR) has polynomial-time strategy extraction on false QBFs.*

Proof. We inductively show that we can compute a winning strategy σ on the current formula $\Pi\Phi$ during our steps of QRAT. σ can be constructed as a circuit with polynomial size in the length of the QRAT proof of $\Pi\Phi$ and consists of Herbrand functions σ_u for each of the universal variables u in the formula, based on existential variables of lower level.

We work backwards in the proof. Initially (i.e. at the end of the proof) we have the empty clause so in the base case our universal strategy sets all u to 0.

For the inductive steps we construct a new strategy σ for $\Pi\Phi$ based on σ' for $\Pi'\Phi'$, which is possible by the Lemmas above. The circuits σ_u constructed for ATA and clause deletion steps are no larger than σ'_u. For UR we have one copy of σ'_u and a circuit to check whether C is satisfied. For QRATU we need a circuit to determine if A and each of the outer clauses are satisfied, and this result with the output of σ'_u determines the final value for u. QRATA is the least obvious case when transforming into a polynomial size circuit. A new circuit is added to decide whether $(A \vee l)$ and the outer clauses are satisfied. The output of this is used to possibly change the input to σ'_u, which can be achieved with a small sub-circuit. Crucially, only one copy of σ'_u is needed.

Eventually we get to our first line and thus provide a strategy for the initial formula which can be constructed as a polynomial circuit.   □

5   Equivalence with QRAT+

In [13] Lonsing and Egly took the QRAT framework and improved on it by relaxing the properties required to add or delete a clause. In QRAT (and the original DRAT) we used the fact that $\Phi \wedge \bar{C}$ is a contradiction that can be checked by unit propagation in order to add C, this is known as Reverse Unit Propagation (RUP) [5]. This works nicely because unit propagation is a polynomial-time procedure and it is done in practice between the main inference steps in a propositional solver. In fact in Conflict-Driven Clause Learning (CDCL) solving it is used exactly in this way, to confirm a conflict whose negation is added as a clause.

However, in the QBF CDCL setting universal reduction is also done in between steps. So Lonsing's and Egly's definition of QRAT+ changed $\vdash_1$ to $\vdash_{1\forall}$, where $\vdash_{1\forall}$ is an inference using unit propagation *and* universal reduction.

In order to make this sound for all the QRAT rules, only some $\forall$-reductions are allowed. To achieve this, an *abstraction* of the quantifier prefix Π is taken.

Definition 9. *Let i be the maximum level of all variables in clause C. Then* $\mathsf{Abs}(\Pi', C)$ *is the quantifier prefix obtained from Π' by setting all quantifiers with level $\leq_{\Pi'} i$ to existential.*

C is a *Quantified* Asymmetric Tautology (QAT) in QBF $\Pi\Phi$ when $\mathsf{Abs}(\Pi', C)\Phi\wedge$ $\bar{C} \vdash_{1\forall} \bot$. We use Π' instead of Π because C could contain variables not in Π.

The asymmetric tautology property used for ATA is replaced by QAT. The property QRAT used for QRATA and QRATU is similarly updated to QRAT+.

Recall clause C has QRAT with respect to literal l in QBF $\Pi\Phi$ if and only if for all $D \in \Phi$ with $\bar{l} \in D$

$$\Phi \wedge \bar{C} \wedge \bar{O}_D \vdash_1 \bot$$

where O_D is the outer clause of D with respect to l. Clause C has QRAT+ with respect to literal l in QBF $\Pi\Phi$ if and only if for all $D \in \Phi$ with $\bar{l} \in D$.

$$\mathsf{Abs}(\Pi', C \vee l)\Phi \wedge \bar{C} \wedge \bar{O}_D \vdash_{1\forall} \bot$$

QATA allows C to be added to QBF $\Pi\Phi$ when C is a Quantified Asymmetric Tautology in $\Pi\Phi$. QRATA+ allows addition of C to $\Pi\Phi$ given there is an existential literal $l \in C$ such that C has QRAT+ with respect to l in $\Pi\Phi$. QRATU+ allows removal of l from $C \vee l$ if C has QRAT+ with respect to l in $\Pi\Phi$. The extended universal reduction rule is not changed in QRAT+.

Lonsing and Egly do not differentiate between a refutational and satisfaction QRAT+. Here we focus on the refutational case where arbitrary clauses can be deleted. Refutational QRAT+ therefore consists of QATA, QRATA+, QRATU+, EUR and clause deletion.

Proposition 1. *Refutationally, QRAT is p-equivalent to QRAT+.*

The key here is that in QRAT we derive some of the intermediate steps used in QRAT+, in particular the clauses that are to be $\forall$-reduced in the $\vdash_{1\forall}$ derivation. From there we can use the universal reduction rule in QRAT to reduce it and continue to use it to get our desired clause using only the rules in QRAT.

Example 2. We use an example from [13] to give a simple illustration of how this method works. Let $\Phi = \forall u_1, u_2 \exists x_1, x_2 \forall u_3 \exists x_3 \bigwedge_{i=0}^{7} C_i$ where

$$\begin{aligned}
C_0 &= (\neg u_2 \vee \neg x_1 \vee \neg x_2) & C_3 &= (u_2 \vee x_1 \vee x_2) & C_6 &= (\neg x_1 \vee x_2 \vee \neg x_3) \\
C_1 &= (\neg u_1 \vee \neg x_1 \vee x_2) & C_4 &= (\neg x_1 \vee \neg x_2 \vee x_3) & C_7 &= (\neg u_3 \vee x_3) \\
C_2 &= (u_1 \vee x_1 \vee \neg x_2) & C_5 &= (u_3 \vee \neg x_3)
\end{aligned}$$

Both u_1 and u_2 can be removed by QRATU+ but not by QRATU. In Φ, C_0 has QRAT+ with respect to $\neg u_2$. The only clause containing u_2 is C_3. Unit propagation of x_1 and x_2 allows us to derive x_3 (from C_4) then we use this to derive u_3 (from C_5) which is then removed by $\forall$-reduction. Therefore we can replace C_0 by $C_0 \backslash \{\neg u_2\}$. This is not possible by QRATU, however it is possible to derive $C = \neg x_1 \vee \neg x_2 \vee u_3$ by ATA since $\Phi \wedge \bar{C} \vdash_1 \bot$. We have unit propagation on x_1 and x_2 as before, then use x_3 to derive u_3 which does not need to be removed by $\forall$-reduction because we can use unit propagation with $\neg u_3$ instead to derive the empty clause. Once C is derived we can safely remove u_3 by $\forall$-reduction to reach the target clause.

Lemma 6. *The QATA step is p-simulated by refutational QRAT(UR).*

Proof. Let Φ be a CNF and let C be a clause not in Φ. Let Π' be a prefix including the variables of C and Φ, and Π a sub-prefix of Π' containing the variables of Φ. QATA derives $\Pi'\Phi \wedge C$ from $\Pi\Phi$ when C has QAT on $\Pi\Phi$. This means $\mathsf{Abs}(\Pi', C)\Phi \wedge \bar{C} \vdash_{1\forall} \bot$, where i is the maximum level of literals in C. However it may not be the case that $\Pi'\Phi \wedge \bar{C} \vdash_1 \bot$ because we may have used a universal reduction step (potentially multiple times).

The idea is that we break the single QAT step into several steps based on what uses unit propagation versus what uses $\forall$-reduction. We deal with unit propagation by using ATA and then use UR or EUR to do the reduction steps.

We single out the clauses that we perform reduction on during the QAT procedure and label them L_i with the literal p_i being reduced (in order, so the first one derived is L_1). So $\Pi'\Phi \wedge \bar{C} \vdash_1 L_1$, and for $i > 1$, $\Pi'\Phi \wedge \bigwedge_{j<i} L_j \backslash \{p_j\} \wedge \bar{C} \vdash_1 L_i$. The condition on reduction is that p_j must be the greatest level literal in L_j, but it must also be at a greater level than any literal in C because we are using the modified prefix so that every variable at a lower level than any variable in C is existentially quantified.

Induction Hypothesis: We can learn $C \vee L_i \backslash \{p_i\}$ in a short proof using only ATA and $\forall$-reduction steps.

Base Case: We need to add $C \vee L_1$ via ATA, we know $\Pi'\Phi \wedge \bar{C} \vdash_1 L_1$ so $\Pi'\Phi \wedge \bar{C} \wedge \bar{L}_1 \vdash_1 \bot$

Inductive Step: From assuming our induction hypothesis we have $\Pi'\Phi \wedge \bigwedge_{j<i}(C \vee L_j \backslash \{p_j\})$ now we need to add $C \vee L_i$ via ATA. For each $j < i$, $(C \vee L_j \backslash \{p_j\}) \wedge \bar{C} \vdash_1 L_j \backslash \{p_j\}$, and $\Phi \wedge \bigwedge_{j<i} L_j \backslash \{p_j\} \wedge \bar{C} \vdash_1 L_i$ so we can join these unit propagation inferences together to get that $\Pi\Phi \wedge \bigwedge_{j<i}(C \vee L_j \backslash \{p_j\}) \wedge \bar{C} \wedge \bar{L}_i \vdash_1 \bot$. We learn $C \vee L_i$ but we now need to remove p_i, which we can do via reduction. Note that per the rules of QRAT+, $\mathrm{lv}(p_i) \geq_{\Pi'} \mathrm{lv}(x)$ for any literal $x \in L_i$ and any x in C. Thus we can reduce p_i.

So we learn all the clauses $C \vee L_i \backslash \{p_i\}$. These new clauses allow us to derive C via ATA without using reduction since those steps are now available. From $\Pi\Phi \wedge \bigwedge_j(C \vee L_j \backslash \{p_j\}) \wedge \bar{C}$ we can derive each $L_j \backslash \{p_j\}$ which gives us every reduced clause we need to derive $\bot$ from $\Phi \wedge \bar{C}$. Hence $\Pi\Phi \wedge \bigwedge_j(C \vee L_j \backslash \{p_j\}) \wedge \bar{C} \vdash_1 \bot$. We can add C via ATA. $\square$

We can take this framework and use it to simulate other rules in the QRAT+ refutation system. We prove this essential lemma about the property QRAT+.

Lemma 7. *If C has QRAT+ in $\Pi\Phi$ with respect to l then for every clause $D \in \Phi$ with $\bar{l} \in D$, the outer resolvent $C \cup \{k \in D \mid \mathrm{lv}(k) \leq_\Pi \mathrm{lv}(l), k \neq \bar{l}\}$ can be added to $\Pi\Phi$ via a sequence of polynomial size iterations of ATA and $\forall$-reduction.*

Proof. Let R_D denote the outer resolvent $C \cup \{k \in D \mid \mathrm{lv}(k) \leq_\Pi \mathrm{lv}(l), k \neq \bar{l}\}$. The property of QRAT+ can be stated as $\mathsf{Abs}(\Pi', C)\Phi \wedge \bar{R}_D \vdash_{1\forall} \bot$ for every $D \in \Phi$ with $\bar{l} \in D$. Let us fix some D and prove we can derive R_D via ATA and $\forall$-reduction steps. Since D is fixed we drop the subscript and let $R = R_D$.

As we did to simulate QATA, we label the clauses that we perform reduction on during the $\vdash_{1\forall}$ procedure as L_i with the literal p_i being reduced (in order, so the first one derived is L_1), we perform induction on this index i.

Induction Hypothesis: We can learn $R \vee L_i \backslash \{p_i\}$ in a short proof using only ATA and $\forall$-reduction steps.

Base Case: We need to add $R \vee L_1$ via ATA, we know $\Phi \wedge \bar{R} \vdash_1 L_1$ so $\Pi'\Phi \wedge \bar{R} \wedge \bar{L}_1 \vdash_1 \bot$. Thus we add $R \vee L_1$. We now reduce p_1 which can be done because $\text{lv}(p_i) \geq_{\Pi'} \text{lv}(x)$ for any literal $x \in L_i$ and any x in C, and since R only adds outer clause literals it works for any $x \in R$ as well.

Inductive Step: From assuming our induction hypothesis we have $\Pi'\Phi \wedge \bigwedge_{j<i}(R \vee L_j \backslash \{p_j\})$ now we need to add $R \vee L_i$ via ATA.

For each $j < i$, $(R \vee L_j \backslash \{p_j\}) \wedge \bar{R} \vdash_1 L_j \backslash \{p_j\}$, and $\Phi \wedge \bigwedge_{j<i} L_j \backslash \{p_j\} \wedge \bar{R} \vdash_1 L_i$ so we can join these unit propagation inferences together to get that $\Pi'\Phi \wedge \bigwedge_{j<i}(R \vee L_j \backslash \{p_j\}) \wedge \bar{R} \wedge \bar{L}_i \vdash_1 \bot$. We learn $R \vee L_i$ but we now need to remove p_i, which we can do via reduction. Note that per the rules of QRAT+, $\text{lv}(p_i) \geq_{\Pi'} \text{lv}(x)$ for any literal $x \in L_i$ and any x in C, and since R only adds outer clause literal it works for any $x \in R$ as well. Thus we can reduce p_i.

Putting this proof together we get all the $(R \vee L_j \backslash \{p_j\})$ clauses we need. $\Pi'\Phi \wedge \bigwedge_{j<i}(R \vee L_j \backslash \{p_j\}) \wedge \bar{R} \vdash_1 L_i$ for every i, thus we also get $\Pi'\Phi \wedge \bigwedge_j (R \vee L_j \backslash \{p_j\}) \wedge \bar{R} \vdash_1 \bot$ and can add R via ATA.

We can then remove the intermediate clauses $R \vee L_j \backslash \{p_j\}$ if necessary. □

Lemma 8. *The QRATA+ step is p-simulated by refutational QRAT(UR).*

Proof. Let Φ be a CNF and let C be a clause not in Φ. Let Π be a prefix including the variables of Φ and Π' be a prefix including the variables of C and Φ. Π is a sub-prefix of Π'. The QRATA+ step derives $\Pi'\Phi \wedge C$ from $\Pi'\Phi$ when C has QRAT+ with respect to l on $\Pi\Phi$.

Let $\Omega = \{O_D \mid O_D \text{ is the outer clause of some clause } D \in \Phi \text{ with } \bar{l} \in D\}$. We aim to add each $C \vee O_D$ via a short proof using ATA and UR rules. This can be done directly via Lemma 7 since $C \vee O_D$ is an outer resolvent and QRATA+ applies. Note that we can continue adding outer resolvents by this method to Φ even after one or more has already been added. The rules ATA and UR that we use in Lemma 7 are not prohibited by the presence of additional clauses.

Now that we have all $C \vee O$ for every $O \in \Omega$, we need to derive C. We can do this via QRATA with respect to l. This is simple, we need to show for each O_D that $\Pi\Phi \wedge \bigwedge_{O\in\Omega}(C \vee O) \wedge \bar{C} \wedge \bar{O}_D \vdash_1 \bot$. We know this is true since we can directly refute the clause $C \vee O_D$ using $\bar{C} \wedge \bar{O}_D$ in each case.

Once we have derived C we can freely delete all clauses from $\bigwedge_{O\in\Omega}(C \vee O)$ if we require to. Note that refutational QRAT does not require you to use QRATE to remove clauses (see Sect. 6.2 in [7]). □

Lemma 9. *The QRATU+ step is p-simulated by refutational QRAT(UR).*

The proof follows similarly to Lemma 8.

Proposition 1 *Refutationally, QRAT is p-equivalent to QRAT+.*

Proof. A QRAT+ proof is a sequence of QATA, QRATA+, QRATU+, EUR and deletion rules. We claim this can be simulated by a QRAT proof, in other words a sequence of ATA, QRATA, QRATU, EUR and deletion steps. EUR and deletion rules remain the same in both systems. In Lemma 6 we showed that QATA steps are simulated by ATA and UR steps, in Lemma 8 we showed that QRATA+ steps are simulated by ATA, QRATA and UR steps and in Lemma 9 we showed that QRATU+ steps are simulated by ATA, QRATU and UR steps. We can simulate UR steps by EUR so we can do all this in refutational QRAT.

QRAT is simulated by QRAT+ also, so they are equivalent.   □

Theorem 3. *Extended Q-Res, refutational QRAT(UR) and refutational QRAT(UR)+ are all equivalent QBF proof systems (modulo NP).*

Proof. Refutational QRAT(UR) simulates refutational QRAT+(UR). This can be seen by putting Lemmas 6, 8 and 9 together to simulate the rules QATA, QRATA+ and QRATU+ using ATA, QRATA, QRATU and UR. Clause deletion and UR are present in both systems.

Extended Q-Res is equivalent QRAT(UR) when both systems have the assistance of an NP oracle. Hence all these systems are equivalent modulo NP.   □

Note that the "modulo NP" here is important because we do not know if Extended Resolution easily proves the reflection principle of QRAT(UR). If it can (as argued in [3]) then they are all equivalent unconditionally. This seems likely as Extended Resolution is known to simulate DRAT [10].

6   Conclusion

We have collapsed QRAT and QRAT+ into the same refutation system. We have also examined the relationships of these systems to Extended Q-Resolution under the framework of an NP oracle and restricting the reduction rule in QRAT.

The NP oracle probably does not make a difference here as Extended Resolution is very powerful and is likely to provide the propositional implications we need anyway, especially when it already simulates DRAT.

Dealing with EUR, in its full power, may prove more tricky, as it will deal with understanding the relationship between strategy extraction and the dependency scheme of [14]. In addition, EUR is used in QRAT to simulate universal expansion and the relationship between strategy extraction and expansion is also opaque at the moment.

We also wish to note that despite the equivalences we show here, it is still our estimation that the QRAT and QRAT+ formats provide practical advantages in QBF solving not covered by complexity theory, just as DRAT continues to be used despite its equivalence to Extended Resolution. It should also be noted that we say nothing on the equivalence of satisfiability proofs for these systems.

Acknowledgements. Research supported by a Postdoctoral Prize Fellowship from EPSRC (1st author).

References

1. Balabanov, V., Jiang, J.H.R.: Unified QBF certification and its applications. Form. Methods Syst. Des. **41**(1), 45–65 (2012)
2. Beyersdorff, O., Hinde, L., Pich, J.: Reasons for hardness in QBF proof systems. Electron. Colloq. Comput. Complexit (ECCC) **24**, 44 (2017). https://eccc. weizmann.ac.il/report/2017/044
3. Chew, L.: Hardness and optimality in QBF proof systems modulo NP. Electron. Colloq. Comput. Complexit (ECCC) **25**, 178 (2018). https://eccc.weizmann.ac.il/report/2018/178
4. Cook, S.A., Reckhow, R.A.: The relative efficiency of propositional proof systems. J. Symb. Log. **44**(1), 36–50 (1979)
5. Goldberg, E., Novikov, Y.: Verification of proofs of unsatisfiability for CNF formulas. In: Proceedings of the Conference on Design, Automation and Test in Europe, DATE 2003, vol. 1. p. 10886. IEEE Computer Society, Washington, DC (2003). http://dl.acm.org/citation.cfm?id=789083.1022836
6. Heule, M., Seidl, M., Biere, A.: Efficient extraction of Skolem functions from QRAT proofs. In: Formal Methods in Computer-Aided Design, FMCAD 2014, Lausanne, Switzerland, 21–24 October 2014, pp. 107–114 (2014). https://doi.org/10.1109/FMCAD.2014.6987602
7. Heule, M.J.H., Seidl, M., Biere, A.: A unified proof system for QBF preprocessing. In: Demri, S., Kapur, D., Weidenbach, C. (eds.) IJCAR 2014. LNCS (LNAI), vol. 8562, pp. 91–106. Springer, Cham (2014). https://doi.org/10.1007/978-3-319-08587-6_7
8. Jussila, T., Biere, A., Sinz, C., Kröning, D., Wintersteiger, C.M.: A first step towards a unified proof checker for QBF. In: Marques-Silva, J., Sakallah, K.A. (eds.) SAT 2007. LNCS, vol. 4501, pp. 201–214. Springer, Heidelberg (2007). https://doi.org/10.1007/978-3-540-72788-0_21
9. Kiesl, B., Heule, M.J.H., Seidl, M.: A little blocked literal goes a long way. In: Proceedings of Theory and Applications of Satisfiability Testing - SAT 2017–20th International Conference, Melbourne, VIC, Australia, 28 August– 1 September 2017, pp. 281–297 (2017). https://doi.org/10.1007/978-3-319-66263-3_18
10. Kiesl, B., Rebola-Pardo, A., Heule, M.J.H.: Extended resolution simulates DRAT. In: Proceedings of the Automated Reasoning - 9th International Joint Conference, IJCAR 2018, Held as Part of the Federated Logic Conference, FloC 2018, Oxford, UK, 14–17 July 2018, pp. 516–531 (2018). https://doi.org/10.1007/978-3-319-94205-6_34
11. Kleine Büning, H., Bubeck, U.: Theory of quantified Boolean formulas. In: Biere, A., Heule, M., van Maaren, H., Walsh, T. (eds.) Handbook of Satisfiability, Frontiers in Artificial Intelligence and Applications, vol. 185, pp. 735–760. IOS Press (2009)
12. Kleine Büning, H., Karpinski, M., Flögel, A.: Resolution for quantified Boolean formulas. Inf. Comput. **117**(1), 12–18 (1995)
13. Lonsing, F., Egly, U.: QRAT+: generalizing QRAT by a more powerful QBF redundancy property. CoRR abs/1804.02908 (2018). http://arxiv.org/abs/1804.02908

14. Slivovsky, F., Szeider, S.: Variable dependencies and Q-resolution. In: Sinz, C., Egly, U. (eds.) SAT 2014. LNCS, vol. 8561, pp. 269–284. Springer, Cham (2014). https://doi.org/10.1007/978-3-319-09284-3_21
15. Wetzler, N., Heule, M.J.H., Hunt, W.A.: DRAT-trim: efficient checking and trimming using expressive clausal proofs. In: Sinz, C., Egly, U. (eds.) SAT 2014. LNCS, vol. 8561, pp. 422–429. Springer, Cham (2014). https://doi.org/10.1007/978-3-319-09284-3_31

A SAT-Based System for Consistent Query Answering

Akhil A. Dixit[1]([✉]) and Phokion G. Kolaitis[1,2]

[1] University of California Santa Cruz, Santa Cruz, USA
akadixit@ucsc.edu
[2] IBM Research - Almaden, San Jose, USA

Abstract. An inconsistent database is a database that violates one or more integrity constraints, such as functional dependencies. Consistent Query Answering is a rigorous and principled approach to the semantics of queries posed against inconsistent databases. The consistent answers to a query on an inconsistent database is the intersection of the answers to the query on every repair, i.e., on every consistent database that differs from the given inconsistent one in a minimal way. Computing the consistent answers of a fixed conjunctive query on a given inconsistent database can be a coNP-hard problem, even though every fixed conjunctive query is efficiently computable on a given consistent database.

We designed, implemented, and evaluated CAvSAT, a SAT-based system for consistent query answering. CAvSAT leverages a set of natural reductions from the complement of consistent query answering to SAT and to Weighted MaxSAT. The system is capable of handling unions of conjunctive queries and arbitrary denial constraints, which include functional dependencies as a special case. We report results from experiments evaluating CAvSAT on both synthetic and real-world databases. These results provide evidence that a SAT-based approach can give rise to a comprehensive and scalable system for consistent query answering.

1 Introduction

Managing inconsistencies in databases is a challenge that arises in several different contexts. Data cleaning is the main approach towards managing inconsistent databases (see the survey [17]). In data cleaning, clustering techniques and/or domain knowledge are used to resolve violations of integrity constraints in a given inconsistent database, thus producing a single consistent database. This approach, however, is often *ad hoc*; for example, if a person has two different social security numbers in a database, which of the two should be kept?

The framework of database repairs and consistent query answering, introduced by Arenas, Bertossi, and Chomicki [3], is an alternative, and arguably more principled, approach to data cleaning. In contrast to data cleaning, the inconsistent database is left as is; instead, inconsistencies are handled at query time by considering all possible repairs of the inconsistent database, where a *repair* of an inconsistent database I is a consistent database J that differs from

© Springer Nature Switzerland AG 2019
M. Janota and I. Lynce (Eds.): SAT 2019, LNCS 11628, pp. 117–135, 2019.
https://doi.org/10.1007/978-3-030-24258-9_8

I in a "minimal" way. The main algorithmic problem in this framework is to compute the *consistent answers* to a query q on a given database I, that is, the tuples that lie in the intersection of the results of q applied on each repair of I (see the monograph [6]). Computing the consistent answers to a query q on I can be computationally harder than evaluating q on I, because an inconsistent database may have exponentially many repairs. By now there is an extensive literature on the computational complexity of the consistent answers for different classes of constraints and queries [7, 18, 22, 29, 30]. For key constraints (the most common constraints) and for conjunctive queries (the most frequently asked queries), the consistent answers appear to exhibit an intriguing trichotomy, namely, the consistent answers of every fixed conjunctive query under key constraints are either first-order rewritable (hence, polynomial-time computable), or are polynomial-time computable but not first-order rewritable, or are coNP-complete. So far, this trichotomy has been proved for self-join free conjunctive queries by Koutris and Wijsen [21, 22]. Moreover, Koutris and Wijsen designed a quadratic algorithm that, given such a conjunctive query and a set of key constraints, determines the side of the trichotomy in which the consistent answers to the query fall. Prior to this work, Fuxman and Miller identified a class of conjunctive queries, called C_{forest}, whose consistent answers are FO-rewritable [12, 14]. Membership in C_{forest}, however, is sufficient but not necessary condition for the FO-rewritability of the consistent answers.

Several academic prototype systems for consistent query answering have been developed [4, 5, 9, 12, 13, 15, 19, 24, 25]. In particular, the ConQuer system [12, 13] is tailored to queries in the class C_{forest}. Other systems use logic programming [5, 15], compact representations of repairs [8], or reductions to solvers. Specifically, the system in [24] uses reductions to answer set programming, while the EQUIP system in [19] uses reductions to binary integer programming and the subsequent deployment of CPLEX. It is fair to say, however, no comprehensive and scalable system for consistent query answering exists at present; this state of affairs has impeded the broader adoption of the framework of repairs and consistent answers as a principled alternative to data cleaning.

In this paper, we report on a SAT-based system for consistent query answering, which we call CAvSAT (Consistent Answers via SAT). The CAvSAT system leverages natural reductions from the complement of consistent query answering to SAT and to WEIGHTED MAXSAT. As such, it can handle the consistent answers to unions of conjunctive queries under *denial* constraints, a broad class of integrity constraints that include functional dependencies (hence also key constraints) as a special case. CAvSAT is the first SAT-based system for consistent query answering. We carried out a preliminary stand-alone evaluation of CAvSAT on both synthetic and real-world databases. The first set of experiments involved the consistent answers of conjunctive queries under key constraints on synthetic databases in which each relation has up to one million tuples. One of the *a priori* unexpected findings is that, for conjunctive queries whose consistent answers are first-order rewritable, CAvSAT had comparable or even better performance to evaluating the first-order rewritings using a database engine, such as PostgreSQL. The second set of experiments involved the consistent answers of (unions of) conjunctive queries under functional dependencies on restaurant inspection records in Chicago and New York with some of the relations exceeding 200000 tuples. The CAvSAT source code is available at the GitHub repository https://github.com/uccross/cavsat via a BSD-style license.

While much more work remains to be done, the experimental finding reported here provide evidence that a SAT-based approach can indeed give rise to a comprehensive and scalable system for consistent query answering.

2 Basic Notions and Background

Databases, Constraints, and Queries. A *relational database schema* $\mathcal{R}$ is a finite collection of relation symbols, each with a fixed positive integer as its arity. The attributes of a relation symbol are names for its columns; attributes can also be identified by their positions, thus $Attr(R) = \{1, ..., n\}$ denotes the set of attributes of R. An $\mathcal{R}$-*database instance* or, simply, an $\mathcal{R}$-*instance* is a collection I of finite relations R^I, one for each relation symbol R in $\mathcal{R}$. An expression of the form $R^I(a_1, ..., a_n)$ is a *fact* of the instance I if $(a_1, ..., a_n) \in R^I$. Every $\mathcal{R}$-instance can be identified with the (finite) set of its facts. The *active domain* of I is the set of all values occurring in facts of I.

Relational database schemas are often accompanied by a set of integrity constraints that impose semantic restrictions on the allowable instances. A *functional dependency* (FD) $x \rightarrow y$ on a relation symbol R is an integrity constraint asserting that if two facts agree on the attributes in x, then they must also agree on the attributes in y. A *key* is a minimal subset x of $Attr(R)$ such that the FD $x \rightarrow Attr(R)$ holds. In this case, the attributes in x are called *key attributes* of R and they are denoted by underlining their corresponding positions; thus, $R(\underline{A}, \underline{B}, C)$ denotes that the attributes A and B form a key of R. Every functional dependency is expressible in first-order logic. For example, the key constraint $A, B \rightarrow C$ in $R(\underline{A}, \underline{B}, C)$ is expressed by the first-order formula

$$\forall x, y, z, z'(R(x, y, z) \wedge R(x, y, z') \rightarrow z = z')$$

Functional dependencies are an important special case of *denial constraints* (DCs), which are expressible by first-order formulas of the form

$$\forall x_1, ..., x_n \neg(\varphi(x_1, ..., x_n) \wedge \psi(x_1, ..., x_n)),$$

or, equivalently,

$$\forall x_1, ..., x_n(\varphi(x_1, ..., x_n) \rightarrow \neg\psi(x_1, ..., x_n)),$$

where $\varphi(x_1, ..., x_n)$ is a conjunction of atomic formulas and $\psi(x_1, ..., x_n)$ is a conjunction of expressions of the form $(x_i \text{ op } x_j)$ with each op a built-in predicate, such as $=, \neq, <, >, \leq, \geq$. In words, a denial constraint prohibits a set of tuples that satisfy certain conditions from appearing together in a database instance.

Let k be a positive integer. A k-*ary query* on a relational database schema $\mathcal{R}$ is a function q that takes an $\mathcal{R}$-instance I as argument and returns a k-relation

$q(I)$ on the active domain of I as value. A *boolean query* on $\mathcal{R}$ is a function that takes an $\mathcal{R}$-instance I as argument and returns true or false as value. As is well known, first-order logic has been successfully used as a query language. In fact, it forms the core of SQL, the main commercial database query language.

A *conjunctive query* is a first-order formula built using the relational symbols, conjunctions, and existential quantifiers. Thus, each conjunctive query is expressible by a first-order formula of the form

$$q(\boldsymbol{z}) := \exists \boldsymbol{w}\ (R_1(\boldsymbol{x_1}) \wedge ... \wedge R_m(\boldsymbol{x_m})),$$

where each $\boldsymbol{x_i}$ is a tuple consisting of variables and constants, $\boldsymbol{z}$ and $\boldsymbol{w}$ are tuples of variables, and the variables in $\boldsymbol{x_1}, ..., \boldsymbol{x_m}$ appear in exactly one of $\boldsymbol{z}$ and $\boldsymbol{w}$. Clearly, a conjunctive query with k free variables $\boldsymbol{z}$ is a k-ary query, while a conjunctive query with no free variables (i.e., all variables are existentially quantified) is a boolean query. Conjunctive queries are also known as *select-project-join* (SPJ) queries and are among the most frequently asked queries in databases. For example, the binary conjunctive query $q(s,t) := \exists c(\text{Enrolls}(s,c) \wedge \text{Teaches}(t,c))$ returns the set of all pairs (s,t) such that student s is enrolled in a course taught by teacher t, while the boolean conjunctive query $q() := \exists x, y, z(E(x,y) \wedge E(y,z) \wedge E(z,x))$ tests whether or not a graph with an edge relation E contains a triangle.

Repairs and Consistent Answers. Let $\mathcal{R}$ be a database schema and let Σ be a set of integrity constraints on $\mathcal{R}$. An $\mathcal{R}$-instance I is *consistent* if $I \models \Sigma$, that is, I satisfies every constraint in Σ; otherwise, I is *inconsistent*. A *repair* of an inconsistent instance I w.r.t. Σ is a consistent instance J that differs from I in a "minimal" way. Different notions of minimality give rise to different types of repairs (see [6] for a comprehensive survey). Here, we focus on *subset repairs*, the most extensively studied type of repairs. An instance J is a *subset repair* of an instance I if $J \subseteq I$ (where I and J are viewed as sets of facts), $J \models \Sigma$, and there exists no instance J' such that $J' \models \Sigma$ and $J \subset J' \subset I$. From now on, by *repair* we mean a subset repair. Arenas, Bertossi, and Chomicki [3] used repairs to give rigorous semantics to query answering on inconsistent databases. Specifically, assume that q is a query, I is an $\mathcal{R}$-instance, and t is a tuple of values. We say that t is a *consistent answer* (also referred as a *certain answer*) to q on I w.r.t. Σ if $t \in q(J)$, for every repair J of I. We write $\text{CONS}(q, I, \Sigma)$ to denote the set of all *consistent answers* to q on I w.r.t. Σ, i.e.,

$$\text{CONS}(q, I, \Sigma) = \bigcap \{q(J) : J \text{ is a repair of } I \text{ w.r.t. } \Sigma\}.$$

If Σ is a fixed set of integrity constraints and q is a fixed query, then the main computational problem associated with the consistent answers is: given an instance I, compute $\text{CONS}(q, I, \Sigma)$. If q is a boolean query, then computing the certain answers becomes the decision problem $\text{CERTAINTY}(q, \Sigma)$: given an instance I, is q true on every repair J of I w.r.t. Σ? When the constraints Σ are understood from the context, we will write $\text{CONS}(q, I)$ and $\text{CERTAINTY}(q)$, instead of $\text{CONS}(q, I, \Sigma)$ and $\text{CERTAINTY}(q, \Sigma)$.

Computational Complexity of Consistent Answers. If Σ is a fixed finite set of denial constraints and q is a k-ary conjunctive query, where $k \geq 1$, then the following problem is in coNP: given an instance I and a tuple t, is t a certain answer to q on I w.r.t. Σ? This is so because to check that t is not a certain answer to q on I w.r.t. Σ, we guess a repair J of I and verify that $t \notin q(J)$ (note that J is a subset of I, evaluating a fixed conjunctive query on a given database is a polynomial-time task, and testing if J is a repair of I w.r.t. denial constraints is a polynomial-time task as well). Similarly, if q is a boolean conjunctive query, then the decision problem CERTAINTY(q, Σ) is in coNP.

Even for key constraints and boolean conjunctive queries, CERTAINTY(q, Σ) exhibits a variety of behaviors within coNP. Indeed, consider the queries

1. PATH$() := \exists x, y, z\ R(\underline{x}, y) \wedge S(\underline{y}, z)$;
2. CYCLE$() := \exists x, y\ R(\underline{x}, y) \wedge S(\underline{y}, x)$;
3. SINK$() := \exists x, y, z\ R(\underline{x}, z) \wedge S(\underline{y}, z)$.

Fuxman and Miller [14] showed that CERTAINTY(PATH) is FO-rewritable, i.e., there is a first-order definable boolean query q' such that CONS(PATH, Σ, I) = $q'(I)$, for every instance I. In fact, q' is $\exists x, y, z\ R(\underline{x}, y) \wedge S(\underline{y}, z) \wedge \forall y'(R(\underline{x}, y') \rightarrow \exists z' S(\underline{y'}, z'))$. Wijsen [28] showed that CERTAINTY(CYCLE) is in P, but it is not FO-rewritable, while Fuxman and Miller [14] showed that CERTAINTY(SINK) is coNP-complete via a reduction from the complement of MONOTONE 3-SAT.

The preceding state of affairs sparked a series of investigations aiming to obtain classification results concerning the computational complexity of the consistent answers (e.g., see [16, 18, 23, 27, 28]). The most definitive result to date is a *trichotomy* theorem, established by Koutris and Wijsen [20–22], for boolean self-join free conjunctive queries, where a conjunctive query is *self-join free* if no relation symbol occurs more than once in the query. This trichotomy theorem asserts that if q is a self-join free conjunctive query with one key per relation symbol, then CERTAINTY(q) is FO-rewritable, or in P but not FO-rewritable, or coNP-complete. Moreover, there is a quadratic algorithm to decide, given such a query, which of the three cases of the trichotomy holds. It remains an open problem whether or not this trichotomy extends to arbitrary boolean conjunctive queries and to arbitrary functional dependencies or denial constraints.

3   Consistent Query Answering for Key Constraints

In this section, we assume that $\mathcal{R}$ is a database schema and Σ is a finite set of primary key constraints on $\mathcal{R}$, i.e., there is one key constraint per each relation of $\mathcal{R}$. We first consider boolean conjunctive queries and, for each fixed boolean conjunctive query q, we give a natural polynomial-time reduction from CERTAINTY(q) to UNSAT. We then extend this reduction to non-boolean conjunctive queries, so that for every fixed non-boolean conjunctive query q, the consistent answers to q can be computed by iteratively solving WEIGHTED MAXSAT instances. In what follows, we heavily use the notions of *key-equal groups* of facts and *minimal witnesses* to a conjunctive query.

Definition 1 *Key-Equal Group.* Let I be an $\mathcal{R}$-instance. We say that two facts of a relation R of I are *key-equal*, if they agree on the key attributes of R. A set S of facts of I is called a *key-equal group* of facts if every two facts in S are key-equal, and no fact in S is key-equal to some fact in $I \backslash S$.

Definition 2 *Minimal Witness.* Let I be an $\mathcal{R}$-instance and let S be a sub-instance of I. We say that S is a *minimal witness* to a conjunctive query q on I, if $S \models q$, and for every proper subset S' of S, we have that $S' \not\models q$.

For each relation R of I, the key-equal groups of R are computed by an SQL query that involves grouping the key attributes of R. Similarly, the set of minimal witnesses to a fixed conjunctive query q on I are computed efficiently as follows. A unique integer *factID* is attached to each fact, by adding an attribute *FactID* to each relation in I that appears in q. Thus, a new instance I' is built, where each relation $R'(FactID, \underline{A}, B)$ in I' is obtained from a relation $R(\underline{A}, B)$ in I. A new non-boolean query q' is constructed, such that each atom $R'(factID_R, \underline{x}, y)$ in q' is constructed from an atom $R(\underline{x}, y)$ of q. The variables of q' that correspond to the *FactID* attributes are not existentially quantified. It is easy to see that each tuple (without duplicate *factIDs*) in $q'(I')$ is in 1-1 correspondence with a minimal witness to q on I.

Boolean Conjunctive Queries. Let q be a fixed boolean conjunctive query over $\mathcal{R}$.

Reduction 1. *Given an $\mathcal{R}$-instance I, we construct a CNF-formula ϕ as follows.*

For each fact f_i of I, introduce a boolean variable x_i, $1 \leq i \leq n$. Let $\mathcal{G}$ be the set of key-equal groups of facts of I, and let $\mathcal{W}$ be the set of minimal witnesses to q on I.

- *For each $G_j \in \mathcal{G}$, construct the clause $\alpha_j = \bigvee\limits_{f_i \in G_j} x_i$.*
- *For each $W_j \in \mathcal{W}$, construct the clause $\beta_j = \bigvee\limits_{f_i \in W_j} \neg x_i$.*
- *Construct the boolean formula $\phi = \left(\bigwedge\limits_{i=1}^{|\mathcal{G}|} \alpha_i \right) \wedge \left(\bigwedge\limits_{j=1}^{|\mathcal{W}|} \beta_j \right)$.*

Proposition 1. *Let ϕ be the CNF-formula constructed using Reduction 1.*

- *The size of ϕ is polynomial in the size of I.*
- *The formula ϕ is satisfiable if and only if $\mathrm{CERTAINTY}(q, \Sigma)$ is false on I.*

The proofs of all propositions are given in the full version of the paper [11].

Non-Boolean Conjunctive Queries. Let q be a fixed non-boolean query on $\mathcal{R}$, i.e., q has one or more free variables. We extend Reduction 1 to Reduction 2, so that one can reason about the certain answers to q on an $\mathcal{R}$-instance I using the satisfying assignments of the CNF-formula ϕ constructed via Reduction 2.

We use the term *potential answers* to refer to the answers to q on I. If a_l is such a potential answer, we write $q[a_l]$ to denote the boolean conjunctive query obtained from q by replacing the free variables in the body of q by corresponding constants from a_l.

Reduction 2. *Given an $\mathcal{R}$-instance I, we construct a CNF-formula ϕ as follows.*

For each fact f_i of I, introduce a boolean variable x_i, $1 \leq i \leq n$, Let $\mathcal{G}$ be the set of key-equal groups of facts of I and let $\mathcal{A}$ be the set of potential answer to q on I. For each $a_l \in \mathcal{A}$, let $\mathcal{W}^l$ denote the set of minimal witnesses to the boolean query $q[a_l]$ on I. For each $a_l \in \mathcal{A}$, introduce a boolean variable p_1, $1 \leq l \leq ..., |\mathcal{A}|$.

- *For each $G_j \in \mathcal{G}$, construct the clause $\alpha_j = \bigvee\limits_{f_i \in G_j} x_i$.*
- *For each $a_l \in \mathcal{A}$ and for each $W_j^l \in \mathcal{W}^l$, construct the clause*
$$\beta_j^l = \left(\bigvee\limits_{f_i \in W_j^l} \neg x_i \right) \vee \neg p_l .$$

- *Construct the boolean formula $\phi = \left(\bigwedge\limits_{i=1}^{|\mathcal{G}|} \alpha_i \right) \wedge \left(\bigwedge\limits_{l=1}^{|\mathcal{A}|} \left(\bigwedge\limits_{j=1}^{|\mathcal{W}^l|} \beta_j^l \right) \right).$*

Proposition 2. *Let ϕ be the CNF-formula constructed using Reduction 2.*

- *The size of ϕ is polynomial in the size I.*
- *There exists a satisfying assignment to ϕ in which a variable p_l is set to 1 if and only if $a_l \notin \mathrm{CONS}(q, I)$.*

Example 1. Consider the flights information database in Table 1. The database schema has three relations, namely, *Airlines*, *Tickets*, and *Flights*; the key attributes of each relation are underlined. This database is inconsistent, as the sets $\{f_1, f_3\}$ and $\{f_8, f_9\}$ of facts violate the key constraints of the relations *Airlines* and *Flights*, respectively.

Suppose we want to find out the codes of the flights that belong to an airline from Canada and fly to the airport OAK. This can be expressed by the unary conjunctive query $q(x) := Flights(x, y, z, p, \text{'OAK'}, q, r) \wedge Airlines(z, \text{'Canada'})$.

Table 1. Flight information records.

Airlines		
Fact	AIRLINE	COUNTRY
f_1	Southwest	United States
f_2	Jazz Air	Canada
f_3	Southwest	Canada

Tickets				
Fact	PNR	CODE	CLASS	FARE
f_4	MJ9C8R	SWA 1568	Economy	430 USD
f_5	KLF88V	MI 471	First	914 USD
f_6	NJ5RT3	SWA 1568	First	112 USD

Flights							
Fact	CODE	DATE	AIRLINE	FROM	TO	DEPARTURE	ARRIVAL
f_7	JZA 8329	01/29/19	Jazz Air	GEG	OAK	16:12 PST	18:00 PST
f_8	SWA 1568	01/29/19	Silkair	YYZ	YAM	18:55 EST	18:44 EST
f_9	SWA 1568	01/29/19	Southwest	LAX	OAK	16:18 PST	17:25 PST

There are two potential answers to q, namely, 'JZA 8329' and 'SWA 1568', so we introduce their corresponding variables p_1 and p_2. Since the facts f_1 and f_3 form a key-equal group, we construct an α-clause $(x_1 \vee x_3)$. Similarly, since the set $\{f_2, f_7\}$ of facts is a minimal witness to $q['$JZA 8329'$]$, we construct the β-clause $(\neg x_2 \vee \neg x_7 \vee \neg p_1)$.

By continuing this way, we obtain the following CNF-formula ϕ:
$$(x_1 \vee x_3) \wedge x_2 \wedge x_4 \wedge x_5 \wedge x_6 \wedge x_7 \wedge (x_8 \vee x_9) \wedge (\neg x_2 \vee \neg x_7 \vee \neg p_1) \wedge (\neg x_3 \vee \neg x_9 \vee \neg p_2).$$
Clauses x_2, x_7, and $(\neg x_2 \vee \neg x_7 \vee \neg p_1)$ force p_1 to take value 0 in each satisfying assignment of ϕ, because the facts f_2 and f_7 appear in every repair of I, thus making 'JZA 8329' a consistent answer to q. In contrast, there is a satisfying assignment of ϕ in which p_2 is set to 1, which implies that 'SWA 1568' is not a consistent answer to q.

Optimizing the Reductions. In real-life applications, a large part of the inconsistent database is consistent. For a boolean query q, if a minimal witness to q is present in the consistent part of the database instance, then we can immediately conclude that CERTAINTY(q, I, Σ) is true. This can be checked with simple SQL queries that involve grouping on the key attributes of each relation. Similarly, for non-boolean queries, the consistent answers coming from the witnesses that belong to the consistent part of the database can be computed efficiently using SQL queries. All additional consistent answers can then be found using the preceding reduction. In this case, we need to introduce boolean variables corresponding to only those facts that contribute to the additional potential answers. This significantly reduces the size of the CNF-formulas produced by Reductions 1 or 2. This optimization has been used earlier in [19], where CONS(q, I, Σ) was reduced to an instance of binary integer programming.

4   Consistent Query Answering Beyond Key Constraints

In this section, we consider the broader class of denial constraints and the more expressive class of unions of conjunctive queries. Note that computing the consistent answers of unions of conjunctive queries under denial constraints is still in coNP, but the consistent answers of a union $Q := q_1 \cup \ldots \cup q_k$ of conjunctive queries $q_1, \ldots, q_k$ is not, in general, equal to the union of the consistent answers of $q_1, \ldots, q_k$.

We give a polynomial-time reduction from CONS(Q, I, Σ) to UNSAT, where Σ is a fixed finite set of denial constraints and Q is a fixed union of non-boolean conjunctive queries. The potential answers to Q are treated in the same way as the potential answers to the conjunctive query q in Reduction 2; to this effect, we introduce a boolean variable for each potential answer. The reduction we give here relies on the notions of *minimal violations* and *near-violations* to the set of denial constraints that we introduce next.

Definition 3. _Minimal violation._ Assume that Σ is a set of denial constraints, I is an $\mathcal{R}$-instance, and S is a sub-instance of I. We say that S is a _minimal violation_ to Σ, if $S \not\models \Sigma$ and for every set $S' \subset S$, we have that $S' \models \Sigma$.

Definition 4. _Near-violation._ Assume that Σ is a set of denial constraints, I is an $\mathcal{R}$-instance, S is a sub-instance of I, and f is a fact of I. We say that S is a _near-violation_ w.r.t. Σ and f, if $S \models \Sigma$ and $S \cup \{f\}$ is a minimal violation to Σ. As a special case, if $\{f\}$ itself is a minimal violation to Σ, then we say that there is exactly one near-violation w.r.t. f, and it is the singleton $\{f_{true}\}$, where f_{true} is an auxiliary fact.

For a fixed finite set Σ of denial constraints, the set of minimal violations to Σ on a given database instance I are computed as follows. The body of a denial constraint $d \in \Sigma$ is treated as a boolean conjunctive query q_d, possibly containing atomic formulas from d that use built-in predicates such as $=$, $\neq$, $<$, $>$, $\leq$, and $\geq$, in addition to the relation symbols. The set of minimal witnesses to q_d on I is computed as described in Sect. 3, which is also, precisely, the set of minimal violations to d. The union of the sets of minimal violations over all denial constraints in Σ gives us the set of minimal violations to Σ. For each fact $f \in I$, the set of near-violations to Σ w.r.t. f can be obtained by removing f from every minimal violation to Σ that contains f.

Let $\mathcal{R}$ be a database schema, let Σ be a fixed finite set of denial constraints on $\mathcal{R}$, and let $Q := q_1 \cup \ldots \cup q_k$ be a union of conjunctive queries $q_1, \ldots, q_k$. Let I be an $\mathcal{R}$-instance, and let Q be the fixed union of k non-boolean conjunctive queries $q_1, \ldots, q_k$.

Reduction 3. _Given an $\mathcal{R}$-instance I, we construct a boolean formula ϕ' as follows._

   Compute the following sets:
   - $\mathcal{V}$: _the set of minimal violations to Σ on I._
   - $\mathcal{N}^i$: _the set of near-violations to Σ, on I, w.r.t. each fact $f_i \in I$._
   - $\mathcal{A}$: _the set of potential answers to Q on I._
   - $\mathcal{W}^l$: _the set of all minimal witnesses to $Q[a_l]$ on I, for each $a_l \in \mathcal{A}$._

For each fact f_i of I, introduce a boolean variable x_i, $1 \leq i \leq n$. For the auxiliary fact f_{true}, introduce a constant $x_{true} = true$. For each $N_j^i \in \mathcal{N}^i$, introduce a boolean variable y_j^i, and for each $a_l \in \mathcal{A}$, introduce a boolean variable p_l.

1. _For each $V_j \in \mathcal{V}$, construct a clause $\alpha_j = \bigvee\limits_{f_i \in V_j} \neg x_i$._

2. _For each $a_l \in \mathcal{A}$ and for each $W_j^l \in \mathcal{W}^l$, construct a clause_
   $$\beta_j^l = \left(\bigvee_{f_i \in W_j^l} \neg x_i \right) \vee \neg p_l.$$

3. *For each $f_i \in I$, construct a clause $\gamma_i = x_i \vee \left(\bigvee_{N_j^i \in \mathcal{N}^i} y_j^i \right)$.*

4. *For each variable y_j^i, construct an expression $\theta_j^i = y_j^i \leftrightarrow \left(\bigwedge_{f_d \in N_j^i} x_d \right)$.*

5. *Construct the following boolean formula ϕ:*

$$\phi' = \left(\bigwedge_{i=1}^{|\mathcal{V}|} \alpha_i \right) \wedge \left(\bigwedge_{l=1}^{|\mathcal{A}|} \left(\bigwedge_{j=1}^{|\mathcal{W}^l|} \beta_j^l \right) \right) \wedge \left(\bigwedge_{i=1}^{|I|} \left(\left(\bigwedge_{j=1}^{|\mathcal{N}^i|} \theta_j^i \right) \wedge \gamma_i \right) \right)$$

Proposition 3. *Let ϕ' be the boolean formula constructed using Reduction 3.*

- *The formula ϕ' can be transformed to an equivalent CNF-formula ϕ whose size is polynomial in the size of I.*
- *There exists a satisfying assignment to ϕ' in which a variable p_l is set to 1 if and only if $a_l \notin \text{CONS}(Q, I, \Sigma)$.*

Example 2. Consider the database instance from Table 1. In addition to the three key constraints from Example 1, suppose the schema now has two additional integrity constraints: (a) if a flight departs from YYZ, then its airline must be Jazz Air; and (b) for Southwest airlines, if two tickets have the same code, then the ticket with an economy class must have lower fare than the one with the first class. These can be expressed as the following denial constraints:

(a) $\forall x, y, z, w, p, q \ \neg(Flights(x, y, z, \text{`YYZ'}, w, p, q) \wedge z \neq \text{`Jazz Air'})$

(b) $\forall x, y, z, w, p, q \ \neg(Flights(x, y, \text{`Southwest'}, z, w, p, q) \wedge Tickets(r, x, \text{`First'}, t)$
$\wedge \ Tickets(r', x, \text{`Economy'}, t') \wedge t \leq t')$

Let us say that we want to find the PNR numbers of the tickets booked with first class, or with Silkair airlines. This can be expressed as the union $Q := q_1 \cup q_2$ of two unary conjunctive queries, where

$q_1(x) := \exists x, y, z \ Tickets(x, y, \text{`First'}, z)$

$q_2(x) := \exists x, y, z, w, p, q, r, s, t \ Tickets(x, y, z, w) \wedge Flights(y, \text{`Silkair'}, p, q, r, s, t)$

The minimal witnesses to Q, the minimal violations to Σ, and the near-violations to Σ w.r.t. each fact of the database are shown in Fig. 1. With these, we construct the α-, β-, γ-clauses, and the θ-expressions of ϕ, as shown in Fig. 2. Even though, for simplicity, it is not mentioned in Reduction 3, we do the following optimization in practice: if $|N_j^i| = 1$, we do not introduce a variable y_j^i, but, we use the x-variable corresponding to the only fact in N_j^i. In each satisfying assignment to ϕ, the variable p_1 must take the value 0. In contrast, this is not the case for p_2 and p_3. By Proposition 3, 'KLF88V' is a consistent answer to Q, but 'MJ9C8R' and 'NJ5RT3' are not.

Minimal violations to Σ:

- $\{f_1, f_3\}, \{f_8\}, \{f_4, f_6, f_9\}$

Minimal witnesses to Q:

- $\{f_5\}, \{f_6\}, \{f_4, f_8\}$

Near-violations to Σ:

- $f_1 : \{f_3\}$ - $f_8 : \{f_{true}\}$
- $f_3 : \{f_1\}$ - $f_9 : \{f_4, f_6\}$
- $f_4 : \{f_6, f_9\}$ - $f_2, f_5, f_7 :$ None
- $f_6 : \{f_4, f_9\}$

Fig. 1. Minimal violations, minimal witnesses, and near-violations in Example 2.

α-clauses: $(\neg x_1 \vee \neg x_3), (\neg x_8), (\neg x_4 \vee \neg x_6 \vee \neg x_9)$

β-clauses: $(\neg x_5 \vee \neg p_1), (\neg x_6 \vee \neg p_2), (\neg x_4 \vee \neg x_8 \vee \neg p_3)$

γ-clauses: $(x_1 \vee y_1^1), (x_2), (x_3 \vee y_1^3), (x_4 \vee y_1^4), (x_5), (x_6 \vee y_1^6), (x_7), (x_8 \vee y_1^8), (x_9 \vee y_1^9)$

θ-expressions: $(y_1^1 \leftrightarrow x_3), (y_1^3 \leftrightarrow x_1), (y_1^4 \leftrightarrow (x_6 \wedge x_9)), (y_1^6 \leftrightarrow (x_4 \wedge x_9)), (y_1^8 \leftrightarrow x_{true}),$

$(y_1^9 \leftrightarrow (x_4 \wedge x_6))$

Fig. 2. The α-, β-, γ-clauses, and the θ-expressions in Example 2.

5 Computing Consistent Answers via WEIGHTED MAXSAT

By Proposition 1, the consistent answer to a boolean conjunctive query over a schema $\mathcal{R}$ with primary key constraints can be computed by solving the UNSAT instance constructed in Reduction 1. For non-boolean queries, however, in a CNF-formula ϕ constructed using Reduction 2 or 3, one needs to identify each variable p_l such that there exists at least one satisfying assignment to ϕ in which p_l gets set to 1. By Proposition 3, the corresponding potential answers can then be discarded for being inconsistent. One way to do this is as follows. Add a clause $(p_1 \vee ... \vee p_{|A|})$ to ϕ, and solve ϕ using a SAT solver. For each p_l that gets set to 1 in the solution of ϕ, remove the literal p_l from ϕ and then solve ϕ again. Repeat this process until ϕ is no longer satisfiable. At the end of this iterative process, the potential answers corresponding to the p-variables that still occur positively in ϕ are precisely the consistent answers to Q on I. This approach, however, requires many SAT instances to be solved when the number of potential answers is large. For this reason, we developed and tested a different method that uses solving WEIGHTED MAXSAT instances. The construction of these WEIGHTED MAXSAT instances is described in Reduction 4.

Reduction 4. *Let the setup be the same as that of Reduction 2 (or Reduction 3).*

1. *Construct a CNF-formula ϕ using Reduction 2 (or Reduction 3).*
2. *Make all clauses in ϕ hard.*
3. *For each $a_l \in A$, construct a unit ϵ-clause $\epsilon_l = (p_l)$.*
4. *Make all ϵ-clauses soft, and of equal weights.*
5. *Construct the WCNF-formula $\psi = \phi \wedge \left(\bigwedge_{l=1}^{|A|} \epsilon_l \right)$.*

Algorithm 1. Eliminating Inconsistent Potential Answers

1: **procedure** ELIMINATEWITHMAXSAT($\psi, \mathcal{A}$)
2:     **let** ANS = **bool array**$[|\mathcal{A}|]$
3:     **for** $l = 1$ to $|\mathcal{A}|$ **do**
4:         ANS$[l] \leftarrow$ true
5:     **let bool** $moreAnswers \leftarrow$ true
6:     **while** $moreAnswers$ **do**
7:         $moreAnswers \leftarrow$ false
8:         **let** $opt \leftarrow$ MAXSAT(ψ)   ▷ Use WEIGHTED MAXSAT solver
9:         **for** $l = 1$ to $|\mathcal{A}|$ **do**
10:             **if** $opt[p_l] = 1$ **then**
11:                 $moreAnswers \leftarrow$ true
12:                 ANS$[l] \leftarrow$ false
13:                 Remove the unit clause (p_l) from ψ
14:                 Remove all clauses containing the literal $\neg p_l$ from ψ
15:                 Add a new unit hard clause ($\neg p_l$) to ψ
16:     **return** ANS

The preceding Algorithm 1 computes the consistent answers by iteratively solving WEIGHTED MAXSAT instances. It takes as inputs the instance ψ constructed using Reduction 4 and the set $\mathcal{A}$ of potential answers. The idea is to eliminate, in each iteration, as many inconsistent answers from $\mathcal{A}$ as possible by solving ψ. After each iteration, ψ is modified in such a way that additional inconsistent answers, if any, can be eliminated in subsequent iterations. In Sect. 6, we carried out experiments in which it turned out that the number of iterations taken by Algorithm 1 is less than 4, even when there is a large number of potential answers.

Proposition 4. *Algorithm 1 returns an array* ANS *such that* $a_l \in$ CONS(Q, I) *if and only if the entry* ANS$[l]$ *is true.*

6   Preliminary Experimental Results

We evaluated the performance of CAvSAT using two different scenarios. First, we experimented with large synthetically generated databases having primary key constraints. We implemented Reduction 2 without the optimization mentioned in Sect. 3. We also implement Reduction 4 and Algorithm 1. We found out that for seven non-boolean FO-rewritable queries that were also used in [19], CAvSAT significantly outperformed the database evaluation of the FO-rewritings obtained using the algorithm from [22]. We also implemented Reduction 2 with the optimization, and evaluated its performance on fourteen additional conjunctive queries whose consistent answers are coNP-complete or are in P but are not FO-rewitable. In the second scenario, we evaluated the performance of CAvSAT using Reduction 3 on a real-world database with functional dependencies. The definitions of the queries used in the experiments, as well as the FO-rewritings of the first seven queries, can be found in the full version of the paper [11].

Experimental Setup. All experiments were carried out on a machine running on Intel Core i7 2.7 GHz, 64 bit Ubuntu 16.04, with 8 GB of RAM. We used PostgreSQL 10.1 as an underlying DBMS, and MaxHS v3.0 solver [10] for solving the WEIGHTED MAXSAT instances. Our system is implemented in Java 9.04.

6.1 Synthetic Data Generation

The synthetic data were generated in two phases: (a) generation of consistent data; (b) injection of inconsistency into consistent data. The parameters used to generate the data were the number of tuples per relation ($rSize$), degree of inconsistency ($inDeg$), and the size of each key-equal group ($kSize$).

Generating Consistent Data. Each relation in the consistent database was generated with the same number of tuples, so that injecting inconsistency with specified $kSize$ and $inDeg$ will make the total number of tuples in the relation equal to $rSize$. For each query used in the experiment, the data was generated so the evaluation of the query on the consistent database results in a relation that has the size 15% to 20% of $rSize$. The values of the third attribute in all of the ternary relations, were chosen from a uniform distribution in the range $[1, rSize/10]$. This was done to simulate a reasonably large number of potential answers. The remaining attributes take values from randomly generated alphanumeric strings of length 10.

Injecting Inconsistency. In each relation, the inconsistency was injected by inserting new tuples to the consistent data, that share the values of the key attributes with some already existing tuples from the consistent data. The parameter $inDeg$ denotes in percentage the number of tuples per relation, that participate in a key violation. We conducted experiments with the varying values for $inDeg$, ranging from 5% to 15%. The values of $kSize$ were uniformly distributed between 2 to 5. The non-key attributes of the newly injected tuples were uniform random alphanumeric strings of length 10.

6.2 Experimental Results

CAvSAT on FO-rewritable Queries. In this set of experiments, we compare the performance of CAvSAT against the FO-rewritings of seven queries over the database with primary key constraints. For queries $q_1, \ldots, q_7$, we computed the FO-rewritings using the algorithm of Koutris and Wijsen in [22]. We refer to these FO-rewritings as *KW-FO-rewritings*. Since the queries $q_1, \ldots, q_4$ happen to be in the class C_{forest}, we computed additional FO-rewritings for them using the algorithm implemented in the consistent query answering system ConQuer [12]; we refer to these rewritings as *ConQuer-FO-rewritings*. Each FO-rewriting was translated into SQL, and fed to PostgreSQL.

Table 2 shows the size of the WCNF-formulas produced by Reduction 4 without optimization (where Reduction 2 is used inside Reduction 4), on these queries over the databases having one million tuples per relation. Figure 3 shows the evaluation time of CAvSAT with these formulas using MaxHS v3.0 solver [10]. The letter E denotes the time required for encoding the problem into a WEIGHTED MAXSAT instance, and the letter S denotes the time taken by Algorithm 1. The percentage adjacent to the letters S and E denotes the degree of inconsistency. Figure 4 (left) shows the significant gain in performance due to the optimization, for databases of size one million tuples per relation and with 10% inconsistency. This is not surprising; since 90% of the data were consistent, it is expected that most of the consistent answers lie in the consistent part of the database. Table 3 shows the size of the WCNF-formulas produced by Reduction 4, with the optimization in place.

Figure 4 (right) shows that for the queries $q_1, \ldots, q_4$ in the class C_{forest}, the performance of CAvSAT is slightly worse, but comparable, to their ConQuer-FO-rewritings. For all seven queries $q_1, \ldots, q_7$, however, CAvSAT significantly outperformed their KW-FO-rewritings, as PostgreSQL hit the two hours timeout while evaluating each KW-FO-rewriting. In fact, this timeout was hit by all seven queries even for databases of size as small as 100 K tuples per relation. For $q_1, \ldots, q_7$, the average number of iterations taken by Algorithm 1 to eliminate all inconsistent potential answers was 2.85.

Table 2. Size of CNF-formula

Query	Variables	Clauses
q_1	2.08M	2.18M
q_2	2.07M	2.12M
q_3	3.15M	3.07M
q_4	3.23M	3.07M
q_5	2.07M	2.12M
q_6	3.3M	3.06M
q_7	3.25M	3.06M

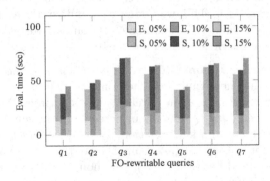

Fig. 3. Evaluation time of CAvSAT without optimization, for 1M tuples/relation.

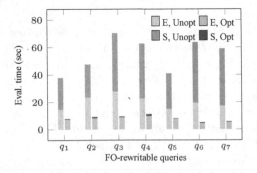
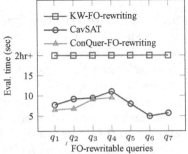

Fig. 4. Evaluation time of CAvSAT with and without optimization (left). Evaluation time of CAvSAT with optimization, in comparison with the KW-FO-rewriting and the ConQuer-FO-rewriting, for 1M tuples/relation with 10% inconsistency (right).

CAvSAT on Harder Queries. In this set of experiments, we considered fourteen additional non-boolean conjunctive queries whose consistent answers are coNP-complete or in P but not FO-rewritable (as mentioned earlier, the definitions of these queries can be found in the full version of the paper). Figure 5 shows that the time required for the optimizing and then constructing the WEIGHTED MAXSAT instance using Reduction 4, dominates over the time taken by Algorithm 1. The solver takes comparatively more time for the queries that have more free variables or more atoms. Table 3 shows the size of the CNF-formulas constructed by Reduction 4 (where Reduction 2 is used inside Reduction 4) in this experiment. The average number of iterations taken by Algorithm 1 to eliminate all inconsistent potential answers to a query was 3.2.

Table 3. The size of the CNF-formulas with optimization, for 1M tuples per relation.

Query	Variables	Clauses	Query	Variables	Clauses	Query	Variables	Clauses
q_1	16.5K	20.9K	q_8	16.6K	16.8K	q_{15}	14.9K	15K
q_2	68.6K	76.0K	q_9	58K	57.7K	q_{16}	58.8K	58.4K
q_3	31.9K	36.8K	q_{10}	31.3K	36.6K	q_{17}	40.1K	41.4K
q_4	117.2K	123.7K	q_{11}	105K	118.1K	q_{18}	107.5K	121.4K
q_5	16.3K	20.6K	q_{12}	116.8K	123.4K	q_{19}	114.4K	120.7K
q_6	32.8K	33.2K	q_{13}	63.2K	65.7K	q_{20}	53.4K	63.7K
q_7	32.5K	33.8K	q_{14}	53.9K	59.2K	q_{21}	170K	199K

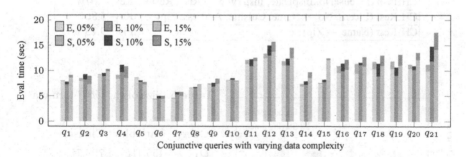

Fig. 5. Evaluation time of CAvSAT for conjunctive queries with varying data complexity, with optimization, over the databases of size 1M tuples/relation.

Results on the Real-World Databases. In this set of experiments, we evaluated the performance of CAvSAT using Reduction 3 on real-world data having key constraints on each relation, along with one functional dependency. The data used are about inspections of food establishments in New York and Chicago, and are taken from [2] and [1]. Part of this data have been previously used for evaluating data cleaning systems, such as HoloClean [26]. Since the structure of the schema or the constraints on the database were not specified by the source, we decomposed the data into four relations, and assumed reasonable key constraints for all relations and also one additional functional dependency, as shown in Table 4. We evaluated the performance of Reduction 3 on six queries (definitions can be found in the full version of the paper [11]). For example, query Q_3 returns the names of the restaurants, such that they are present in both New York and Chicago, and they were inspected on the same day. Figure 6 shows that the solver took the most amount of time to compute answers to this query. Not surprisingly, the evaluation time increases as the number of atoms or the number of free variables in the query grow. Table 5 shows the size of the CNF-formulas produced by Reduction 4 (where Reduction 3 is used inside Reduction 4). No optimization was implemented in this set of experiments.

Table 4. The schema and the constraints of the real-world database.

Relation	# Tuples
NY_Insp (<u>LicenseNo</u>, Risk, InspDate, InspType, Result)	229K
NY_Rest (Name, <u>LicenseNo</u>, Cuisine, Address, Zip)	26.5K
CH_Insp (<u>LicenseNo</u>, Risk, InspDate, InspType, Result)	167K
CH_Rest (Name, <u>LicenseNo</u>, Facility, Address, Zip)	31.1K

Constraint	Type	Violations
NY_Insp (LicenseNo, InspDate, InspType → Risk, Result)	Key	25.6%
NY_Rest (LicenseNo → Name, Cuisine, Address, Zip)	Key	0%
CH_Insp (LicenseNo, InspDate, InspType → Risk, Result)	Key	0.07%
CH_Rest (LicenseNo → Name, Cuisine, Address, Zip)	Key	5.86%
CH_Rest (Name → Zip)	FD	9.73%

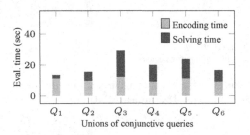

Fig. 6. Evaluation time of CAvSAT on real data.

Table 5. Size of the CNF-formula.

Query	Variables	Clauses
Q_1	455.1K	793.7K
Q_2	456.5K	794K
Q_3	455.1K	671.5K
Q_4	476K	861.5K
Q_5	486.7K	836.2K
Q_6	455.5K	1.12M

7 Concluding Remarks

We designed and implemented CAvSAT, the first SAT-based system for consistent query answering. Our preliminary stand-alone evaluation shows that a SAT-based approach can give rise to a scalable system for consistent query answering. We note that, on queries with first-order rewritable consistent answers, CAvSAT had comparable or even better performance to evaluating the first-order rewritings using a database engine. This finding suggests a potential difference between theory and practice, since the study of first-order rewritability of the consistent answers was motivated from having an efficient evaluation of consistent answers using the database engine alone.

The next step in this investigation is to carry out an extensive comparative evaluation of CAvSAT with other systems for consistent query answering and, in particular, with systems that use reduction-based methods [19,24].

Acknowledgments. Dixit is supported by the Center for Research in Open Source Software (CROSS) at UC Santa Cruz. Kolaitis is supported by NSF Grant IIS:1814152.

References

1. Food Inspections, City of Chicago, August 2011. https://data.cityofchicago.org/Health-Human-Services/Food-Inspections/4ijn-s7e5
2. New York City Restaurant Inspection Results, Department of Health and Mental Hygiene (DOHMH), August 2014. https://data.cityofnewyork.us/Health/DOHMH-New-York-City-Restaurant-Inspection-Results/43nn-pn8j
3. Arenas, M., Bertossi, L., Chomicki, J.: Consistent query answers in inconsistent databases. In: Proceedings of the Eighteenth ACM SIGMOD-SIGACT-SIGART Symposium on Principles of Database Systems, PODS 1999, pp. 68–79. ACM, New York (1999). https://doi.org/10.1145/303976.303983
4. Arenas, M., Bertossi, L.E., Chomicki, J.: Answer sets for consistent query answering in inconsistent databases. TPLP **3**(4–5), 393–424 (2003). https://doi.org/10.1017/S1471068403001832
5. Barceló, P., Bertossi, L.E.: Logic programs for querying inconsistent databases. In: Proceedings Practical Aspects of Declarative Languages, 5th International Symposium, PADL 2003, New Orleans, LA, USA, 13–14 January 2003, pp. 208–222 (2003). https://doi.org/10.1007/3-540-36388-2_15
6. Bertossi, L.E.: Database Repairing and Consistent Query Answering. Synthesis Lectures on Data Management, Morgan & Claypool Publishers (2011). https://doi.org/10.2200/S00379ED1V01Y201108DTM020
7. ten Cate, B., Fontaine, G., Kolaitis, P.G.: On the data complexity of consistent query answering. In: International Conference on Database Theory (ICDT), pp. 22–33 (2012)
8. Chomicki, J., Marcinkowski, J., Staworko, S.: Computing consistent query answers using conflict hypergraphs. In: Proceedings of the Thirteenth ACM International Conference on Information and Knowledge Management, CIKM 2004, pp. 417–426. ACM, New York (2004). https://doi.org/10.1145/1031171.1031254
9. Chomicki, J., Marcinkowski, J., Staworko, S.: Hippo: a system for computing consistent answers to a class of SQL queries. In: Bertino, E., Christodoulakis, S., Plexousakis, D., Christophides, V., Koubarakis, M., Böhm, K., Ferrari, E. (eds.)

EDBT 2004. LNCS, vol. 2992, pp. 841–844. Springer, Heidelberg (2004). https://doi.org/10.1007/978-3-540-24741-8_53

10. Davies, J., Bacchus, F.: Solving MAXSAT by solving a sequence of simpler SAT instances. In: Lee, J. (ed.) CP 2011. LNCS, vol. 6876, pp. 225–239. Springer, Heidelberg (2011). https://doi.org/10.1007/978-3-642-23786-7_19

11. Dixit, A.A., Kolaitis, P.G.: A SAT-based system for consistent query answering. abs/1905.02828 (2019). http://arxiv.org/abs/1905.02828

12. Fuxman, A., Fazli, E., Miller, R.J.: ConQuer: efficient management of inconsistent databases. In: Proceedings of the 2005 ACM SIGMOD International Conference on Management of Data, SIGMOD 2005, pp. 155–166. ACM, New York (2005). https://doi.org/10.1145/1066157.1066176

13. Fuxman, A., Fuxman, D., Miller, R.J.: ConQuer: a system for efficient querying over inconsistent databases. In: Proceedings of the 31st International Conference on Very Large Data Bases, VLDB 2005, pp. 1354–1357. VLDB Endowment (2005). http://dl.acm.org/citation.cfm?id=1083592.1083774

14. Fuxman, A., Miller, R.J.: First-order query rewriting for inconsistent databases. J. Comput. Syst. Sci. **73**(4), 610–635 (2007)

15. Greco, G., Greco, S., Zumpano, E.: A logical framework for querying and repairing inconsistent databases. IEEE Trans. Knowl. Data Eng. **15**(6), 1389–1408 (2003). https://doi.org/10.1109/TKDE.2003.1245280

16. Grieco, L., Lembo, D., Rosati, R., Ruzzi, M.: Consistent query answering under key and exclusion dependencies: algorithms and experiments. In: Proceedings of the 14th ACM International Conference on Information and Knowledge Management, CIKM 2005, pp. 792–799. ACM, New York (2005). https://doi.org/10.1145/1099554.1099742

17. Ilyas, I.F., Chu, X.: Trends in cleaning relational data: consistency and deduplication. Found. Trends Databases **5**(4), 281–393 (2015). https://doi.org/10.1561/1900000045

18. Kolaitis, P.G., Pema, E.: A dichotomy in the complexity of consistent query answering for queries with two atoms. Inf. Process. Lett. **112**(3), 77–85 (2012). https://doi.org/10.1016/j.ipl.2011.10.018

19. Kolaitis, P.G., Pema, E., Tan, W.: Efficient querying of inconsistent databases with binary integer programming. PVLDB **6**(6), 397–408 (2013). https://doi.org/10.14778/2536336.2536341

20. Koutris, P., Wijsen, J.: The data complexity of consistent query answering for self-join-free conjunctive queries under primary key constraints. In: Proceedings of the 34th ACM SIGMOD-SIGACT-SIGAI Symposium on Principles of Database Systems, PODS 2015, pp. 17–29. ACM, New York (2015). https://doi.org/10.1145/2745754.2745769

21. Koutris, P., Wijsen, J.: Consistent query answering for primary keys. SIGMOD Rec. **45**(1), 15–22 (2016). https://doi.org/10.1145/2949741.2949746

22. Koutris, P., Wijsen, J.: Consistent query answering for self-join-free conjunctive queries under primary key constraints. ACM Trans. Database Syst. **42**(2), 9:1–9:45 (2017). https://doi.org/10.1145/3068334

23. Lembo, D., Rosati, R., Ruzzi, M.: On the first-order reducibility of unions of conjunctive queries over inconsistent databases. In: Grust, T., et al. (eds.) EDBT 2006. LNCS, vol. 4254, pp. 358–374. Springer, Heidelberg (2006). https://doi.org/10.1007/11896548_28

24. Manna, M., Ricca, F., Terracina, G.: Consistent query answering via ASP from different perspectives: theory and practice. CoRR abs/1107.4570 (2011). http://arxiv.org/abs/1107.4570

25. Marileo, M.C., Bertossi, L.E.: The consistency extractor system: answer set programs for consistent query answering in databases. Data Knowl. Eng. **69**(6), 545–572 (2010). https://doi.org/10.1016/j.datak.2010.01.005
26. Rekatsinas, T., Chu, X., Ilyas, I.F., Ré, C.: HoloClean: holistic data repairs with probabilistic inference. Proc. VLDB Endow. **10**(11), 1190–1201 (2017). https://doi.org/10.14778/3137628.3137631
27. Wijsen, J.: Consistent query answering under primary keys: a characterization of tractable queries. In: Proceedings of the 12th International Conference on Database Theory, ICDT 2009, pp. 42–52. ACM, New York (2009). https://doi.org/10.1145/1514894.1514900
28. Wijsen, J.: A remark on the complexity of consistent conjunctive query answering under primary key violations. Inf. Process. Lett. **110**(21), 950–955 (2010). https://doi.org/10.1016/j.ipl.2010.07.021
29. Wijsen, J.: Certain conjunctive query answering in first-order logic. ACM Trans. Database Syst. **37**(2), 9:1–9:35 (2012). https://doi.org/10.1145/2188349.2188351
30. Wijsen, J.: Charting the tractability frontier of certain conjunctive query answering. In: Proceedings of the 32nd ACM SIGMOD-SIGACT-SIGAI Symposium on Principles of Database Systems, PODS 2013, pp. 189–200. ACM, New York (2013). https://doi.org/10.1145/2463664.2463666

Incremental Inprocessing in SAT Solving

Katalin Fazekas[1(✉)], Armin Biere[1], and Christoph Scholl[2]

[1] Johannes Kepler University, Linz, Austria
{katalin.fazekas,armin.biere}@jku.at
[2] Albert–Ludwigs–University, Freiburg, Germany
scholl@informatik.uni-freiburg.de

Abstract. Incremental SAT is about solving a sequence of related SAT problems efficiently. It makes use of already learned information to avoid repeating redundant work. Also preprocessing and inprocessing are considered to be crucial. Our calculus uses the most general redundancy property and extends existing inprocessing rules to incremental SAT solving. It allows to automatically reverse earlier simplification steps, which are inconsistent with literals in new incrementally added clauses. Our approach to incremental SAT solving not only simplifies the use of inprocessing but also substantially improves solving time.

1 Introduction

Solving a sequence of related SAT problems incrementally [1–4] is crucial for the efficiency of SAT based model checking [5–8], and important in many domains [9–12]. Utilizing the effort already spent on a SAT problem by keeping learned information (such as variable scores and learned clauses) can significantly speed-up solving similar problems. Equally important are formula simplification techniques such as variable elimination, subsumption, self-subsuming resolution, and equivalence reasoning [13–16].

These simplifications are not only applied before the problem solving starts (*preprocessing*), but also periodically during the actual search (*inprocessing*) [17]. In this paper we focus on how to efficiently combine simplification techniques with incremental SAT solving.

Consider the SAT problem $F^0 = (a \lor b) \land (\neg a \lor \neg b)$. Both clauses are redundant and can be eliminated by for instance variable or blocked clause elimination [14, 16]. The resulting empty set of clauses is of course satisfiable and the SAT solver could for example simply just assign *false* to both variable as a solution. That is of course not a satisfying assignment of F^0, but can be transformed into one by solution reconstruction [17,18], taking eliminated clauses into account. As we will see later, this would set the truth value of either a or b to true.

Now consider the SAT problem $F^1 = (a \lor b) \land (\neg a \lor \neg b) \land (\neg a) \land (\neg b)$ which is actually an extension of F^0 with the clauses $(\neg a)$ and $(\neg b)$. Simply adding them to our simplified F^0 (i.e. to the empty set of clauses) would result in a formula that again is satisfied by assigning false to each variable. However, using solution

© Springer Nature Switzerland AG 2019
M. Janota and I. Lynce (Eds.): SAT 2019, LNCS 11628, pp. 136–154, 2019.
https://doi.org/10.1007/978-3-030-24258-9_9

reconstruction on that assignment leads to the same solution as before, one that satisfies $(a \lor b)$, and thus would actually falsify $(\neg a)$ or $(\neg b)$. The solver would incorrectly report that F^1 is satisfiable, and even return an invalid solution. Thus naively using inprocessing in an incremental setting is not sound.

Obviously one can just give up on incrementality and simply solve F^1 from scratch but with pre- and inprocessing. Another trivial approach is to use learned information from solving F^0, but then disable inprocessing. A compromise is to disallow inprocessing partially by *freezing* [7] those variables that are not allowed to be involved in simplifications ("Don't Touch" variables in [8]). This is rather error-prone and cumbersome for the user, and even often impossible [19].

Our approach benefits from most inprocessing techniques, without freezing any variables. It identifies potential problems between an eliminated clause, such as $(a \lor b)$ in the example, and new clauses, such as $(\neg a)$ and $(\neg b)$. In such a case it moves back the eliminated clause to the formula before adding the new clauses. This greatly simplifies the way how incremental SAT solvers can be used.

The specialized approach in [19] focuses on three preprocessing techniques (variable elimination, subsumption and self-subsumption of [14]). It applies a preprocessing phase before each incremental SAT call. Instead of that, we adapt and extend the framework of [17] and present a generic calculus which allows to combine a much broader set of pre- and inprocessing techniques with incremental SAT solving. Actually, we use the *most general redundancy property* [20, 21] that covers not only all techniques in [19], but also provides optimized procedures for equivalence literal reasoning [13] and even blocked clause elimination [16]. However, we do not yet support techniques that remove models, such as blocked clause addition [17, 22–24] (neither does [19]).

Our approach is also more precise than [19] since it allows to distinguish simplification steps applied on different phases of variables, i.e. we provide a literal- and not just variable-based approach. On the practical side, beyond enabling a wider range of pre- and inprocessing techniques, we present a simple algorithm, which yields an efficient implementation as confirmed by our experiments. Using dedicated techniques for inprocessing under assumptions, as [25] extends [19] based on [26], is orthogonal to the approach presented in this paper.

After preliminaries we present our new rules for incremental SAT solving in Sect. 3 which are proven correct in Sect. 4. We discuss implementation details in Sect. 5 followed by experimental results in Sect. 6 before we conclude in Sect. 7.

2   Preliminaries

Satisfiability. A *literal* is either a Boolean variable (v), or its negation $(\neg v)$. A *clause* is a disjunction of literals, and a *formula* in conjunctive normal form (CNF) is a conjunction of clauses. If convenient, we consider a clause as a set of literals and a formula as a set of clauses. A (partial) *truth assignment* τ is a consistent set of literals assigning truth values to variables as follows. In case $v \in \tau$, then v is assigned *true* by τ (denoted as $\tau(v) = \top$), while if $\neg v \in \tau$, then v is assigned *false* $(\tau(v) = \bot)$. A truth assignment *satisfies* a literal ℓ (denoted as $\tau(\ell) = \top$) if $\ell \in \tau$ and it *falsifies* it (denoted as $\tau(\ell) = \bot$) if $\neg\ell \in \tau$, where $\neg\ell = \neg v$ if $\ell = v$ and $\neg\ell = v$ if $\ell = \neg v$. Neither satisfied nor falsified literals are *undefined*. A clause is a *tautology* if it contains both a literal and the negation of it. The application of a truth assignment τ to an arbitrary formula F, denoted as $\tau(F)$ or $F|_\tau$, is defined as usual. When it is convenient, we will use sets of literals directly as truth assignments. We further use $\tau_1 \circ \tau_2$ to denote the composition of truth assignments τ_1 and τ_2 in the natural way, i.e., $(\tau_1 \circ \tau_2)(F) = \tau_1(\tau_2(F))$.

The *satisfiability problem* (SAT) for a CNF asks whether there is a truth assignment such that all clauses contain at least one satisfied literal. A truth assignment satisfying a formula is also called a *model*. Formulas F_1, F_2 are *logically equivalent*, denoted as $F_1 \equiv F_2$, if they are satisfied by exactly the same truth assignments, while they are *satisfiability equivalent*, denoted as $F_1 \equiv_{sat} F_2$, if both of them are satisfiable or both of them are unsatisfiable.

Incremental SAT Problems. An incremental SAT problem $\mathcal{F}$ is a sequence of clause sets $\langle \Delta_0, \dots, \Delta_n \rangle$. In phase $i = 0, \dots, n$ the task is to determine the satisfiability of $F^i = \wedge_{s=0\dots i}\Delta_s$, the conjunction of all added clauses up to this point. If F^i is unsatisfiable, then F^j for all $j > i$ is unsatisfiable as well, as each iteration just augments the set of clauses. The focus of this paper is on the case where F^i is satisfiable. We rely on the common approach to always choose the given assumptions, literals that are assumed to be true in a phase, as first decisions during search and thus w.l.o.g. do not need to consider assumptions in this paper explicitly. See Minisat [3] for implementation details or [27,28] for abstract solvers following that approach. However, the variables of assumptions are not allowed to be eliminated or occur in witnesses (e.g., as blocking literal in blocked clauses [16]), i.e., they have to be considered *frozen* [7,8] internally.

Example 1. Consider the incremental SAT problem $\mathcal{F} = \langle \{(a \vee b)\}, \{a, b\} \rangle$. It consists of two SAT queries: $F^0 = (a \vee b)$ and $F^1 = F^0 \wedge a \wedge b = (a \vee b) \wedge a \wedge b$.

Redundancy in SAT. Inprocessing in SAT solving relies on the concept of adding and removing *redundant* clauses. To simplify matters, in this paper we use the most general redundancy notion [20,21]. It covers most techniques used in current SAT solvers including resolution asymmetric tautology (RAT), which was used in the original work on inprocessing [17]. As [20,21] points out, any clause redundancy can produce a "witness", e.g. a blocking literal in case of a blocked clause, which allows polynomial solution reconstruction. The following two essential definitions are adapted from [20,21]:

Definition 1 (Witness Labelled Clause). *A set of literals ω and a clause C such that $\omega \cap C \neq \emptyset$ is called a witness labelled clause and written as $(\omega : C)$.*

A witness is a set of literals and can be interpreted as a partial truth assignment. With this interpretation, the truth assignment α which falsifies a given clause C but is undefined otherwise is also written as $\alpha = \neg C$.

Definition 2 (Clause Redundancy). *A witness labelled clause $(\omega : C)$ is redundant with respect to a formula F if $\omega(C) = \top$ and $F|_\alpha \models F|_\omega$ for $\alpha = \neg C$. This is also denoted as $F \wedge C \equiv_{sat}^\omega F$.*

As has been shown in [20,21], this is the most general notion of redundancy and allows to simulate all other types of clause redundancy. The corresponding proof (of Theorem 1 in [20,21]) allows to "fix" an assignment using the witness. We formalize that part of the proof and extend it to *partial* truth assignments, which allows to use partial truth assignments in the witness reconstruction process satisfying only the simplified formula and is further useful to produce a partial satisfying assignment after reconstruction (used for instance in [29]).

Proposition 1. *Assume* $F \wedge C \equiv^{\omega}_{sat} F$ *as above. Let* τ *be a (partial) truth assignment with* $\tau(F) = \top$ *and* $\tau(C) \neq \top$. *Then* $\gamma(F \wedge C) = \top$ *with* $\gamma = \tau \circ \omega$.

Proof. Clearly $\gamma(C) = \omega(C) = \top$. We need to show $\gamma(D) = \top$ for all $D \in F$. Observe $\alpha \circ \tau = \tau \circ \alpha$ with $\alpha = \neg C$ since $\tau(C) \neq \top$ and α and τ are consistent. Thus, $\top = \tau(F) = (\alpha \circ \tau)(F) = (\tau \circ \alpha)(F) = (\tau \circ \alpha)(F \wedge \neg C) = \tau(F|_\alpha)$ since $\alpha(\neg C) = \top$. Using $F|_\alpha \models F|_\omega$ and because $F|_\omega$ remains satisfied for all extensions of τ, we get $\top = (\beta \circ \tau)(F|_\omega) = (\beta \circ \tau \circ \omega)(F) = (\beta \circ \gamma)(F)$, where $\beta = \neg D$ is the truth assignment falsifying the clause $D \in F$, which in particular gives $(\beta \circ \gamma)(D) = \top$. Since $\beta(D) = \bot$ we obtain $\top = (\beta \circ \gamma)(D) = \gamma(D)$.   □

Inprocessing. Our goal is to adjust and extend the abstract framework of [17] such that incremental SAT solving with inprocessing can be handled. The derivation performed by an inprocessing SAT solver is modelled as a sequence of abstract states. Each state consist of three components: a set of irredundant clauses φ that the solver aims to satisfy, a set of redundant clauses ρ that can be removed without changing the satisfiability of the formula under consideration and an ordered sequence of witness labelled clauses σ (that are actually just literal-clause pairs in [17]), to keep track of eliminated clauses.

To make the paper more self contained Fig. 1 lists the original rules of [17], together with the proposed RAT instantiation of side conditions ♯ and ♭. Rule STRENGTHEN strengthens the irredundant set of clauses, by moving a clause from the redundant set into it, while rule FORGET allows to eliminate a redundant clause from ρ. Rule LEARN introduces a new clause C into the redundant set of clauses in case C has RAT w.r.t. $\varphi \wedge \rho$. Rule WEAKEN simplifies the irredundant set by eliminating a clause C from it if C has RAT on a literal l of C w.r.t. φ. The eliminated clause is moved to the end of the literal-clause pair sequence σ.

$$\frac{\varphi\,[\rho]\,\sigma}{\varphi\,[\rho \wedge C]\,\sigma}\;♯ \qquad \frac{\varphi\,[\rho \wedge C]\,\sigma}{\varphi\,[\rho]\,\sigma} \qquad \frac{\varphi\,[\rho \wedge C]\,\sigma}{\varphi \wedge C\,[\rho]\,\sigma} \qquad \frac{\varphi \wedge C\,[\rho]\,\sigma}{\varphi\,[\rho \wedge C]\,\sigma \cdot (l:C)}\;♭$$
$$\quad\text{LEARN} \qquad\qquad \text{FORGET} \qquad\quad \text{STRENGTHEN} \qquad\quad \text{WEAKEN}$$

where ♯ is "C has RAT w.r.t. $\varphi \wedge \rho$" and ♭ is "C has RAT on l w.r.t. φ".

Fig. 1. Instantiated (with RAT) inprocessing rules as introduced in [17]

Model Reconstruction. One challenge of using inprocessing is to guarantee that a satisfying assignment of the final formula can be transformed to a satisfying assignment of the original, non-processed formula. A sequence of witness labelled clauses σ is used as part of the abstract state to keep track of clauses eliminated by WEAKEN during inprocessing. The process of solution reconstruction described through pseudo code in [17] can be formalized as follows:

Definition 3 (Reconstruction Function). *Given a truth assignment τ and a sequence of witness labelled clauses σ, the reconstruction function is defined as*

$$\mathcal{R}(\tau, \varepsilon) = \tau, \qquad \mathcal{R}(\tau, \sigma \cdot (\omega : D)) = \begin{cases} \mathcal{R}(\tau, \sigma) & \text{if } \tau(D) = \top \\ \mathcal{R}((\tau \circ \omega), \sigma) & \text{otherwise.} \end{cases}$$

The reconstruction function takes a (partial) truth assignment τ and a sequence of witness labelled clauses σ as inputs. It traverses σ in reverse order and sets truth values of those literals in τ to true that are witnesses of not yet satisfied clauses in σ. We are now ready to formalize the central concept of this paper:

Definition 4 (Reconstruction Property). *A sequence of witness labelled clauses σ satisfies the reconstruction property w.r.t. a formula F iff for all truth assignments τ satisfying F, the result of the reconstruction function $\mathcal{R}$ on τ and σ is a satisfying assignment for $F \wedge \sigma$. An abstract state $\varphi\,[\rho]\,\sigma$ satisfies the reconstruction property iff σ satisfies the reconstruction property w.r.t. φ.*

For the expression $F \wedge \sigma$ in this definition we interpret σ as a set of its clauses.

3   Inprocessing Rules for Incremental Solving

Our first goal is to determine how information, such as learned clauses, can be transferred from one incremental solving phase to the next, utilizing that the sub-problem F^{i+1} is an extension of the previously solved sub-problem F^i. Thus, instead of solving F^{i+1} from scratch, previously learned facts are reused to avoid repeated work. This is sound if the incremental approach gives the same answer (satisfiable or unsatisfiable) as solving from scratch.

More formally, it is crucial that $\varphi^i_{k_i} \wedge \Delta_{i+1} \equiv_{sat} F^{i+1}$ holds, where $\varphi^i_{k_i}$ is the set of irredundant clauses at the end of the evaluation of F^i. We also need to make sure that $F^{i+1} \models \rho^i_{k_i}$, i.e. the redundant clause set at the end of the evaluation of F^i can be reused. Furthermore, we need to guarantee that a model for F^{i+1} can be resconstructed from any satisfying assignment of $\varphi^{i+1}_{k_{i+1}}$.

To establish notation and to emphasize what we would like to improve in this paper, we briefly describe how inprocessing in a non-incremental solver (as in e.g. [30] with cloning) would look like using only the original inprocessing rules of [17] (shown in Fig. 1). Each phase $i = 0, \dots, n$ of solving an incremental problem $\mathcal{F}$ consists of a derivation of a formula $\varphi^i_{k_i} \wedge \rho^i_{k_i}$ as a sequence of states $\langle \varphi^i_0\,[\rho^i_0]\,\sigma^i_0, \dots, \varphi^i_{k_i}\,[\rho^i_{k_i}]\,\sigma^i_{k_i} \rangle$, where (for all $j = 1, \dots, k_i$)

(a) $\varphi_0^0 = F^0$, $\rho_0^0 = \emptyset$, $\sigma_0^0 = \varepsilon$
(b) $\varphi_{j-1}^i [\rho_{j-1}^i] \sigma_{j-1}^i$ results in $\varphi_j^i [\rho_j^i] \sigma_j^i$ as application of a rule in Fig. 1
(c) $\varphi_0^{i+1} = F^{i+1}$, $\rho_0^{i+1} = \emptyset$, $\sigma_0^{i+1} = \varepsilon$.

The *initial state* defined in (a) starts the derivation with F^0 as irredundant set of clauses, with an empty σ and without any redundant clause. Then following (b) the solver applies the rules of Fig. 1 until it reaches a state $\varphi_{k_i}^i [\rho_{k_i}^i] \sigma_{k_i}^i$ in which satisfiability of $\varphi_{k_i}^i \wedge \rho_{k_i}^i$ is determined. The new phase starts by adding a set of clauses to the problem, as described by (c). Such a derivation only relies on the original rules of [17], so each phase has to restart with completely empty ρ and σ and no information learned from solving F^i can be reused to solve F^{i+1}.

To capture inprocessing in an incremental solver we have to extend and modify the calculus of [17] (in Fig. 1). The initial state in (a) and the components of abstract states remain the same as in [17] (see Sect. 2), except that σ is more general. In our new calculus it consists of witness labelled clauses instead of literal-clause pairs, which allows to capture any redundancy property (not just RAT). We will refer on that component of a state as *reconstruction stack*.

Next sections describe the derivations of $\varphi_{j+1}^i [\rho_{j+1}^i] \sigma_{j+1}^i$ from $\varphi_j^i [\rho_j^i] \sigma_j^i$ for each $0 \leq j < k_i$ in each phase $i = 0, \ldots, n$ and show a sound way to start a new phase $i + 1$ from state $\varphi_{k_i}^i [\rho_{k_i}^i] \sigma_{k_i}^i$ when adding Δ_{i+1} to $\varphi_{k_i}^i$.

3.1   Constrained Learning

The side condition $\boxed{\sharp}$ of rule LEARN in Fig. 1 allows to learn clauses that remove models of the current formula. However, as the following example demonstrates, this is not correct in the context of incremental solving.

Example 2. Consider the incremental SAT problem $\mathcal{F} = \langle \{(a \vee b)\}, \{a, b\} \rangle$. First in phase $i = 0$ the evaluation of F^0 starts from the initial state $(a \vee b) [\emptyset] \varepsilon$. Now the clause $(\neg a \vee \neg b)$ can be learned since it has the RAT property w.r.t. $(a \vee b)$ (this is $\boxed{\sharp}$ in Fig. 1). Then, rule STRENGTHEN can be applied on $(\neg a \vee \neg b)$ which yields state $(a \vee b) \wedge (\neg a \vee \neg b) [\emptyset] \varepsilon$, with a satisfiable set of irredundant clauses. In the next phase $i = 1$ we add Δ_1 and target to solve the formula $F^1 = (a \vee b) \wedge a \wedge b$, which still is satisfiable. However, conjoining Δ_1 to the irredundant clause set of the last state of the previous phase leads to the state $(a \vee b) \wedge (\neg a \vee \neg b) \wedge a \wedge b [\emptyset] \varepsilon$ with an unsatisfiable irredundant clause set.

Thus in our calculus the precondition of learning (LEARN⁻) is $\varphi \wedge \rho \models C$, i.e. we allow to learn only *implied* clauses. Compared to [17] our new rule LEARN⁻ is weaker due to this stronger side condition. It still covers most learning techniques in current SAT solvers, except forms of extended resolution such as blocked clause addition [17,22–24]. Learned clauses can be forgotten (FORGET) or moved to the irredundant formula (STRENGTHEN) as in [17].

3.2 Stronger Weakenings

We decompose the original weakening rule (WEAKEN in Fig. 1) of [17] into two rules: WEAKEN$^+$ and DROP (see Fig. 2). WEAKEN$^+$, as the name suggests, weakens the current formula by eliminating a clause C from the irredundant set while pushing it to the reconstruction stack. The DROP rule allows to weaken the current formula by eliminating an implied clause from the irredundant set. Removal of implied clauses from φ does not introduce (nor remove) models and so it is not necessary to save these clauses on the reconstruction stack. In our implementation the DROP rule is also used for more advanced equivalence-literal reasoning techniques [13,31,32]. Further, in current implementations weakening is always immediately followed by a forget step (simulating WEAKEN$^+$).

$$\frac{\varphi \wedge C\,[\rho]\,\sigma}{\varphi\,[\rho]\,\sigma \cdot (\omega : C)}\;\boxed{\flat} \qquad\qquad \frac{\varphi \wedge C\,[\rho]\,\sigma}{\varphi\,[\rho]\,\sigma}\;\boxed{\varnothing}$$

$$\text{WEAKEN}^+ \qquad\qquad\qquad \text{DROP}$$

where $\boxed{\flat}$ is $\varphi \wedge C \equiv^\omega_{sat} \varphi$ and $\boxed{\varnothing}$ is $\varphi \models C$

Fig. 2. New weakening and dropping rules

3.3 Incremental Clause Addition

The main feature of incremental SAT solving is the possibility to extend the previously solved formula with a set of new clauses. In non-incremental SAT solving, clauses determined to be redundant, always remain redundant. In incremental SAT solving arbitrary clauses can be added and thus previous simplifications might need to be reconsidered and potentially reversed.

Example 3. Consider the incremental SAT problem $\mathcal{F} = \langle \{F^0\}, \{(\neg a \vee b)\}\rangle$, where $F^0 = (a \vee b) \wedge (\neg a \vee \neg b) \wedge (a \vee \neg b)$ and $F^1 = F^0 \wedge (\neg a \vee b)$. Phase $i = 0$ starts from the state $(a \vee b) \wedge (\neg a \vee \neg b) \wedge (a \vee \neg b)\,[\emptyset]\,\varepsilon$. Resolving the first clause on a always produces tautological resolvents (i.e. it is blocked [16]). Thus WEAKEN$^+$ can be applied with witness a. Afterwards no other irredundant clause contains literal b and so both remaining irredundant clauses are blocked on $\neg b$. Thus they can be eliminated by WEAKEN$^+$ too, which results in state $\emptyset\,[\emptyset]\,(a : (a \vee b)) \cdot (\neg b : (\neg a \vee \neg b)) \cdot (\neg b : (a \vee \neg b))$, without irredundant clauses left, and the solver concludes F^0 to be satisfiable. Adding the new clause $(\neg a \vee b)$ to incrementally solve F^1 yields a state with a satisfiable set of irredundant clauses. But F^1 is actually unsatisfiable, so just adding $(\neg a \vee b)$ is not sound.

There are different ways to avoid unsoundness. An obvious way is to simply disallow simplifications over variables (or actually literals in our calculus) that might occur in later phases. In essence, this is the solution implemented through freezing in current SAT solvers [7], which ensures that the reconstruction stack does not contain frozen variables as witnesses. These frozen variables are then the only variables of the current formula that are allowed to reoccur in new clauses. We capture this property as follows.

Definition 5 (Clean Clause). *A clause C is clean w.r.t. a sequence of witness labelled clauses σ iff for all $(\omega : D) \in \sigma$ we have that $\neg C \cap \omega = \emptyset$.*

Example 4. The clause $(a \vee b)$ is not clean w.r.t. $(\neg b : (\neg a \vee \neg b)) \cdot (\neg b : (a \vee \neg b))$ because $\neg(a \vee b) \cap (\neg b) \neq \emptyset$. On the other hand, $(\neg a \vee \neg b)$ is clean w.r.t. the witness labelled clause sequence $(\neg b : (a \vee \neg b))$ since $\neg(\neg a \vee \neg b) \cap (\neg b) = \emptyset$.

$$\frac{\varphi\,[\rho]\,\sigma}{\varphi \wedge \Delta\,[\rho]\,\sigma}\;\boxed{\mathcal{I}}$$

ADDCLAUSES

where $\boxed{\mathcal{I}}$ is that each clause of Δ is clean w.r.t. σ

Fig. 3. New rule to capture clause set augmentation

With this definition the freezing approach guarantees that every added clause is clean w.r.t. the reconstruction stack. Building on that observation, we can now introduce clause addition (ADDCLAUSES in Fig. 3), where the side condition requires that each new clause in Δ is clean w.r.t. the reconstruction stack σ. If the added clauses are clean w.r.t. the reconstruction stack, then every assignment satisfying them will remain satisfying after applying the reconstruction function:

Lemma 1. *If a clause C is clean w.r.t. a sequence of witness labelled clauses σ, then for all truth assignments τ with $\tau(C) = \top$ we have that $\mathcal{R}(\tau, \sigma)(C) = \top$.*

Proof. By induction on the length of σ. The base case $\sigma = \varepsilon$ is trivial. Now consider $\sigma \cdot (\omega : D)$ and $\tau' = \tau$ if $\tau(D) = \top$, $\tau' = \tau \circ \omega$ otherwise. If $\tau(D) = \top$, then $\tau'(C) = \tau(C) = \top$. For $\tau(D) \neq \top$ there is $\ell \in C$ with $\tau(\ell) = \top$. As C is clean w.r.t. $(\omega : D)$, i.e., $\neg C \cap \omega = \emptyset$, we have $\neg \ell \notin \omega$ and so $\tau'(\ell) = (\tau \circ \omega)(\ell) = \tau(\ell) = \top$. This also holds if $\ell \in \omega$, since then $\omega(\ell) = \top$. Now it follows by induction applied to τ' and σ: $\top = \mathcal{R}(\tau', \sigma)(C) = \mathcal{R}(\tau, \sigma \cdot (\omega : D))(C)$. $\quad\square$

Thus, as long as all our clause elimination steps are based on witnesses that never occur in new clauses, we can add clauses without any problem in new incremental calls. However, this approach requires to know in advance in every phase i every literal of every Δ_j with $j > i$. Beyond that, it allows less clauses to be eliminated. Fortunately we can do better.

Instead of prohibiting simplifications, we allow arbitrary inprocessing as in a non-incremental SAT solver, but later reverse simplifications inconsistent with new clauses. It would be easy to just reverse all simplifications by reintroducing all eliminated clauses, but this is costly (as our experiments show). Therefore, it would be desirable to reverse a minimal subset of simplifications, but such a minimal set is in general difficult to identify.

As compromise we try to cheaply identify a sufficient subset of problematic simplifications as follows. If a new clause is not clean w.r.t. the reconstruction stack, we reverse those simplifications which have a negated literal of the new clause in the witness. Reversing all these problematic steps yields a clean reconstruction stack for all new clauses that in turn allows to apply rule ADDCLAUSES.

3.4 Reversing Weakening

The side condition of rule ADDCLAUSES identifies which simplifications need to be reversed in order to add a set of new clauses to the formula. What is missing is a rule to actually reverse these steps. The challenge with reversing clause eliminations is that many simplification steps are dependent on each other, e.g., in F^0 of Example 3 the last two clauses became blocked only after the first simplification step. Therefore one can not just arbitrarily reverse simplifications:

Example 5. Consider again the inprocessing of F^0 in Example 3, with the final state $\emptyset \, [\emptyset] \, (a : (a \vee b)) \cdot (\neg b : (\neg a \vee \neg b)) \cdot (\neg b : (a \vee \neg b))$. Assume we reverse the first simplification step, i.e. we move $(a \vee b)$ from the reconstruction stack to the irredundant clauses. The truth assignment $\tau = \{\neg a, b\}$ would satisfy the irredundant clauses of the resulting state $(a \vee b) \, [\emptyset] \, (\neg b : (\neg a \vee \neg b)) \cdot (\neg b : (a \vee \neg b))$. The reconstruction function on that assignment and the current stack would be $\mathcal{R}(\tau, (\neg b : (\neg a \vee \neg b)) \cdot (\neg b : (a \vee \neg b)))$. Since $\tau(a \vee \neg b) \neq \top$, it first updates τ with the witness of that clause and becomes $\tau' = (\tau \circ \{\neg b\}) = \{\neg a, \neg b\}$. Then, τ' satisfies the next clause of the stack and so $\mathcal{R}(\tau', (\neg b : (\neg a \vee \neg b))) = \mathcal{R}(\tau', \varepsilon) = \tau'$. However, $\tau'(a \vee b) = \bot$. Thus, reversing only the first simplification step led to a state where we failed to reconstruct a solution for F^0.

$$\frac{\varphi \, [\rho] \, \sigma \cdot (\omega : C) \cdot \sigma'}{\varphi \wedge C \, [\rho] \, \sigma \cdot \sigma'} \boxed{\partial}$$

RESTORE

where $\boxed{\partial}$ is "C is clean w.r.t. σ'"

Fig. 4. New rule to reverse a weakening step

Our main contribution is the rule RESTORE in Fig. 4 which provides a sound way to reintroduce *selected* clauses from the stack back to the formula using the concept of clean clauses of Definition 5 as precondition.

Example 6. Consider again formula F^0 of Example 3. Example 4 shows that the first clause of the stack is not clean w.r.t. its suffix $((a \vee b)$ w.r.t. $(\neg b : (\neg a \vee \neg b)) \cdot$

$(\neg b : (a \vee \neg b)))$, but the second and third clauses are both clean $((\neg a \vee \neg b)$ w.r.t. $(\neg b : (a \vee \neg b))$ and $(a \vee \neg b)$ w.r.t. $\varepsilon)$. Restoring the second clause leads to the state $(\neg a \vee \neg b) [\emptyset] (a : (a \vee b)) \cdot (\neg b : (a \vee \neg b))$. A satisfying assignment of $(\neg a \vee \neg b)$ is $\tau = \{\neg a, \neg b\}$. The reconstruction function on τ and the current stack would be then $\mathcal{R}(\tau, (a : (a \vee b)) \cdot (\neg b : (a \vee \neg b))) = \mathcal{R}(\tau, (a : (a \vee b)))$, since $\tau(a \vee \neg b) = \top$. Because $\tau(a \vee b) \neq \top$, τ needs to be updated with the witness a, $\tau' = \tau \circ \{a\} = \{a, \neg b\}$. Then $\mathcal{R}(\tau, (a : (a \vee b))) = \mathcal{R}(\tau', \varepsilon) = \tau'$. The resulting assignment τ' satisfies not just the irredundant formula but each clause of the stack as well. Similarly, starting from any other satisfying assignment of $(\neg a \vee \neg b)$, the result of the reconstruction function satisfies all clauses.

3.5   Incremental Inprocessing Rules

The final and complete version of our calculus is shown in Fig. 5. To keep the notation simple the precise indexing of the states were so far omitted. Following the convention introduced at the beginning of this section, each single-line rule allows to derive a state $\varphi^i_{j+1} [\rho^i_{j+1}] \sigma^i_{j+1}$ from a state $\varphi^i_j [\rho^i_j] \sigma^i_j$, with $0 \leq i \leq n$ and $0 \leq j < k_i$, while our double-line rule ADDCLAUSES transits from a state $\varphi^i_{k_i} [\rho^i_{k_i}] \sigma^i_{k_i}$ to state $\varphi^{i+1}_0 [\rho^{i+1}_0] \sigma^{i+1}_0$.

$$\frac{\varphi [\rho] \sigma}{\varphi [\rho \wedge C] \sigma} \boxed{\sharp} \qquad \frac{\varphi [\rho \wedge C] \sigma}{\varphi [\rho] \sigma} \qquad \frac{\varphi [\rho \wedge C] \sigma}{\varphi \wedge C [\rho] \sigma} \qquad \frac{\varphi [\rho] \sigma}{\varphi \wedge \Delta [\rho] \sigma} \boxed{\mathcal{I}}$$

$$\text{LEARN}^- \qquad\qquad \text{FORGET} \qquad\qquad \text{STRENGTHEN} \qquad\qquad \text{ADDCLAUSES}$$

$$\frac{\varphi \wedge C [\rho] \sigma}{\varphi [\rho] \sigma \cdot (\omega : C)} \boxed{b} \qquad \frac{\varphi \wedge C [\rho] \sigma}{\varphi [\rho] \sigma} \boxed{\emptyset} \qquad \frac{\varphi [\rho] \sigma \cdot (\omega : C) \cdot \sigma'}{\varphi \wedge C [\rho] \sigma \cdot \sigma'} \boxed{\partial}$$

$$\text{WEAKEN}^+ \qquad\qquad \text{DROP} \qquad\qquad \text{RESTORE}$$

where $\boxed{\sharp}$ is $\varphi \wedge \rho \models C$, $\boxed{b}$ is $\varphi \wedge C \equiv^\omega_{sat} \varphi$, $\boxed{\emptyset}$ is $\varphi \models C$,

$\boxed{\partial}$ is C is clean w.r.t. σ' and $\boxed{\mathcal{I}}$ is that each clause in Δ is clean w.r.t. σ

Fig. 5. Incremental inprocessing rules

4   Formal Correctness

First we show that learned clauses are still valid in the next phase, and then prove that solutions can be reconstructed in each satisfiable state. In these proofs the set of irredundant clauses are always considered in *combination* together with the clauses on the reconstruction stack, i.e., $\varphi^i_j \wedge \sigma^i_j$. An important finding of our paper is that these combined formulas always imply the redundant clauses.

Proposition 2. *In any derivation in our calculus starting from the initial state the property* $\varphi^i_j \wedge \sigma^i_j \models \rho^i_j$ *holds for each phase* $i = 0 \ldots n$ *and* j *with* $0 \leq j \leq k_i$.

Proof. In the initial state $\varphi_0^0 \wedge \sigma_0^0 \models \rho_0^0$ trivially holds because ρ_0^0 is empty. Assume that $\varphi_j^i \wedge \sigma_j^i \models \rho_j^i$ holds (for any i and j s.t. $0 \leq i \leq n$ and $0 \leq j < k_i$). We show that any transition maintains the property. In case rule FORGET or STRENGTHEN is applied, ρ_{j+1}^i is weaker than ρ_j^i. In case of FORGET, $\varphi_{j+1}^i = \varphi_j^i$ and $\sigma_{j+1}^i = \sigma_j^i$, while in case of STRENGTHEN φ_{j+1}^i is even stronger than φ_j^i, and thus $\varphi_{j+1}^i \wedge \sigma_{j+1}^i \models \rho_{j+1}^i$ trivially follows in both cases. Rules WEAKEN$^+$ and RESTORE only move a clause between φ_j^i and σ_j^i and so $\varphi_j^i \wedge \sigma_j^i$ remains unchanged. Due to $\boxed{\varnothing}$, in case of DROP, $\varphi_j^i \equiv \varphi_{j+1}^i$, and so it also trivially maintains the property. When LEARN$^-$ transits from state j to $j + 1$, we get from the inductive assumption that $\varphi_j^i \wedge \sigma_j^i \models \varphi_j^i \wedge \rho_j^i$ and due to $\boxed{\sharp}$, we know that $\varphi_j^i \wedge \rho_j^i \models C$, and so $\varphi_j^i \wedge \sigma_j^i \models \rho_j^i \wedge C = \rho_{j+1}^i$. When starting a new phase (i.e. moving from i to $i + 1$ where $0 \leq i < n$ and j is k_i) only new clauses are added to $\varphi_{k_i}^i$ by ADDCLAUSES, and so $\varphi_0^{i+1} \wedge \sigma_0^{i+1} \models \rho_0^{i+1}$ clearly holds. $\square$

With this proposition we can now prove that the combined formulas remain logically equivalent during a derivation, unless new clauses are added.

Proposition 3. *In any derivation starting from the initial state, the property $\varphi_j^i \wedge \sigma_j^i \equiv \varphi_{j+1}^i \wedge \sigma_{j+1}^i$ holds for phase $i = 0 \ldots n$ and each j with $0 \leq j < k_i$.*

Proof. Only the rules STRENGTHEN and DROP change the combined formula. However, STRENGTHEN strengthens with an implied clause (due to Proposition 2), while DROP guarantees logical equivalence due to its side condition. $\square$

From that follows that at any point of a derivation within one phase the combined formula is logically equivalent to the incremental sub-problem:

Corollary 1. $F^i \equiv \varphi_0^i \wedge \sigma_0^i \equiv \varphi_1^i \wedge \sigma_1^i \equiv \cdots \equiv \varphi_{k_i}^i \wedge \sigma_{k_i}^i$.

Proof. $F^0 \equiv \varphi_0^0 \wedge \varepsilon \equiv \cdots \equiv \varphi_{k_0}^0 \wedge \sigma_{k_0}^0$. By an inductive argument and Proposition 3: $F^{i+1} = F^i \wedge \Delta_{i+1} \equiv \varphi_{k_i}^i \wedge \sigma_{k_i}^i \wedge \Delta_{i+1} = \varphi_0^{i+1} \wedge \sigma_0^{i+1} \equiv \cdots \equiv \varphi_{k_{i+1}}^{i+1} \wedge \sigma_{k_{i+1}}^{i+1}$. $\square$

Moreover, an important practical consequence of Corollary 1 and Proposition 2 is that it is sound to keep the learned clauses of the solver when new clauses are added:

Corollary 2. $F^{i+1} \models \rho_{k_i}^i$.

Before we can prove that we can reconstruct a model for the original incremental problem from a model of the current irredundant clauses using the reconstruction stack we need the following lemma.

Lemma 2. *For a given truth assignment τ and a sequence of witness labelled clauses $\sigma \cdot \sigma'$ we have $\mathcal{R}(\tau, \sigma \cdot \sigma') = \mathcal{R}(\mathcal{R}(\tau, \sigma'), \sigma)$.*

Proof. By induction over the length of σ'. The base case $\sigma' = \varepsilon$ is trivial. Now consider $\sigma' = \sigma'' \cdot (\omega : C)$ and let $\tau' = \tau$ if $\tau(C) = \top$, $\tau' = \tau \circ \omega$ otherwise. Since $\mathcal{R}(\tau, \sigma \cdot \sigma') = \mathcal{R}(\tau', \sigma \cdot \sigma'')$ and $\mathcal{R}(\tau, \sigma') = \mathcal{R}(\tau', \sigma'')$, $\mathcal{R}(\tau, \sigma \cdot \sigma') = \mathcal{R}(\mathcal{R}(\tau, \sigma'), \sigma)$ follows from the induction hypothesis applied to τ' and $\sigma \cdot \sigma''$. $\square$

Theorem 1 (Reconstructiveness). *In any derivation starting from the initial state, every state satisfies the reconstruction property of Definition 4.*

Proof. In the initial state the reconstruction stack is empty, and so for any satisfying assignment τ of F^0, $\mathcal{R}(\tau, \varepsilon)(F^0) = \top$. To simplify notation, we first consider only a single phase i (with $0 \leq i \leq n$), and omit the superscript i. Assume that in a state j (where $0 \leq j < k_i$), the reconstruction property holds. Let τ be a truth assignment with $\tau(\varphi_j) = \top$. Then $\mathcal{R}(\tau, \sigma_j)(\varphi_j \wedge \sigma_j) = \top$ follows by induction. In case LEARN$^-$ or FORGET was applied to state j, we have $\varphi_{j+1} = \varphi_j$ and $\sigma_{j+1} = \sigma_j$, thus the reconstruction property remains true. Rule STRENGTHEN moves a clause C from ρ_j to φ_{j+1} and so $\varphi_{j+1} = \varphi_j \wedge C$ and $\sigma_{j+1} = \sigma_j$. In case $\tau(\varphi_{j+1}) = \top$ we have $\mathcal{R}(\tau, \sigma_{j+1})(\varphi_j \wedge \sigma_{j+1}) = \mathcal{R}(\tau, \sigma_j)(\varphi_j \wedge \sigma_j) = \top$ by induction. Then Proposition 2 gives $\varphi_j \wedge \sigma_j \models C$, thus $\mathcal{R}(\tau, \sigma_{j+1})(\varphi_j \wedge C \wedge \sigma_{j+1}) = \top$. From the side condition of DROP we know that $\tau(\varphi_{j+1} \wedge C) = \top$ whenever $\tau(\varphi_{j+1}) = \top$, and thus $\mathcal{R}(\tau, \sigma_{j+1})(\varphi_{j+1} \wedge \sigma_{j+1}) = \top$ again by induction. When WEAKEN$^+$ is applied, a redundant clause C is removed from φ_j and pushed to σ_{j+1} (i.e. $\varphi_j = \varphi_{j+1} \wedge C$) witnessed by ω. Assume $\tau(\varphi_{j+1}) = \top$. We apply the induction hypothesis to the truth assignments τ and $(\tau \circ \omega)$ to get:

$$\tau(\varphi_{j+1} \wedge C) = \top \implies \mathcal{R}(\tau, \sigma_j)(\varphi_{j+1} \wedge C \wedge \sigma_j) = \top \tag{1}$$

$$(\tau \circ \omega)(\varphi_{j+1} \wedge C) = \top \implies \mathcal{R}((\tau \circ \omega), \sigma_j)(\varphi_{j+1} \wedge C \wedge \sigma_j) = \top. \tag{2}$$

If $\tau(C) = \top$, then $\mathcal{R}(\tau, \sigma_j \cdot (\omega : C))(\varphi_{j+1} \wedge C \wedge \sigma_j) = \top$ due to (1). Furthermore, assuming the side condition of WEAKEN$^+$, we know that $(\omega : C)$ is redundant w.r.t. φ_{j+1}. If $\tau(C) \neq \top$, then $(\tau \circ \omega)(\varphi_{j+1} \wedge C) = \top$ using Proposition 1. And with (2) we also get $\mathcal{R}(\tau, \sigma_j \cdot (\omega : C))(\varphi_{j+1} \wedge C \wedge \sigma_j) = \top$ if $\tau(C) \neq \top$. When we restore a clause C by RESTORE, we know that if $\tau(\varphi_j \wedge C) = \top$ then $\tau(C) = \top$. Further, we know from the side condition of RESTORE that C is clean w.r.t. σ', and so with Lemma 1, we obtain $\mathcal{R}(\tau, \sigma')(C) = \top$. From that and from Lemma 2 it follows that $\mathcal{R}(\tau, \sigma \cdot (\omega : C) \cdot \sigma') = \mathcal{R}(\mathcal{R}(\tau, \sigma'), \sigma \cdot (\omega : C)) = \mathcal{R}(\mathcal{R}(\tau, \sigma'), \sigma) = \mathcal{R}(\tau, \sigma \cdot \sigma')$, where $\varphi_j \wedge \sigma \wedge C \wedge \sigma'$ evaluates to true due to the induction hypothesis. When a new phase starts (i.e. $0 \leq i < n$ and $j = k_i$) as Δ_{i+1} is added to $\varphi^i_{k_i}$ by ADDCLAUSES, each new clause is clean w.r.t. $\sigma^i_{k_i}$. Thus, due to Lemma 1, the reconstruction function does not destroy any satisfying assignment of Δ_{i+1}. □

Theorem 2 (Correctness). *In any derivation starting from the initial state, for each phase $i = 0 \ldots n$ we have $F^i \equiv_{sat} \varphi^i_j \wedge \rho^i_j$ for all j with $0 \leq j \leq k_i$.*

Proof. From Proposition 2 and Theorem 1 it follows, that φ^i_j is unsatisfiable if $\varphi^i_j \wedge \rho^i_j$ is unsatisfiable. In this case also F^i is unsatisfiable using Corollary 1. Otherwise, if $\varphi^i_j \wedge \rho^i_j$ is satisfiable, then F^i is satisfiable due to Theorem 1 and again Corollary 1. □

To summarize, our calculus fulfills all the desiderata listed at the beginning of Sect. 3: (*i*) we can reuse the gained information of previous iterations (including learned clauses), (*ii*) we can continue with incremental solving in a satisfiability preserving way, and (*iii*) the reconstruction property guarantees that we can get a solution to the original problem in case of satisfiability.

5   Implementation

Based on our new approach we added incremental inprocessing to the SAT solver CaDiCaL [33]. Rule WEAKEN$^+$ is defined in our calculus based on the most general redundancy property and so it allows to employ every clause elimination procedure implemented in CaDiCaL including variable elimination [14], vivification [34,35], equivalent-literal substitution [31,32], hyper-binary resolution [13], (self-)subsumption [14] and blocked clause elimination [16]. Combining DROP with WEAKEN$^+$ allows efficient equivalence literal substitution, since only two binary clauses have to be stored on the stack for each literal in a strongly connected component [31,32] instead of all clauses with that literal. Similarly, gate-based variable elimination [14] only requires to save gate clauses.

At the heart of our new calculus are the RESTORE and ADDCLAUSES rules. They allow to reverse problematic simplification steps and add new clauses. In practice, SAT solvers are used via an interface (e.g. IPASIR [2] in CaDiCaL) to add new clauses Δ and then asked to solve the extended formula $F \wedge \Delta$. Before solving $F \wedge \Delta$, our approach first performs a sequence of RESTORE steps in order to make each clause in Δ clean w.r.t. the reconstruction stack σ using the algorithm RestoreAddClauses in Fig. 6. Then the new and restored clauses are added to the irredundant clauses and a new incremental solving phase starts.

RestoreAddClauses (new clauses Δ, reconstruction stack σ)

1    $(\omega_1 : C_1) \cdots (\omega_n : C_n) := \sigma$
2   **for** i **from** 1 **to** n
3   **if** exists $\ell \in \omega_i$ where $\neg\ell$ occurs in Δ **then**
4            $\Delta := \Delta \cup C_i, \quad \sigma := \sigma \setminus (\omega_i : C_i)$
5   **return** $\langle \Delta, \sigma \rangle$

Fig. 6. Algorithm RestoreAddClauses to identify and restore all tainted clauses.

The algorithm in Fig. 6 presents a simple implementation that identifies a sufficient set of clauses to restore in order to make Δ clean. It follows the idea of "taint-checking", commonly used to reason about information-flow (see e.g. [36]). First consider every clause that comes from the user as tainted, because it potentially leads to problems. Then check whether these tainted clauses (actually literals of these clauses) trigger any clause on the stack to be restored. In that case the literals of the restored clause become tainted as well and recursively might trigger further clauses. However, restored clauses only need to be clean w.r.t. the reconstruction stack after them (see RESTORE in Fig. 4), while the clauses in Δ need to be clean w.r.t. the whole reconstruction stack. Therefore, the need for restoring is checked by traversing the stack from bottom to top (left to right). If a clause has to be restored, it can only trigger to restore clauses to its right. Thus, already processed clauses on the left do not have to be reconsidered.

The method takes the new clauses Δ and the current stack σ as input and checks each previous simplification step from left to right (see Line 1–2). Whenever the witness of a simplification has a literal that occurs negated in Δ, the simplification is reversed by restoring the eliminated clause from the stack. The check in Line 3 is actually asking whether there is a clause in Δ (i.e. in the set of new or already restored clauses) that is not clean w.r.t. $(\omega_i : C_i)$. To implement this check efficiently, we mark literals in Δ as tainted and in σ as witness. If the check succeeds, we need to restore the problematic C_i so that at the end we have a clean stack. In Line 4 the restored clause is added to Δ and removed from the stack. At the end of the procedure, Δ contains all the new and restored clauses, which added to the formula together with the new σ achieves the same effect as applying a sequence of RESTORE steps and a final ADDCLAUSES.

6 Experiments

We implemented a new bounded model checker called CaMiCaL for AIGER models [37], as used in the hardware model checking competition (HWMCC) [38]. Unrolling is simulated symbolically through substitution [39] in combination with structural hashing [40,41] and local low-level AIG optimizations [42]. As back-end different configurations of our SAT solver CaDiCaL [33] and other state-of-the-art incremental SAT solvers are used. The model checker was run on all the 300 models of the single safety property track of HWMCC'17 [38] up to bound 1000 with a time limit of 3600 s (for each model) and memory limit of 8 GB on our cluster with Intel Xeon E5-2620 v4 @ 2.10 GHz CPUs.

Results are presented in a similar way as the well-known cactus plots of the SAT Competition, except that we do not measure the overall running time of the model checker, but the time needed for one (incremental) call to the SAT solver. Figure 7 shows the distribution of these solving times. For example, if the model checker finished proving unsatisfiability for bound 41 after 110 s and then proved unsatisfiability for the consecutive bound 42 at 125 s then the time difference of 15 s is accounted for bound 42 on this instance. At the end each instance contributes as many solving times as bounds for it are solved.

As expected, worst performance is observed when the SAT solver is used in a completely non-incremental way (cadical-non-incremental), even with pre- and inprocessing enabled. It improves, if the model checker is allowed to assume earlier bounds to be good (cadical-non-incremental-assume-good-earlier). Incremental SAT solving is better as configuration cadical-restore-all-clauses shows, which employs pre- and inprocessing, but at the beginning of incremental calls restores all weakened clauses. However, disabling pre- and inprocessing completely during incremental SAT solving (cadical-no-inprocessing) is even better.

Configuration cadical-freeze can use variables for simplification which are not frozen. This again improves performance and there is no need to restore clauses. In bounded model checking (BMC) only variables encoding the next state are used in future calls and freezing them is sufficient. However, it required substantial programming effort to identify the set of frozen variables. Further,

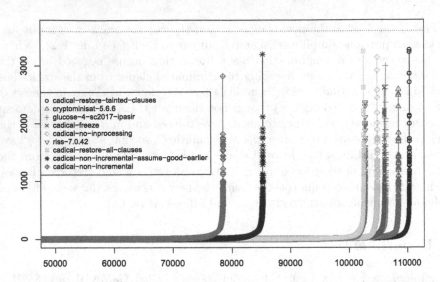

Fig. 7. Experimental results on all the 300 instances of the single safety property track of HWMCC'17. The x-axis corresponds to all bounds solved over all models sorted by the time needed for the SAT call for each bound, which is on the y-axis. The dotted horizontal line at 3600 s shows the time limit for solving all bounds of each model.

optimizations during CNF encoding, including structural hashing [40,41] across time frames or local two-level AIG optimizations [42], make it difficult to predict future use of variables. In other cases freezing might not even be possible [19].

Giving up on freezing makes use of our framework and gave the best solving times as configuration cadical-restore-tainted-clauses shows. This not only simplifies the way the solver is used through the API (no need to freeze variables) but also improves solving time. We measured the time spent in RestoreAddClauses to be less than 1% of the overall running time: 0.14% for our best configuration cadical-restore-tainted-clauses and 0.33% for cadical-restore-all-clauses. Our best configuration only restored 17% of the clauses. Restoring all clauses also lead to 3.4 times more eliminated clauses (applications of WEAKEN$^+$) in total.

Note that one can not get rid of freezing completely, since assumptions (for the "bad" state property in BMC) have to be frozen internally. Keeping freezing in the API might for instance also be useful for CNF simplification [8].

We also have similar results using freezing (as it is necessary for the solver Lingeling [30]) versus restoring tainted clauses for CaDiCaL as SAT solver back-end of our SMT solver Boolector [43]. We solved more benchmarks and decreased solving time significantly with the consequence that CaDiCaL is likely to replace Lingeling as incremental SAT solver back-end in the future.

We also considered other highly ranked SAT solvers in incremental tracks of the SAT Competition [2,44,45]: Glucose 4 [46], CryptoMiniSAT 5.6.6 [45,47] and Riss 7.0.42 [48]. CryptoMiniSAT performs significantly better than the other two external solvers. It is the only external solver which performs inprocessing during solving, including distillation [49]. Even though CryptoMiniSAT implements the same solution as [19] for incremental bounded variable elimination (BVE), this feature cannot be enabled through the API, and is disabled in our experiments. According to Mate Soos (private communication) scheduling BVE efficiently for incremental SAT solving is difficult for CryptoMiniSAT. We simply schedule BVE in CaDiCaL in the same way as during stand-alone SAT solving, with a persistent schedule across incremental invocations. Note that CaDiCaL only tries to eliminate variables and clauses which are newly added (or restored).

Source code of CaDiCaL and CaMiCaL and experimental data related to Fig. 7 can be found at http://fmv.jku.at/incrinpr including additional plots.

7 Conclusion

This paper presents a calculus that extends the framework of [17] to capture incremental SAT solving. It uses the most general clause redundancy property and is able to simulate most simplifications implemented in state-of-the-art SAT solvers. Our proposed approach is simple, eases the burden of using SAT solvers, can be implemented efficiently, and also reduces solving time substantially. As future work we want to support techniques which remove models, such as blocked clause addition, and techniques for simplifying under assumptions.

Acknowledgments. This research has been supported by the Austrian Science Fund (FWF) under projects W1255-N23 and S11408-N23. We thank Mathias Preiner and Aina Niemetz for their help in experimenting with Boolector and Håkan Hjort for providing feedback on using an incremental version of CaDiCaL.

References

1. Audemard, G., Lagniez, J.-M., Simon, L.: Improving glucose for incremental SAT solving with assumptions: application to MUS extraction. In: Järvisalo, M., Van Gelder, A. (eds.) SAT 2013. LNCS, vol. 7962, pp. 309–317. Springer, Heidelberg (2013). https://doi.org/10.1007/978-3-642-39071-5_23
2. Balyo, T., Biere, A., Iser, M., Sinz, C.: SAT race 2015. Artif. Intell. **241**, 45–65 (2016)
3. Eén, N., Sörensson, N.: An extensible SAT-solver. In: Giunchiglia, E., Tacchella, A. (eds.) SAT 2003. LNCS, vol. 2919, pp. 502–518. Springer, Heidelberg (2004). https://doi.org/10.1007/978-3-540-24605-3_37
4. Hooker, J.N.: Solving the incremental satisfiability problem. J. Log. Program. **15**(1&2), 177–186 (1993)
5. Biere, A., Cimatti, A., Clarke, E., Zhu, Y.: Symbolic model checking without BDDs. In: Cleaveland, W.R. (ed.) TACAS 1999. LNCS, vol. 1579, pp. 193–207. Springer, Heidelberg (1999). https://doi.org/10.1007/3-540-49059-0_14
6. Bradley, A.R.: SAT-based model checking without unrolling. In: Jhala, R., Schmidt, D. (eds.) VMCAI 2011. LNCS, vol. 6538, pp. 70–87. Springer, Heidelberg (2011). https://doi.org/10.1007/978-3-642-18275-4_7
7. Eén, N., Sörensson, N.: Temporal induction by incremental SAT solving. Electr. Notes Theor. Comput. Sci. **89**(4), 543–560 (2003)

8. Kupferschmid, S., Lewis, M.D.T., Schubert, T., Becker, B.: Incremental preprocessing methods for use in BMC. Form. Methods Syst. Des. **39**(2), 185–204 (2011)
9. Gocht, S., Balyo, T.: Accelerating SAT based planning with incremental SAT solving. In: Barbulescu, L., Frank, J., Mausam, Smith, S.F. (eds.) Proceedings of the 27th International Conference on Automated Planning and Scheduling, ICAPS 2017, pp. 135–139. AAAI Press (2017)
10. Martins, R., Joshi, S., Manquinho, V.M., Lynce, I.: On using incremental encodings in unsatisfiability-based MaxSAT solving. JSAT **9**, 59–81 (2014)
11. Nadel, A.: Boosting minimal unsatisfiable core extraction. In: Bloem, R., Sharygina, N. (eds.) Proceedings of 10th International Conference on Formal Methods in Computer-Aided Design, FMCAD 2010, pp. 221–229. IEEE (2010)
12. Sebastiani, R.: Lazy satisability modulo theories. JSAT **3**(3–4), 141–224 (2007)
13. Bacchus, F., Winter, J.: Effective preprocessing with hyper-resolution and equality reduction. In: Giunchiglia, E., Tacchella, A. (eds.) SAT 2003. LNCS, vol. 2919, pp. 341–355. Springer, Heidelberg (2004). https://doi.org/10.1007/978-3-540-24605-3_26
14. Eén, N., Biere, A.: Effective preprocessing in SAT through variable and clause elimination. In: Bacchus, F., Walsh, T. (eds.) SAT 2005. LNCS, vol. 3569, pp. 61–75. Springer, Heidelberg (2005). https://doi.org/10.1007/11499107_5
15. Heule, M.J.H., Järvisalo, M., Biere, A.: Efficient CNF simplification based on binary implication graphs. In: Sakallah, K.A., Simon, L. (eds.) SAT 2011. LNCS, vol. 6695, pp. 201–215. Springer, Heidelberg (2011). https://doi.org/10.1007/978-3-642-21581-0_17
16. Järvisalo, M., Biere, A., Heule, M.J.H.: Blocked clause elimination. In: Esparza, J., Majumdar, R. (eds.) TACAS 2010. LNCS, vol. 6015, pp. 129–144. Springer, Heidelberg (2010). https://doi.org/10.1007/978-3-642-12002-2_10
17. Järvisalo, M., Heule, M.J.H., Biere, A.: Inprocessing rules. In: Gramlich, B., Miller, D., Sattler, U. (eds.) IJCAR 2012. LNCS (LNAI), vol. 7364, pp. 355–370. Springer, Heidelberg (2012). https://doi.org/10.1007/978-3-642-31365-3_28
18. Järvisalo, M., Biere, A.: Reconstructing solutions after blocked clause elimination. In: Strichman, O., Szeider, S. (eds.) SAT 2010. LNCS, vol. 6175, pp. 340–345. Springer, Heidelberg (2010). https://doi.org/10.1007/978-3-642-14186-7_30
19. Nadel, A., Ryvchin, V., Strichman, O.: Preprocessing in incremental SAT. In: Cimatti, A., Sebastiani, R. (eds.) SAT 2012. LNCS, vol. 7317, pp. 256–269. Springer, Heidelberg (2012). https://doi.org/10.1007/978-3-642-31612-8_20
20. Heule, M.J.H., Kiesl, B., Biere, A.: Short proofs without new variables. In: de Moura, L. (ed.) CADE 2017. LNCS (LNAI), vol. 10395, pp. 130–147. Springer, Cham (2017). https://doi.org/10.1007/978-3-319-63046-5_9
21. Heule, M.J.H., Kiesl, B., Biere, A.: Strong extension-free proof systems. J. Autom. Reason. (2019). https://doi.org/10.1007/s10817-019-09516-0
22. Audemard, G., Katsirelos, G., Simon, L.: A restriction of extended resolution for clause learning sat solvers. In: Proceedings of the 24th AAAI Conference on Artificial Intelligence (AAAI 2010). AAAI Press (2010)
23. Kullmann, O.: On a generalization of extended resolution. Discret. Appl. Math. **96–97**, 149–176 (1999)
24. Manthey, N., Heule, M.J.H., Biere, A.: Automated reencoding of Boolean formulas. In: Biere, A., Nahir, A., Vos, T. (eds.) HVC 2012. LNCS, vol. 7857, pp. 102–117. Springer, Heidelberg (2013). https://doi.org/10.1007/978-3-642-39611-3_14
25. Nadel, A., Ryvchin, V., Strichman, O.: Ultimately incremental SAT. In: Sinz, C., Egly, U. (eds.) SAT 2014. LNCS, vol. 8561, pp. 206–218. Springer, Cham (2014). https://doi.org/10.1007/978-3-319-09284-3_16

26. Nadel, A., Ryvchin, V.: Efficient SAT solving under assumptions. In: Cimatti, A., Sebastiani, R. (eds.) SAT 2012. LNCS, vol. 7317, pp. 242–255. Springer, Heidelberg (2012). https://doi.org/10.1007/978-3-642-31612-8_19

27. Blanchette, J.C., Fleury, M., Lammich, P., Weidenbach, C.: A verified SAT solver framework with learn, forget, restart, and incrementality. J. Autom. Reason. **61**(1–4), 333–365 (2018)

28. Fazekas, K., Bacchus, F., Biere, A.: Implicit hitting set algorithms for maximum satisfiability modulo theories. In: Galmiche, D., Schulz, S., Sebastiani, R. (eds.) IJCAR 2018. LNCS (LNAI), vol. 10900, pp. 134–151. Springer, Cham (2018). https://doi.org/10.1007/978-3-319-94205-6_10

29. Balyo, T., Fröhlich, A., Heule, M.J.H., Biere, A.: Everything you always wanted to know about blocked sets (but were afraid to ask). In: Sinz, C., Egly, U. (eds.) SAT 2014. LNCS, vol. 8561, pp. 317–332. Springer, Cham (2014). https://doi.org/10.1007/978-3-319-09284-3_24

30. Biere, A.: Yet another local search solver and Lingeling and friends entering the SAT competition 2014. In: Balint, A., Belov, A., Heule, M.J.H., Järvisalo, M. (eds.) SAT Competition 2014. Department of Computer Science Series of Publications B, pp. 39–40. University of Helsinki (2014)

31. Aspvall, B., Plass, M.F., Tarjan, R.E.: A linear-time algorithm for testing the truth of certain quantified Boolean formulas. Inf. Process. Lett. **8**(3), 121–123 (1979)

32. Brafman, R.I.: A simplifier for propositional formulas with many binary clauses. In: Nebel, B. (ed.) Proceedings of the Seventeenth International Joint Conference on Artificial Intelligence, IJCAI 2001, Morgan Kaufmann, pp. 515–522 (2001)

33. Biere, A.: CaDiCaL, Lingeling, Plingeling, Treengeling and YalSAT entering the SAT competition 2018. In: Heule, M.J.H., Järvisalo, M., Suda, M. (eds.) Proceedings of SAT Competition 2018 - Solver and Benchmark Descriptions. Volume B-2018-1 of Department of Computer Science Series of Publications B, pp. 13–14. University of Helsinki (2018)

34. Luo, M., Li, C., Xiao, F., Manyà, F., Lü, Z.: An effective learnt clause minimization approach for CDCL SAT solvers. In: Sierra, C. (ed.) Proceedings of the 26th International Joint Conference on Artificial Intelligence, IJCAI 2017, pp. 703–711. ijcai.org (2017)

35. Piette, C., Hamadi, Y., Sais, L.: Vivifying propositional clausal formulae. In: Ghallab, M., Spyropoulos, C.D., Fakotakis, N., Avouris, N.M. (eds.) Proceedings of the 18th European Conference on Artificial Intelligence, ECAI 2008. Volume 178 of Frontiers in Artificial Intelligence and Applications, pp. 525–529. IOS Press (2008)

36. Schwartz, E.J., Avgerinos, T., Brumley, D.: All you ever wanted to know about dynamic taint analysis and forward symbolic execution (but might have been afraid to ask). In: 31st IEEE Symposium on Security and Privacy, S&P 2010, 16–19 May 2010, pp. 317–331. IEEE Computer Society, Berleley/Oakland (2010)

37. Biere, A., Heljanko, K., Wieringa, S.: AIGER 1.9 and beyond. Technical report, FMV reports series, Institute for Formal Models and Verification, Johannes Kepler University, Altenbergerstr. 69, 4040 Linz, Austria (2011)

38. Biere, A., van Dijk, T., Heljanko, K.: Hardware model checking competition 2017. In: Stewart, D., Weissenbacher, G. (eds.) Formal Methods in Computer Aided Design, FMCAD 2017, p. 9. IEEE (2017)

39. Jussila, T., Biere, A.: Compressing BMC encodings with QBF. Electr. Notes Theor. Comput. Sci. **174**(3), 45–56 (2007)

40. Heule, M.J.H., Järvisalo, M., Biere, A.: Revisiting hyper binary resolution. In: Gomes, C., Sellmann, M. (eds.) CPAIOR 2013. LNCS, vol. 7874, pp. 77–93. Springer, Heidelberg (2013). https://doi.org/10.1007/978-3-642-38171-3_6

41. Kuehlmann, A., Paruthi, V., Krohm, F., Ganai, M.K.: Robust Boolean reasoning for equivalence checking and functional property verification. IEEE Trans CAD Integr. Circ. Syst. **21**(12), 1377–1394 (2002)
42. Brummayer, R., Biere, A.: Local two-level and-inverter graph minimization without blowup. In: Proceedings of the 2nd Doctoral Workshop on Mathematical and Engineering Methods in Computer Science (MEMICS 2006) (2006)
43. Niemetz, A., Preiner, M., Wolf, C., Biere, A.: BTOR2, BtorMC and Boolector 3.0. In: Chockler, H., Weissenbacher, G. (eds.) CAV 2018. LNCS, vol. 10981, pp. 587–595. Springer, Cham (2018). https://doi.org/10.1007/978-3-319-96145-3_32
44. Balyo, T., Heule, M.J.H., Järvisalo, M. (eds.): Proceedings of SAT Competition 2016 - Solver and Benchmark Descriptions. Volume B-2016-1 of Department of Computer Science Series of Publications B. University of Helsinki (2016)
45. Balyo, T., Heule, M.J.H., Järvisalo, M., (eds.): Proceedings of SAT Competition 2017 - Solver and Benchmark Descriptions. Volume B-2017-1 of Department of Computer Science Series of Publications B. University of Helsinki (2017)
46. Audemard, G., Simon, L.: Glucose and syrup in the SAT 2017. In: Balyo, T., Heule, M.J.H., Järvisalo, M. (eds.) Proceedings of SAT Competition 2017 - Solver and Benchmark Descriptions. Volume B-2017-1 of Department of Computer Science Series of Publications B, pp. 16–17. University of Helsinki (2017)
47. Soos, M., Nohl, K., Castelluccia, C.: Extending SAT solvers to cryptographic problems. In: Kullmann, O. (ed.) SAT 2009. LNCS, vol. 5584, pp. 244–257. Springer, Heidelberg (2009). https://doi.org/10.1007/978-3-642-02777-2_24
48. Manthey, N.: Riss 7. In Balyo, T., Heule, M.J.H., Järvisalo, M., (eds.) Proceedings of SAT Competition 2017 - Solver and Benchmark Descriptions. Volume B-2017-1 of Department of Computer Science Series of Publications B, p. 29. University of Helsinki (2017)
49. Han, H., Somenzi, F.: Alembic: an efficient algorithm for CNF preprocessing. In: Proceedings of the 44th Design Automation Conference, DAC 2007, pp. 582–587. IEEE (2007)

Local Search for Fast Matrix Multiplication

Marijn J. H. Heule[1]([✉]), Manuel Kauers[2], and Martina Seidl[3]

[1] Department of Computer Science, The University of Texas, Austin, USA
marijn@heule.nl
[2] Institute for Algebra, J. Kepler University, Linz, Austria
[3] Institute for Formal Models and Verification, J. Kepler University,
Linz, Austria

Abstract. Laderman discovered a scheme for computing the product of two 3×3 matrices using only 23 multiplications in 1976. Since then, some more such schemes were proposed, but nobody knows how many such schemes there are and whether there exist schemes with fewer than 23 multiplications. In this paper we present two independent SAT-based methods for finding new schemes using 23 multiplications. Both methods allow computing a few hundred new schemes individually, and many thousands when combined. Local search SAT solvers outperform CDCL solvers consistently in this application.

1 Introduction

Matrix multiplication is a fundamental operation with applications in nearly any area of science and engineering. However, after more than 50 years of work on matrix multiplication techniques (see, e.g., [3,5,11,13]), the complexity of matrix multiplication is still a mystery. Even for small matrices, the problem is not completely understood, and understanding these cases better can provide valuable hints towards more efficient algorithms for large matrices.

The naive way for computing the product C of two 2×2 matrices A, B requires 8 multiplications:

$$\begin{pmatrix} a_{11} & a_{12} \\ a_{21} & a_{22} \end{pmatrix} \begin{pmatrix} b_{11} & b_{12} \\ b_{21} & b_{22} \end{pmatrix} = \begin{pmatrix} a_{11}b_{11} + a_{12}b_{21} & a_{11}b_{12} + a_{21}b_{22} \\ a_{21}b_{11} + a_{22}b_{21} & a_{21}b_{12} + a_{22}b_{22} \end{pmatrix} = \begin{pmatrix} c_{11} & c_{12} \\ c_{21} & c_{22} \end{pmatrix}$$

Strassen observed 50 years ago that C can also be computed with only 7 multiplications [15]. His scheme proceeds in two steps. In the first step he introduces auxiliary variables $M_1, \ldots, M_7$ which are defined as the product of certain linear

M. J. H. Heule is supported by NSF grant CCF-1813993 and AFRL Award FA8750-15-2-0096.
M. Kauers is supported by the Austrian FWF grants P31571-N32 and F5004.
M. Seidl is supported by the Austrian FWF grant NFN S11408-N23 and the LIT AI Lab funded by the State of Upper Austria.

M. Janota and I. Lynce (Eds.): SAT 2019, LNCS 11628, pp. 155–163, 2019.
https://doi.org/10.1007/978-3-030-24258-9_10

combinations of the entries of A and B. In the second step the entries of C are obtained as certain linear combinations of the M_i:

$$M_1 = (a_{11} + a_{22})(b_{11} + b_{22}) \qquad c_{11} = M_1 + M_4 - M_5 + M_7$$
$$M_2 = (a_{21} + a_{22})(b_{11}) \qquad c_{12} = M_3 + M_5$$
$$M_3 = (a_{11})(b_{12} - b_{22}) \qquad c_{21} = M_2 + M_4$$
$$M_4 = (a_{22})(b_{21} - b_{11}) \qquad c_{22} = M_1 - M_2 + M_3 + M_6$$
$$M_5 = (a_{11} + a_{12})(b_{22})$$
$$M_6 = (a_{21} - a_{11})(b_{11} + b_{12})$$
$$M_7 = (a_{12} - a_{22})(b_{21} + b_{22})$$

Recursive application of this scheme gave rise to the first algorithm for multiplying arbitrary $n \times n$ matrices in subcubic complexity. Winograd [16] showed that Strassen's scheme is optimal in the sense that there does not exist a similar scheme with fewer than 7 multiplications, and de Groote [7] showed that Strassen's scheme is essentially unique.

Less is known about 3×3 matrices. The naive scheme requires 27 multiplications, and in 1976 Laderman [10] found one with 23. Similar as Strassen, he defines $M_1, \dots, M_{23}$ as products of certain linear combinations of the entries of A and B. The entries of $C = AB$ are then obtained as linear combinations of $M_1, \dots, M_{23}$. It is not known whether 23 is optimal (the best lower bound is 19 [2]). It is known however that Laderman's scheme is *not* unique. A small number of intrinsically different schemes have been found over the years. Of particular interest are schemes in which all coefficients in the linear combinations are $+1$, -1, or 0. The only four such schemes (up to equivalence) we are aware of are due to Laderman, Smirnov [14], Oh et al. [12], and Courtois et al. [6].

While Smirnov and Oh et al. found their multiplication schemes with computer-based search using non-linear numerical optimization methods, Courtois found his multiplication scheme using a SAT solver. This is also what we do here. We present two approaches which allowed us to generate more than 13,000 mutually inequivalent new matrix multiplication schemes for 3×3 matrices, using altogether about 35 years of CPU years. We believe that the new schemes are of interest to the matrix multiplication community. We therefore make them publicly available in various formats and grouped by invariants at

http://www.algebra.uni-linz.ac.at/research/matrix-multiplication/.

2   Encoding and Workflow

To search for multiplication schemes of 3×3 matrices having the above form, we define the M_i as product of linear combination of all entries of A and B with undetermined coefficients $\alpha_{ij}^{(\ell)}, \beta_{ij}^{(\ell)}$:

$$M_1 = (\alpha_{11}^{(1)} a_{11} + \dots + \alpha_{33}^{(1)} a_{33})(\beta_{11}^{(1)} b_{11} + \dots + \beta_{33}^{(1)} b_{33})$$
$$\vdots$$
$$M_{23} = (\alpha_{11}^{(23)} a_{11} + \dots + \alpha_{33}^{(23)} a_{33})(\beta_{11}^{(23)} b_{11} + \dots + \beta_{33}^{(23)} b_{33})$$

Similarly, we define the c_{ij} as linear combinations of the M_i with undetermined coefficients $\gamma_{i,j}^{(\ell)}$:

$$c_{11} = \gamma_{11}^{(1)} M_1 + \cdots + \gamma_{11}^{(23)} M_{23}, \quad \ldots, \quad c_{33} = \gamma_{33}^{(1)} M_1 + \cdots + \gamma_{33}^{(23)} M_{23}$$

Comparing the coefficients of all terms $a_{i_1 i_2} b_{j_1 j_2} c_{k_1 k_2}$ in the equations $c_{ij} = \sum_k a_{ik} b_{kj}$ leads to the polynomial equations

$$\sum_{\ell=1}^{23} \alpha_{i_1 i_2}^{(\ell)} \beta_{j_1 j_2}^{(\ell)} \gamma_{k_1 k_2}^{(\ell)} = \delta_{i_2 j_1} \delta_{i_1 k_1} \delta_{j_2 k_2}$$

for $i_1, i_2, j_1, j_2, k_1, k_2 \in \{1, 2, 3\}$. These 729 cubic equations with 621 variables are also known as *Brent equations* [4]. The δ_{uv} on the right are Kronecker-deltas, i.e., $\delta_{uv} = 1$ if $u = v$ and $\delta_{uv} = 0$ otherwise. Each solution of the system of these equations corresponds to a matrix multiplication scheme. The equations become slightly more symmetric if we flip the indices of the γ_{ij}, and since this is the variant mostly used in the literature, we will also adopt it from now on.

Another view on the Brent equations is as follows. View the $\alpha_{i_1 i_2}^{(\ell)}, \beta_{j_1 j_2}^{(\ell)}, \gamma_{k_1 k_2}^{(\ell)}$ as variables, as before, and regard $a_{i_1 i_2}, b_{j_1 j_2}, c_{k_1 k_2}$ as polynomial indeterminants. Then the task consists of instantiating the variables in such a way that

$$\sum_{\ell=1}^{23} (\alpha_{11}^{(\ell)} a_{11} + \cdots)(\beta_{11}^{(\ell)} b_{11} + \cdots)(\gamma_{11}^{(\ell)} c_{11} + \cdots) = \sum_{i=1}^{3} \sum_{j=1}^{3} \sum_{k=1}^{3} a_{ij} b_{jk} c_{ki}$$

holds as equation of polynomials in the variables $a_{i_1 i_2}, b_{j_1 j_2}, c_{k_1 k_2}$. Expanding the left hand side and equating coefficients leads to the Brent equations as stated before (but with indices of γ flipped, as agreed). In other words, expanding the left hand side, all terms have to cancel except for the terms on the right. We found it convenient to say that a term $a_{i_1 i_2} b_{j_1 j_2} c_{k_1 k_2}$ has "type m" if $m = \delta_{i_2 j_1} + \delta_{j_2 k_1} + \delta_{k_2 i_1}$. With this terminology, all terms of types 0, 1, 2 have to cancel each other, and all terms of type 3 have to survive. Note that since all 27 type 3 terms must be produced by the 23 summands on the left, some summands must produce more than one type 3 term.

For solving the Brent equations with a SAT solver, we use $\mathbb{Z}_2$ as coefficient domain, so that multiplication translates into 'and' and addition translates into 'xor'. When, for example, the variable $\alpha_{i_1 i_2}^{(\ell)}$ is true in a solution of the corresponding SAT instance, this indicates that the term $a_{i_1 i_2}$ occurs in M_ℓ, and likewise for the b-variables. If $\gamma_{k_1 k_2}^{(\ell)}$ is true, this means that M_ℓ appears in the linear combination for $c_{k_1 k_2}$. We call $\alpha_{i_1 i_2}^{(\ell)}, \beta_{j_1 j_2}^{(\ell)}$, and $\gamma_{k_1 k_2}^{(\ell)}$ the *base variables*.

In order to bring the Brent equations into CNF, we use Tseitin transformation, i.e., we introduce definitions for subformulas to avoid exponential blow-up. To keep the number of fresh variables low, we do not introduce one new variable for every cube but only for pairs of literals, i.e., we encode a cube $(\alpha \wedge \beta \wedge \gamma)$ as $u \leftrightarrow (\alpha \wedge \beta)$ and $v \leftrightarrow (u \wedge \gamma)$. In this way, we can reuse u. Furthermore, a sum $v_1 \oplus \cdots \oplus v_m$ with $m \geq 4$ is encoded by $w \leftrightarrow (v_1 \oplus v_2 \oplus v_3)$ and $v_4 \oplus \cdots \oplus v_m \oplus w$, with the latter sum being encoded recursively. This encoding seems to require the smallest sum of the number of variables and the number of clauses—a commonly used optimality heuristic. The used scripts are available at

https://github.com/marijnheule/matrix-challenges/tree/master/src.

The generation of new schemes proceeds in several steps, with SAT solving being the first and main step. If the SAT solver finds a solution, we next check whether it is equivalent to any known or previously found solution modulo de Groote's symmetry group [7]. If so, we discard it. Otherwise, we next try to simplify the new scheme by searching for an element in its orbit which has a smaller number of terms. The scheme can then be used to initiate a new search. In the fourth step, we use Gröbner bases to lift the scheme from the coefficient domain $\mathbb{Z}_2$ to $\mathbb{Z}$. Finally, we cluster large sets of similar solutions into parameterized families.

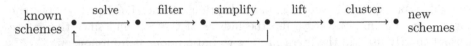

In the present paper, we give a detailed description of the first step in this workflow. The subsequent steps use algebraic techniques unrelated to SAT and will be described in [9].

3   Solving Methods

The *core* of a scheme is the pairing of the type 3 terms. Our first method focuses on finding schemes with new cores, while our second method searches for schemes that are similar to an existing one and generally has the same core. For all experiments we used the local search SAT solver `yalsat` [1] as this solver performed best on instances from this application. We also tried solving these instances using CDCL solvers, but the performance was disappointing. We observed a possible explanation: The runtime of CDCL solvers tends to be exponential in the average backtrack level (ABL) on unsatisfiable instances. For most formulas arising from other applications, ABL is small (< 50), while on the matrix multiplication instances ABL is large (> 100).

3.1   Random Pairings of Type 3 Terms and Streamlining

Two of the known schemes, those of Smirnov [14] and Courtois et al. [6], have the property that each type 3 term occurs exactly once and at most two type 3 terms occur in the same summand. We decided to use this pattern to search for new schemes: randomly pair four type 3 terms and assign the remaining type 3 terms to the other 19 summands. Only in very rare cases, random pairing could be extended to a valid scheme in reasonable time, say a few minutes. In the other cases it is not known whether the pairing cannot be extended to a valid scheme or whether finding such a scheme is very hard.

Since the number of random pairings that could be extended to a valid scheme was very low, we tried adding *streamlining constraints* [8] to formulas. A streamlining constraint is a set of clauses that guides the solver to a solution, but these clauses may not (and generally are not) implied by the formula. Streamlining constraints are usually patterns observed in solutions of a given problem, potentially of smaller sizes. We experimented with various streamlining constraints, such as enforcing that each type 0, type 1, and type 2 term occurs either zero times or twice in a scheme (instead of an even number of times). The most effective streamlining constraint that we came up with was observed in the Smirnov scheme: for each summand that is assigned a single type 3 term, enforce that (i) one matrix has either two rows, two columns or a row and a column fully assigned to zero and (ii) another matrix has two rows and two columns assigned to zero, i.e., the matrix has a single nonzero entry. This streamlining constraint reduced the runtime from minutes to seconds. Yet some random pairings may only be extended to a valid scheme that does not satisfy the streamlining constraint.

3.2 Neighborhood Search

The second method is based on neighborhood search: we select a scheme, randomly fix some the corresponding base variables, and search for an assignment for the remaining base variables. This simple method turned out to be remarkably effective to find new schemes. The only parameter for this method is the number of base variables that will be fixed. The lower the number of fixed base variables, the higher the probability to find a different scheme and the higher the costs to find an assignment for the remaining base variables. We experimented with various values and it turned out that fixing 2/3 of the 621 base variables (414) is effective to find many new schemes in a reasonable time.

The neighborhood search is able to find many new schemes, but in almost all cases they have the same pairing of type 3 terms. Only in some rare cases the pairing of type 3 terms is different. Figure 1 shows such an example: scheme A has term $a_{13}b_{31}c_{11}$ in summand 22 and term $a_{23}b_{33}c_{32}$ in summand 23, while the neighboring scheme B has term $a_{13}b_{33}c_{31}$ in summand 22 and terms $a_{13}b_{31}c_{11}$, $a_{23}b_{33}c_{32}$, and $a_{13}b_{33}c_{31}$ in summand 23.

4 Evaluation and Analysis

The methods presented in Sect. 3 enabled us to find several hundreds of solutions individually, but they were particularly effective when combined. The first method allows finding schemes that can be quite different compared to the known schemes. However, finding a scheme using that method may require a few CPU hours as most pairings of type 3 terms cannot be extended to a valid scheme that satisfies the streamlining constraints. The second method can find schemes that are very similar to known ones with a second. The neighborhood of known solutions turned out to be limited to a few hundred of new schemes.

1	$(a_{11} + a_{13} + a_{21} + a_{22} + a_{23})(b_{13})(c_{22} + c_{32})$
2	$(a_{11} + a_{13} + a_{23})(b_{13} + b_{32})(c_{11} + c_{22} + c_{31} + c_{32})$
3	$(a_{11} + a_{13})(b_{32})(c_{21} + c_{22} + c_{31} + c_{32})$
4	$(a_{11} + a_{31})(b_{11} + b_{12} + b_{13})(c_{23})$
5	$(a_{11} + a_{33})(b_{11} + b_{13} + b_{32})(c_{11} + c_{23})$
6	$(a_{12} + a_{13} + a_{23})(b_{13} + b_{33})(c_{11} + c_{31})$
7	$(a_{12} + a_{22} + a_{32})(b_{21} + b_{22} + b_{23})(c_{33})$
8	$(a_{12} + a_{31} + a_{32} + a_{33})(b_{22})(c_{23} + c_{33})$
9	$(a_{12} + a_{33})(b_{13} + b_{21} + b_{33})(c_{11} + c_{33})$
10	$(a_{12})(b_{13} + b_{23} + b_{33})(c_{31} + c_{33})$
11	$(a_{21} + a_{31} + a_{33})(b_{11})(c_{12} + c_{22})$
12	$(a_{21})(b_{11} + b_{12} + b_{13})(c_{22})$
13	$(a_{22} + a_{31} + a_{33})(b_{13} + b_{22})(c_{12} + c_{13} + c_{22} + c_{33})$
14	$(a_{22} + a_{32} + a_{33})(b_{21})(c_{13} + c_{33})$
15	$(a_{22})(b_{13} + b_{21} + b_{22})(c_{12} + c_{13})$
16	$(a_{22})(b_{13} + b_{23})(c_{32} + c_{33})$
17	$(a_{23})(b_{31})(c_{11} + c_{12} + c_{31} + c_{32})$
18	$(a_{31} + a_{33})(b_{11} + b_{13} + b_{22})(c_{12} + c_{13} + c_{22} + c_{23})$
19	$(a_{33})(b_{11} + b_{21} + b_{31})(c_{11} + c_{13})$
20A	$(a_{12})(b_{22})(c_{21} + c_{23})$
21A	$(a_{11})(b_{12} + b_{32})(c_{21} + c_{23})$
22A	$(a_{13} + a_{33})(b_{31} + b_{32} + b_{33})(c_{11})$
23A	$(a_{23})(b_{31} + b_{32} + b_{33})(c_{11} + c_{31} + c_{32})$
20B	$(a_{11} + a_{12})(b_{22})(c_{21} + c_{23})$
21B	$(a_{11})(b_{12} + b_{22} + b_{32})(c_{21} + c_{23})$
22B	$(a_{13} + a_{33})(b_{31} + b_{32} + b_{33})(c_{31} + c_{32})$
23B	$(a_{13} + a_{23} + a_{33})(b_{31} + b_{32} + b_{33})(c_{11} + c_{31} + c_{32})$

Fig. 1. Two neighboring schemes with 19 identical summands and 4 different ones.

In contrast, some of the schemes found using the first method have a large neighborhood. We approximated the size of the neighborhood of a scheme using the following experiment: Start with a given scheme S and find a neighboring scheme by randomly fixing 2/3 of the base variables. Once a neighboring scheme S' is found, find a neighboring scheme of S', etc. We ran this experiment on a machine with 48 cores of the Lonestar 5 cluster of Texas Advanced Computing Center. We started each experiment using 48 threads with each thread assigned a different seed. Figure 2 shows the number of different schemes (after sorting) found in 1000 seconds when starting with one of the four known schemes and a scheme that was computed from the streamlining method. Some of these different schemes are new, while others are equivalent to each other or known ones. We only assure here that they are not identical after sorting the summands.

Observe that the number of different schemes found in 1000 seconds depends a lot on the starting scheme. No different neighboring scheme was found for

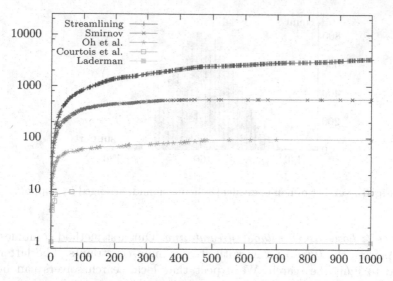

Fig. 2. The number of different schemes (vertical axis in logscale) found within a period of time (horizontal axis in seconds) during a random walk in the neighborhood of a given scheme.

Laderman's scheme, only 9 different schemes were found for the scheme of Courtois et al., 94 different schemes were found for the scheme of Oh et al., 561 new schemes were found for Smirnov's scheme, and 3359 schemes were found using a randomly selected new scheme obtained with the streamlining method.

In view of the large number of solutions we found, it is also interesting to compare them with each other. For example, if we define the *support* of a solution as the number of base variables set to 1, we observe that the support seems to follow a normal distribution with mean around 160, see Fig. 3 for a histogram. We can also see that the Laderman scheme differs in many ways from all the other solutions. It is, for example, the only scheme whose core consists of four quadruples of type 3 terms. In 89% of the solutions, the core consists of four pairs of type 3 terms, about 10% of the solution have three pairs and one quadrupel, and less than 1% of the schemes have cores of the form 2-2-2-2-3 or 2-2-2-3-4.

5   Challenges

The many thousands of new schemes that we found may still be just the tip of the iceberg. However, we also observed that the state-of-the-art SAT solving techniques are unable to answer several other questions. This section provides four challenges for SAT solvers with increasing difficulty. For each challenge we constructed one or more formulas that are available at

https://github.com/marijnheule/matrix-challenges.

The challenges are hard, but they may be doable in the coming years.

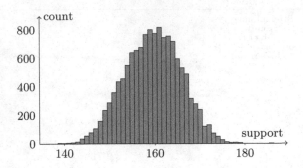

Fig. 3. Number of non-equivalent schemes found, arranged by support.

Challenge 1: Local search without streamlining. Our first method combines randomly pairing the type 3 terms with streamlining constraints. The latter was required to limit the search. We expect that local search solvers can be optimized to efficiently solve the formulas without the streamlining constraints. This may result in schemes that are significantly different compared to ones we found. We prepared ten satisfiable formulas with hardcoded pairings of type 3 terms. Five of these formulas can be solved using `yalsat` in a few minutes. All of these formulas appear hard for CDCL solvers (and many local search solvers).

Challenge 2: Prove unsatisfiability of subproblems. We observed that complete SAT solvers performed weakly on our matrix multiplication instances. It seems therefore unlikely that one could prove any optimality results for the product of two 3×3 matrices using SAT solvers in the near future. A more realistic challenge concerns proving unsatisfiability of some subproblems. We prepared ten formulas with 23 multiplications and hardcoded pairings of type 3 terms. We expect that these formulas are unsatisfiable.

Challenge 3: Avoiding a type 3 term in a summand. All known schemes have the following property: each summand has at least one type 3 term. We do not know whether there exists a scheme with 23 multiplications such that one of the summands contains no type 3 term. The challenge problem blocks the existence of a type 3 term in the last summand and does not have any additional (streamlining) constraints.

Challenge 4: Existence of a scheme with 22 multiplications. The main challenge concerns finding a scheme with only 22 multiplications. The hardness of this challenge strongly depends on whether there exists such a scheme. The repository contains a plain formula for a scheme with 22 multiplications.

Acknowledgments. The authors acknowledge the Texas Advanced Computing Center at The University of Texas at Austin for providing HPC resources that have contributed to the research results reported within this paper.

References

1. Biere, A.: CaDiCaL, Lingeling, Plingeling, Treengeling and YalSAT entering the SAT competition 2018. In: Proceedings of the SAT Competition 2018 – Solver and Benchmark Descriptions. Department of Computer Science Series of Publications B, vol. B-2018-1, pp. 13–14. University of Helsinki (2018)
2. Bläser, M.: On the complexity of the multiplication of matrices of small formats. J. Complex. **19**(1), 43–60 (2003)
3. Bläser, M.: Fast Matrix Multiplication. Number 5 in Graduate Surveys. Theory of Computing Library (2013)
4. Brent, R.P.: Algorithms for matrix multiplication. Technical report, Department of Computer Science, Stanford (1970)
5. Bürgisser, P., Clausen, M., Shokrollahi, M.A.: Algebraic Complexity Theory, vol. 315. Springer, Heidelberg (2013)
6. Courtois, N., Bard, G.V., Hulme, D.: A new general-purpose method to multiply 3×3 matrices using only 23 multiplications. CoRR, abs/1108.2830 (2011)
7. de Groote, H.F.: On varieties of optimal algorithms for the computation of bilinear mappings I. The isotropy group of a bilinear mapping. Theor. Comput. Sci. **7**(1), 1–24 (1978)
8. Gomes, C., Sellmann, M.: Streamlined constraint reasoning. In: Wallace, M. (ed.) CP 2004. LNCS, vol. 3258, pp. 274–289. Springer, Heidelberg (2004)
9. Heule, M.J.H., Kauers, M., Seidl, M.: New ways to multiply 3×3 matrices (in preparation)
10. Laderman, J.D.: A noncommutative algorithm for multiplying 3×3 matrices using 23 multiplications. Bull. Am. Math. Soc. **82**(1), 126–128 (1976)
11. Landsberg, J.M.: Geometry and Complexity Theory, vol. 169. Cambridge University Press, Cambridge (2017)
12. Oh, J., Kim, J., Moon, B.-R.: On the inequivalence of bilinear algorithms for 3×3 matrix multiplication. Inf. Process. Lett. **113**(17), 640–645 (2013)
13. Pan, V.Y.: Fast feasible and unfeasible matrix multiplication. CoRR, abs/1804.04102 (2018)
14. Smirnov, A.V.: The bilinear complexity and practical algorithms for matrix multiplication. Comput. Math. Math. Phys. **53**(12), 1781–1795 (2013)
15. Strassen, V.: Gaussian elimination is not optimal. Numer. Math. **13**(4), 354–356 (1969)
16. Winograd, S.: On multiplication of 2×2 matrices. Linear Algebra Appl. **4**(4), 381–388 (1971)

Speeding Up Assumption-Based SAT

Randy Hickey[✉] and Fahiem Bacchus[✉]

Department of Computer Science, University of Toronto, Toronto, Canada
{rhickey,fbacchus}@cs.toronto.edu

Abstract. Assumption based SAT solving is an essential tool in many applications of SAT solving, especially in incremental SAT solving. For example, assumption based SAT solving is used when solving MaxSat, when computing minimal unsatisfiable subsets and minimal correction sets, and in various inductive verification applications. The MiniSat SAT solver introduced a simple technique for extending a SAT solver to allow it to handle assumptions by forcing the SAT solver to make the assumed literals its initial decisions. This approach persists in almost all current SAT solvers making it the most commonly used technique for handling assumptions. In this paper we explain some deficiencies in this approach that can hinder its efficiency, and provide a very simple modification that fixes these deficiencies. We show that our modification makes a non-trivial difference in practice, e.g., allowing two tested state of the art MaxSat solvers to solve 50+ new instances. This improvement is particularly useful since our modification is extremely simple to implement. We also examine the issue of repeated work when the solver backtracks over the assumptions, e.g., on restarts or when a new unit clause is learnt, and develop a new method for avoiding this repeated work that addresses some deficiencies of prior approaches.

1 Introduction

A wide range of applications of SAT solving rely on assumption-based incremental SAT solving. This includes algorithms for bounded model checking, e.g., [10,11,14,16]; minimum unsatisfiable set (MUSes) extraction, e.g., [6–8,20], computing minimal correction sets (MCSes), e.g., [5,6,22,27]; and solving maximum satisfiability (MaxSat), e.g., [2,3,13,21,23,26,28].

Assumption-based SAT involves requesting the SAT solver to find a solution that also satisfies a specified set of assumptions, encoded as a conjunction of literals. Assumptions are particularly useful in *incremental* SAT solving. In incremental SAT solving the SAT solver is called on a sequence of formulas that are closely related to each other. Each formula could be solved by invoking a new instance of the SAT solver; but then information computed during one solving episode (e.g., learnt clauses) cannot easily be exploited in subsequent solving episodes. The idea of incremental SAT solving is to use only one instance of the SAT solver for all of the formulas so that all information computed when solving the previous formulas can be retained to make solving the next formula

© Springer Nature Switzerland AG 2019
M. Janota and I. Lynce (Eds.): SAT 2019, LNCS 11628, pp. 164–182, 2019.
https://doi.org/10.1007/978-3-030-24258-9_11

more efficient. In this case, we can monotonically add clauses to the SAT solver or use different sets of assumptions to specify each new formula to be solved. With assumptions, e.g., we can add or remove certain clauses by adding to those clauses a new literal ℓ. When we assume $\neg\ell$ these clauses become active (added to the formula), and when we assume ℓ these clauses become inactive (removed from the formula).

The most common technique for supporting assumptions in SAT solvers was originally proposed in [16] and implemented in the MiniSat solver [15]. This technique involves forcing the SAT solver to make as its initial decisions the assumed literals. This approach is still used in the most commonly available SAT solvers handling assumptions, including MiniSat [15], Glucose [4], Lingeling [9], and CryptoMiniSat [30]. However, as we will explain below, this approach can suffer from unnecessary overhead when enforcing the assumptions. One of the main contributions of this paper is an extremely simple and light-weight technique for eliminating this overhead.

There has been previous work on improving assumption-based SAT solving [4,19,24,25]. However, in contrast to the techniques presented in [19,24,25] the techniques we present in this paper are much more light-weight. By this we mean two things: (1) our techniques are much easier to implement, e.g., no major new data-structures or algorithms are needed; and (2) our techniques yield good performance improvements on relatively easy instances that the SAT solver can solve in two hundred seconds or less. In contrast the heavy-weight techniques presented in [19,24,25] are more complex to implement and the empirical evidence presented in these papers indicate that they often slow the SAT solver down on easier instances. These heavy-weight techniques do, however, often pay off (sometimes dramatically) on harder instances on which the SAT solver needs many hundreds or even thousands of seconds. So although our techniques continue to improve SAT solver performance on harder instances, the improvements they yield on such instances are unlikely to be as great as these heavy-weight techniques.

This is an important contrast as assumption-based SAT solving is applied in a diverse set of application areas. In model-checking applications the instances are often very large and very hard, and as shown in [25] on these types of instances the heavy-weight techniques they describe can be very effective. The application area we are most interested in, however, is MaxSat solving. In that domain some solvers, e.g., MaxHS use SAT solving to solve quite simple SAT instances [12], and the key to performance is SAT solver throughput, i.e., solving many instances quickly. Other MaxSat solvers like RC2 [18] solve harder SAT instances than MaxHS, but most of these instances still take less than a few hundred seconds. We demonstrate that our techniques speed up both MaxHS and RC2, enabling both state of the art solvers to solve more instances. Our techniques also speed up the MUS extraction Muser tool [8].

In particular, we present two light-weight techniques for speeding up assumption-based SAT solving. Our first technique is to enqueue all of the assumptions at once in one decision level rather than the standard technique of sequentially making each assumption a decision and performing unit-propagation after each decision. We show that the standard technique can suffer from unnecessary overhead that enqueueing all at once eliminates. We also provide an extensive empirical verification of the effectiveness of this simple idea. Our second technique is to develop a way of enhancing trail-savings in the presence of assumptions during the same SAT solve and between different SAT solves. Although our techniques are not technically sophisticated they have a significant advantage in that they are very **cost effective**: they are easy to implement and provide a non-trivial performance improvement.

In the rest of the paper we will first give some necessary background. Then we motivate our first technique by demonstrating that the standard way of dealing with assumptions can incur overheads that are easily fixed by our approach. We then describe prior work on trail savings as applied to assumption based SAT solving, and show how a simple method can provide savings both on restarts and when unit clauses are learnt. We also show how trail savings can be realized between two SAT solves using different sets of assumptions. Finally, we present empirical results that demonstrate that our techniques provide non-trivial performance improvements.

2   Background

When given an input CNF formula F, a SAT solver can produce either a satisfying truth assignment, or conclude that F is unsatisfiable. Assumption-based SAT solving extends the capacity of the SAT solver by asking it to solve F subject to a set of assumptions A which must be a set of literals. Now the SAT solver must either find a truth assignment satisfying $F \wedge A$ (i.e., a truth assignment satisfying F that also makes all of the literals in A true), or it must conclude that $F \wedge A$ is unsatisfiable. Furthermore, and most critical in many applications, when $F \wedge A$ is unsatisfiable the SAT solver must return a clause C such that (1) C contains only negated literals of A, i.e., $A \models \neg C$ and (2) $F \models C$. Putting (1) and (2) together we obtain $F \models \neg A$. We call any clause C that satisfies (1) and (2) a **conflict clause for** A. For any such clause C every model of F must falsify at least one of the literals of A whose negation is contained in C.

It should be noted that the conflict clause C returned by the SAT solver need not be minimal. That is, there could be another clause $C' \subsetneq C$ satisfying the above two conditions, but the SAT solver did not compute it. In many applications the size of the returned conflict is important for performance: the shorter the returned conflict clause is the more effective it tends to be for the application.

We assume the reader is familiar with conflict-driven clause learning (CDCL) SAT solvers [29] and the concepts of unit-propagation and conflict analysis. Some familiarity with the MiniSat or Glucose code base [4,15] might also be helpful. Modern SAT solvers use a two-watch-literal scheme to effect efficient unit-propagation. Treating a clause C as an indexed array of literals, the MiniSat scheme is to delegate the first two literals in the clause, $C[0]$ and $C[1]$, as the watch literals. Associated with every literal ℓ is a list of all clauses that ℓ is currently a watcher for, $watchlist(\ell)$. Hence, for clause C, $C[0] = \ell$ or $C[1] = \ell$ if and only if $C \in watchlist(\ell)$.

CDCL SAT solvers utilize a trail containing the current path of the search tree being explored. This path consists of the set of literals currently assigned $true$. Let $val(\ell)$ denote the assigned truth value ($true$ or $false$) of literal ℓ. Hence the trail contains a sequence of literals $\ell_1, \ldots, \ell_k$ such that $val(\ell_i) = true$ for $1 \le i \le k$. The literals on the trail are divided up into decision levels starting at zero. Decision level zero contains literals implied by the input formula F. Subsequent decision levels are started by finding a new unvalued literal d that the SAT solver decides to make $true$ (a decision literal). The trail's decision level is then incremented and that literal is then **enqueued**; i.e., it is assigned the value $true$ and it is added to the end of the trail. Whenever a literal is enqueued its decision level is the trail's current decision level so the decision literal d starts a new decision level in the trail.

The SAT solver keeps a unit-prop pointer to the last literal on the trail that has been unit propagated. The literals on the trail between the unit-prop pointer and the end of the trail have not yet been unit propagated. So whenever new literals are added to the trail, the suffix of un-propagated literals grows. Before the next decision is made the SAT solver unit propagates every un-propagated literal on the trail moving the unit-prop pointer forward. This process might enqueue new literals, but it eventually terminates (by finding a conflict, or by running out of literals to unit propagate). After all literals have been unit-propagated the SAT solver starts a new decision level. Hence, all literals enqueued by unit propagation are at the same level as the most recent decision. If a conflict (a clause falsified by the trail) is found, unit propagation is terminated and the solver backtracks after learning a new clause. If all variables have been valued the SAT solver terminates returning "satisfiable".

The process of unit propagating a literal ℓ involves examining all clauses in $watchlist(\neg\ell)$ to determine if any of them have become falsified or unit (i.e., all but one literal in the clause has been falsified). A clause cannot become unit unless one of its watches has become false, hence it is sufficient to check only the clauses in $watchlist(\neg\ell)$ when ℓ is made $true$. If the clause $C \in watchlist(\neg\ell)$ has become falsified, then unit propagation can stop and clause learning and backtracking can occur. If C has become unit with sole remaining unfalsified literal x, then x can be enqueued if it is not already on the trail. Determining if C has become falsified or unit requires examining the non-watch literals in C (i.e., from $C[2]$ onwards) until a non-false non-watch literal x is found. In the worst case this requires examining all literals in C (except for $C[0]$ and $C[1]$). If such a literal x is found, it replaces ℓ as one of C's watches. That is, x and ℓ are swapped with each other in C, and C is removed from $watchlist(\neg\ell)$ and placed on $watchlist(\neg x)$. If no such x exists, then if C's other watch is unvalued it is the sole remaining non-false literal in C and it can be enqueued, else if C's other watch is already false C has become falsified.[1]

3   Equeueing All Assumptions at once

3.1   Standard Approach

Let the input formula be F and the assumptions $A = \{a_1, \ldots, a_k\}$. The standard technique of supporting assumptions [16] used in most modern SAT solvers is to require that the SAT solver choose a_i as the decision literal whenever it starts decision level i for $1 \leq i \leq k$. If a_i is already *true* the SAT solver increments the decision level and continues to the $i+1$-th decision, i.e., an empty decision level is added to the trail. If a_i is already *false*, then it is the case that the previous $i-1$ assumption decisions were sufficient to imply $\neg a_i$. So $C = (\neg a_1, \ldots, \neg a_i)$ is already a clause such that $F \models (\neg a_1, \ldots, \neg a_i)$ and $A \models \neg C$. That is, C is conflict clause for A. However, by conflict analysis (described in more detail below) a shorter conflict than C can typically be computed. Otherwise a_i is still unvalued, and it is enqueued as a new decision. After all assumptions are enqueued the SAT solver is free to make decisions according to its normal heuristics.

Since A always becomes a prefix of the trail, if a satisfying model is found that model must satisfy $F \wedge A$. Otherwise eventually a conflict will be found that forces $\neg a_j$ for some j at some level i where $i < j$. In that case the SAT solver will backtrack to level i and add $\neg a_j$ as a new unit implicant. Then when the SAT solver descends again to level j it will find that a_j is already *false* and will invoke conflict analysis to compute and return a conflict clause over A. Note that irrespective of the satisfiability status of $F \wedge A$, the SAT solver during its search might learn any number of new clauses that cause new unit-implicants to be added into the assumption levels (levels $i \leq k$). Each such new implicant at level i will cause the SAT solver to undo the decisions $a_{i+1}, \ldots, a_k$, after which it will have to make those decisions again as it re-descends the search tree.

This backtracking into the various assumption levels to add new implied literals and then having to redo the remaining assumption decisions is the **first inefficiency** of the standard technique. In some applications, particularly in MaxSat solving, there can often be thousands of assumptions, so this inefficiency can have a significant impact. The following example shows that this inefficiency can potentially induce an overhead that is quadratic in the number of assumptions.

Example 1. Let the assumptions $A = \{a_1, \ldots, a_n\}$ be processed by the SAT solver in this order, and the input formula F consist of three sets of clauses: (1) $\{(\neg a_1, z_i^1, z_i^2) | i = 1 \ldots n\}$, (2) $\{(\neg y_i, \neg z_i^1) | i = 1 \ldots n\}$, and (3) $\{(z_i^1, \neg z_i^2) | i = 1 \ldots n\}$. Furthermore, assume that after setting the assumptions the SAT solver's

---

[1] Quick checks to determine if C is already satisfied can be made first by checking data in the watch data structure (the blocking literal) and checking if the clause's other watch is *true*.

branching heuristic selects variables $y_1, \ldots, y_n$ in that order before any of the variables $z_1^1, z_1^2, \ldots, z_n^1, z_n^2$, and follows a default phase of first setting each literal to *true*.

On its first descent the SAT solver will assume a_1, ..., a_n in the first n decision levels. None of these decisions cause any unit propagations. Then it will select y_1 as its $n+1$-th decision. The literal $\neg z_1^1$ will be implied by unit propagation from the clause $(\neg y_1, \neg z_1^1)$ in set (2). Then the literal $\neg z_1^2$ will be implied from the clause $(z_1^1, \neg z_1^2)$ in set (3). This will falsify the clause $(\neg a_1, z_1^1, z_1^2)$ in set (1). The 1-UIP clause $(\neg a_1, z_1^1)$ will be generated from resolving the conflict $(\neg a_1, z_1^1, z_1^2)$ against the reason clause $(z_1^1, \neg z_1^2)$ for $\neg z_1^2$. This will cause the SAT solver to backtrack to level 1, where both z_1^1 and $\neg y_1$ will be unit implied.

After this the SAT solver will repeat setting $a_2, \ldots, a_n$ as assumptions, and at decision level $n+1$ will select y_2 as the decision literal. Applying the same reasoning as before, except with the clauses $(\neg y_2, \neg z_2^1)$, $(z_2^1, \neg z_1^2)$, and $(\neg a_1, z_2^1, z_2^2)$, we see that again the SAT solver will backtrack to level 1 where it will add the unit implicants z_2^1 and $\neg y_2$. This would continue to happen n times, giving rise to the SAT solver making the assumption decisions $O(n^2)$ times before concluding that all clauses of F are satisfied.   ∎

The **second inefficiency** of the standard technique arises from the observation by Gent [17] that the unit propagation scheme used in most modern CDCL solvers can visit the literals of the **same** clause $O(n^2)$ times, where n is the length of the clause, when descending a single branch. This second inefficiency arises from having to move a clause from one watch list to another $O(n)$ times and each time having to scan $O(n)$ literals in the clause.

Example 2. Let the assumptions $A = \{a_1, \ldots, a_n\}$ be processed by the SAT solver in this order. Consider the length n clause $C = (x, \neg a_1, \neg a_2, \ldots, \neg a_{n-1})$. The two watch literals are x (at $C[0]$) and a_1 (at $C[1]$). The assumption a_1 is enqueued first by the SAT solver, and when unit propagated $C \in watchlist(\neg a_1)$ will be checked from $C[2]$ onwards for a new non-false non-watch literal. This search will find $\neg a_2$ at position $C[2]$ and make it a new watch by swapping $\neg a_1$ (at $C[1]$) and $\neg a_2$ (at $C[2]$) and placing C in $watchlist(\neg a_2)$. The assumption a_2 will be enqueued next, and $C \in watchlist(\neg a_2)$ will be searched again, this time from $C[2]$ to $C[3]$, before a new watch, $\neg a_3$ is found at $C[3]$. Literals $\neg a_2$ (at $C[1]$) and $\neg a_3$ (at $C[3]$) will be swapped and C placed in $watchlist(\neg a_3)$. When the i-th assumption a_i is made, C will be on $watchlist(\neg a_i)$, and positions $C[2]$ to $C[i+1]$ will be searched until finding $\neg a_{i+1}$ as a new watch at position $C[i+1]$. Literals $\neg a_i$ (at $C[1]$) and $\neg a_{i+1}$ (at $C[i+1]$) will be swapped and C placed in $watchlist(\neg a_{i+1})$. In total, in making the first n assumptions the SAT solver will have to visit $O(n^2)$ literals in C to finally conclude that x is unit implied by the assumptions.[2] Figure 1 illustrates this process.   ∎

---

[2] This description follows the MiniSat and Glucose schemes for watch literals, but this particular type of implementation is not necessary. Scanning $O(n^2)$ literals in the clause down a single branch occurs with any implementation that stores no information about the previous scan [17].

$$\begin{array}{cccccc} & U & F & U & U & & U \\ \text{Propagating } \neg a_1: & x & \neg a_1 & \neg a_2 & \neg a_3 & \cdots & \neg a_{n-1} \\ & & & \vdash\!\!\!-\!\!\!\dashv & & & \end{array}$$

$$\begin{array}{cccccc} & U & F & F & U & & U \\ \text{Propagating } \neg a_2: & x & \neg a_2 & \neg a_1 & \neg a_3 & \cdots & \neg a_{n-1} \\ & & & \vdash\!\!\!-\!\!\!-\!\!\!-\!\!\!\dashv & & & \end{array}$$

$$\vdots$$

$$\begin{array}{cccccc} & U & F & F & F & & F \\ \text{Propagating } \neg a_{n-1}: & x & \neg a_{n-1} & \neg a_1 & \neg a_2 & \cdots & \neg a_{n-2} \\ & & & \vdash\!\!\!-\!\!\!-\!\!\!-\!\!\!-\!\!\!-\!\!\!-\!\!\!-\!\!\!-\!\!\!-\!\!\!\dashv & & & \end{array}$$

Fig. 1. The propagate routine searching for a new watcher for clause C of Example 2 when each assumptions is made at a separate decision level and unit propagated before moving to the next assumption. The first two literals are the current watched literals and the line shows the span of literals traversed in searching for a replacement watcher. Truth values are above each literal; "F" represents false, "T" represents true, and "U" represents unassigned.

3.2 New Approach

The technique we propose for fixing both of these inefficiencies is very simple. *After all literals at level zero have been unit propagated, the SAT solver increments the decision level and enqueues all assumptions at decision level 1, after which it performs unit propagation.* So all assumptions are placed on the trail at the top of level 1, and unit propagation is performed only after all assumption literals are on the trail and have been assigned the value *true*.

If when processing the assumptions the SAT solver finds one a_i that is already *true*, then a_i must have been made true at level 0 since unit propagation has not yet been run at level 1. That is, a_i is already on the trail in level 0 and we do not need to enqueue it again so we can skip it. Similarly if a_i is already *false* then $\neg a_i$ must have been made true at level 0, so $F \models (\neg a_i)$ and the SAT solver can return the conflict clause $(\neg a_i)$.

Enqueueing all assumptions at level 1 also necessitates a change to the SAT solver's **analyzeFinal** routine which is called in the standard technique to compute a conflict clause when an assumption a_i is about to be enqueued as the i-th decision and is discovered to be *false*. We will describe those changes below, but first we explain how our technique fixes the two inefficiencies identified above.

With our technique all assumptions are at level 1, so if a new unit implicant of the assumptions is found during search, that clause will cause a backtrack to level 1. None of the assignments in level 1 will be undone, instead the new unit will be added to the bottom of level 1 and search will continue after unit propagation is run on the new unit. This resolves the first inefficiency.

Example 3 (Example 1 continued). With the same set of assumptions A and input formula F as Example 1 our technique would operate as follows. First all assumption literals $a_1, \ldots a_n$ would be enqueued at level 1 (in this example F has no initial unit clauses so level 0 will be empty). As before unit propagation will not find any implied units. Then the SAT solver will increment the decision level to 2 and select y_1 as the next decision. Unit propagation operates in the same way as in Example 1, and the learnt 1-UIP clause will again be $(\neg a_1, z_1^1)$. This will cause the SAT solver to backtrack to level 1, where it will add z_1^1 and $\neg y_1$ at the bottom of level 1 without disturbing any of the assumption literals on the trail. The SAT solver will then make a new decision, y_2 and in the same way a backtrack to level 1 will be generated where the new units z_2^1 and $\neg y_2$ will be added at the bottom of level 1 with out disturbing the previously added units z_1^1 and $\neg y_1$. This will continue n times until all y_i are set and all clauses are satisfied. So this process will require making only n instead of $O(n^2)$ assumption decisions.   ∎

Our technique also addresses the second inefficiency. In particular, since all assumption literals are valued before unit propagation starts, no clause will ever be moved to the watch list of a negated assumption literal as all of these literals are already *false*.

Propagating all	U	F	F	F		F
assumptions at once:	x	$\neg a_1$	$\neg a_2$	$\neg a_3$	...	$\neg a_{n-1}$

Fig. 2. The propagate routine searching for a new watcher when all assumptions are enqueued before unit propagation on the clause from Example 2. Line shows span of literals traversed in searching for a replacement watch.

Example 4 (Example 2 continued). With the same set of assumptions A and clause $C = (x, \neg a_1, \neg a_2, \ldots, \neg a_{n-1})$ as Example 2 our technique would operate as shown in Fig. 2. Since all of the assumption literals in the clause have already been made false, unit propagation will visit the literals from $C[2]$ to $C[n-1]$ only once concluding that x is unit implied by the assumptions. That is, with our technique only $O(n)$ literals of C will be examined instead of $O(n^2)$ with the standard technique.   ∎

3.3   Implementation

Our new approach of enqueueing all assumptions at once is very simple to implement, and we illustrate this in the framework of the MiniSat code base. It should be equally easy to implement our approach in non-MiniSat based SAT solvers. Two routines need to be altered: (1) the main **search()** routine that makes new decisions, invokes unit propagation, and performs clause learning when a conflict is detected; and (2) the **analyzeFinal()** routine that computes the final conflict clause. The needed changes to the **search()** routine are shown in Algorithm 1. If a conflict occurs at decision level 1 (where the assumptions are enqueued) we must convert it into a conflict for the assumptions (line 6); and if we are making a decision at level 0 we enqueue all assumptions at level 1 (lines 19–26).

Algorithm 1. Code changes for a MiniSat based SAT Solver to implement enqueueing all assumptions at once. Line 6 is added and lines 11–17 are replaced with lines 19–17.

```
1  search ()
2  while true do
3  |    confl = unitPropagate()
4  |    if confl then
5  |    |    if decisionLevel() ≡ 0 then return false;
6  |    |    if decisionLevel() ≡ 1 then              /* ADD THIS LINE*/
7  |    |    |    analyzeFinal(confl, conflict); return false;
8  |    |    analyze conflict and backtrack
9  |    else                                          /* No conflict */
10 |    |    begin /* REMOVE THIS BLOCK /*
11 |    |    |    while assumptions remain do
12 |    |    |    |    aᵢ = nextAssumption();
13 |    |    |    |    if value(aᵢ) ≡ false then  analyzeFinal(¬aᵢ, conflict);
14 |    |    |    |    else if value(aᵢ) ≡ true then  newDecisionLevel();
15 |    |    |    |    else nextDecision = aᵢ; break;
16 |    |    |    if nextDecision ≡ NIL then  nextDecision = heuristic();
17 |    |    |    newDecisionLevel(); enqueue(nextDecision);
18 |    |    begin /* ADD THIS BLOCK */
19 |    |    |    if decisionLevel() == 0 then
20 |    |    |    |    newDecisonLevel()
21 |    |    |    |    forall the Assumptions aᵢ do
22 |    |    |    |    |    if value(aᵢ) ≡ false then  conflict = {aᵢ}; return false;
23 |    |    |    |    |    if value(aᵢ) ≢ true then  enqueue(aᵢ);
24 |    |    |    else
25 |    |    |    |    nextDecision = heuristic()
26 |    |    |    |    newDecisionLevel(); enqueue(nextDecision)
```

The routine **analyzeFinal()** also changes. In the standard approach it is passed an assumption literal that has been falsified by a prior set of assumption decisions (line 13), whereas in the new approach it is passed the conflict clause found at level 1 by unit propagation (line 6). In both cases the routine must return the computed conflict in the passed **conflict** vector.

The standard MiniSat implementation starts with C equal to $\neg a_i$'s reason clause (i.e., the clause that became unit implying $\neg a_i$). Then while there exists a literal ℓ in C not equal to $\neg a_i$ and such that $\neg \ell$ has been unit implied by reason clause $reason(\neg \ell)$ we replace C by the resolution of C and $reason(\neg \ell)$. The end result is a conflict clause that contains $\neg a_i$ and the negation of decision literals. Since, $\neg a_i$ was implied at a level where all decisions are assumptions, the computed clause contains only negated assumption literals as required.

The new implementation is very similar. It starts with C equal to the passed conflict. Then while there exists a literal ℓ in C such that $\neg\ell$ has been unit implied by reason clause $reason(\neg\ell)$ we replace C by the resolution of C and $reason(\neg\ell)$. Since the conflict occurs at level 1 only the assumption literals have no reason clause, so this process removes all non-assumption literals from C and the final result contains only negated assumption literals as required.

It should be noted however that these two different approaches can produce different conflict clauses even if the initial conflict and trail are similar. In some cases the standard approach might produce a shorter clause and in other cases our new approach might produce a shorter clause.

Example 5. The following table illustrates a case where the standard technique will learn a shorter conflict clause.

Standard			Enqueue All		
Level	Lit	Reason	Level	Lit	Reason
1	a_1	nil	1	a_1	nil
1	a_2	$(a_2, \neg a_1)$	1	a_2	nil
1	a_3	$(a_3, \neg a_1)$	1	a_3	nil
1	$\neg a_4$	$(\neg a_4, \neg a_3, \neg a_2, \neg a_1)$	1	a_4	nil
2	empty level		Conflict at level 1: $(\neg a_4, \neg a_3, \neg a_2, \neg a_1)$		
3	empty level				
4	Conflict at level 4: $\neg a_4$				

The table shows the trail when a conflict is found. The left three columns show the trail for the standard approach, and the right three show the trail for our proposed approach of enqueueing all assumptions at level 1. The table indicates the decision level, the literal on the trail and the reason clause. Literals with non-nil reasons are implied literals.

The standard approach detects a conflict at level 4 when it tries to make a_4 true as the next decision and finds that a_4 is already false. The conflict it learns starts with the reason clause for $\neg a_4$, $(\neg a_4, \neg a_3, \neg a_2, \neg a_1)$, and proceeds to resolve away all implied literals from this clause except for $\neg a_4$. This involves resolving away a_3 and a_2 to obtain the conflict $(\neg a_4, \neg a_1)$.

Our new technique, on the other hand, will enqueue all assumptions at level 1, and then discover that the clause $(\neg a_4, \neg a_3, \neg a_2, \neg a_1)$ is falsified. In the new technique none of these literals are implied so no resolution steps are performed. Hence it will return this longer clause as the conflict.

On the other hand, the following table illustrates a case where the standard technique learns a longer conflict clause.

Standard			Enqueue All		
Level	Lit	Reason	Level	Lit	Reason
1	a_1	nil	1	a_1	nil
2	a_2	nil	1	a_2	nil
3	a_3	nil	1	a_3	nil
3	a_4	$(a_4, \neg a_3, \neg a_2, \neg a_1)$	1	a_4	nil
3	$\neg a_5$	$(\neg a_5, \neg a_4)$	1	a_5	nil
4	empty level		Conflict at level 1: $(\neg a_5, \neg a_4)$		
5	Conflict at level 5: $\neg a_5$				

The standard technique discovers a conflict at level 5 when it finds that a_5 is already falsified. Starting with the reason clause $(\neg a_5, \neg a_4)$, it will resolve away the implied literal $\neg a_4$ to obtain the conflict clause $(\neg a_5, \neg a_3, \neg a_2, \neg a_1)$.

Our new technique, on the other hand, will return the shorter clause $(\neg a_5, \neg a_4)$ as the conflict, as again none of these literals are implied so no resolution steps will be performed. ∎

As we will demonstrate later, although our technique could return longer conflicts in the same context, it generally returns shorter ones than the standard technique.

3.4   Previous Work

Gent [17] proposed an alternate method for eliminating the second inefficiency where a clause moves from the watch list of one assumption to another many times and each time has many of its literals scanned. In particular, he proposed keeping track of, for each clause, the location where the previous search for a non-falsified literal stopped. Then when a new search is made, the search starts at this previous location and if necessary wraps around. Gent showed that this reduces the worst case number of literals that could be visited in a single clause along a single branch to $2n$ down from $O(n^2)$ (n is the clause length). Unlike our approach Gent's method is more intrusive, requiring a change to the clause data structure (to track the location the previous search stopped). However, it accounts for non-assumption literals as well as assumption literals. Nevertheless, Gent also showed that his technique did not yield any significant gains in SAT solver performance on the instances he experimented with, whereas our technique does yield performance gains, perhaps because it fixes both inefficiencies.

Audemard et al. [4] proposed four lightweight techniques for improving assumption-based SAT solving in Glucose. First, they ignored the assumption literals when computing the LBD score of a learnt clause. Our method comes close to achieving the same improvement: with it all assumption literals are at the same level, so they contribute only 1 (instead of 0) to the LBD score. Second, they store all of the assumption literals at the end of each learnt clause so as to avoid having to scan these literals on LBD score updates. Third, they check only the two watch literals of a clause to see if it is satisfied to avoid scanning the entire clause.[3] Our technique does not achieve the second nor the third improvement.

Fourth, during unit propagation when scanning a clause for a new non-false watcher, they continue searching until they find a non-false non-assumption literal. If none exist they use the last non-false literal found (even if it is an assumption literal). This technique addresses some of the second inefficiency, but not all of it. In particular, if in Example 2 the assumptions A are set in the order $a_1, a_{n-1}, a_{n-2}, \ldots, a_2$ then the clause $(x, \neg a_1, \neg a_2, \ldots, \neg a_{n-1})$ will still have its literals scanned $O(n^2)$ times when each assumption is a separate decision. For example, when a_1 is made true, all of the literals $\neg a_2, \ldots, \neg a_{n-1}$ will be scanned, skipping over non-false assumption literals, until finally returning the last non-false assumption literal $\neg a_{n-1}$ as the new watcher.

It can also be noted that none of the techniques of [4, 17] address the first inefficiency.

4   Trail Savings

When a restart returns the SAT solver to level zero, the solver can often proceed to reproduce the same initial sequence of decisions that were on the trail before the restart. Redoing these decisions is redundant work and methods for saving this work by making the SAT solver backtrack only to the point where the restart causes a divergence in the decisions made have been developed [31]. Similarly, [4] proposes only backtracking to the bottom of the assumption levels on restart, as all of the assumption decisions will be redone by the solver.

However, these techniques do not help when the SAT solver learns a new unit clause. Unit clauses are added to level 0 so the solver must backtrack across the assumptions to insert the new unit into the trail. After this it must redo the assumptions. Our trail savings method is based on the observation that after a new unit is added to level 0, the set of literals forced to be true at level 0 and 1 can only either (a) be contradictory or (b) be a superset of the previous set of literals forced at level 0 and 1.

In particular, let L_0 and L_1 be the literals at level 0 and level 1 before a new unit clause (ℓ) is found. The new unit clause will cause a backtrack to the end of level 0 where ℓ will be added and unit propagated. Let L'_0 be the new literals at level 0 after this process. If no conflict is found we have that $L_o \subset L'_0$. Let L'_1 be the new level 1 generated by enqueueing all the assumptions not already in L'_0 and then performing unit propagation. If L'_1 does not generate a contradiction, we must have that $L_1 \subseteq L'_0 \cup L'_1$. If $x \in L_1$ is an assumption then $x \in L'_0 \cup L'_1$ as L'_1 starts with all assumptions not already at level 0. Considering the unit implicants $x \in L_1$ in the order they appeared in the trail, we can conclude inductively that for every other literal l in x's reason clause we

---

[3] It is not clear if this third technique is an improvement outside of the context of MUS extraction.

have that $\neg l \in L'_0 \cup L'_1$. Hence, x must also be a unit implicant at level 1 or 0. Therefore, $L_1 \subset L'_o \cup L'_1$.

Our new technique based on this observation is as follows. When a new unit is learnt we save a copy of the level 1 trail (all literals and clause reasons). Then we backtrack the trail to the end of level 0, add the new unit, and perform unit propagation. If no contradiction is found we then enqueue each literal from our copy of the level 1 preserving the order these literals previously appeared on the trail (so assumptions are enqueued first). If any of these literals is already *true*, we skip enqueueing it. If any of these literals is already *false*, we can compute a conflict for the assumptions. If the falsified literal is an assumption $\neg a_i$ then the conflict is $(\neg a_i)$, otherwise the conflict is computed by passing the stored reason clause associated with the falsified literal (this reason clause has become falsified at level 1) to **analyzeFinal** to compute the conflict. After all saved literals have been added to level 1, we invoke unit propagation from the top of level 1. Note that although we can restore the literals from level 1 we still have to unit propagate them once again. However, this unit propagation process is sped up by the fact that many literals have already been made *true* at level 1. Hence, the second inefficiency of moving a clause from one watch literal to another will occur less frequently.

We can further extend trail savings to provide a head start for the SAT solver when it is called again with a different set of assumptions. Note that level 0 is always preserved between calls to the SAT solver as this level does not depend on the assumptions. Our extension is to also try to preserve as much of level 1 as is possible between SAT calls. The method is to save all of the literals implied at level 1, and their reason clauses, at the time the SAT solver exits. Then when the SAT solver is called again with a new set of assumptions, we enqueue all of these assumptions at level 1 as normal. After the new assumptions are enqueued, and before they are unit propagated, we check all of the saved literals previously implied at level 1, in the order they were previously on the trail. If their reasons are still unit clauses under the new trail, we enqueue them on the trail with the same reason clause.[4] Note that by examining these implied literals in trail order we can detect preserved implied literals that rely on previously preserved implied literals. For example, say x and y were at level 1 at the end of the previous SAT solve with y appearing after x, and with reason clauses $(x, \neg a1)$ and $(y, \neg x)$. Then if the new SAT call includes the previous assumption a_1 our technique will detect that x is still unit implied with the same reason clause, and x will be added to the trail. Then y will also be detected to still be unit because x has already been added to the trail.

---

[4] We save a reference to the reason clause, so before checking to see if the reason is still unit we must ensure that the references haven't been changed by garbage collection, and that the implied literal is still at position zero in the reason clause (this is a MiniSat invariant for reason clauses).

5   Experiments and Results

We implemented our techniques in the MiniSat 2.2 and Glucose 3.0 SAT solvers. In Glucose 3.0 we also preserved the already implemented incremental techniques of [4]. We then used these modified SAT solvers in the MaxHS [12] and RC2 [18] MaxSat solvers, both of which are state-of-the-art MaxSat solvers. MaxHS uses MiniSat while RC2 uses Glucose.

We then ran these solvers using both the modified and original SAT solvers on 7439 benchmark instances (4627 unweighted and 2812 weighted) collected from the 2008 to 2018 MaxSat Evaluations [1]. This benchmark set includes all non-random instances used and submitted to these evaluations, excluding 825 "abrame-habet" random maxcut instances that were categorized as "crafted" instances (none of these are solvable by either MaxHS nor RC2). We also removed duplicate instances that had different names but were the same except for comment lines. The experiments were run on 2.4 GHz Intel cores with 30 min CPU time and 5.24 GB memory limits.

	Total				Unweighted		Weighted	
	MaxHS	+/-	RC2	+/-	MaxHS	RC2	MaxHS	RC2
original	6052	0/0	6030	0/0	3940	3993	2112	2037
enqueue assumptions as set	6131	114/35	6081	99/48	3995	4032	2136	2049
enqueue assumptions as set + save literals after learnt units	6136	120/36	6079	95/46	3991	4030	2145	2049
enqueue assumptions as set + save literals after learnt units + save literals from last invocation	6138	115/29	6080	92/42	3989	4030	2149	2050

Fig. 3. Number of MaxSat instances solved by MaxHS and RC2 using different extensions of the underlying SAT solver. The $+/-$ column shows the number of instances gained/lost vs the original.

Figure 3 shows that our first technique of enqueueing all assumptions at once is surprisingly effective yielding 79 newly solved instances for MaxHS and 51 newly solved instances for RC2. It should be noted that both of these solvers are state-of-the-art and techniques that allow them to solve this many new instances are not easy to find. Trail savings within the same SAT solver call is a less successful improvement. It gains 5 more problems for MaxHS but loses 2 problems for RC2. Trail savings across different SAT solver calls, is even less impactful. Hence, in terms of number of instances solved trail savings do not seem to be either positively or negatively significant, and the remaining experiments use only the enqueueing technique.

Figure 4 shows a cactus plot of the two solvers with and without enqueueing. The plot shows that enqueueing provides a general speedup for both solvers with more instances generally being solved at every time bound in the plot. Note the first 5000 instances were solved by both solvers in ≤ 50 s per instance, so we truncated that part of the plot to show more detail on the harder instances.

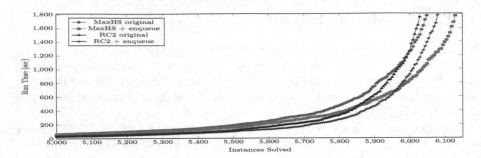

Fig. 4. Cactus plot comparing total instances solved within a given time bound for MaxHS and RC2 with and without our assumption enqueueing techniques. The first 5000 instances were solved in less than 50 s each, so that part of the plot is truncated.

The run times shown in Fig. 4 are affected both by the speed of the SAT solver and by the sequence of conflicts returned by the different SAT solvers. That is, the SAT calls the MaxSat solver performs diverges as the instance is solved. Hence, to get a more precise picture of the SAT solver speedup and the quality of conflicts obtained by our technique, we changed RC2 so that it always invokes both the original Glucose solver and then the modified Glucose solver with enqueueing. However, RC2 always uses the conflict returned by the original Glucose solver in its further processing. In this way, each version of the SAT solver is solving an identical sequence of SAT calls during the processing of each MaxSat instance. We ran this modified version of RC2 on the same suite of 7439 MaxSat instances, giving it 3600 s per instance to account for the doubled up SAT solving. From this setup we obtained a sequence of matched pairs of SAT solver calls where the standard and equeueing versions of the SAT solver are both invoked to solve the same formula subject to the same set of assumptions and with the same history of previous calls.

We measured the CPU time each SAT solver call took in each matched pair, discarding those pairs where both solvers took less than 0.1 s. In particular, we did not compare the CPU times of SAT solver calls that were so fast that they were likely to be too noisy. This yielded 22,810 matched pairs of SAT solver calls. There is quite a bit of variance in the runtimes, so a scatter plot of these points was not very informative. Instead for each pair we computed $\log_2(\frac{\text{Original Glucose CPU}}{\text{Enqueueing Glucose CPU}})$. This number is positive when original Glucose is slower and symmetric but negative when original Glucose is faster. Hence, the absolute value of the negative numbers represent instances that had that $\log_2$ speedup ratio, while the positive numbers represent instances that had that $\log_2$ slowdown ratio. Figure 5 left shows a side-by-side histogram of the number of instances that had similar amounts of $\log_2$ speedup and slowdown ratio. The Figure shows that although there is considerable variance among the instances— some were sped up while others were slowed down—there is a general trend that for all bands of speedup/slowdown factors more instances had a speedup of that factor than had a slowdown of that factor.

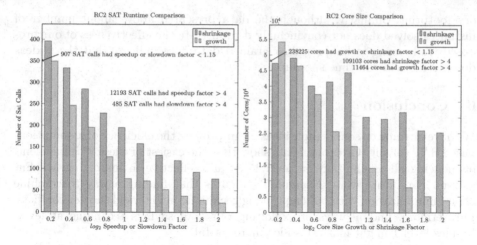

Fig. 5. Left: Comparison of number of instances that had corresponding $\log_2$ speedups or slowdowns. **Right:** Individual core size (right) $\log_2$ comparison.

Our setup also yielded 949,003 matched pairs of conflicts for each solver. In each matched pair one conflict was produced by original Glucose and the other one by enqueueing Glucose, both from solving the identical SAT problem under an identical history of previous calls. As with the run times, we show in Fig. 5 a side-by-side histogram of the $\log_2(\frac{\text{Size of conflict from Original Glucose}}{\text{Size of conflict from enqueueing Glucose}})$ growth and shrinkage ratios: shrinkage indicates that enqueueing Glucose produces a shorter conflict, while growth indicates that it produces a longer conflict. As with the run times there is a considerable variance in core sizes among these identical SAT calls—on some calls enqueueing Glucose produced larger conflicts, and on others it produced smaller conflicts. Nevertheless, the general trend is that for any band of growth/shrinkage ratio, more conflicts produced by enqueueing Glucose had that amount of shrinkage than that amount of growth. The only divergence from this trend was that there were more cores grown by a $\log_2$ factor between 0.1 to 0.3 (the band centered at 0.2) than shrunk by this factor. Most notably, however, almost 10 times as many cores were more than a factor of 4 smaller than were more than a factor of 4 larger.

This experiment shows that the overall better performance of enqueueing Glucose in RC2 is likely a product of both faster SAT calls and smaller conflicts. Together these two effects tend to make RC2 more effective. Although we do not have similar data for MiniSat used in MaxHS, we expect that similar results hold since MaxHS is also more effective with enqueueing.

Although we do not have space to show the data, we also experimented with MUS extraction in the Muser tool [8], and showed that our techniques speed up Muser and allowed it to produce smaller MUSes. In the benchmark suite we used for Muser we were not able, however, to solve any additional instances. The heavier-weight techniques of [19] (involving introducing new abbreviation literals) were able solve additional instances mainly by reducing Muser's memory footprint. Similarly, although the data presented above about number of instances solved does not convincingly demonstrate the effectiveness of our proposed trail saving techniques, finer grained data does indicate that these ideas do tend to yield run time speedups.

6   Conclusion

We have introduced some simple ideas for improving the efficiency of assumption-based SAT solving. Or experiments show that the easiest of these to implement, enqueueing all assumptions at level 1, is quite effective in improving MaxSat solvers. For future work, Example 5 indicates that finding a way to combine clause minimization with our enqueueing technique might yield shorter conflicts. Furthermore, finding ways of exploiting our trail savings technique at levels besides level 1 might make this idea more useful.

References

1. Maxsat evaluation series: 2006–2016 http://www.maxsat.udl.cat/, 2017–2018 https://maxsat-evaluations.github.io/
2. Alviano, M., Dodaro, C., Ricca, F.: A MaxSat algorithm using cardinality constraints of bounded size. In: Proceedings of the Twenty-Fourth International Joint Conference on Artificial Intelligence, IJCAI 2015, Buenos Aires, Argentina, 25–31 July 2015, pp. 2677–2683 (2015). http://ijcai.org/Abstract/15/379
3. Ansótegui, C., Didier, F., Gabàs, J.: Exploiting the structure of unsatisfiable cores in MaxSat. In: Proceedings of the Twenty-Fourth International Joint Conference on Artificial Intelligence, IJCAI 2015, Buenos Aires, Argentina, 25–31 July 2015, pp. 283–289 (2015). http://ijcai.org/Abstract/15/046
4. Audemard, G., Lagniez, J.-M., Simon, L.: Improving glucose for incremental SAT solving with assumptions: application to MUS extraction. In: Järvisalo, M., Van Gelder, A. (eds.) SAT 2013. LNCS, vol. 7962, pp. 309–317. Springer, Heidelberg (2013). https://doi.org/10.1007/978-3-642-39071-5_23
5. Bacchus, F., Davies, J., Tsimpoukelli, M., Katsirelos, G.: Relaxation search: a simple way of managing optional clauses. In: Proceedings of the Twenty-Eighth AAAI Conference on Artificial Intelligence, 27–31 July 2014, Québec City, Québec, Canada, pp. 835–841 (2014). http://www.aaai.org/ocs/index.php/AAAI/AAAI14/paper/view/8618
6. Bacchus, F., Katsirelos, G.: Using minimal correction sets to more efficiently compute minimal unsatisfiable sets. In: Kroening, D., Păsăreanu, C.S. (eds.) CAV 2015. LNCS, vol. 9207, pp. 70–86. Springer, Cham (2015). https://doi.org/10.1007/978-3-319-21668-3_5
7. Belov, A., Lynce, I., Marques-Silva, J.: Towards efficient MUS extraction. AI Commun. 25(2), 97–116 (2012). https://doi.org/10.3233/AIC-2012-0523
8. Belov, A., Marques-Silva, J.: Muser2: an efficient MUS extractor. JSAT 8(3/4), 123–128 (2012). https://satassociation.org/jsat/index.php/jsat/article/view/101
9. Biere, A.: Cadical, Lingeling, Plingeling, Treengeling and YalSAT entering the sat competition 2018. In: Heule, M.J.H., Järvisalo, M., Suda, M. (eds.) Proceedings of SAT COMPETITION 2018 Solver and Benchmark Descriptions. University of Helsinki (2018)

10. Cabodi, G., Lavagno, L., Murciano, M., Kondratyev, A., Watanabe, Y.: Speeding-up heuristic allocation, scheduling and binding with sat-based abstraction/refinement techniques. ACM Trans. Design Autom. Electr. Syst. **15**(2), 121–1234 (2010). https://doi.org/10.1145/1698759.1698762
11. Claessen, K., Sörensson, N.: A liveness checking algorithm that counts. In: Formal Methods in Computer-Aided Design, FMCAD 2012, Cambridge, UK, 22–25 October 2012, pp. 52–59 (2012). http://ieeexplore.ieee.org/document/6462555/
12. Davies, J., Bacchus, F.: Solving MAXSAT by solving a sequence of simpler SAT instances. In: Proceedings Principles and Practice of Constraint Programming - CP 2011–17th International Conference, CP 2011, Perugia, Italy, 12–16 September 2011, pp. 225–239 (2011). https://doi.org/10.1007/978-3-642-23786-7_19
13. Davies, J., Bacchus, F.: Postponing optimization to speed up MAXSAT solving. In: Proceedings of the Principles and Practice of Constraint Programming - 19th International Conference, CP 2013, Uppsala, Sweden, 16–20 September 2013, pp. 247–262 (2013). https://doi.org/10.1007/978-3-642-40627-0_21
14. Eén, N., Mishchenko, A., Amla, N.: A single-instance incremental SAT formulation of proof- and counterexample-based abstraction. In: Proceedings of 10th International Conference on Formal Methods in Computer-Aided Design, FMCAD 2010, Lugano, Switzerland, 20–23 October, pp. 181–188 (2010).http://ieeexplore.ieee.org/document/5770948/
15. Eén, N., Sörensson, N.: An extensible SAT-solver. In: Giunchiglia, E., Tacchella, A. (eds.) SAT 2003. LNCS, vol. 2919, pp. 502–518. Springer, Heidelberg (2004). https://doi.org/10.1007/978-3-540-24605-3_37
16. Eén, N., Sörensson, N.: Temporal induction by incremental SAT solving. Electr. Notes Theor. Comput. Sci. **89**(4), 543–560 (2003). https://doi.org/10.1016/S1571-0661(05)82542-3
17. Gent, I.P.: Optimal implementation of watched literals and more general techniques. J. Artif. Intell. Res. **48**, 231–251 (2013). https://doi.org/10.1613/jair.4016
18. Ignatiev, A., Morgado, A., Marques-Silva, J.: RC2: a python-based MaxSat solver. In: Bacchus, F., Järvisalo, M., Martins, R. (eds.) MaxSAT Evaluation 2018 Solver and Benchmark Descriptions. University of Helsinki (2018)
19. Lagniez, J.-M., Biere, A.: Factoring out assumptions to speed up MUS extraction. In: Järvisalo, M., Van Gelder, A. (eds.) SAT 2013. LNCS, vol. 7962, pp. 276–292. Springer, Heidelberg (2013). https://doi.org/10.1007/978-3-642-39071-5_21
20. Liffiton, M.H., Previti, A., Malik, A., Marques-Silva, J.: Fast, flexible MUS enumeration. Constraints **21**(2), 223–250 (2016). https://doi.org/10.1007/s10601-015-9183-0
21. Martins, R., Manquinho, V., Lynce, I.: Open-WBO: a modular MaxSAT solver'. In: Sinz, C., Egly, U. (eds.) SAT 2014. LNCS, vol. 8561, pp. 438–445. Springer, Cham (2014). https://doi.org/10.1007/978-3-319-09284-3_33
22. Mencía, C., Previti, A., Marques-Silva, J.: Literal-based MCS extraction. In: Proceedings of the Twenty-Fourth International Joint Conference on Artificial Intelligence, IJCAI 2015, Buenos Aires, Argentina, 25–31 July 2015, pp. 1973–1979 (2015). http://ijcai.org/Abstract/15/280
23. Morgado, A., Dodaro, C., Marques-Silva, J.: Core-guided MaxSAT with soft cardinality constraints. In: Proceedings of the Principles and Practice of Constraint Programming - 20th International Conference, CP 2014, Lyon, France, 8–12 September 2014, pp. 564–573 (2014). https://doi.org/10.1007/978-3-319-10428-7_41
24. Nadel, A., Ryvchin, V.: Efficient SAT Solving under assumptions. In: Cimatti, A., Sebastiani, R. (eds.) SAT 2012. LNCS, vol. 7317, pp. 242–255. Springer, Heidelberg (2012). https://doi.org/10.1007/978-3-642-31612-8_19

25. Nadel, A., Ryvchin, V., Strichman, O.: Ultimately incremental SAT. In: Sinz, C., Egly, U. (eds.) SAT 2014. LNCS, vol. 8561, pp. 206–218. Springer, Cham (2014). https://doi.org/10.1007/978-3-319-09284-3_16
26. Narodytska, N., Bacchus, F.: Maximum satisfiability using core-guided MaxSat resolution. In: Proceedings of the Twenty-Eighth AAAI Conference on Artificial Intelligence, Québec City, Québec, Canada, 27–31 July 2014, pp. 2717–2723 (2014). http://www.aaai.org/ocs/index.php/AAAI/AAAI14/paper/view/8513
27. Previti, A., Mencía, C., Järvisalo, M., Marques-Silva, J.: Improving MCS enumeration via caching. In: Proceedings of the Theory and Applications of Satisfiability Testing - SAT 2017–20th International Conference, Melbourne, VIC, Australia, 28 August–1 September 2017, pp. 184–194 (2017). https://doi.org/10.1007/978-3-319-66263-3_12
28. Saikko, P., Berg, J., Järvisalo, M.: LMHS: A SAT-IP hybrid MaxSAT solver. In: Creignou, N., Le Berre, D. (eds.) SAT 2016. LNCS, vol. 9710, pp. 539–546. Springer, Cham (2016). https://doi.org/10.1007/978-3-319-40970-2_34
29. Silva, J.P.M., Lynce, I., Malik, S.: Conflict-driven clause learning SAT solvers. In: Handbook of Satisfiability, pp. 131–153. IOS Press (2009). https://doi.org/10.3233/978-1-58603-929-5-131
30. Soos, M.: The cryptominisat 5.5 set of solvers at the sat competition 2018. In: Heule, M.J.H., Järvisalo, M., Suda, M. (eds.) Proceedings of SAT COMPETITION 2018 Solver and Benchmark Descriptions. University of Helsinki (2018)
31. van der Tak, P., Ramos, A., Heule, M.: Reusing the assignment trail in CDCL solvers. JSAT 7(4), 133–138 (2011). https://satassociation.org/jsat/index.php/jsat/article/view/89

Simplifying CDCL Clause Database Reduction

Sima Jamali[✉] and David Mitchell

Simon Fraser University, Burnaby, Canada
sima_jamali@sfu.ca, mitchell@cs.sfu.ca
http://www.cs.sfu.ca

Abstract. CDCL SAT solvers generate many "learned" clauses, so effective clause database reduction strategies are important to performance. Over time reduction strategies have become complex, increasing the difficulty of evaluating particular factors or introducing new refinements. At the same time, it has been unclear if the complexity is necessary. We introduce a simple online clause reduction scheme, which involves no sorting. We instantiate this scheme with simple mechanisms for taking into account clause activity and LBD within the winning solver from the 2018 SAT Solver Competition, obtaining performance comparable to the original. We also present empirical data on the effects of simple measures of clause age, activity and LBD on performance.

Keywords: Clause deletion · CDCL · Clause database reduction

1 Introduction

CDCL SAT solvers generate a very large number of new "learned" clauses, so clause management methods are central to solver performance [2,13]. In particular, most learned clauses must be deleted to keep the clause database of practical size, and the clause database reduction scheme is one of a small number of key heuristic mechanisms in a CDCL solver [3,14]. Typical clause maintenance strategies involve two stores of learned clauses, which we will call Core and Local. Clauses placed in Core are retained for the entire run. The size of Core is limited by being selective about which clauses are added. The large majority of learned clauses are placed in Local. The size of Local is limited by periodic deletion of "low quality" clauses, which are deemed unlikely to be of high future utility. The quality measure is typically a combination of size, age, literal block distance (LBD) and some measure of usage or activity [3,8–11,14,15].

 Major changes to the general scheme are rare, but over time many refinements have combined to make the overall mechanism in the best recent solvers quite complex. Most details have intuitive explanations, and were chosen based on empirical performance. At the same time, the complexity seems perhaps a bit much relative to our understanding of "clause quality". This complexity makes

© Springer Nature Switzerland AG 2019
M. Janota and I. Lynce (Eds.): SAT 2019, LNCS 11628, pp. 183–192, 2019.
https://doi.org/10.1007/978-3-030-24258-9_12

it hard to evaluate the contributions of individual elements, and is an obstacle to adding new features or refined quality measures.

There are two main aspects to a clause deletion strategy. The first is a method to categorize clauses as likely to be useful (high quality), or not (low quality). The second is implementation of an algorithmic method to remove low quality clauses efficiently. In an idealized scheme, we might have a clause quality measure Q, and keep the clauses in a heap so that the lowest quality clause(s) can be removed when the clause database is deemed too large. Conventional wisdom is that using a heap would be too inefficient. It also seems unlikely that spending time to obtain the very worst clause is necessary. Thus, fast heuristics are desired. One scheme, which we call Delete-Half, is to periodically sort the clauses of Local and delete the half with lowest quality. This scheme has been very widely used for many years, but there are many other possible schemes. While some solvers use other schemes (e.g., [4,16]), we think much more investigation is justified. Regarding clause quality, we expect a very good clause quality measure to involve a combination of many factors. The dominant current quality measure uses VSIDS-like clause "activities". Unfortunately, the way activities are computed and maintained in practice makes it hard to combine activity with other measures of quality in a simple and meaningful way.

The goal of this work is to identify simple methods that might largely account for effectiveness of the best current schemes. We make the following contributions.

1. We introduce a new "online" clause deletion scheme which is simple to implement and maintains the size of Local at a desired value. It does not use sorting and in many natural instantiations takes constant time per conflict. The scheme is presented in Sect. 2.
2. We show that a simple instantiation of this scheme performs comparably to the state of the art. In particular, we implemented the scheme within MapleLCMDistChronoBT, the first-place solver from the 2018 SAT Competition [1,12]. This instantiation takes into account clause usage and LBD using very simple mechanisms. The resulting solver (Online-RU-T2Flag) and its performance are described in Sect. 4.
3. To aid in understanding the degree to which the particular methods play a role in solver performance, we present data from a number of experiments measuring performance or other properties. These appear throughout remaining sections.

Our performance evaluations are carried out using the 400 formulas from the main track of the 2018 SAT Solver Competition, with a 5000 second cut off. Our baseline solver for performance evaluation is MapleLCMDistChronoBT, winner of the competition and all other solvers are modified versions of it. The computations were performed on the Cedar compute cluster [6] on 32-core, 128 GB nodes with Intel "Broadwell" CPUs running at 2.1 Ghz.

1.1 MapleLCMDistChronoBT Clause Database Management

Many top-performing solvers in recent SAT Solver Competitions have been variants or derivatives of MapleSAT [11]. For simplicity, we focus on the first-place solver from the 2018 competition, MapleLCMDistChronoBT [12], which uses the deletion scheme introduced in COMiniSatPS [14,15].

This scheme has three clause databases, called Core, Tier2 and Local. The decision of where to store a newly learned clause in is based on its LBD: Core if LBD ≤ 3, Tier2 if $4 \leq$ LBD ≤ 6 and Local if $6 <$ LBD. If after 100,000 conflicts there are not enough clauses in Core, the core threshold is changed from 3 to 5. A clause may be moved from one DB to another based on LBD or usage. The LBD of each clause is recomputed whenever it is used in conflict analysis or the clause simplifying procedure [3]. If the LBD of a clause is sufficiently reduced, it is moved from Local to Tier2 or Core, or from Tier2 to Core. Every 10,000 conflicts, every clause in Tier2 that has not been used during the last 30,000 conflicts is moved to Local. Every 15,000 conflicts, all the clauses in Local are sorted by their activity and MapleLCMDistChronoBT deletes half of the clauses with lower activities. Clauses that are a reason for a current assignment and clauses with recent improvement in LBD are saved from deletion [3,14].

2 Online Clause Deletion

Our online clause deletion scheme is as follows. The clauses of Local are maintained in a circular list L with an index variable i that traverses the list in one direction. The index identifies the current "deletion candidate" L_i. We have a clause quality measure Q, and some threshold quality value q. When a new learned clause C needs to be stored in Local, we select a "low quality" clause in the list to be replaced with C by sequential search. As long as $Q(L_i) \geq q$, we increment i ("saving" clause L_i for one more "round"); The first time $Q(L_i) < q$, we replace L_i with C (deleting the "old" L_i). The clause quality measure threshold must be chosen so that there are always sufficiently many "low-quality" clauses in the list. There are algorithmic methods to ensure this (for example, using a feedback control mechanism) but it is not hard to obtain good practical performance without them.

Relating Delete-Half and Online Deletion. Consider a Delete-Half scheme with a sort-and-reduce phase every k conflicts. Roughly speaking (ignoring some details for simplicity) each clause is inspected every k conflicts, deleted if its quality is below the median of the current clauses in Local. If we instantiate our online scheme with $S = 2k$, and keep q sufficiently close to the median, we expect each clause to be inspected every k conflicts and deleted if its quality is below the median of the current clauses in Local. In this sense, the two schemes can be made quite close: we trade off sorting for dynamically estimating the median. In doing so, we get a clause database of uniform size, rather than one that significantly grows and shrinks.

Age-Based Deletion. A trivially implemented version of our scheme assumes $Q(C) < q$ for every clause C. This results in a pure age-based scheme: Each new learned clause replaces the oldest clause in Local. This very low-cost scheme works surprisingly well. Figure 1 shows a "cactus-plot" comparison of default MapleLCMDistChronoBT with 3 variants using online deletion. (The Local size limit is set to 80,000 clauses in all solvers using online deletion reported here.)

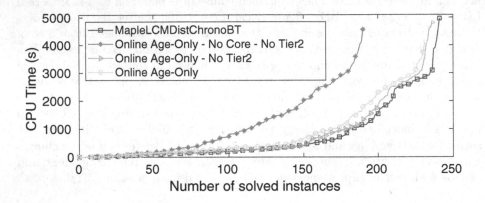

Fig. 1. Simple online deletion performance.

Online Age-Only - No Core, No Tier2. This has no permanent store at all, just pure age-based deletion of all learned clauses.

Online Age-Only - No Tier2. This version keeps clauses with LBD ≤ 3 or Size ≤ 4 permanently in Core, and uses pure age-based deletion from Local.

Online Age-Only. In this version we use Core and Tier2 just as in MapleL-CMDistChronoBT, but use age-based online deletion from Local. If a clause is moved from Tier2 to Local, it replaces the oldest clause in Local.

Figure 1 shows that Core and Tier2 are important to the performance of MapleLCMDistChronoBT. It also shows that in the presence of Core and Tier2 a simple pure age-based deletion scheme for Local gives quite good performance.

We make two observations regarding this second point. First, in online deletion with Local of size S, if the probability of saving a clause is as most 0.5 (see Fig. 4), then every learned clause is kept for at least $S/2$ conflicts, giving it substantial time to be used. Delete-Half schemes generally do not ensure this. Second, age is highly correlated with usage rate, and can account for a large fraction of decisions that would be made based on clause activities. This is illustrated by Fig. 2, which shows the average usage rates of clauses that have been in Local for at least 10K conflicts, at different ages. The usage rate of most clauses drops very quickly.

3 Clause Usage

MiniSAT and many of its successors, including MapleLCMDistChronoBT, use
clause "activity" scores in their clause deletion schemes [8,11,15]. If a clause is
used in conflict analysis, its activity is "bumped", meaning it's activity score is
increased by a reward value. The reward is initialized to 1 and divided by 0.999
(the decay factor) at each conflict, to similate decay of activities. To prevent
activity overflow, when the activity of any clause reaches 1e20, all activity values
and the reward value are divided by 1e−20 [5,8].

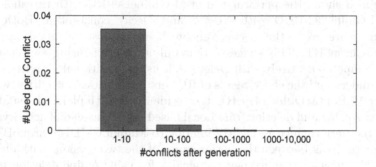

Fig. 2. Rate of use of clauses in Local at different ages.

This scheme, with many variations, has been widely used, but it also has
inconvenient aspects as discussed above. We anticipated that, in the presence
of Core, much simpler usage measures might be effective. Here we report two
that we have considered. Both are extremely simple to implement. We follow
their descriptions with reports of three experiments that may shed light on the
performance of the RU measures.

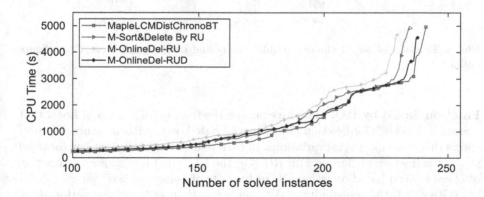

Fig. 3. Online deletion with recent usage

M-OnlineDel-RU. In this version, the measure Q of quality (or activity) is just the number of times the clause was used in conflict analysis during the last "round". That is, we count uses and reset the count to zero if the clause becomes the candidate for deletion but is saved. We denote this measure RU, for Recently-Used. If the threshold value is q (denoted RU = q), a clause will be saved if it was used q or more times in the last round.

M-OnlineDel-RUD. This is similar to M-OnlineDel-RU, but instead of resetting RU to 0 when a clause is saved, we decay it by dividing it by a constant. We call this measure RUD.

Figure 3 shows the performance of M-OnlineDel-RU with threshold RU = 2 and M-OnlineDel-RUD with RU = 2 and Decay constant 4. Both versions perform quite well, the decay version being almost as good as MapleL-CMDistChronoBT. This suggests that online deletion using simple measures might compete effectively with Delete-Half using traditional activities.

To understand the effectiveness of RU versus traditional activities, we created a solver **M-Sort&Delete By RU** that is identical to MapleLCMDistChronoBT but does sorting and deletion from Local based on RU instead of activity. Figure 3 shows the performance is slightly inferior to MapleLCMDistChronoBT on our benchmark, lying between the performance of the two versions with online deletion. This suggests that we pay no penalty for using online deletion instead of the Delete-Half scheme, and confirms that in the presence of Core and Tier2 a simple usage measure can be almost as useful as traditional clause activities.

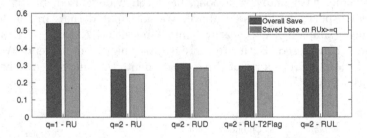

Fig. 4. Fraction of saved clauses in different online deletion schemes (Color figure online)

Fraction Saved by RU. Here we examine the fraction of clauses in Local that become candidates for deletion but are saved based on the RU measure. Figure 4 shows this value for several variations. In each pair of bars, the right bar (orange) shows the fraction of clauses with RU $\geq$ q; the left bar (blue) shows the fraction of clauses saved based on either RU or because of being "locked" [8].

With q = 1, the probability of deletion is less than 1/2, and the performance of the solver is poor. In contrast, with q = 2, about three quarters of clauses are deleted, and the performance is quite good as shown in Fig. 3. In the remainder of the paper, all solvers using RU have q is set to 2.

Table 1. Commonality among high-activity clauses and recently-used clauses.

Formula	Local size	RU ≥ 1 (%)	RU ≥ 2 (%)	RUD ≥ 2 (%)	RUL ≥ 2 (%)
201	28470	12150 (100)	2162 (100)	2265 (97)	3591 (100)
CNP-5-200	28699	13177 (98)	4122 (100)	4923 (87)	2915 (99)
Karatsuba	25251	11091 (86)	1730 (90)	2111 (80)	5571 (89)
T62.2.0.cnf	7097	2474 (100)	521 (100)	618 (86)	1574 (100)
ae_rphp	30535	12586 (87)	6575 (94)	8004 (78)	7278 (94)
apn-sbox6	29422	15459 (87)	6423 (92)	7129 (85)	6282 (90)
cms-scheel	21828	8602 (100)	2454 (100)	2698 (92)	3752 (100)
courses	13869	3241 (100)	854 (100)	1092 (83)	1610 (100)
cz-alt-3-7	26577	11276 (99)	2346 (100)	2639 (93)	7247 (99)
dist9.c	26274	15150 (84)	4182 (92)	4614 (87)	6651 (90)
Average	**23802**	**10521 (93)**	**3137 (97)**	**3609 (87)**	**3395 (96)**

Clauses Saved by RU and Activity. We examined the clauses in Local just before the 10th clause deletion in MapleLCMDistChronoBT, and measured their RU and activity values to see what fraction of clauses would be saved by our RU-based schemes. Table 1 shows the results for one formula from each of 10 families. The first column is the number of clauses in Local just before deletion. Other columns show the number of clauses that would be saved due to RU ≥ q, and the fraction (in percent) of these clauses that have high enough activity to be saved by Delete-Half. On average this fraction is between 87 and 97%, suggesting that simple RU counters can account for a significant fraction of decisions based on activities.

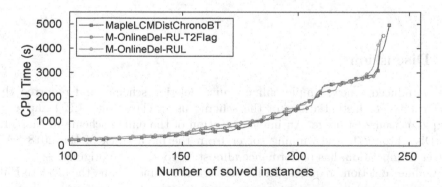

Fig. 5. Online deletion with usage and LBD

4   Clause LBD and Tier2

LBD is used in MapleLCMDistChronoBT for initial placement of a learned clause, and to move clauses between stores if the LBD changes. Here we report two simple methods to take into account LBD changes in a solver with online deletion and no Tier2. Figure 5 show the resulting performance.

M-OnlineDel-RU-T2Flag. Here we replace Tier2 with a rough simulation, by adding a "Tier 2 flag" to clauses in Local. We set the flag true if MapleL-CMDistChronoBT would move it from Local to Tier2, and false for the reverse direction. Clauses with this flag true are always saved. This is not an accurate Tier2 simulation, because the size of the clause DB does not change appropriately. Nonetheless, the resulting performance is very close to the original solver.

M-OnlineDel-RUL. Here we take LBD into account by modifying the usage scoring. Instead of incrementing RU by 1 each time a clause is used, we increment by c/LBD, for a constant c. We call this RUL, for RU with LBD. The RUL values are re-set to zero when a clause is saved. The curve in Fig. 5 is the performance with $c = 20$.

Table 2. Performance on satisfiable *vs* unsatisfiable formulas.

Solver	# Solved	SAT	UNSAT
MapleLCMDistChronoBT	**241**	138	**103**
M-OnlineDel-RU	230	132	98
M-OnlineDel-RUL	237	140	97
M-OnlineDel-RUD	238	**141**	97
M-OnlineDel-RU-T2Flag	238	139	99

5   Discussion

We introduced a new, simple online clause deletion scheme, and reported the performance of instantiations of the scheme using clause age, LBD and very simple measures of usage. An implementation of the online scheme in MapleL-CMDistChronoBT, the winning solver from the main track of the 2018 SAT Solver Competition, has performance almost as good as the original.

Online deletion requires less computation time than the Delete-Half scheme. However, the fraction of run time consumed by deletion in MapleL-CMDistChronoBT is small, so this is not a major performance factor.

The online deletion schemes in this paper use age or age modified by a fixed quality threshold. A dynamic threshold may be more desirable, in which case we may use a feedback control scheme to ensure the threshold is such that the fraction of saved clauses is suitable (*cf* Fig. 4).

We continue to investigate more refined versions of our scheme, in particular with regard to clause quality measures and clause database size. Table 2 shows that our modified solvers are biased toward Satisfiable instances, and we will work on shifting this bias.

Acknowledgement. This research was supported and enabled in part by the Natural Sciences and Engineering Research Council of Canada (NSERC), WestGrid (www.westgrid.ca) and Compute Canada (www.computecanada.ca) [7].

References

1. The international SAT competitions web page. http://www.satcompetition.org
2. Ansótegui, C., Giráldez-Cru, J., Levy, J., Simon, L.: Using community structure to detect relevant learnt clauses. In: Heule, M., Weaver, S. (eds.) SAT 2015. LNCS, vol. 9340, pp. 238–254. Springer, Cham (2015). https://doi.org/10.1007/978-3-319-24318-4_18
3. Audemard, G., Simon, L.: Predicting learnt clauses quality in modern SAT solvers. In: Proceedings of the 21st International Jiont Conference on Artificial Intelligence, IJCAI 2009, pp. 399–404. Morgan Kaufmann Publishers Inc., San Francisco (2009)
4. Biere, A.: Pre, icosat@sc'09. solver description for SAT competition 2009. SAT Competitive Event Booklet (2009)
5. Biere, A., Fröhlich, A.: Evaluating CDCL Variable scoring schemes. In: Heule, M., Weaver, S. (eds.) SAT 2015. LNCS, vol. 9340, pp. 405–422. Springer, Cham (2015). https://doi.org/10.1007/978-3-319-24318-4_29
6. Cedar, A Compute Canada Cluster. https://docs.computecanada.ca/wiki/Cedar
7. Compute Canada: Advanced Research Computing (ARC) Systems. https://www.computecanada.ca/
8. Eén, N., Sörensson, N.: An extensible SAT-solver. In: Giunchiglia, E., Tacchella, A. (eds.) SAT 2003. LNCS, vol. 2919, pp. 502–518. Springer, Heidelberg (2004). https://doi.org/10.1007/978-3-540-24605-3_37
9. Goldberg, E., Novikov, Y.: BerkMin: A fast and robust SAT-solver. In: Lauwereins, R., Madsen, J. (eds.) Design, Automation, and Test in Europe, pp. 465–478. Springer, Dordrecht (2008). https://doi.org/10.1007/978-1-4020-6488-3_34
10. Jamali, S., Mitchell, D.: Centrality-based improvements to CDCL heuristics. In: Beyersdorff, O., Wintersteiger, C.M. (eds.) SAT 2018. LNCS, vol. 10929, pp. 122–131. Springer, Cham (2018). https://doi.org/10.1007/978-3-319-94144-8_8
11. Liang, J.H., Ganesh, V., Poupart, P., Czarnecki, K.: Learning rate based branching heuristic for SAT solvers. In: Creignou, N., Le Berre, D. (eds.) SAT 2016. LNCS, vol. 9710, pp. 123–140. Springer, Cham (2016). https://doi.org/10.1007/978-3-319-40970-2_9
12. Nadel, A., Ryvchin, V.: Chronological backtracking. In: Beyersdorff, O., Wintersteiger, C.M. (eds.) SAT 2018. LNCS, vol. 10929, pp. 111–121. Springer, Cham (2018). https://doi.org/10.1007/978-3-319-94144-8_7
13. Newsham, Z., Ganesh, V., Fischmeister, S., Audemard, G., Simon, L.: Impact of community structure on SAT solver performance. In: Sinz, C., Egly, U. (eds.) SAT 2014. LNCS, vol. 8561, pp. 252–268. Springer, Cham (2014). https://doi.org/10.1007/978-3-319-09284-3_20
14. Oh, C.: Between SAT and UNSAT: the fundamental difference in CDCL SAT. In: Heule, M., Weaver, S. (eds.) SAT 2015. LNCS, vol. 9340, pp. 307–323. Springer, Cham (2015). https://doi.org/10.1007/978-3-319-24318-4_23

15. Oh, C.: Improving SAT solvers by exploiting empirical characteristics of CDCL. Ph.D. thesis, New York University (2016)
16. Soos, M., Nohl, K., Castelluccia, C.: Extending SAT solvers to cryptographic problems. In: Kullmann, O. (ed.) SAT 2009. LNCS, vol. 5584, pp. 244–257. Springer, Heidelberg (2009). https://doi.org/10.1007/978-3-642-02777-2_24

QRAT Polynomially Simulates
∀-Exp+Res

Benjamin Kiesl[1,2(✉)] and Martina Seidl[3]

[1] Institute of Logic and Computation, TU Wien, Vienna, Austria
benjamin.kiesl@gmail.com
[2] CISPA Helmholtz Center for Information Security, Saarbrücken, Germany
[3] Institute for Formal Models and Verification, JKU Linz, Linz, Austria

Abstract. The proof system ∀-Exp+Res formally captures expansion-based solving of quantified Boolean formulas (QBFs) whereas the QRAT proof system captures QBF preprocessing. From previous work it is known that certain families of formulas have short proofs in QRAT but not in ∀-Exp+Res. However, it was not known if the two proof systems were incomparable (i.e., if there also existed QBFs with short ∀-Exp+Res proofs but without short QRAT proofs), or if QRAT polynomially simulates ∀-Exp+Res. We close this gap of the QBF-proof-complexity landscape by presenting a polynomial simulation of ∀-Exp+Res in QRAT. Our simulation shows how definition introduction combined with extended-universal reduction can mimic the concept of universal expansion.

1 Introduction

Proof systems for quantified Boolean formulas (QBFs) have been extensively studied to obtain a better understanding of the strengths and limitations of different QBF-solving approaches (e.g., [3,5,8,10,13,25]). Much is known about instantiation-based proof systems [4,5,16], which provide the foundation for expansion-based solvers [7,15], and about Q-resolution systems [1,2,6,13,14,20, 23,26,28], which provide the foundation for search-based solvers [21,22]. There is, however, one other practically useful proof system that is quite different from the aforementioned ones and whose exact place in the complexity landscape is still unclear: the QRAT proof system [12].

The QRAT proof system is a generalization of DRAT [27] (the de-facto standard for proofs in practical SAT solving) that has its strengths when it comes to preprocessing: Many QBF solvers benefit from preprocessing techniques to simplify a QBF before they actually evaluate its truth. With the QRAT system, it is possible to certify the correctness of virtually all preprocessing simplifications performed by state-of-the-art QBF solvers and preprocessors. Recently, it has been shown that QRAT can polynomially simulate the long-distance-resolution

This work has been supported by the Austrian Science Fund (FWF) projects W1255-N23 and S11408-N23 and the LIT AI Lab funded by the State of Upper Austria.

© Springer Nature Switzerland AG 2019
M. Janota and I. Lynce (Eds.): SAT 2019, LNCS 11628, pp. 193–202, 2019.
https://doi.org/10.1007/978-3-030-24258-9_13

calculus, a strictly stronger extension of the Q-Resolution calculus [18]. So far, however, it has not been known if QRAT can also polynomially simulate the instantiation-based calculus ∀-Exp+Res [16]. In this short paper, we show that this is indeed the case by providing a simulation whose resulting QRAT proof is only linear in the size of the original ∀-Exp+Res proof.

2   Preliminaries

We consider *quantified Boolean formulas* in *prenex conjunctive normal form* (PCNF), which are of the form $\mathcal{Q}.\psi$, where $\mathcal{Q}$ is a *quantifier prefix* and ψ, called the *matrix* of the QBF, is a propositional formula in conjunctive normal form (CNF); we define propositional formulas and quantifier prefixes in the following.

Propositional formulas in CNF are built from variables and logical operators as follows. A *literal* is either a variable x (a *positive literal*) or the negation $\bar{x}$ of a variable x (a *negative literal*). The *complement* $\bar{l}$ of a literal l is defined as $\bar{l} = \bar{x}$ if $l = x$ and $\bar{l} = x$ if $l = \bar{x}$. A *clause* is a finite disjunction of the form $(l_1 \vee \cdots \vee l_n)$ where $l_1, \ldots, l_n$ are literals. We denote the empty clause by $\bot$. A clause with exactly one literal is a *unit clause*. A *formula* is a finite conjunction of the form $C_1 \wedge \cdots \wedge C_m$ where $C_1, \ldots, C_m$ are clauses. Clauses can be viewed as sets of literals, and formulas can be viewed as sets of clauses. For an expression (i.e., a literal, formula, etc.) E, we denote the set of variables occurring in E by $var(E)$. If $var(E)$ is a singleton set, we sometimes treat it like a variable.

A *quantifier prefix* has the form $\mathcal{Q}_1 X_1 \ldots \mathcal{Q}_q X_q$ where all the X_i are mutually disjoint sets of variables, $\mathcal{Q}_i \in \{\forall, \exists\}$, and $\mathcal{Q}_i \neq \mathcal{Q}_{i+1}$. The quantifier of a literal l is $\mathcal{Q}_i$ if $var(l) \in X_i$. Given a literal l with quantifier $\mathcal{Q}_i$ and a literal k with quantifier $\mathcal{Q}_j$, we write $l \leq_{\mathcal{Q}} k$ if $i \leq j$, and $l <_{\mathcal{Q}} k$ if $i < j$. We sometimes write $l \leq k$ instead of $l \leq_{\mathcal{Q}} k$, and we write $l < k$ instead of $l <_{\mathcal{Q}} k$ if $\mathcal{Q}$ is clear from the context. If $l \leq k$, we say that l occurs *left of* k.

Given a literal l and a propositional formula ψ, we define $\psi[l]$ to be the formula obtained from ψ by first removing all clauses that contain l and then removing $\bar{l}$ from all remaining clauses. The result of applying the *unit-clause rule* to ψ is the formula $\psi[l]$ where (l) is a unit clause in F. The iterated application of the unit-clause rule, until either the empty clause is derived or no unit clauses are left, is called *unit propagation*. In case unit propagation derives the empty clause, we say that unit propagation derived a *conflict* on ψ.

Given a propositional formula ψ and a clause $(l_1 \vee \cdots \vee l_k)$, we say that ψ implies $(l_1 \vee \cdots \vee l_k)$ via unit propagation—denoted by $\psi \vdash_1 (l_1 \vee \cdots \vee l_k)$— if unit propagation derives a conflict on $\psi \wedge (\bar{l}_1) \wedge \cdots \wedge (\bar{l}_k)$. For example, the formula $(\bar{x} \vee z) \wedge (\bar{y} \vee \bar{z})$ implies the clause $(\bar{x} \vee \bar{y})$ via unit propagation since unit propagation derives a conflict on $(\bar{x} \vee z) \wedge (\bar{y} \vee \bar{z}) \wedge (x) \wedge (y)$.

A QBF $\exists x \mathcal{Q}.\psi$ is true if at least one of $\mathcal{Q}.\psi[x]$ and $\mathcal{Q}.\psi[\bar{x}]$ is true, otherwise it is false. Respectively, a QBF $\forall x \mathcal{Q}.\psi$ is true if both $\mathcal{Q}.\psi[x]$ and $\mathcal{Q}.\psi[\bar{x}]$ are true, otherwise it is false. If the matrix ψ of a QBF $\mathcal{Q}.\psi$ is the empty formula, then $\mathcal{Q}.\psi$ is true. If ϕ contains the empty clause, then $\mathcal{Q}.\psi$ is false.

An *assignment* is a function from variables to the truth values 1 (*true*) and 0 (*false*). We denote assignments by the sequences of literals they satisfy. E.g., $x\bar{y}$ denotes the assignment that assigns 1 to x and 0 to y.

Finally, for the formal definition of polynomial simulations between proof systems we refer to Cook and Reckhow [9]. An informal summary is this: A proof system f polynomially simulates a proof system g if there exists a polynomial-time procedure that transforms g-proofs into f-proofs.

3 The QRAT Proof System

Here, we introduce the basics of the QRAT proof system [12]. The two main concepts behind QRAT are QRAT *literals* and universal reduction via the *reflexive-resolution-path dependency scheme* [11].

The definition of QRAT literals is based on the notion of an *outer resolvent*. Given two clauses $C \vee l, D \vee \bar{l}$ of a QBF $Q.\psi$, the outer resolvent $C \vee l \bowtie_Q^l D \vee \bar{l}$ of $C \vee l$ with $D \vee \bar{l}$ upon l is the clause consisting of all literals in C together with those literals of D that occur left of l, i.e., the clause $C \cup \{k \mid k \in D \text{ and } k \leq_Q l\}$. If all outer resolvents upon a literal are implied via unit propagation, then that literal is a QRAT *literal* [12]:

Definition 1. *A literal l is a* QRAT *literal in a clause $C \vee l$ with respect to a QBF $Q.\psi$ if, for every clause $D \vee \bar{l} \in \psi \setminus \{C \vee l\}$, it holds that $\psi \vdash_1 C \vee l \bowtie_Q^l D \vee \bar{l}$.*

Example 1. Let $C = (b \vee x \vee y)$ and let $\phi = \exists ab \forall x Q y \exists c.(\bar{b} \vee \bar{y} \vee c) \wedge (a \vee \bar{y} \vee c) \wedge (a \vee b \vee x)$, where $Q \in \{\exists, \forall\}$. The literal y is a QRAT literal in C with respect to ϕ since there are two outer resolvents: the tautology $(b \vee \bar{b} \vee x)$, obtained by resolving with $(\bar{b} \vee \bar{y} \vee c)$, and the clause $(a \vee b \vee x)$, obtained by resolving with $(a \vee \bar{y} \vee c)$. The matrix of ϕ implies both outer resolvents via unit propagation.

Let $\phi = Q.\psi$ be a QBF. If a universal literal u is a QRAT literal in a clause $C \in \psi$, the removal of u from C is called QRAT-*literal elimination*. If, after adding a universal literal u to a clause $C \in \psi$, u becomes a QRAT literal, then this addition is called QRAT-*literal addition*. If a clause contains an existential QRAT literal, it is called a QRAT clause (or simply a QRAT) with respect to ϕ; its addition to a QBF is called QRAT *addition* and its removal is called QRAT *elimination*. It can be shown that QRAT-literal addition and elimination as well as QRAT-clause addition and elimination preserve the truth value of a QBF.

The introduction of definition clauses of the form $(\bar{x} \vee y), (x \vee \bar{y})$ (where x is a fresh variable not occurring in ϕ), is an instance of QRAT addition if we put x

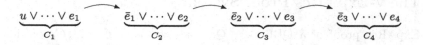

Fig. 1. A resolution path from u to e_4.

into the same quantifier block as y: $(\bar{x} \vee y)$ is a QRAT since x is fresh and thus there are no outer resolvents upon $\bar{x}$; $(x \vee \bar{y})$ is then a QRAT since the only outer resolvent upon x is the tautology $(\bar{y} \vee y)$, obtained by resolving with $(\bar{x} \vee y)$.

The reflexive-resolution-path dependency scheme (short, $\mathcal{D}^{\mathrm{rrs}}$) is based on the notion of a *resolution path* [11]. Intuitively, a QBF contains a resolution path between a universal literal u and an existential literal e if we can start with a clause that contains u and perform a number of resolution steps over existential literals that occur right of u to obtain a clause that contains both u and e. An example of a resolution path is given in Fig. 1.

Definition 2. *Given a QBF $\phi = \mathcal{Q}.\psi$, a universal literal u, and an existential literal e_n, ϕ contains a* resolution path *from u to e_n if there exists a sequence $C_1, \ldots, C_n$ of clauses and a sequence $e_1, \ldots, e_{n-1}$ of existential literals such that*

(1) $u \in C_1$ *and* $e_n \in C_n$,
(2) $e_1, \ldots, e_n$ *occur right of u,*
(3) $e_i \in C_i$, $\bar{e}_i \in C_{i+1}$, *for $i \in 1, \ldots, n-1$, and*
(4) $var(e_i) \neq var(e_{i+1})$ *for $i \in 1, \ldots, n-1$.*

The reflexive-resolution-path dependency scheme defines that a literal e *depends* on a literal u if and only if e is existential, u is universal, and at least one of the following conditions holds: (1) There exist resolution paths from u to e and from $\bar{u}$ to $\bar{e}$. (2) There exist resolution paths from u to $\bar{e}$ and from $\bar{u}$ to e.

Next we define the QRAT proof system. In the QRAT proof system, a *derivation* for a QBF $\phi = \mathcal{Q}.\psi$ is a sequence $M_1, \ldots, M_n$ of proof steps. Starting with $\phi_0 = \phi$, every M_i modifies ϕ_{i-1} in one of the following five ways, which results in a new formula $\phi_i = \mathcal{Q}_i.\psi_i$, which we call the *accumulated formula* at step i:

(1) Add a clause that is implied by ψ_{i-1} via unit propagation.
(2) Add a clause that is a QRAT clause with respect to ϕ_{i-1}.
(3) Remove an arbitrary clause from ϕ_{i-1}.
(4) Remove a QRAT literal from a clause in ϕ_{i-1}.
(5) Remove a universal literal u from a clause $C \vee u \in \phi_{i-1}$ where all $l \in C$ are independent of u according to $\mathcal{D}^{\mathrm{rrs}}$ ("extended universal reduction").

A QRAT derivation $M_1, \ldots, M_n$ thus derives new formulas $\phi_1, \ldots, \phi_n$ from ϕ. If the final formula ϕ_n contains $\bot$, then the derivation is a (*refutation*) *proof* of ϕ. To simplify the presentation, we do not specify in detail how the modification steps M_i are represented syntactically, but it should be clear that their size needs to be at most linear with respect to the involved clauses and literals. Note that certain proof steps can modify the quantifier prefix.

4   The ∀-Exp+Res Proof System

A ∀-Exp+Res proof of a QBF $\phi = \mathcal{Q}.\psi$ is a sequence $C_1, \ldots, C_n$ of clauses where each clause is obtained either via the *axiom rule* or the *resolution rule*. The axiom rule is as follows:

$$\frac{C}{\{l^{\tau_l} \mid l \in C, l \text{ is existential}\}} \text{ (Ax)}$$

Here, C is a clause of ψ, τ is an assignment that falsifies all universal literals of C, and τ_l denotes the assignment τ restricted to the universal variables u with $u < l$. Intuitively, τ_l can be seen as an annotation of the literal l. For example, the axiom rule allows us to use the assignment $\tau = u\bar{v}$ for deriving the clause $x^u \vee \bar{y}^{u\bar{v}}$ from the formula $\forall u \exists x \forall v \exists y.(\bar{u} \vee x \vee v \vee \bar{y})$. The resolution rule of ∀-Exp+Res is just the usual resolution rule from propositional logic—it derives a new clause C_k from two earlier clauses C_i, C_j with $i, j < k$:

$$\frac{C \vee l^\tau \qquad D \vee \bar{l}^\tau}{C \vee D} \text{ (Res)}$$

We next illustrate the intuition behind the simulation of ∀-Exp+Res by QRAT.

5   Simulating ∀-Exp+Res by QRAT: Intuition

To simulate ∀-Exp+Res by QRAT, we need to find a way to simulate applications of the axiom rule. Intuitively, the axiom rule introduces multiple instantiations of a single existential variable because, in satisfying assignments of the formula, this variable might take different truth values depending on the truth values of the universal variables that occur left of it. We can introduce these instantiations in QRAT by first adding definitions of the new variables and then eliminating the superfluous universal variables with extended-universal reduction. Once this is done, we can just straightforwardly perform the remaining resolution steps in QRAT since resolvents are implied via unit propagation. Assume a ∀-Exp+Res proof uses the axiom rule as follows, where the quantifier prefix is $\forall u \exists x \forall v \exists y$:

$$\frac{\bar{u} \vee x \vee v \vee y}{x^u \vee y^{u\bar{v}}} \text{ (Ax)}$$

We simulate the derivation of $(x^u \vee y^{u\bar{v}})$ in QRAT as follows:

(1) Add definitions for the new variables x^u and $y^{u\bar{v}}$, where x^u goes to the same quantifier block as x and $y^{u\bar{v}}$ goes to the same quantifier block as y. The new clauses are $(\bar{x} \vee x^u), (x \vee \bar{x}^u), (\bar{y} \vee y^{u\bar{v}}), (y \vee \bar{y}^{u\bar{v}})$.
(2) Add a clause that is similar to the original clause $(\bar{u} \vee x \vee v \vee y)$, with the only difference that we now use the new annotated variables instead of the original ones. Observe that we can resolve $(\bar{x} \vee x^u)$ with $(\bar{u} \vee x \vee v \vee y)$ to replace x by x^u; likewise for y and $y^{u\bar{v}}$. Because of this, $(\bar{u} \vee x \vee v \vee y)$ and the definition clauses together imply the new clause, $(\bar{u} \vee x^u \vee v \vee y^{u\bar{v}})$, via unit propagation.
(3) Eliminate $(\bar{u} \vee x \vee v \vee y)$ and the definition clauses introduced in step 1.
(4) Eliminate the universal literals $\bar{u}$ and v from $(\bar{u} \vee x^u \vee v \vee y^{u\bar{v}})$ by extended universal reduction, resulting in the clause $(x^u \vee y^{u\bar{v}})$.

The correctness of the fourth step is a consequence of Lemma 1, which we prove in the next section, where we define our simulation.

6   Simulating ∀-Exp+Res by QRAT

We start with a QBF $\mathcal{Q}.\psi$ and a ∀-Exp+Res proof π of $\mathcal{Q}.\psi$. We then construct a QRAT proof Π of $\mathcal{Q}.\psi$ as follows:

Step 1 (Introduction of Definitions): For each annotated variable x^τ in the ∀-Exp+Res proof π, we introduce a definition of the form $(\bar{x} \vee x^\tau), (x \vee \bar{x}^\tau)$. We also put x^τ into the same quantifier block as x. Note that each annotated variable must have been obtained by an application of the axiom rule. The definition introductions are QRAT additions, as explained on page 3. We denote the resulting accumulated formula by $\mathcal{Q}'.\psi_1$.

Step 2 (Introduction of Annotated Clauses): For each clause $C^\tau \in \pi$ that was obtained from a clause $C \in \psi$ by applying the axiom rule with the assignment τ, we add the clause $C^\tau \vee \bar{u}_1 \vee \cdots \vee \bar{u}_k$. Since C and the definitions of the annotated literals of C^τ are in ψ_1, the clause $C^\tau \vee \bar{u}_1 \vee \cdots \vee \bar{u}_k$ is implied via unit propagation and thus it can be added as a QRAT. We denote the accumulated formula after performing all these QRAT additions by $\mathcal{Q}'.\psi_2$

Step 3 (Elimination of Input Clauses and Definitions): We now eliminate all clauses of ψ as well as the definitions introduced in step 1 since we don't need them anymore. Note that QRAT allows the elimination of arbitrary clauses. We thus obtain the accumulated formula $\mathcal{Q}'.\psi_3$ with $\psi_3 := \psi_2 \setminus \psi_1$.

Step 4 (Removal of Universal Literals): We now remove all universal literals from the clauses in ψ_3. We start by removing the occurrences of the right-most variable u and apply extended universal reduction on all clauses in which it occurs. Once u is eliminated, we move on to the new right-most variable and eliminate it. We also remove eliminated variables from the quantifier prefix. We repeat this for all universal literals and denote the resulting accumulated formula by $\mathcal{Q}''.\psi_4$. It remains to show that all the removal steps are valid extended-universal-reduction steps. This is a consequence of the following lemma:

Lemma 1. *If $\mathcal{Q}'.\psi_3$ contains a resolution path from u to e, then e must be an annotated literal of the form l^τ where the assignment τ falsifies u.*

Proof. Suppose there exists a resolution path $C_1, \ldots, C_n$ from u to e. We show by induction on n that e is of the form l^τ where τ falsifies u.

BASE CASE ($n = 1$): C_1 contains both u and e. Hence, all the existential literals of C_1 must have been obtained by instantiating with an assignment that falsifies all universal literals of C_1. Moreover, by the definition of resolution paths, e must occur right of u. Hence, e must be of the form l^τ where τ falsifies u.

INDUCTION STEP ($n > 1$): Since $C_1, \ldots, C_n$ is a resolution path from u to e, we know that $e \in C_n$ and that $C_1, \ldots, C_{n-1}$ is a resolution path from u to some literal e_{n-1} such that $e_{n-1} \in C_{n-1}$ and $\bar{e}_{n-1} \in C_n$. By the induction hypothesis, e_{n-1} is of the form l_{n-1}^τ where τ falsifies u. But then, since $\bar{e}_{n-1} \in C_n$, we know that C_n must have been obtained by instantiating it with an assignment that falsifies u. It follows that e is of the form l^τ where τ falsifies u.   □

$$\dfrac{\dfrac{a \vee x \vee b \vee y \vee c \qquad a \vee x \vee b \vee y \vee \bar{c}}{\dfrac{a \vee b^{\bar{x}} \vee c^{\bar{x}\bar{y}} \qquad a \vee b^{\bar{x}} \vee \bar{c}^{\bar{x}\bar{y}}}{\dfrac{a \vee b^{\bar{x}}}{a}}}}{\dfrac{\dfrac{x \vee \bar{b}}{\bar{b}^{\bar{x}}} \quad \dfrac{\bar{y} \vee c}{c^{xy}} \quad \dfrac{\bar{a} \vee \bar{x} \vee b \vee \bar{c}}{\bar{a} \vee b^{x} \vee \bar{c}^{xy}}}{\dfrac{\bar{a} \vee b^{x}}{b^{x}} \qquad \dfrac{\bar{x} \vee \bar{b}}{\bar{b}^{x}}} \atop \bot}$$

Fig. 2. Example of a ∀-Exp+Res refutation.

Thus, whenever we eliminate a universal literal u from a clause C in step 4, then $\mathcal{D}^{\mathrm{rrs}}$ defines each existential literal $e \in C$ that occurs right of u to be independent of u (literals to the left of u are trivially independent of u): Since $e \in C$, we know that e is of the form l^{τ} where τ falsifies u. Thus, there cannot exist resolution paths from $\bar{u}$ to e or to $\bar{e}$, for otherwise Lemma 1 would tell us that e is of the form l^{σ} where σ falsifies $\bar{u}$. Hence, e is independent of u according to $\mathcal{D}^{\mathrm{rrs}}$. Note that, strictly speaking, Lemma 1 would only guarantee that the first elimination of a universal literal is a valid extended-universal-reduction step (because the elimination modifies the formula $\mathcal{Q}'.\psi_3$). However, since the elimination of universal literals does not introduce additional resolution paths, all eliminations of universal literals are valid extended-universal-reduction steps.

Step 5 (Resolution Proof): In this last step, we perform all resolution steps of π as QRAT additions to derive the empty clause. This is possible since ψ_4 contains all clauses that are involved in the resolution proof. We thus conclude:

Theorem 2. Π is a QRAT refutation of $\mathcal{Q}.\psi$.

We illustrate our simulation on an example before showing that it is polynomial:

Example 2. Fig. 2 shows a ∀-Exp+Res refutation of $\exists a \forall x \exists b \forall y \exists c.\psi$ with

$$\psi = (a \vee x \vee b \vee y \vee c) \wedge (a \vee x \vee b \vee y \vee \bar{c}) \wedge (x \vee \bar{b}) \wedge (\bar{y} \vee c) \wedge (\bar{a} \vee \bar{x} \vee b \vee \bar{c}) \wedge (\bar{x} \vee \bar{b}).$$

For simulating this proof in QRAT, we proceed as follows.

(1) We introduce definitions of the annotated variables by adding the following eight QRAT clauses:

$$\begin{array}{llll} (\bar{b} \vee b^{\bar{x}}) & (\bar{b} \vee b^{x}) & (\bar{c} \vee c^{\bar{x}\bar{y}}) & (\bar{c} \vee c^{xy}) \\ (b \vee \bar{b}^{\bar{x}}) & (b \vee \bar{b}^{x}) & (c \vee \bar{c}^{\bar{x}\bar{y}}) & (c \vee \bar{c}^{xy}) \end{array}$$

(2) We then introduce the following QRAT clauses, which correspond to applications of the axiom rule in ∀-Exp+Res:

$$\begin{array}{lll} (a \vee x \vee b^{\bar{x}} \vee y \vee c^{\bar{x}\bar{y}}) & (\bar{y} \vee c^{xy}) & (x \vee \bar{b}^{\bar{x}}) \\ (a \vee x \vee b^{\bar{x}} \vee y \vee \bar{c}^{\bar{x}\bar{y}}) & (\bar{a} \vee \bar{x} \vee b^{x} \vee \bar{c}^{xy}) & (\bar{x} \vee \bar{b}^{x}) \end{array}$$

(3) We remove the original clauses and the clauses introduced in step 1. Only the clauses introduced in step 2 remain.

(4) From the remaining clauses, we first remove all occurrences of y and then all occurrences of x via extended universal reduction. We obtain the clauses introduced by applications of the axiom rule in the ∀-Exp+Res proof.

(5) Finally, we can simply perform the resolution steps of the ∀-Exp+Res proof to obtain a QRAT refutation of the input formula.

This concludes the example.   □

It remains to show that the simulation is polynomial. We first bound the size (measured by the number of symbols) of the resulting QRAT proof:

Lemma 3. Π *is linear in the size of* π.

Proof. In step 1 (definition introduction), we perform two QRAT additions for each annotated variable in the ∀-Exp+Res proof π. The size of the corresponding QRAT derivation is clearly linear with respect to π. In step 2, we perform one QRAT addition for each application of the axiom rule in π, again resulting in a linear-size QRAT derivation. In step 3, we eliminate clauses of ψ and definitions— also clearly linear. In step 4, we remove universal literals from existing clauses. All these universal literals are contained in π as their respective clauses are involved in applications of the axiom rule. Hence, also this step yields a QRAT derivation of linear size. Finally, the resolution proof derived in step 5 is part of π and thus also of linear size with respect to π. We conclude that the size of the final QRAT proof is linear with respect to the size of π.   □

It should now be clear that our simulation can be performed in polynomial time:

Theorem 4. QRAT *polynomially simulates* ∀-*Exp+Res.*

7   Conclusion

We filled an empty spot in the QBF-proof-complexity landscape by showing that QRAT polynomially simulates universal expansion in general, and the proof system ∀-Exp+Res in particular. Our approach is similar to the approach in [12], which mimics the expansion of inner-most universal variables in QRAT.

There are, however, some subtle but important differences to [12]. First, in [12] the universal variables are fully expanded, which could potentially duplicate the whole formula. In contrast, we expand arbitrary variables and focus only on the clauses that are used as axioms in the ∀-Exp+Res proof. By only deriving these clauses (using new definitions), we can ensure that the resulting proof is small. Second, in [12] the QRAT proof is generated during proof search, when it is still unclear if the formula is true or false. In our simulation here, the proof of unsatisfiability is given as input and therefore we know from the beginning that the formula is false. This allows us to delete clauses eagerly (deletion doesn't have to preserve satisfiability), which is not the case in [12].

A closer look at our simulation shows that the only features of QRAT needed for the simulation are Q-resolution, definition introduction, and extended universal reduction. A system that uses only these features could be seen as *extended*

$Q(\mathcal{D}^{rrs})$-*resolution* in the dependency framework of Slivovsky and Szeider [24]. In propositional logic, we know that extended resolution polynomially simulates DRAT [19] but it is not known if extended Q-resolution [17] or extended $Q(\mathcal{D}^{rrs})$-resolution can polynomially simulate QRAT. It is also still unclear how QRAT is related to the stronger expansion-based systems IR-calc and IRM-calc [4]. Finally, since there exist efficient proof checkers for QRAT and since the size increase induced by our simulation is only linear, our simulation could be used in practice to validate the results of expansion-based solvers.

References

1. Balabanov, V., Jiang, J.R.: Unified QBF certification and its applications. Formal Methods Syst. Des. **41**(1), 45–65 (2012)
2. Balabanov, V., Widl, M., Jiang, J.-H.R.: QBF resolution systems and their proof complexities. In: Sinz, C., Egly, U. (eds.) SAT 2014. LNCS, vol. 8561, pp. 154–169. Springer, Cham (2014). https://doi.org/10.1007/978-3-319-09284-3_12
3. Beyersdorff, O., Bonacina, I., Chew, L.: Lower bounds: from circuits to QBF proof systems. In: Proceedings of the 2016 ACM Conference on Innovations in Theoretical Computer Science (ITCS 2016), pp. 249–260. ACM (2016)
4. Beyersdorff, O., Chew, L., Janota, M.: On unification of QBF resolution-based calculi. In: Csuhaj-Varjú, E., Dietzfelbinger, M., Ésik, Z. (eds.) MFCS 2014. LNCS, vol. 8635, pp. 81–93. Springer, Heidelberg (2014). https://doi.org/10.1007/978-3-662-44465-8_8
5. Beyersdorff, O., Chew, L., Janota, M.: Proof complexity of resolution-based QBF calculi. In: Proceedings of the 32nd International Symposium on Theoretical Aspects of Computer Science (STACS 2015). LIPIcs, vol. 30, pp. 76–89. Schloss Dagstuhl - Leibniz-Zentrum für Informatik (2015)
6. Beyersdorff, O., Chew, L., Mahajan, M., Shukla, A.: Are short proofs narrow? QBF resolution is not simple. In: Proceedings of the 33rd Symposium on Theoretical Aspects of Computer Science (STACS 2016). LIPIcs, vol. 47, pp. 15:1–15:14. Schloss Dagstuhl - Leibniz-Zentrum fuer Informatik (2016)
7. Bloem, R., Braud-Santoni, N., Hadzic, V., Egly, U., Lonsing, F., Seidl, M.: Expansion-based QBF solving without recursion. In: Proceedings of the International Conference on Formal Methods in Computer Aided Design (FMCAD 2018), pp. 1–10. IEEE (2018)
8. Chen, H.: Proof complexity modulo the polynomial hierarchy: understanding alternation as a source of hardness. In: Proceedings of the 43rd International Colloquium on Automata, Languages, and Programming (ICALP 2016). LIPIcs, vol. 55, pp. 94:1–94:14. Schloss Dagstuhl - Leibniz-Zentrum fuer Informatik (2016)
9. Cook, S.A., Reckhow, R.A.: The relative efficiency of propositional proof systems. J. Symb. Logic **44**(1), 36–50 (1979)
10. Egly, U.: On stronger calculi for QBFs. In: Creignou, N., Le Berre, D. (eds.) SAT 2016. LNCS, vol. 9710, pp. 419–434. Springer, Cham (2016). https://doi.org/10.1007/978-3-319-40970-2_26
11. Gelder, A.: Variable independence and resolution paths for quantified boolean formulas. In: Lee, J. (ed.) CP 2011. LNCS, vol. 6876, pp. 789–803. Springer, Heidelberg (2011). https://doi.org/10.1007/978-3-642-23786-7_59
12. Heule, M.J.H., Seidl, M., Biere, A.: Solution validation and extraction for QBF preprocessing. J. Autom. Reasoning **58**(1), 1–29 (2016)

13. Janota, M.: On Q-resolution and CDCL QBF solving. In: Creignou, N., Le Berre, D. (eds.) SAT 2016. LNCS, vol. 9710, pp. 402–418. Springer, Cham (2016). https://doi.org/10.1007/978-3-319-40970-2_25

14. Janota, M., Grigore, R., Marques-Silva, J.: On QBF proofs and preprocessing. In: McMillan, K., Middeldorp, A., Voronkov, A. (eds.) LPAR 2013. LNCS, vol. 8312, pp. 473–489. Springer, Heidelberg (2013). https://doi.org/10.1007/978-3-642-45221-5_32

15. Janota, M., Klieber, W., Marques-Silva, J., Clarke, E.M.: Solving QBF with counterexample guided refinement. Artif. Intell. **234**, 1–25 (2016)

16. Janota, M., Marques-Silva, J.: On propositional QBF expansions and Q-resolution. In: Järvisalo, M., Van Gelder, A. (eds.) SAT 2013. LNCS, vol. 7962, pp. 67–82. Springer, Heidelberg (2013). https://doi.org/10.1007/978-3-642-39071-5_7

17. Jussila, T., Biere, A., Sinz, C., Kröning, D., Wintersteiger, C.M.: A first step towards a unified proof checker for QBF. In: Marques-Silva, J., Sakallah, K.A. (eds.) SAT 2007. LNCS, vol. 4501, pp. 201–214. Springer, Heidelberg (2007). https://doi.org/10.1007/978-3-540-72788-0_21

18. Kiesl, B., Heule, M.J.H., Seidl, M.: A little blocked literal goes a long way. In: Gaspers, S., Walsh, T. (eds.) SAT 2017. LNCS, vol. 10491, pp. 281–297. Springer, Cham (2017). https://doi.org/10.1007/978-3-319-66263-3_18

19. Kiesl, B., Rebola-Pardo, A., Heule, M.J.H.: Extended resolution simulates DRAT. In: Galmiche, D., Schulz, S., Sebastiani, R. (eds.) IJCAR 2018. LNCS (LNAI), vol. 10900, pp. 516–531. Springer, Cham (2018). https://doi.org/10.1007/978-3-319-94205-6_34

20. Kleine Büning, H., Karpinski, M., Flögel, A.: Resolution for quantified boolean formulas. Inf. Comput. **117**(1), 12–18 (1995)

21. Lonsing, F., Egly, U.: DepQBF 6.0: a search-based QBF solver beyond traditional QCDCL. In: de Moura, L. (ed.) CADE 2017. LNCS (LNAI), vol. 10395, pp. 371–384. Springer, Cham (2017). https://doi.org/10.1007/978-3-319-63046-5_23

22. Peitl, T., Slivovsky, F., Szeider, S.: Dependency learning for QBF. In: Gaspers, S., Walsh, T. (eds.) SAT 2017. LNCS, vol. 10491, pp. 298–313. Springer, Cham (2017). https://doi.org/10.1007/978-3-319-66263-3_19

23. Slivovsky, F., Szeider, S.: Variable dependencies and Q-resolution. In: Sinz, C., Egly, U. (eds.) SAT 2014. LNCS, vol. 8561, pp. 269–284. Springer, Cham (2014). https://doi.org/10.1007/978-3-319-09284-3_21

24. Slivovsky, F., Szeider, S.: Soundness of Q-resolution with dependency schemes. Theor. Comput. Sci. **612**, 83–101 (2016)

25. Tentrup, L.: On expansion and resolution in CEGAR based QBF solving. In: Majumdar, R., Kunčak, V. (eds.) CAV 2017. LNCS, vol. 10427, pp. 475–494. Springer, Cham (2017). https://doi.org/10.1007/978-3-319-63390-9_25

26. Gelder, A.: Contributions to the theory of practical quantified boolean formula solving. In: Milano, M. (ed.) CP 2012. LNCS, pp. 647–663. Springer, Heidelberg (2012). https://doi.org/10.1007/978-3-642-33558-7_47

27. Wetzler, N., Heule, M.J.H., Hunt, W.A.: DRAT-trim: efficient checking and trimming using expressive clausal proofs. In: Sinz, C., Egly, U. (eds.) SAT 2014. LNCS, vol. 8561, pp. 422–429. Springer, Cham (2014). https://doi.org/10.1007/978-3-319-09284-3_31

28. Zhang, L., Malik, S.: Conflict driven learning in a quantified boolean satisfiability solver. In: Proceedings of the 2002 IEEE/ACM International Conference on Computer-aided Design (ICCAD 2002), pp. 442–449. ACM/IEEE Computer Society (2002)

QRATPre+: Effective QBF Preprocessing via Strong Redundancy Properties

Florian Lonsing[1(✉)] and Uwe Egly[2]

[1] Computer Science Department, Stanford University, Stanford, CA 94305, USA
lonsing@cs.stanford.edu
[2] Institute of Logic and Computation, TU Wien, 1040 Vienna, Austria

Abstract. We present version 2.0 of QRATPre+, a preprocessor for quantified Boolean formulas (QBFs) based on the QRAT proof system and its generalization QRAT$^+$. These systems rely on strong redundancy properties of clauses and universal literals. QRATPre+ is the first implementation of these redundancy properties in QRAT and QRAT$^+$ used to simplify QBFs in preprocessing. It is written in C and features an API for easy integration in other QBF tools. We present implementation details and report on experimental results demonstrating that QRATPre+ improves upon the power of state-of-the-art preprocessors and solvers.

1 Introduction

The application of preprocessing prior to the actual solving process is crucial for the performance of most of the quantified Boolean formula (QBF) solvers [17,20,21]. Preprocessors aim at decreasing the complexity of a given formula with respect to the number of variables, the number of clauses, or the number of quantifier blocks. Contrary to complete QBF solvers, preprocessors are incomplete but can detect redundant parts of the given formula by applying resource-restricted reasoning. Bloqqer [6] and HQSpre [26] are leading preprocessors which show their power in the yearly QBFEVAL competitions[1] in potentially almost doubling the number of instances solved by certain state-of-the-art solvers. These tools apply a diverse set of redundancy elimination techniques.

We present QRATPre+ 2.0,[2] a QBF preprocessor based on the QRAT$^+$ proof system [18]. QRATPre+ processes QBFs in prenex conjunctive normal form (PCNF) and eliminates redundant clauses and universal literals. Redundancy checking relies on redundancy properties defined by the QRAT$^+$ proof system, which is a generalization of the QRAT (quantified resolution asymmetric tautology) proof system [7,9]. QRAT is a lifting of (D)RAT [11,25] from the propositional to the QBF level, and it simulates virtually all simplification rules applied

---

Part of this work was carried out while the first author was employed at the Institute of Logic and Computation, TU Wien, Austria. This work is supported by the Austrian Science Fund (FWF) under grant S11409-N23.

[1] http://www.qbflib.org/index_eval.php.

[2] QRATPre+ is licensed under GPLv3: https://lonsing.github.io/qratpreplus/.

M. Janota and I. Lynce (Eds.): SAT 2019, LNCS 11628, pp. 203–210, 2019.
https://doi.org/10.1007/978-3-030-24258-9_14

in state-of-the-art QBF preprocessors. This is made possible by strong redundancy properties.

However, the strong redundancy properties of QRAT have not been applied to preprocess QBFs so far. With QRATPre+ we close this gap and leverage the power of QRAT and QRAT$^+$ for QBF preprocessing. Compared to the initially released version 1.0, version 2.0 comes with a more modularized code base and a C API that allows to easily integrate QRATPre+ in other tools. QRATPre+ currently applies only rewrite rules of the QRAT and QRAT$^+$ proof systems that remove either clauses or universal literals from a PCNF. That is, it does not attempt to add redundant parts with the aim to potentially enable further simplifications later on. Despite this fact, experimental results with benchmarks from QBFEVAL'18 clearly indicate the effectiveness of QRATPre+. It improves on state-of-the-art preprocessors, such as Bloqqer and HQSpre, and solvers in terms of formula size reduction and solved instances, respectively.

2   QRAT$^+$ Redundancy Checking for Preprocessing

QRATPre+ eliminates redundant clauses and universal literals within clauses from a QBF in PCNF. Redundancy checking in QRATPre+ relies on redundancy properties defined by the QRAT$^+$ proof system. We present QRAT$^+$ only informally and refer to related work instead [18].

Let $\phi := \Pi.(\psi \wedge (C' \cup \{l\}))$ be a PCNF with *prefix* $\Pi := Q_1 B_1 \ldots Q_i B_i \ldots \ldots Q_n B_n$, where $Q_i B_i$ are *quantifier blocks (qblocks)* consisting of a quantifier $Q_i \in \{\forall, \exists\}$ and a block (i.e., set) B_i of variables. We write $Q(B_1 \ldots B_n)$ for $Q(B_1 \cup \ldots \cup B_n)$. Index i is the *nesting level* of qblock $Q_i B_i$ and of the variables in B_i. Formula $\psi \wedge (C' \cup \{l\})$ is in CNF, where $(C' \cup \{l\})$ is a clause in ϕ containing literal l. We consider only PCNFs without tautological clauses of the form $(C \cup \{p\} \cup \{\bar{p}\})$ for some propositional variable p.

Given a clause $C := (C' \cup \{l\})$ in ϕ, checking whether C or the literal $l \in C$ is redundant requires to consider all clauses D_i in the *resolution neighborhood* $\mathsf{RN}(C, l) := \{D_i \mid D_i \in \phi, \bar{l} \in D_i\}$ of C with respect to l, cf. [11–13]. This is illustrated in Fig. 1. Given C and some $D_j \in \mathsf{RN}(C, l)$, we first compute the *outer resolvent* $\mathsf{OR}_j \subset (C \cup D_j)$ of C and D_j on literal l [9]. Then we determine the maximum nesting level $i := max(levels(\Pi, \mathsf{OR}_j))$ of variables appearing in OR_j. Based on i we construct the PCNF $Abs(\Pi.\psi, i) := \exists (B_1 \ldots B_i) Q_{i+1} B_{i+1} \ldots Q_n B_n.\psi$ by converting all universal quantifiers in the subprefix $Q_1 B_1 \ldots Q_i B_i$ to existential ones. The resulting PCNF $Abs(\Pi.\psi, i)$ is an *abstraction* of $\Pi.\psi$.

We add the negation $\overline{\mathsf{OR}_j}$ of the outer resolvent OR_j, which is a set of unit clauses, to the abstraction $Abs(\Pi.\psi, i)$ to obtain the formula $Abs(\Pi.(\psi \wedge \overline{\mathsf{OR}_j}), i)$. Finally, we check whether a conflict, i.e., the empty clause $\emptyset$, is derived by applying *QBF unit propagation (QBCP)* [3,5,15,27] to $Abs(\Pi.(\psi \wedge \overline{\mathsf{OR}_j}), i)$. If so, which we denote by $Abs(\Pi.(\psi \wedge \overline{\mathsf{OR}_j}), i) \vdash_\mathsf{v} \emptyset$, then the current outer resolvent OR_j has the *quantified asymmetric tautology (QAT)* [18] redundancy property.

If *all possible* outer resolvents of $C := (C' \cup \{l\})$ and clauses D in the resolution neighborhood $\mathsf{RN}(C, l)$ of C with respect to literal l have the QAT property,

$$D_1 = D_1' \cup \{\bar{l}\} \quad \cdots \quad \boxed{D_j = D_j' \cup \{\bar{l}\}} \quad \cdots \quad D_n = D_n' \cup \{\bar{l}\}$$

$$OR_j \subset (C \cup D_j)$$
$$i := max(levels(\Pi, OR_j))$$
$$\Pi := Q_1 B_1 \ldots Q_i B_i \ldots Q_n B_n$$
$$Abs(\Pi.\phi, i) := \exists (B_1 \ldots B_i) Q_{i+1} B_{i+1} \ldots Q_n B_n.\phi$$
$$\text{Check for conflict:} \quad Abs(\Pi.(\phi \wedge \overline{OR_j}), i) \vdash_{\scriptscriptstyle \mathsf{v}} \emptyset?$$

$$\boxed{C = C' \cup \{l\}}$$

Fig. 1. Redundancy checking of clause $C := (C' \cup \{l\})$ based on QRAT^+. The resolution neighborhood $\mathsf{RN}(C, l)$ of C consists of the clauses D_i with $\bar{l} \in D_i$ shown on top. The boxes indicate the pair of clauses for which the current outer resolvent OR_j is computed.

then clause C has the QRAT^+ *redundancy property* on literal l. In this case, either C or l is redundant, depending on whether l is existential or universal, respectively. Note that in general, for every outer resolvent OR_j, the index i for which the abstraction $Abs(\Pi.(\psi \wedge \overline{OR_j}), i)$ is constructed may be different.

Eliminating redundant clauses or universal literals via the above workflow is denoted by QRATE^+ and QRATU^+, respectively. In QRATPre+, we apply the QRATE^+ and QRATU^+ rewrite rules for preprocessing. The QRAT redundancy property [9] differs from QRAT^+ [18] in that for QRAT always a full abstraction $Abs(\Pi.(\psi \wedge \overline{OR_j}), i)$ with $i := n$ is constructed, regardless of the actual maximum nesting level of variables in the current outer resolvent OR_j. Moreover, QRAT^+ relies on QBCP which includes the universal reduction operation [14] to temporarily shorten clauses during propagation. This way, potentially more conflicts are derived. In contrast to that, QRAT applies propositional unit propagation to abstractions where all variables are existential. Due to that, the QRAT^+ redundancy property is more general and stronger than QRAT.

3 Implementation Details

Algorithm 1 shows the high-level workflow implemented in QRATPre+. To limit the computational costs of the relative expensive techniques QRATE^+ and QRATU^+, which are based on QBCP, we apply cheaper ones first to reduce the formula size upfront. In *quantified blocked clause elimination (QBCE)* [1], which is a restriction of QRATE^+, it is checked whether every outer resolvent (cf. Fig. 1) contains a pair of complementary literals. *Blocked literal elimination (BLE)* [8] is a restriction of QRATU^+ and, like QBCE, checks for complementary literals in outer resolvents. Before checking whether some clause C has the QRAT^+ property, we check whether it has the QAT property. This is done analogously to checking whether an outer resolvent has the QAT property in QRAT^+ testing (cf. Sect. 2). QAT checking is necessary in the workflow since it is not subsumed by QRATE^+.

During clause elimination, all clauses found redundant in the current PCNF ϕ' are cleaned up lazily in one pass after an application of a technique. In literal elimination, redundant literals are cleaned up eagerly since clauses are shortened, which increases chances to detect conflicts in QBCP.

If the QRAT$^+$ redundancy check fails for some clause C with literal $l \in C$ and clause $D_i \in RN(C, l)$ in the resolution neighborhood of C, we mark D_i as a *witness* for that failure. If a witness D is found redundant then all clauses in $RN(D, l)$, for all literals

Input: PCNF ϕ.
Output: Simplified PCNF ϕ'.
1 $\phi' = \phi$;
2 **do**
       // clause elimination
3      $\phi' := QBCE(\phi')$;
4      $\phi' := QAT(\phi')$;
5      $\phi' := QRATE^+(\phi')$;
       // literal elimination
6      $\phi' := BLE(\phi')$;
7      $\phi' := QRATU^+(\phi')$;
8 **while** ϕ' *changed*;

Algorithm 1: QRATPre+ workflow.

$l \in D$, are scheduled for being checked in the next iteration as D has prevented at least one of these clauses from being detected before. This witness-based scheduling potentially avoids superfluous redundancy checks. In our experiments the median number of clauses being checked in a run of QRATPre+ on a given PCNF was only by a factor of 3.3 larger than the initial number of clauses in the PCNF.

We maintain the index i indicating the maximum nesting level of variables in the current outer resolvent being tested (cf. Fig. 1). Any variable with an index smaller than i is treated as an existential one during QBCP. This way, abstractions $Abs(\Pi.\psi, i)$ are constructed implicitly. For QBCP, we implemented standard two-literal watching [4]. However, when assignments are retracted after deriving a conflict, then in general the literal watchers have to be restored to literals which are existential in the input PCNF rather than in the current abstraction. Restoring literal watchers in our implementation is necessary to maintain certain invariants, in contrast to, e.g., QBCP in QCDCL solvers.

Compared to version 1.0, version 2.0 of QRATPre+ comes with a C API that allows for easy integration in other tools. The API provides functions to import and export PCNFs and to configure the preprocessing workflow. Options include switches to toggle the individual redundancy tests in the main loop, user-defined limits, and shuffling the orderings in which clauses are tested. Shuffling may affect the result of preprocessing since the QAT, QRATE$^+$ and QRATU$^+$ rewrite rules are not confluent. In the default configuration, QRATPre+ does not shuffle clauses and applies rewrite rules until saturation.

4   Experiments

QRATPre+ improves on state-of-the-art preprocessors and solvers in terms of formula size reduction and solved instances. We ran experiments with the preprocessors Bloqqer v37 [6] and HQSpre 1.3 [26], and the solvers CAQE (commit-ID 9b95754 on GitHub) [23,24], RAReQS 1.1 [10], ljtihad v2 [2], Qute 1.1 [22], and DepQBF 6.03 [16]. The solvers implement different solving paradigms, e.g., QCDCL (Qute and DepQBF), expansion (RAReQS and Ijtihad), and clausal abstraction (CAQE).

The following experiments were run on a cluster of Intel Xeon CPUs (E5-2650v4, 2.20 GHz) running Ubuntu 16.04.1. We used the 463 instances from the PCNF track of QBFEVAL'18. In all our experiments, we allowed 600 s CPU time and 7 GB of memory for each call of Bloqqer, HQSpre, or QRATPre+. For formulas where Bloqqer or HQSpre exceeded these limits, we considered the original, unpreprocessed formula. In contrast to Bloqqer and HQSpre, we implemented a soft time limit in QRATPre+ which, when exceeding the limit, allows to print the preprocessed formula with identified redundant parts being removed.

Table 1 shows the effect of preprocessing by QRATPre+ (Q), Bloqqer (B), HQSpre (H), and combinations, where QRATPre+ is called before (QB, QH) or after (BQ, HQ) Bloqqer or HQSpre. The table shows *average numbers* of clauses (#cl), qblocks (#qb), existential (#∃l) and universal literals (#∀l) as a percentage relative to the original benchmark

Table 1. Effect of preprocessing.

	Q	B	H	QB	BQ	QH	HQ
#cl	79	78	23	70	69	18	17
#qb	92	22	17	21	22	59	17
#∃l	82	85	27	80	77	20	21
#∀l	73	95	74	82	86	44	53

set. Except for qblocks, QRATPre+ considerably further reduces the formula size when combined with both Bloqqer (B vs. QB and BQ) and HQSpre (H vs. QH and HQ). These results include solved instances: QRATPre+, Bloqqer, and HQSpre solve 18, 74, and 158 original instances, respectively. The preprocessors timed out on 12 (HQSpre), 38 (Bloqqer), and 59 (QRATPre+, soft time limit) original instances.

The application of computationally inexpensive techniques like QBCE and BLE to shrink the PCNF before applying more expensive ones like QRATE$^+$ and QRATU$^+$ pays off. With the original workflow (Algorithm 1 and column Q in Table 1), QRATPre+ spends 96 s on average per instance, compared to 120 s when disabling QBCE and BLE, where it exceeds the time limit on 77 instances.

Shuffling the ordering of clauses before applying QRATE$^+$ and QRATU$^+$ in Algorithm 1 based on five different random seeds hardly had any effect on the aggregate data in column Q in Table 1, except for an increase in eliminated universal literals by one percent. When using the redundancy property of QRAT, which is weaker than the one of QRAT$^+$, by constructing full abstractions (cf. Sect. 2), we observed a moderate decrease of all four metrics by one percent each.

Tables 2a to f show the numbers of instances solved after preprocessing with different combinations of QRATPre+, Bloqqer, and HQSpre. We used a limit of 1800s CPU time and 7 GB of memory for solving. Times for preprocessing are not included in the times reported in the tables. Preprocessing by QRATPre+ increases the number of solved instances in most cases, except for Ijtihad on instances preprocessed with HQSpre and QRATPre+ (Tables 2e and f). Similar to formula size reduction shown in Table 1, the ordering of whether to apply QRAT-Pre+ before or after Bloqqer or HQSpre has an impact on solving performance. Interestingly, for all solvers the combination BQ (Bloqqer before QRATPre+)

Table 2. QBFEVAL'18: solved instances (S), unsatisfiable ($\bot$), satisfiable ($\top$), and uniquely solved ones (U), and total CPU time in kiloseconds (K) including time outs.

(a) Original instances.

Solver	S	$\bot$	$\top$	U	Time
CAQE	151	107	44	11	586K
DepQBF	149	87	62	50	592K
RAReQS	147	115	32	4	588K
Ijtihad	132	111	21	2	609K
Qute	98	79	19	6	665K

(b) QRATPre+ only (Q).

Solver	S	$\bot$	$\top$	U	Time
CAQE	211	135	76	29	487K
RAReQS	178	120	58	9	533K
DepQBF	165	83	82	33	562K
Ijtihad	156	111	45	1	562K
Qute	137	92	45	9	598K

(c) Bloqqer only (B).

Solver	S	$\bot$	$\top$	U	Time
CAQE	269	150	119	24	383K
RAReQS	258	159	99	7	399K
Ijtihad	200	128	72	4	482K
DepQBF	198	99	99	21	501K
Qute	189	114	75	2	512K

(d) QRATPre+ and Bloqqer (QB).

Solver	S	$\bot$	$\top$	U	Time
CAQE	290	169	121	37	347K
RAReQS	260	163	97	5	390K
Ijtihad	216	140	76	2	456K
DepQBF	210	107	103	21	481K
Qute	197	119	78	2	493K

(e) HQSpre only (H).

Solver	S	$\bot$	$\top$	U	Time
CAQE	322	183	139	21	290K
RAReQS	292	180	112	2	317K
DepQBF	270	160	110	20	364K
Qute	253	161	92	3	390K
Ijtihad	249	167	82	0	394K

(f) HQSpre and QRATPre+ (HQ).

Solver	S	$\bot$	$\top$	U	Time
CAQE	325	188	137	14	279K
RAReQS	303	189	114	3	304K
DepQBF	271	158	113	20	362K
Qute	263	170	93	2	377K
Ijtihad	245	166	79	0	407K

results in a decrease of solved instances compared to QB (QRATPre+ before Bloqqer). We made similar observations for combinations with HQSpre (QH and HQ).[3]

5   Conclusion

We presented version 2.0 of QRATPre+, a preprocessor for QBFs in PCNF that is based on strong redundancy properties of clauses and universal literals defined by the QRAT$^+$ proof system [18]. QRAT$^+$ is a generalization of the QRAT proof system [7,9]. QRATPre+ is the first implementation of the QRAT and QRAT$^+$ redundancy properties for applications in QBF preprocessing. As such, the techniques implemented in QRATPre+ are orthogonal to techniques applied in state-of-the-art preprocessors like Bloqqer and HQSpre. Our experiments demonstrate a considerable performance increase of preprocessing and solving. QRATPre+ comes with a C API that allows easy integration into other tools.

---

[3] We refer to an online appendix [19] for results with combinations BQ and QH.

We observed a sensitivity of solvers to the ordering in which QRATPre+ is coupled with other preprocessors. To better understand the interplay between redundancy elimination and the proof systems implemented in solvers, we want to further analyze this phenomenon. We used a simple but effective witness-based scheduling to avoid superfluous redundancy checks. However, with more sophisticated watched data structures the run time of QRATPre+ could be optimized. To enhance the power of redundancy checking, it could be beneficial to selectively add redundant formula parts to enable additional simplifications afterwards.

References

1. Biere, A., Lonsing, F., Seidl, M.: Blocked clause elimination for QBF. In: Bjørner, N., Sofronie-Stokkermans, V. (eds.) CADE 2011. LNCS (LNAI), vol. 6803, pp. 101–115. Springer, Heidelberg (2011). https://doi.org/10.1007/978-3-642-22438-6_10
2. Bloem, R., Braud-Santoni, N., Hadzic, V., Egly, U., Lonsing, F., Seidl, M.: Expansion-based QBF solving without recursion. In: FMCAD, pp. 1–10. IEEE (2018)
3. Cadoli, M., Giovanardi, A., Schaerf, M.: An algorithm to evaluate quantified boolean formulae. In: AAAI, pp. 262–267. AAAI Press/The MIT Press (1998)
4. Gent, I., Giunchiglia, E., Narizzano, M., Rowley, A., Tacchella, A.: Watched data structures for QBF solvers. In: Giunchiglia, E., Tacchella, A. (eds.) SAT 2003. LNCS, vol. 2919, pp. 25–36. Springer, Heidelberg (2004). https://doi.org/10.1007/978-3-540-24605-3_3
5. Giunchiglia, E., Narizzano, M., Tacchella, A.: Clause/term resolution and learning in the evaluation of quantified boolean formulas. JAIR **26**, 371–416 (2006). https://doi.org/10.1613/jair.1959
6. Heule, M., Järvisalo, M., Lonsing, F., Seidl, M., Biere, A.: Clause elimination for SAT and QSAT. JAIR **53**, 127–168 (2015). https://doi.org/10.1613/jair.4694
7. Heule, M., Seidl, M., Biere, A.: A unified proof system for QBF preprocessing. In: Demri, S., Kapur, D., Weidenbach, C. (eds.) IJCAR 2014. LNCS (LNAI), vol. 8562, pp. 91–106. Springer, Cham (2014). https://doi.org/10.1007/978-3-319-08587-6_7
8. Heule, M.J.H., Seidl, M., Biere, A.: Blocked literals are universal. In: Havelund, K., Holzmann, G., Joshi, R. (eds.) NFM 2015. LNCS, vol. 9058, pp. 436–442. Springer, Cham (2015). https://doi.org/10.1007/978-3-319-17524-9_33
9. Heule, M., Seidl, M., Biere, A.: Solution validation and extraction for QBF preprocessing. J. Autom. Reasoning **58**(1), 97–125 (2017)
10. Janota, M., Klieber, W., Marques-Silva, J., Clarke, E.: Solving QBF with counterexample guided refinement. Artif. Intell. **234**, 1–25 (2016)
11. Järvisalo, M., Heule, M.J.H., Biere, A.: Inprocessing rules. In: Gramlich, B., Miller, D., Sattler, U. (eds.) IJCAR 2012. LNCS (LNAI), vol. 7364, pp. 355–370. Springer, Heidelberg (2012). https://doi.org/10.1007/978-3-642-31365-3_28
12. Kiesl, B., Seidl, M., Tompits, H., Biere, A.: Super-blocked clauses. In: Olivetti, N., Tiwari, A. (eds.) IJCAR 2016. LNCS (LNAI), vol. 9706, pp. 45–61. Springer, Cham (2016). https://doi.org/10.1007/978-3-319-40229-1_5
13. Kiesl, B., Seidl, M., Tompits, H., Biere, A.: Blockedness in propositional logic: are you satisfied with your neighborhood? In: IJCAI, pp. 4884–4888 (2017). www.ijcai.org

14. Kleine Büning, H., Karpinski, M., Flögel, A.: Resolution for quantified boolean formulas. Inf. Comput. **117**(1), 12–18 (1995). https://doi.org/10.1006/inco.1995.1025

15. Letz, R.: Lemma and model caching in decision procedures for quantified boolean formulas. In: Egly, U., Fermüller, C.G. (eds.) TABLEAUX 2002. LNCS (LNAI), vol. 2381, pp. 160–175. Springer, Heidelberg (2002). https://doi.org/10.1007/3-540-45616-3_12

16. Lonsing, F., Egly, U.: DepQBF 6.0: a search-based QBF solver beyond traditional QCDCL. In: de Moura, L. (ed.) CADE 2017. LNCS (LNAI), vol. 10395, pp. 371–384. Springer, Cham (2017). https://doi.org/10.1007/978-3-319-63046-5_23

17. Lonsing, F., Egly, U.: Evaluating QBF solvers: quantifier alternations matter. In: Hooker, J. (ed.) CP 2018. LNCS, vol. 11008, pp. 276–294. Springer, Cham (2018). https://doi.org/10.1007/978-3-319-98334-9_19

18. Lonsing, F., Egly, U.: QRAT$^+$: generalizing QRAT by a more powerful QBF redundancy property. In: Galmiche, D., Schulz, S., Sebastiani, R. (eds.) IJCAR 2018. LNCS (LNAI), vol. 10900, pp. 161–177. Springer, Cham (2018). https://doi.org/10.1007/978-3-319-94205-6_12

19. Lonsing, F., Egly, U.: QRATPre+: effective QBF preprocessing via strong redundancy properties. CoRR abs/1904.12927 (2019). https://arxiv.org/abs/1904.12927. SAT 2019 proceedings version with appendix

20. Lonsing, F., Seidl, M., Van Gelder, A.: The QBF gallery: behind the scenes. Artif. Intell. **237**, 92–114 (2016)

21. Marin, P., Narizzano, M., Pulina, L., Tacchella, A., Giunchiglia, E.: Twelve years of QBF evaluations: QSAT Is PSPACE-hard and it shows. Fundam. Inform. **149**(1–2), 133–158 (2016)

22. Peitl, T., Slivovsky, F., Szeider, S.: Dependency learning for QBF. In: Gaspers, S., Walsh, T. (eds.) SAT 2017. LNCS, vol. 10491, pp. 298–313. Springer, Cham (2017). https://doi.org/10.1007/978-3-319-66263-3_19

23. Rabe, M.N., Tentrup, L.: CAQE: a certifying QBF solver. In: FMCAD, pp. 136–143. IEEE (2015)

24. Tentrup, L.: On expansion and resolution in CEGAR based QBF solving. In: Majumdar, R., Kunčak, V. (eds.) CAV 2017. LNCS, vol. 10427, pp. 475–494. Springer, Cham (2017). https://doi.org/10.1007/978-3-319-63390-9_25

25. Wetzler, N., Heule, M.J.H., Hunt, W.A.: DRAT-trim: efficient checking and trimming using expressive clausal proofs. In: Sinz, C., Egly, U. (eds.) SAT 2014. LNCS, vol. 8561, pp. 422–429. Springer, Cham (2014). https://doi.org/10.1007/978-3-319-09284-3_31

26. Wimmer, R., Reimer, S., Marin, P., Becker, B.: HQSpre – an effective preprocessor for QBF and DQBF. In: Legay, A., Margaria, T. (eds.) TACAS 2017. LNCS, vol. 10205, pp. 373–390. Springer, Heidelberg (2017). https://doi.org/10.1007/978-3-662-54577-5_21

27. Zhang, L., Malik, S.: Conflict driven learning in a quantified boolean satisfiability solver. In: ICCAD, pp. 442–449. ACM/IEEE Computer Society (2002)

On Computing the Union of MUSes

Carlos Mencía[1]($^{\boxtimes}$), Oliver Kullmann[2], Alexey Ignatiev[3,4], and Joao Marques-Silva[3]

[1] University of Oviedo, Gijón, Spain
`menciacarlos@uniovi.es`
[2] Swansea University, Swansea, UK
`o.kullmann@swansea.ac.uk`
[3] Faculty of Science, University of Lisbon, Lisbon, Portugal
`{aignatiev,jpms}@ciencias.ulisboa.pt`
[4] ISDCT SB RAS, Irkutsk, Russia

Abstract. This paper considers unsatisfiable CNF formulas and addresses the problem of computing the union of the clauses included in some minimally unsatisfiable subformula (MUS). The union of MUSes represents a useful notion in infeasibility analysis since it summarizes all the causes for the unsatisfiability of a given formula. The paper proposes a novel algorithm for this problem, developing a refined recursive enumeration of MUSes based on powerful pruning techniques. Experimental results indicate the practical suitability of the approach.

1 Introduction

A growing number of practical applications of Boolean Satisfiability (SAT) solvers is related to the analysis of overconstrained formulas, with different instantiations of model-based diagnosis (MBD) representing one of the most visible classes of applications. Examples of diagnosis problems include software fault localization [11, 12], spreadsheet debugging [8, 9], design debugging [28, 31], type error debugging [32], and axiom pinpointing [29], among others. In most settings, the core reasoning services used in the analysis of overconstrained formulas include extraction and enumeration of minimally unsatisfiable subformulas (MUSes), and also minimal correction subsets (MCSes). Tightly related with these core services, there is also interest in smallest (or minimum cost) MCSes (i.e. the maximum satisfiability (MaxSAT) problem) and, in some settings, the smallest MUS problem (SMUS). It is well-known that the SMUS problem is hard for the second level of the polynomial hierarchy (Σ_2^P). Similarly, MCS and MUS membership problems are hard for Σ_2^P, and the specification of preferences of clauses to include in MUSes and MCSes are also hard for Σ_2^P [18].

Practical evidence suggests that complete enumeration of MUSes and MCSes is often beyond the reach of existing technologies, and a practical solution is the partial enumeration of either MUSes or MCSes [1, 6, 19, 24–26]. A possible drawback of these

This research is supported by the Spanish Government under project TIN2016-79190-R and by the Principality of Asturias under grant IDI/2018/000176. This work is also supported by FCT grants ABSOLV (PTDC/CCI-COM/28986/2017), FaultLocker (PTDC/CCI-COM/29300/2017), SAFETY (SFRH/BPD/120315/2016), and SAMPLE (CEECIND/04549/2017), and by EPSRC grant EP/S015523/1.

M. Janota and I. Lynce (Eds.): SAT 2019, LNCS 11628, pp. 211–221, 2019.
https://doi.org/10.1007/978-3-030-24258-9_15

approaches is that there is no guarantee that the actual sources of inconsistency in the given context are reported. An alternative solution is to compute the union of all sources of inconsistency, i.e. the union of all MUSes (UMU). Information about UMU can restrict the number of clauses (or components in MBD) to analyze, with the guarantee that the source of inconsistency is present in the reported set.

Since MCS or MUS membership queries can be answered with a Σ_2^P oracle [10], the union of MUSes or MCSes can be obtained with a linear (but still large) number of calls to that oracle. However, it is often unrealistic in practice to call a Σ_2^P oracle say tens of thousands of times or more. Yet another approach consists in computing the intersection of all maximal satisfiable subsets (MSSes) and, from this compute the union of MUSes [14]. However, as hinted above, it is often unrealistic to enumerate all MCSes (and so MSSes). In contrast, this paper describes a novel algorithm for the direct computation of the union of MUSes. The experimental results provide evidence that the proposed algorithm offers a viable alternative to existing approaches.

2   Preliminaries

We consider propositional formulas in *conjunctive normal form* (CNF), defined as a conjunction, or set, of clauses $\mathcal{F} = \{c_1, c_2, ..., c_m\}$ over a set of variables $V(\mathcal{F}) = \{x_1, x_2, ..., x_n\}$, where a clause is a disjunction of literals, and a literal is a variable x or its negation $\neg x$.

A truth assignment, or interpretation, is a mapping $\mu : V(\mathcal{F}) \rightarrow \{0, 1\}$. If μ satisfies a formula $\mathcal{F}$, μ is referred to as a *model* of $\mathcal{F}$. Given two formulas $\mathcal{F}$ and $\mathcal{G}$, $\mathcal{F}$ entails $\mathcal{G}$ (written $\mathcal{F} \vDash \mathcal{G}$) iff all the models of $\mathcal{F}$ are models of $\mathcal{G}$. A formula $\mathcal{F}$ is satisfiable (written $\mathcal{F} \nvDash \perp$) iff there exists a model for it, and unsatisfiable (written $\mathcal{F} \vDash \perp$) otherwise. SAT is the NP-complete problem [4] of deciding the satisfiability of a formula.

The following definitions characterize two dual notions in the analysis of unsatisfiable CNF formulas:

Definition 1 (MUS). $\mathcal{M} \subseteq \mathcal{F}$ *is a* minimally unsatisfiable subformula *(MUS) if and only if* $\mathcal{M} \vDash \perp$ *and for all* $\mathcal{M}' \subsetneq \mathcal{M}$, $\mathcal{M}' \nvDash \perp$.

Definition 2 (MCS). $\mathcal{C} \subseteq \mathcal{F}$ *is a* minimal correction subset *(MCS) if and only if* $(\mathcal{F} \setminus \mathcal{C}) \nvDash \perp$ *and for all* $\mathcal{C}' \subsetneq \mathcal{C}$, $(\mathcal{F} \setminus \mathcal{C}') \vDash \perp$.

MUSes represent minimal explanations of unsatisfiability, while MCSes represent irreducible subsets of clauses whose removal renders $\mathcal{F}$ satisfiable. Every MUS (resp. MCS) is a minimal hitting set of the set of all MCSes (resp. MUSes) [3,27]. There can be a worst-case exponential number of MUSes and MCSes [20].

Among the clauses in the MUSes of $\mathcal{F}$, a special case is that of *necessary* clauses. A clause c is necessary for (the unsatisfiability of) $\mathcal{F}$ iff $(\mathcal{F} \setminus \{c\}) \nvDash \perp$. Necessary clauses belong to all the MUSes of a formula. In a broader sense, a clause is said to be *potentially necessary* [15] (*pn-clause* in short) iff it belongs to some MUS of $\mathcal{F}$, since it can become necessary after the removal of other clauses. The problem of deciding whether a given clause is potentially necessary is known to be Σ_2^P-complete (as the proof of Theorem 4 in [18] shows).

Input: $\mathcal{F}_0$ an unsatisfiable CNF formula
Output: CUMU, the union of MUSes of $\mathcal{F}_0$

1 $\mathcal{F}_0 \leftarrow \text{LeanKernel}(\mathcal{F}_0)$
2 $\mathcal{M} \leftarrow \text{ComputeMUS}(\mathcal{F}_0, \emptyset)$
3 CUMU $\leftarrow \mathcal{M}$
4 nec $\leftarrow \text{ComputeNecessaryClauses}(\mathcal{F}_0, \mathcal{M})$
5 **if** nec $\neq \mathcal{M}$ **then**
6      $\mathcal{F}_0 \leftarrow \mathcal{F}_0 \setminus \{c \in (\mathcal{F}_0 \setminus \text{nec}); \text{nec} \vDash \{c\}\}$
7      $\text{UMU\_rec}(\mathcal{F}_0, \text{nec}, \text{nec})$
8 **end**
9 **return** CUMU

Algorithm 1. Main procedure

This paper addresses the computation of the set of all pn-clauses of an unsatisfiable formula $\mathcal{F}$, i.e., the union of its MUSes, denoted $\text{UMU}(\mathcal{F})$. By the aforementioned hitting set duality relationship, the union of all MCSes of $\mathcal{F}$ is the same as $\text{UMU}(\mathcal{F})$.

A useful notion for *efficiently* over-approximating $\text{UMU}(\mathcal{F})$ is the *lean kernel* of $\mathcal{F}$, defined as the set of clauses used in some resolution refutation of $\mathcal{F}$, or, equivalently, those clauses which are not touched by any autarky (see [13] for the background, and see [16] for a recent paper on computing the lean kernel). A set of variables $A \subseteq V(\mathcal{F})$ is an *autarky* (or an *autarky varset*) iff there exists a truth assignment to the variables in A that satisfies all the clauses in $\mathcal{F}$ containing literals in the variables of A. There exists a unique largest autarky (varset), since the union of autarky (varsets) is again an autarky (varset). The lean kernel of $\mathcal{F}$ is obtained by removing the clauses containing some literal in the variables of the largest autarky of $\mathcal{F}$. In [15], the clauses in the lean kernel of $\mathcal{F}$ are referred to as *useful* clauses. Although some of these might not belong to $\text{UMU}(\mathcal{F})$, they can participate in short extended-resolution refutations of $\mathcal{F}$.

Example 1. Consider $\mathcal{F}_{ex} = \{c_1: (x_1), c_2: (\neg x_1), c_3: (\neg x_2), c_4: (\neg x_1 \vee x_2), c_5: (\neg x_2 \vee x_1), c_6: (\neg x_3 \vee x_4), c_7: (\neg x_4 \vee x_3)\}$. The largest autarky is $A = \{x_3, x_4\}$, so the lean kernel of $\mathcal{F}_{ex}$ is the clause-set $\{c_1, ..., c_5\}$. $\mathcal{F}_{ex}$ has the two MUSes $\mathcal{M}_1 = \{c_1, c_2\}$, $\mathcal{M}_2 = \{c_1, c_3, c_4\}$; and the three MCSes $\mathcal{C}_1 = \{c_1\}, \mathcal{C}_2 = \{c_2, c_3\}, \mathcal{C}_3 = \{c_2, c_4\}$. The union of MUSes is $\text{UMU}(\mathcal{F}_{ex}) = \{c_1, c_2, c_3, c_4\}$.

3 Computing the Union of MUSes

Consider an unsatisfiable formula $\mathcal{F}_0$. The main procedure for computing $\text{UMU}(\mathcal{F}_0)$ is shown in Algorithm 1. The core of our algorithm is recursive, with the initial recursive call in Line 7, and the recursive function UMU_rec given in Algorithm 2. Our approach develops a recursive enumeration of MUSes, keeping track of all the clauses found to be in $\text{UMU}(\mathcal{F}_0)$ along the search in the *global* under-approximation CUMU $\subseteq \text{UMU}(\mathcal{F}_0)$. Variable CUMU is always a subset of $\text{UMU}(\mathcal{F}_0)$, and equality holds upon termination (which is guaranteed). Each node of the recursion tree is characterized by a subformula $\mathcal{F} \subseteq \mathcal{F}_0$, from which an MUS $\mathcal{M} \subseteq \mathcal{F}$ is extracted. The MUS $\mathcal{M}$ is then used for splitting, removing one clause $c_i \in \mathcal{M}$ from $\mathcal{F}$ in each new branch. So, along

Global: CUMU, under-approximation of the union of MUSes
   $\mathcal{F}_0$, initial formula (for early termination checks)
Input: $\mathcal{F}$, nec, forced, clause sets
Output: Boolean value, indicating if CUMU $= \mathcal{F}_0$

```
1   F ← LeanKernel(F)
2   if (F ⊆ CUMU) or (forced ⊄ F) then  return false
3   M ← ComputeMUS(F, nec)
4   CUMU ← CUMU ∪ M
5   if CUMU = F₀ then return true
6   if (M = F) or (F ⊆ CUMU) then return false
7   for c ∈ (M \ forced) do
8       if SAT(F \ {c}) then
9           nec ← (nec ∪ {c})
10      else
11          if UMU_rec(F \ {c}, nec, forced) then return true
12          if F ⊆ CUMU then return false
13      end
14      forced ← (forced ∪ {c})
15      if not SAT(forced) then return false
16  end
17  return false
```

Algorithm 2. Recursive function UMU_rec

any path of the recursion tree, the current $\mathcal{F}$ is shrinking, by removing clauses, while CUMU $\subseteq$ UMU$(\mathcal{F}_0)$ is growing, over the whole search. So at any point we can abort the computation (say, due to a time limit), and can use the current CUMU as an under-approximation of UMU$(\mathcal{F}_0)$.

Some preprocessing steps are taken in Algorithm 1. In Line 1 the input formula $\mathcal{F}_0$ is reduced to its lean kernel, potentially removing a number of clauses not in UMU$(\mathcal{F}_0)$. Then in Line 2 an MUS $\mathcal{M} \subseteq \mathcal{F}_0$ is extracted, which serves to initialize CUMU, and to compute the set nec of all necessary clauses of $\mathcal{F}_0$ in Line 4 (note nec $\subseteq \mathcal{M}$). If nec $= \mathcal{M}$ then the algorithm terminates, since $\mathcal{F}_0$ only contains one MUS (due to different MUSes being incomparable regarding set-inclusion). Otherwise, clauses in $\mathcal{F}_0 \setminus$ nec entailed by nec are removed in Line 6 (no MUS of $\mathcal{F}_0$ contains such a clause, since it must contain nec as well), and the recursive procedure is invoked (parameters are passed by value). Note that CUMU can be initialized with better under-approximations of UMU$(\mathcal{F}_0)$ than a single MUS (e.g. the union of some MCSes or MUSes known).

Now we come to UMU_rec (Algorithm 2). First we need to explain the structure of the splitting (the loop from Line 7 to Line 16). We split on any MUS $\mathcal{M} \subseteq \mathcal{F}$, computed in Line 3, where nec is now (only) some set of necessary clauses of $\mathcal{F}$, possibly missing some (only at the root we spend the effort to compute all). We add the clauses in $\mathcal{M}$ to CUMU (Line 4), and return (i.e., abort this branch of the recursion tree) if CUMU is all of the original $\mathcal{F}_0$ (Line 5; here indeed everything is finished), or when $\mathcal{F}$ cannot possibly contain any new clauses for CUMU (Line 6). Let $\mathcal{M} = \{c_1, \ldots, c_k\}$; the basic splitting idea is that branch i *excludes* clause c_i for any MUS considered in that branch — no MUS yielding some new clause, other than what we got from $\mathcal{M}$, can contain all of $\mathcal{M}$.

Now we make the branches *possibly* disjoint ("morally" one might say), by *possibly including* in branch i the clauses $c_1, \ldots, c_{i-1}$. We note that the search is complete: Consider another MUS $\mathcal{M}'$ of $\mathcal{F}$ which contributes a new clause to CUMU. So there is some i with $c_i \notin \mathcal{M}'$, and for the smallest such i we have $c_1, \ldots, c_{i-1} \in \mathcal{M}'$.

The exclusion is lazily noted by adding the excluded clause to clause-set forced. We say "possibly exclude", which means the following: The branch can be aborted, if it is determined (by some incomplete, but correct argument — we prune "as good as we can") that every MUS $\mathcal{M} \subseteq \mathcal{F}$ with forced $\subset \mathcal{M}$ fulfils $\mathcal{M} \subseteq$ CUMU. On the other side, we can include into CUMU in Line 4 whatever we want (as long as it belongs to UMU($\mathcal{F}_0$)), the algorithm will always be correct.

The clause-set forced is the third argument of function UMU_rec, set in the original invocation to nec (Algorithm 1, Line 7), and updated in Line 14 of Algorithm 2. The second argument is the current value of nec, which is updated in Line 9: if $\mathcal{F}$ without c is satisfiable, then c is a necessary clause for $\mathcal{F}$, and considering the branch without c (Line 11 and Line 12) is useless.

It remains to discuss the remaining simplifications and shortcuts. In Line 1, $\mathcal{F}$ is reduced to its lean kernel (recall that no clause satisfied by some autarky can be element of any MUS). If $\mathcal{F}$ cannot contribute new clauses, or falls outside of forced, then we return (Line 2). The return value true of UMU_rec means that we are done completely, since all of $\mathcal{F}_0$ (the lean kernel of the original formula minus clauses entailed by the necessary clauses) has been covered by CUMU.

We always maintain the invariants nec $\subseteq$ forced, and that forced is satisfiable (while irredundancy holds initially, but may be lost along a path — experimentation showed the irredundancy test to be rather expensive). We conclude by summarizing the argumentation of the section:

Theorem 1. *Algorithm 1 correctly computes* UMU($\mathcal{F}_0$) *and terminates for all inputs* $\mathcal{F}_0$.

3.1 Additional Procedures

Each node in the search tree defined by Algorithm 2 involves a number of computations requiring SAT oracles, besides the tests in Line 8 and Line 15.

The procedure ComputeMUS($\mathcal{F}$, nec) computes an MUS $\mathcal{M} \subseteq \mathcal{F}$, where nec is a subset of the necessary clauses of $\mathcal{F}$. It follows a deletion-based approach with clause-set refinement [2]. In short, the procedure maintains two sets of clauses: $\mathcal{M}$, initialized with the clauses in nec, and $\mathcal{R}$, a *reference* set of clauses initially set to $\mathcal{F} \setminus$ nec. As an invariant, $(\mathcal{M} \cup \mathcal{R}) \vDash \bot$. Iteratively, it tests the satisfiability of $(\mathcal{M} \cup \mathcal{R} \setminus \{c\})$, for some $c \in \mathcal{R}$. If satisfiable, c is moved to $\mathcal{M}$. Otherwise, $\mathcal{R}$ is updated to $\mathcal{C} \setminus \mathcal{M}$, with $\mathcal{C}$ an unsatisfiable core produced by the SAT solver. The procedure terminates when $\mathcal{R}$ becomes empty, returning the MUS $\mathcal{M}$, requiring $|\mathcal{F}| - |$nec$|$ satisfiability tests at most.

Given an MUS $\mathcal{M} \subseteq \mathcal{F}$, the procedure ComputeNecessaryClauses($\mathcal{F}$, $\mathcal{M}$) identifies all the necessary clauses of $\mathcal{F}$. For each clause $c \in \mathcal{M}$ it tests the satisfiability of $\mathcal{F} \setminus \{c\}$. The clause c is necessary iff the latter formula is satisfiable. This procedure is only used in Algorithm 1 before the invocation to the recursive procedure, where (some) necessary clauses along a path are identified by a call to the SAT solver in the splitting step (Line 8 in Algorithm 2).

On the other hand, the procedure $\texttt{LeanKernel}(\mathcal{F})$ computes the lean kernel of $\mathcal{F}$ by first computing its largest autarky following the approach in [21]. Since this step is performed at every node (on different subsets of $\mathcal{F}_0$) we use a dedicated instantiation of a SAT solver for this task, which operates on an extension of the encoding Γ_3 proposed in [21] for representing largest autarkies. The encoding is extended with a selector variable s_i for each clause $c_i \in \mathcal{F}$, so that subformulas of $\mathcal{F}$ can be *activated* via assumptions, enabling reusing the encoding across the computation of the lean kernel of different subformulas. Initially, a formula referred to as $\mathcal{F}_{aut}$ is built from $\mathcal{F}_0$. Each variable $x_i \in V(\mathcal{F}_0)$ results in two variables $x_i^0, x_i^1 \in V(\mathcal{F}_{aut})$. Then, for each clause $c_i \in \mathcal{F}_0$ the following clauses are added, where $P(c_i)$ (resp. $N(c_i)$) denotes the positive (resp. negative) literals in c_i: (i) each $x_j \in P(c_i)$ results in the clause $s_i \rightarrow (x_j^0 \rightarrow \bigvee_{x_k \in P(c_i) \setminus \{x_j\}} x_k^1 \vee \bigvee_{x_k \in N(c_i)} x_k^0)$; (ii) each $x_j \in N(c_i)$ results in the clause $s_i \rightarrow (x_j^1 \rightarrow \bigvee_{x_k \in P(c_i)} x_k^1 \vee \bigvee_{x_k \in N(c_i) \setminus \{x_j\}} x_k^0)$. Finally, AtMost1 constraints of the form $(\neg x_i^0 \vee \neg x_i^1)$ for all $x_i \in V(\mathcal{F}_0)$ are added to $\mathcal{F}_{aut}$. The resulting encoding has $2n + m$ variables and $L + n$ clauses, where n is the number of variables, m is the number of clauses and L is the number of literals in $\mathcal{F}$.

Then, the computation of the largest autarky of any given $\mathcal{F} \subseteq \mathcal{F}_0$ amounts to computing a model of the formula $\mathcal{F}_{aut} \wedge \bigwedge_{\{c_i \in \mathcal{F}\}} s_i$ that maximizes the number of variables x_i^0 and x_i^1 set to true. This model is unique, so it corresponds to a maximal model (w.r.t. variables x_i^0 and x_i^1), that can be computed by reducing the problem to the extraction of an MCS of a formula considering the clauses in $\mathcal{F}_{aut} \wedge \bigwedge_{\{c_i \in \mathcal{F}\}} s_i$ as *hard* and a *soft* unit clause for each variable x_i^0 and x_i^1. For this purpose, we use the *Clause D* algorithm, as proposed in [21]. The set of variables appearing in the complement of the computed MCS corresponds to the largest autarky of $\mathcal{F}$. The lean kernel of $\mathcal{F}$ is obtained by removing all the clauses containing a variable in the computed autarky.

4   Experimental Results

This section presents an experimental assessment of the proposed approach to computing the union of MUSes, $\texttt{UMU}(\mathcal{F})$, of an unsatisfiable formula $\mathcal{F}$. The experiments were performed on a Linux machine (Intel Xeon 2.26 GHz, 128GByte). The time limit for each process was set to 900 s and the memory limit to 4GByte.

A prototype implementing the proposed algorithm was written in C++ on top of the known *caching* MCS enumerator *mcscache* proposed in [25]. In the following, the prototype is referred to as *umuser*. Among the existing alternative state-of-the-art MCS and MUS enumerators [6, 24–26], we opted to compare the prototype against *mcscache* as they share the same code base and the interface to a SAT solver. The underlying SAT solver used in both tools is MiniSat 2.2 [5]. Additionally and following the ideas of Sect. 3, we considered a combination of both tools connected in the following setup: (i) first, *mcscache* was used to enumerate MCSes within 3 min and (ii) second, *umuser* was bootstrapped with the clauses in these MCSes as an initial approximation of $\texttt{UMU}(\mathcal{F})$. This approach is referred to as *umuser**.

The considered benchmark suite includes two sets of instances. The first one is derived from the MUS track of the SAT competition 2011[1]. These benchmarks

---

[1] https://satcompetition.org/2011/.

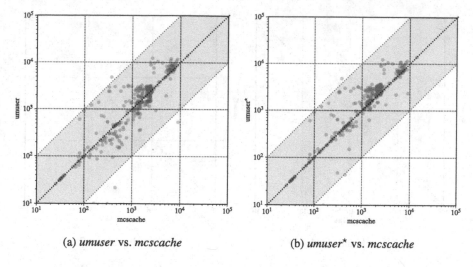

(a) *umuser* vs. *mcscache*   (b) *umuser** vs. *mcscache*

Fig. 1. The number of clauses in $\mathrm{UMU}(\mathcal{F})$ computed within 900 s.

were widely used in prior work on MUS and MCS computation and enumeration [2,19,22,23]. Since computing all clauses participating in an MUS/MCS is computationally hard, for our evaluation purposes, we took only instances with at most 20000 clauses from this benchmark set. The second set of instances comprises the benchmarks previously studied in a number of settings, including MCS [22] and SMUS (smallest MUS) [7] computation. These include automotive product configuration benchmarks [30] and circuit diagnosis. The total number of benchmark instances considered in both sets is 427.

First of all, it should be noted that *mcscache* terminates for 98 instances while the default setup of *umuser* can finish only for 41 of them by the time limit. This does not come as a surprise provided that the recursive procedure of *umuser* is computationally more expensive than simple MCS enumeration. In general, when targeting instances with few MCSes, an MCS enumerator may be the best approach to use, while *umuser* should be reasonable to apply when exhaustive enumeration of MCSes or MUSes is infeasible. Also note that the combined variant *umuser** terminates for 94 benchmark instances, several of them solved during the MCS enumeration phase. Similar results were observed with a longer time limit of 30 min, with only 7 more instances solved.

To illustrate the power of the proposed approach w.r.t. under-approximations of $\mathrm{UMU}(\mathcal{F})$, the second and the most important part of the experiment is to evaluate the number of clauses reported to participate in $\mathrm{UMU}(\mathcal{F})$. Figure 1 shows the comparison between *umuser/umuser** and *mcscache* in terms of the number of clauses in $\mathrm{UMU}(\mathcal{F})$ found within 900 s. As can be seen in Fig. 1a, apart from a number of outliers that can be exhaustively solved by *mcscache*, the basic version of *umuser* tends to compute more clauses than what can be achieved by *mcscache*, indicating its effectiveness at conducting the search towards new clauses in $\mathrm{UMU}(\mathcal{F})$. Furthermore, as shown in Fig. 1b, *umuser** strengthens the approach more, which enables it to find even more clauses. Concretely, the average number of clauses in $\mathrm{UMU}(\mathcal{F})$ per instance computed

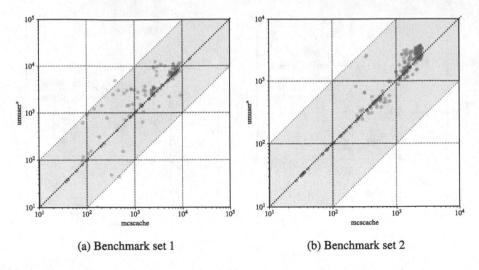

(a) Benchmark set 1   (b) Benchmark set 2

Fig. 2. The number of clauses in $\mathrm{UMU}(\mathcal{F})$ computed for the two benchmark sets.

by *mcscache* within 900 s is 2089.58, while *umuser* and *umuser** compute 2421.87 and 2623.14, respectively. A more detailed comparison between *umuser** and *mcscache* is shown in Fig. 2, where the two considered benchmark sets are analyzed separately. Both Fig. 2a and Fig. 2b confirm that *umuser** has an advantage over *mcscache* in terms of the number of clauses identified to belong to $\mathrm{UMU}(\mathcal{F})$.

Computing lean kernels allows the identification of clauses not in $\mathrm{UMU}(\mathcal{F})$. At the root of the recursion tree, on average 242 such clauses were identified, with an average reduction of 15.9% w.r.t. the number of clauses of the formula. The extent of this reduction depends on each instance, ranging from 0% (62 instances) to 97.32%, and a median value of 3.32%. These results encourage future research on strategies for activating and deactivating the computation of lean kernels on demand, depending on its effectiveness.

5   Conclusions

Identifying clauses that take part in the unsatisfiability of a formula represents a useful task in infeasibility analysis. In this context, clauses might be classified in three levels regarding their *importance*. The extreme cases are the set of necessary clauses, which belong to all MUSes, and the lean kernel, containing all the clauses that can be used in a resolution refutation. In the middle, the union of MUSes, $\mathrm{UMU}(\mathcal{F})$, constitutes a concise (and accurate) explanation of all the causes of unsatisfiability. This paper addresses the Σ_2^P-hard problem of computing $\mathrm{UMU}(\mathcal{F})$ and proposes a novel algorithm for this task. Based on a lazy splitting along MUSes (similar to splitting on a clause in SAT solving) and a global collection of clauses found for $\mathrm{UMU}(\mathcal{F})$, the algorithm uses a variety of NP oracles for powerful pruning techniques, obtaining a shortened implicit enumeration of MUSes. These pruning steps can be relaxed or strengthened in various ways, giving rise to a general framework for computing $\mathrm{UMU}(\mathcal{F})$, and opening a wide space of promising possibilities for the future. Experimental results show that under a given time limit the new algorithm produces (much) better approximations of $UMU(\mathcal{F})$ than an approach based on MCS enumeration. The results also reveal potential benefits of combining both approaches, encouraging further research on alternative methods.

References

1. Bacchus, F., Katsirelos, G.: Finding a collection of MUSes incrementally. In: Quimper, C.-G. (ed.) CPAIOR 2016. LNCS, vol. 9676, pp. 35–44. Springer, Cham (2016). https://doi.org/10.1007/978-3-319-33954-2_3
2. Belov, A., Lynce, I., Marques-Silva, J.: Towards efficient MUS extraction. AI Commun. **25**(2), 97–116 (2012). https://doi.org/10.3233/AIC-2012-0523
3. Birnbaum, E., Lozinskii, E.L.: Consistent subsets of inconsistent systems: structure and behaviour. J. Exp. Theor. Artif. Intell. **15**(1), 25–46 (2003). https://doi.org/10.1080/0952813021000026795
4. Cook, S.A.: The complexity of theorem-proving procedures. In: Harrison, M.A., Banerji, R.B., Ullman, J.D. (eds.) Proceedings of the 3rd Annual ACM Symposium on Theory of Computing, Shaker Heights, Ohio, USA, 3–5 May 1971. pp. 151–158. ACM (1971). https://doi.org/10.1145/800157.805047
5. Eén, N., Sörensson, N.: An extensible SAT-solver. In: Giunchiglia, E., Tacchella, A. (eds.) SAT 2003. LNCS, vol. 2919, pp. 502–518. Springer, Heidelberg (2004). https://doi.org/10.1007/978-3-540-24605-3_37
6. Grégoire, É., Izza, Y., Lagniez, J.: Boosting MCSes enumeration. In: Lang [17], pp. 1309–1315. https://doi.org/10.24963/ijcai.2018/182
7. Ignatiev, A., Janota, M., Marques-Silva, J.: Quantified maximum satisfiability. Constraints **21**(2), 277–302 (2016). https://doi.org/10.1007/s10601-015-9195-9
8. Jannach, D., Schmitz, T.: Model-based diagnosis of spreadsheet programs: a constraint-based debugging approach. Autom. Softw. Eng. **23**(1), 105–144 (2016). https://doi.org/10.1007/s10515-014-0141-7
9. Jannach, D., Schmitz, T., Hofer, B., Schekotihin, K., Koch, P.W., Wotawa, F.: Fragment-based spreadsheet debugging. Autom. Softw. Eng. **26**(1), 203–239 (2019). https://doi.org/10.1007/s10515-018-0250-9
10. Janota, M., Marques-Silva, J.: On deciding MUS membership with QBF. In: Lee, J. (ed.) CP 2011. LNCS, vol. 6876, pp. 414–428. Springer, Heidelberg (2011). https://doi.org/10.1007/978-3-642-23786-7_32
11. Jose, M., Majumdar, R.: Bug-assist: assisting fault localization in ANSI-C programs. In: Gopalakrishnan, G., Qadeer, S. (eds.) CAV 2011. LNCS, vol. 6806, pp. 504–509. Springer, Heidelberg (2011). https://doi.org/10.1007/978-3-642-22110-1_40
12. Jose, M., Majumdar, R.: Cause clue clauses: error localization using maximum satisfiability. In: Hall, M.W., Padua, D.A. (eds.) Proceedings of the 32nd ACM SIGPLAN Conference on Programming Language Design and Implementation. PLDI 2011, San Jose, CA, USA, 4–8 June 2011. pp. 437–446. ACM (2011). https://doi.org/10.1145/1993498.1993550
13. Kleine Büning, H., Kullmann, O.: Minimal unsatisfiability and autarkies. In: Biere, A., Heule, M., van Maaren, H., Walsh, T. (eds.) Handbook of Satisfiability, Frontiers in Artificial Intelligence and Applications, vol. 185, pp. 339–401. IOS Press (2009). https://doi.org/10.3233/978-1-58603-929-5-339
14. Kullmann, O.: An application of matroid theory to the SAT problem. In: Proceedings of the 15th Annual IEEE Conference on Computational Complexity, Florence, Italy, 4–7 July 2000, p. 116. IEEE Computer Society (2000). https://doi.org/10.1109/CCC.2000.856741

15. Kullmann, O., Lynce, I., Marques-Silva, J.: Categorisation of clauses in conjunctive normal forms: minimally unsatisfiable sub-clause-sets and the lean kernel. In: Biere, A., Gomes, C.P. (eds.) SAT 2006. LNCS, vol. 4121, pp. 22–35. Springer, Heidelberg (2006). https://doi.org/10.1007/11814948_4

16. Kullmann, O., Marques-Silva, J.: Computing maximal autarkies with few and simple oracle queries. In: Heule, M., Weaver, S. (eds.) SAT 2015. LNCS, vol. 9340, pp. 138–155. Springer, Cham (2015). https://doi.org/10.1007/978-3-319-24318-4_11

17. Lang, J. (ed.): Proceedings of the Twenty-Seventh International Joint Conference on Artificial Intelligence. IJCAI 2018, Stockholm, Sweden, 13–19 July 2018 (2018). www.ijcai.org

18. Liberatore, P.: Redundancy in logic I: CNF propositional formulae. Artif. Intell. **163**(2), 203–232 (2005). https://doi.org/10.1016/j.artint.2004.11.002

19. Liffiton, M.H., Previti, A., Malik, A., Marques-Silva, J.: Fast, flexible MUS enumeration. Constraints **21**(2), 223–250 (2016). https://doi.org/10.1007/s10601-015-9183-0

20. Liffiton, M., Sakallah, K.: Searching for autarkies to trim unsatisfiable clause sets. In: Kleine Büning, H., Zhao, X. (eds.) SAT 2008. LNCS, vol. 4996, pp. 182–195. Springer, Heidelberg (2008). https://doi.org/10.1007/978-3-540-79719-7_18

21. Marques-Silva, J., Ignatiev, A., Morgado, A., Manquinho, V.M., Lynce, I.: Efficient autarkies. In: Schaub, T., Friedrich, G., O'Sullivan, B. (eds.) ECAI 2014–21st European Conference on Artificial Intelligence, 18–22 August 2014, Prague, Czech Republic - Including Prestigious Applications of Intelligent Systems (PAIS 2014). Frontiers in Artificial Intelligence and Applications, vol. 263, pp. 603–608. IOS Press (2014). https://doi.org/10.3233/978-1-61499-419-0-603

22. Mencía, C., Ignatiev, A., Previti, A., Marques-Silva, J.: MCS extraction with sublinear oracle queries. In: Creignou, N., Le Berre, D. (eds.) SAT 2016. LNCS, vol. 9710, pp. 342–360. Springer, Cham (2016). https://doi.org/10.1007/978-3-319-40970-2_21

23. Mencía, C., Previti, A., Marques-Silva, J.: Literal-based MCS extraction. In: Yang, Q., Wooldridge, M.J. (eds.) Proceedings of the Twenty-Fourth International Joint Conference on Artificial Intelligence. IJCAI 2015, 25–31 July 2015, Buenos Aires, Argentina, pp. 1973–1979. AAAI Press (2015). http://ijcai.org/Abstract/15/280

24. Narodytska, N., Bjørner, N., Marinescu, M., Sagiv, M.: Core-guided minimal correction set and core enumeration. In: Lang [17], pp. 1353–1361. https://doi.org/10.24963/ijcai.2018/188

25. Previti, A., Mencía, C., Järvisalo, M., Marques-Silva, J.: Improving MCS enumeration via caching. In: Gaspers, S., Walsh, T. (eds.) SAT 2017. LNCS, vol. 10491, pp. 184–194. Springer, Cham (2017). https://doi.org/10.1007/978-3-319-66263-3_12

26. Previti, A., Mencía, C., Järvisalo, M., Marques-Silva, J.: Premise set caching for enumerating minimal correction subsets. In: McIlraith, S.A., Weinberger, K.Q. (eds.) Proceedings of the Thirty-Second AAAI Conference on Artificial Intelligence, (AAAI 2018), the 30th innovative Applications of Artificial Intelligence (IAAI 2018), and the 8th AAAI Symposium on Educational Advances in Artificial Intelligence (EAAI 2018), New Orleans, Louisiana, USA, 2–7 February 2018. pp. 6633–6640. AAAI Press (2018). https://www.aaai.org/ocs/index.php/AAAI/AAAI18/paper/view/17328

27. Reiter, R.: A theory of diagnosis from first principles. Artif. Intell. **32**(1), 57–95 (1987). https://doi.org/10.1016/0004-3702(87)90062-2

28. Safarpour, S., Mangassarian, H., Veneris, A.G., Liffiton, M.H., Sakallah, K.A.: Improved design debugging using maximum satisfiability. In: Proceedings of the 7th International Conference on Formal Methods in Computer-Aided Design, FMCAD 2007, Austin, Texas, USA, 11–14 November 2007, pp. 13–19. IEEE Computer Society (2007). https://doi.org/10.1109/FAMCAD.2007.26

29. Schlobach, S., Huang, Z., Cornet, R., van Harmelen, F.: Debugging incoherent terminologies. J. Autom. Reasoning **39**(3), 317–349 (2007). https://doi.org/10.1007/s10817-007-9076-z

30. Sinz, C., Kaiser, A., Küchlin, W.: Formal methods for the validation of automotive product configuration data. AI EDAM **17**(1), 75–97 (2003). https://doi.org/10.1017/S0890060403171065

31. Smith, A., Veneris, A.G., Ali, M.F., Viglas, A.: Fault diagnosis and logic debugging using Boolean satisfiability. IEEE Trans. CAD Integr. Circ. Syst. **24**(10), 1606–1621 (2005). https://doi.org/10.1109/TCAD.2005.852031

32. Stuckey, P.J., Sulzmann, M., Wazny, J.: Interactive type debugging in Haskell. In: Proceedings of the ACM SIGPLAN Workshop on Haskell, Haskell 2003, Uppsala, Sweden, 28 August 2003. pp. 72–83. ACM (2003). https://doi.org/10.1145/871895.871903

Revisiting Graph Width Measures
for CNF-Encodings

Stefan Mengel[1] and Romain Wallon[1,2（✉）]

[1] CRIL-CNRS UMR 8188, 62307 Lens, France
{mengel,wallon}@cril.fr
[2] Université d'Artois, Lens, France

Abstract. We consider bounded width CNF-formulas where the width is measured by popular graph width measures on graphs associated to CNF-formulas. Such restricted graph classes, in particular those of bounded treewidth, have been extensively studied for their uses in the design of algorithms for various computational problems on CNF-formulas. Here we consider the expressivity of these formulas in the model of clausal encodings with auxiliary variables. We first show that bounding the width for many of the measures from the literature leads to a dramatic loss of expressivity, restricting the formulas to such of low communication complexity. We then show that the width of optimal encodings with respect to different measures is strongly linked: there are two classes of width measures, one containing primal treewidth and the other incidence cliquewidth, such that in each class the width of optimal encodings only differs by constant factors. Moreover, between the two classes the width differs at most by a factor logarithmic in the number of variables. Both these results are in stark contrast to the setting without auxiliary variables where all width measures we consider here differ by more than constant factors and in many cases even by linear factors.

1 Introduction

Graph width measures like treewidth and cliquewidth have been studied extensively in the context of propositional satisfiability. The general idea is to assign graphs to CNF-formulas and compute their width with respect to different width measures. Then, if the resulting width is small, there are algorithms that solve SAT, but also more complex problems like #SAT or MAX-SAT or even QBF efficiently, see e.g. [10,13,25,28,29,32] for this line of work. There is also a considerable body of work on reasoning problems from artificial intelligence restricted to knowledge encoded by CNF-formulas with restricted underlying graphs: for example, treewidth restrictions have been studied for abduction, closed world reasoning, circumscription, disjunctive logic programming [19] and answer set programming [20]. There is thus by now a large body of work on how problems can be solved on bounded width CNF-formulas for different graph width measures.

© Springer Nature Switzerland AG 2019
M. Janota and I. Lynce (Eds.): SAT 2019, LNCS 11628, pp. 222–238, 2019.
https://doi.org/10.1007/978-3-030-24258-9_16

Curiously, however, there seems to be very little work on the natural question what we can actually encode with these restricted CNF-formulas. This question is pertinent because good algorithms for problems are less attractive if they cannot deal with interesting instances. We make two main contributions on the expressivity of bounded width CNF-formulas here.

First, we show one can give lower bounds for the width of any encoding of a function by means of communication complexity. This was known for tree-width [8], but we extend it for many different width measures, in particular (signed and unsigned) cliquewidth [13,32], modular treewidth [25] and MIM-width [28]. As a consequence, in a sense, for all these measures, formulas of bounded width can only encode simple functions.

All these lower bounds do not only work for *representations* of functions as CNF-formulas but also on *clausal encodings*, i.e. CNF-formulas using auxiliary variables. It is folklore that adding auxiliary variables can decrease the size of an encoding: for example the parity function has no subexponential CNF-representations but there is an easy linear size encoding using auxiliary variables. We here observe a similar effect for the example of treewidth: we show that any CNF-representation of the **AtMostOne**$_n$-function of n inputs without auxiliary variables has primal treewidth $n - 1$ which is the highest possible. But when authorizing the use of auxiliary variables, **AtMostOne**$_n$ can be computed with formulas of bounded treewidth easily. This shows that lower bounds for clausal encodings are far stronger than those of CNF-representations. Considering that **AtMostOne**$_n$ is arguably a very easy function, we feel that encodings with auxiliary variables are the more interesting notion in our setting so we focus on them here.

In a second main contribution, we focus on the *relative* expressive power of different graph width measures for clausal encodings. For the graph width measures studied in the literature, it is known that without auxiliary variables the expressivity of bounded width CNF-formulas is different for all notions and they form a partial order with so-called MIM-width as the most general notion, see e.g. [7, Sect. 5]. Somewhat surprisingly, the situation changes completely when one allows auxiliary variables: in this setting, the commonly considered width notions are all up to constant factors either equivalent to primal treewidth or to incidence cliquewidth. This is true for every individual function. We remark that for the parameters primal treewidth, dual treewidth and incidence treewidth, it was already known that the width of encodings minimizing the respective width measures differs only by constant factors [8,23,30]. All other relationships are new.

We also show that, assuming that an optimal encoding of a function has at least primal treewidth $\log(n)$ where n is the number of variables, incidence cliquewidth and primal treewidth differ exactly by a factor of $\Theta(\log(n))$ for optimal encodings. So, up to a logarithmic scaling, in fact all the width measures in [13,25,28,29,32] coincide when allowing auxiliary variables. Note that this scaling exactly corresponds to the runtime differences of many algorithms: while treewidth based algorithms often have runtimes of the type $2^{O(k)}n^c$ for treewidth k and a constant c, cliquewidth based algorithms typically give run-times roughly $n^{O(k')}$ for cliquewidth k'. These runtimes coincide exactly when treewidth and cliquewidth differ by a logarithmic factor which, as we show here, they do generally for encodings with auxiliary variables.

We finally use our main results for several applications. In particular, we answer an open question of [8] on the cliquewidth of the permutation function $\mathbf{PERM}_n$, see Sect. 6 for details.

Most of our results use machinery recently developed in the area of knowledge compilation. In particular, we use a combination of the algorithm in [4], the width notion for DNNF developed in [9] and the lower bound techniques from [5,27]. Relying on these building blocks, most of our proofs become rather simple.

Due to lack of space, some of the proofs are given only in the full version of this paper.

2   Preliminaries

2.1   CNF-Formulas and Their Graphs

We use standard notations for CNF-formulas as it can e.g. be found in [3]. Let X be a set of variables. A *representation* of a Boolean function f in variables X is a CNF-formula F on the variable set X that has as models exactly the assignments on which f evaluates to true. A *clausal encoding* of f is a CNF-formula F' on a variable set $X \cup Y$ such that

- for every assignment $a : X \to \{0, 1\}$ on which f evaluates to true, there is an extension a' of a to Y that is a model of F', and
- for every assignment $a : X \to \{0, 1\}$ on which f evaluates to false, no extension a' of a to Y is a model of F'.

The variables in Y are called *auxiliary variables*. An auxiliary variable y is called *dependent* if and only if in the first item above all extensions a' satisfying F' take the same value on y [17]. We say that a clausal encoding has *dependent auxiliary variables* if all its auxiliary variables are dependent. Note that for such an encoding the extension a' is unique.

We use standard notations from graph theory and assume the reader to have a basic background in the area [12]. By $N(v)$ we denote the open neighborhood of a vertex in a graph.

We will also in some parts of this paper deal with Boolean circuits. We assume that the reader is familiar with basic definitions in the area. As it is common when considering circuits with structurally restricted underlying graphs, we assume that every input variable appears in only one input gate. This property is sometimes called the *read-once property*.

To every CNF-formula F, we assign two graphs. The *primal graph* of F has as vertices the variables of F and two variables x, y are connected by an edge if and only if there is a clause C such that a literal in x and a literal in y appear in C. The *incidence graph* of F has as vertex set the union of the variable set and the clause set of F. Edges in the incidence graph are exactly the pairs x, C where x is a variable and C a clause that contains a literal in x.

A *tree decomposition* $(T, (B_t)_{t \in V(T)})$ of a graph $G = (V, E)$ consists of a tree T and, for every node t of T, a set $B_t \subseteq V$ called *bag* such that:

- $\bigcup_{v \in V(T)} B_t = V$,
- for every edge $uv \in E$ there is a bag B_t such that $\{u, v\} \subseteq B_t$, and
- for every $v \in V$, the set $\{t \in V(T) \mid v \in B_t\}$ is connected in T.

The *width* of a tree decomposition is defined as $\max\{|B_t| \mid t \in V(T)\} - 1$. The *treewidth* $\mathsf{tw}(G)$ of G is defined as the minimum width taken over all tree decompositions of G. The *primal treewidth* $\mathsf{tw_p}(F)$ of a CNF-formula F is defined as the treewidth of its primal graph and the *incidence treewidth* $\mathsf{tw_i}(F)$ of F is defined as that of the incidence graph.

We say that two vertices u, v in a graph $G = (V, E)$ have the same neighborhood type if and only if $N(u) \setminus \{v\} = N(v) \setminus \{u\}$. It can be shown that having the same neighborhood type is an equivalence relation on V. A generalization of treewidth is *modular treewidth* which is defined as follows: from a graph G we construct a new graph G' by contracting all vertices sharing a neighborhood type, i.e., from every equivalence class we delete all vertices but one. The modular treewidth of G is then defined to be the treewidth of G'. The modular treewidth $\mathsf{mtw}(F)$ of a CNF-formula F is defined as the modular treewidth of its incidence graph.

We will deal with several other graph width measures for a CNF-formula in the remainder of this paper, in particular dual treewidth $\mathsf{tw_d}(F)$, signed incidence cliquewidth $\mathsf{scw}(F)$, incidence cliquewidth $\mathsf{cw}(F)$, and MIM-width $\mathsf{mimw}(F)$. Since for those notions we will only use some of their properties, we will refrain from overwhelming the reader by giving their definitions and refer to the literature, e.g. [13, 28, 29, 32, 33].

We also consider the treewidth $\mathsf{tw}(C)$ and the cliquewidth $\mathsf{cw}(C)$ of Boolean circuits C.

2.2 Communication Complexity

We here give a very basic introduction to communication complexity. For more details, the reader is referred to the very readable textbook [22].

Let X be a set of variables and $\Pi = (Y, Z)$ a partition of X. A *combinatorial rectangle* respecting Π is a function $r(X) = r_1(Y) \wedge r_2(Z)$. For a Boolean function f on X, a *rectangle cover of size s* respecting Π is defined to be a representation

$$f(X) = \bigvee_{i=1}^{s} r_1^i(Y) \wedge r_2^i(Z),$$

where all $r_1^i(Y) \wedge r_2^i(Z)$ are combinatorial rectangles respecting Π. The non-deterministic communication complexity $\mathsf{cc}(f, \Pi) = \mathsf{cc}(f, (Y, Z))$ of f is defined as $\log(s_{\min})$ where $s_{\min}$ is the minimum size of any rectangle cover of f respecting Π.

The best-case non-deterministic communication complexity with $\frac{1}{3}$-balance $cc_{best}^{1/3}(f)$ is defined as $cc_{best}^{1/3}(f) := \min_{\Pi}(cc(f, \Pi))$ where the minimum is over all partitions $\Pi = (Y, Z)$ of X with $\min(|Y|, |Z|) \geq |X|/3$.

2.3 Structured Deterministic DNNF

Out of the rich landscape of representations from knowledge compilation, see e.g. [11, 26], we only introduce one that we will use in the remainder of this paper. For all circuits in this section, we assume that $\wedge$-gates have exactly two inputs while the number of $\vee$-gates may be arbitrary.

A *v-tree* T for a variable set X is a full binary tree whose leaves are in bijection with X. We call the variable assigned by this bijection to a leaf v the *label* of v. For a node $t \in T$, we denote by T_t the subtree of T that has t as its root and by $var(T_t)$ the variables that are labels of leaves in T_t.

We give some definitions from [9]. A *complete structured DNNF D* structured by a v-tree T is a Boolean circuit with the following properties: there is a labeling μ of the nodes in T with subsets of gates of D such that:

- For every gate g of D there is a unique node t_g of T with $g \in \mu(t_g)$.
- If t is a leaf labeled by a variable x, then $\mu(t)$ may only contain x and $\neg x$. Moreover, for every input gate g, the node t_g is a leaf.
- For every $\vee$-gate g, all inputs are $\wedge$-gates in $\mu(t_g)$.
- Every $\wedge$-gate g has exactly two inputs g_1, g_2 that are both $\vee$-gates or input gates. Moreover, t_{g_1} and t_{g_2} are the children of t_g in T and in particular $t_{g_1} \neq t_{g_2}$.

The *width* $wi(D)$ of D is defined as the maximal number of $\vee$-gates in any set $\mu(t)$.

We often speak of complete structured DNNF without mentioning the v-tree by which it is structured in cases where the form of the v-tree is unsubstantial.

Intuitively, a complete structured DNNF is a Boolean circuit in negation normal form in which the gates are organized into blocks $\lambda(t)$ which form a tree shape. In every block one then computes a 2-DNF whose inputs are gates from the blocks that are the children of $\lambda(t)$ in the tree shape.

A complete structured DNNF is called *deterministic* if and only if for every assignment and for every $\vee$-gate, at most one input evaluates to true.

Note that we do not allow constant input gates here. We remark that if we allowed those, we could always get rid of them in the circuit by propagation without changing any other properties of the circuit, see [9, Sect. 4]. We also remark that in a complete structured DNNF D, we can *forget* a variable x, i.e., construct a complete structured DNNF D' computing $\exists x D$, by setting all occurrences of x and $\neg x$ to 1 and propagating the constants in the obvious way. This operation does not increase the width, see [9]. However, if D is deterministic, this is generally not the case for D'.

3   The Effect of Auxiliary Variables

In this section, we will show that introducing auxiliary variables may arbitrarily reduce the treewidth of encodings. Note that this is not very surprising since it is not too hard to see that CNF-representations of, say, the parity function, are of high treewidth. However, in this case the *size* of the representation is exponential, so in a sense parity is a hard function for CNF-representations anyway. Here we will show that even for functions that have small CNF-representations there can be a large gap between the treewidth of representations and clausal encodings with auxiliary variables. To this end, consider the **AtMostOne**$_n$-function on variables $x_1, \ldots, x_n$ which accepts exactly those assignments in which at most one variable is assigned to 1. There is an obvious quadratic size representation as

$$\textbf{AtMostOne}_n = \bigwedge_{i,j \in [n], i < j} \neg x_i \vee \neg x_j.$$

However, this representation has as primal graph the clique K_n which is of treewidth $n - 1$. In the next lemma, we will see that in fact there is no representation of **AtMostOne**$_n$ that is of smaller primal treewidth.

Lemma 1. *Any CNF-representation of the **AtMostOne**$_n$-function of n inputs without auxiliary variables has primal treewidth $n - 1$.*

Proof. Let $x_1, \ldots, x_n$ be the variables of **AtMostOne**$_n$. We proceed with two claims.

Claim. Every non-tautological clause C of any CNF representation of **AtMostOne**$_n$ must contain at least the negation of two variables from $x_1, \ldots, x_n$.

Proof. Suppose that a clause C does not contain two such literals. Then, there are two possible cases: either C contains no negated variables or exactly one.

In the first case, the model of **AtMostOne**$_n$ setting all variables to 0 does not satisfy C, so C cannot be part of the CNF representation.

In the second case, let x_i be the (only) variable of **AtMostOne**$_n$ appearing negatively in C. Then, the model of **AtMostOne**$_n$ setting only x_i to 1 and all other variables to 0 does not satisfy C, so C cannot be part of the CNF representation, either.

Hence, at least two negated variables must appear in C.   □

From this claim, we will deduce that all pairs of variables must appear conjointly in at least one clause.

Claim. For each pair of variables x_i, x_j from $x_1, \ldots, x_n$ with $i \neq j$, there is a clause in the CNF representation of **AtMostOne**$_n$ containing both $\neg x_i$ and $\neg x_j$.

Proof. Suppose that, for a pair x_i, x_j such a clause does not exist. Let a be the assignment that sets exactly the variables x_i, x_j to 1 and all other variables to 0.

Let C be a clause from the CNF representation. By our previous claim, C contains two negated variables from $x_1, \ldots, x_n$. Because of our assumption, at least one of these literals is neither $\neg x_i$ nor $\neg x_j$, and this literal is satisfied by a. Thus C is satisfied by a.

Since this is true for every clause C, it follows that a satisfies all the clauses of the representation, so it is one of its models. However, a is not a model of **AtMostOne**$_n$. As a consequence, a clause containing both $\neg x_i$ and $\neg x_j$ must exist, which is also true for every pair x_i, x_j.   □

This last claim shows that for each pair of variables, there is a clause containing both of them. It follows that all variables are connected to all other variables in the primal graph of the representation. So the primal graph is a clique which has treewidth $n - 1$.   □

If we allow the use of auxiliary variables, we may decrease the treewidth dramatically.

Lemma 2. *There is a clausal encoding of* **AtMostOne**$_n$ *of primal treewidth 2.*

Proof. We use the well-known ladder encoding from [16], see also [3, Sect. 2.2.5]. We introduce the auxiliary variables $y_0, \ldots, y_n$. The encoding consists of the following clauses:

- the validity clauses $\neg y_{i-1} \lor y_i$, and
- clauses representing the constraint $x_i \leftrightarrow (\neg y_{i-1} \land y_i)$

for every $i \in [n]$. It is easy to see that this encoding is correct.

Concerning the treewidth bound, we construct for every index $i \in [n]$ the bag $B_i := \{y_{i-1}, y_i, x_i\}$. Then $(P_n, (B_i)_{i \in [n]})$ where P_n has nodes $[n]$ and edges $\{(i, i+1) \mid i \in [n-1]\}$ is a tree decomposition of width 2.   □

4   Width vs. Communication

In this section, we show that from communication complexity we get lower bounds for the various width notions of Boolean functions. The main building block is the following result that is an application of the main result of [27] to complete structured DNNF.

Theorem 1. *Let D be a complete structured DNNF structured by a v-tree T computing a function f in variables X. Let t be a node of T and let $Y := \mathsf{var}(T_t)$ and $Z = X \setminus \mathsf{var}(T_t)$. Finally, let ℓ be the number of $\lor$-gates in $\mu(t)$. Then there is a rectangle cover of f respecting (Y, Z) of size at most ℓ.*

Note that in [27] the considered models are structured DNNF that are not necessarily complete, a slightly more general model than ours. Thus the statement in [27] is slightly different. However, it is easy to see that in our restricted setting, their proof shows the statement we give above, see also the discussion in [5, Sect. 5].

Since Theorem 1 is somewhat technical, it will be more convenient here to use the following easy consequence that one gets directly with the definitions.

Proposition 1. *Let D be a complete structured DNNF structured by a v-tree T computing a function f in variables X. Let t be a node of T and let $Y := \mathsf{var}(T_t)$ and $Z = X \setminus \mathsf{var}(T_t)$. Then*

$$\log(\mathsf{wi}(D)) \geq \mathsf{cc}(f, (Y, Z)).$$

In many cases, instead of considering explicit v-trees, it is more convenient to simply use best-case communication complexity.

Corollary 1. *Let f be a Boolean function in variables X. Then, for every complete structured DNNF computing f, we have*

$$\mathsf{wi}(D) \geq 2^{\mathsf{cc}_{\mathsf{best}}^{1/3}(f)}.$$

Proof. Note that for every v-tree with X on the leaves, there is a node t such that $|X|/3 \leq |\mathsf{var}(T_t)| \leq 2|X|/3$. Plugging this into Proposition 1 directly yields the result.   □

We will use Corollary 1 to turn compilation algorithms that produce complete structured DNNF based on a parameter of the input as in [1, 6] into inexpressivity bounds based on this parameter. We first give an abstract version of this result that we will instantiate for concrete measures later on.

Theorem 2. *Let $\mathcal{C}$ be a (fully expressive) representation language for Boolean functions. Let p be a parameter $\mathsf{p} : \mathcal{C} \to \mathbb{N}$. Assume that there is for every Boolean function f and every $C \in \mathcal{C}$ that encodes f a complete structured DNNF with*

$$\mathsf{wi}(D) \leq 2^{\mathsf{p}(C)}.$$

Then we have

$$\mathsf{p}(C) \geq \mathsf{cc}_{\mathsf{best}}^{1/3}(f).$$

Proof. From the assumption, we get $\mathsf{p}(C) \geq \log(\mathsf{wi}(D))$. Then we apply Corollary 1 to directly get the result.   □

Intuitively, it is exactly the algorithmic usefulness of parameters that makes the resulting instances inexpressive. Note that it is not surprising that instances whose expressiveness is severely restricted allow for good algorithmic properties. However, here we see that the inverse of this statement is also true in a quite harsh way: if a parameter has good algorithmic properties allowing efficient compilation into DNNF, then this parameter puts strong restrictions on the complexity of the expressible functions.

Note that instead of Corollary 1 we could have used Proposition 1 in the proof of Theorem 2 to get a slightly stronger result. We chose to go with a simpler statement here but note that we will use the extended strength of Proposition 1 later on in Sect. 6.

From Theorem 2, we directly get lower bounds for the width measures studied in [13, 24, 28, 29, 32]. The first result considers the parameters with respect to which SAT is fixed-parameter tractable.

Corollary 2. *There is a constant $b > 0$ such that for every Boolean function f and every CNF C encoding f we have*

$$\min\{\mathsf{tw_i}(C), \mathsf{tw_p}(C), \mathsf{tw_d}(C), \mathsf{scw}(C)\} \geq b \cdot \mathsf{cc}_{\mathsf{best}}^{1/3}(f)$$

Proof. This follows directly from Theorem 2 and the fact that for all these parameters we have compilation algorithms of width $2^{O(k)}$.   □

Using the compilation algorithm from [1, 2], we get essentially the same result for circuit representations.

Corollary 3. *There is a constant $b > 0$ such that for every Boolean function f and every circuit C encoding f we have*

$$\min\{\mathsf{tw}(C), \mathsf{cw}(C)\} \geq b \cdot \mathsf{cc}_{\mathsf{best}}^{1/3}(f)$$

We remark that for treewidth 1 the circuits of Corollary 3 boil down to so-called read-once functions which have been studied extensively, see e.g. [18].

Finally, we give a version for parameters that allow polynomial time algorithms when fixed but no fixed-parameter algorithms.

Corollary 4. *There is a constant $b > 0$ such that for every Boolean function f in n variables and every CNF C encoding f we have*

$$\min\{\mathsf{mimw}(C), \mathsf{cw}(C), \mathsf{mtw}(C)\} \geq b \cdot \frac{\mathsf{cc}_{\mathsf{best}}^{1/3}(f)}{\log(n)}$$

Proof. All of the width measures in the statement allow compilation into complete structured DNNF of size – and thus also width – $n^{O(k)}$ for parameter value k and n variables [4]. Thus, with Theorem 2, for each measure there is a constant b' with $\log(n^k) = k\log(n) \geq b'\mathsf{cc}_{\mathsf{best}}^{1/3}(f)$ which completes the proof.   □

Note that the bounds of Corollary 4 are lower by a factor of $\log(n)$ than those of Corollary 2. We will see in the next section that in a sense this difference is unavoidable.

5   Relations Between Different Width Measures of Encodings

In this section, we will show that the different width measures for optimal clausal encodings are strongly related. We will start by proving that primal treewidth bounds imply bounds for modular treewidth and cliquewidth.

Theorem 3. *Let k be a positive integer and f be a Boolean function of n variables that has a CNF-encoding F of primal treewidth at most $k \log(n)$. Then f also has a CNF-encoding F' of modular incidence treewidth and cliquewidth $O(k)$. Moreover, if F has dependent auxiliary variables, then so has F'.*

Due to space limitations, the proof of Theorem 3 is given in the full version of this paper. Let us here discuss this result a little. It is well known that the modular treewidth and the cliquewidth of a CNF formula can be much smaller than its treewidth [32]. Theorem 3 strengthens this by saying essentially that for *every* function we can gain a factor logarithmic in the number of variables.

In particular, this shows that the lower bounds we can show with Corollary 4 are the best possible: the maximal lower bounds we can show are of the form $n/\log(n)$ and since there is always an encoding of every function of treewidth n, by Theorem 3 there is always an encoding of cliquewidth roughly $n/\log(n)$. Thus the maximal lower bounds of Corollary 4 are tight up to constants.

Note that for Theorem 3, it is important that we are allowed to change the encoding. For example, the primal graph of the formula $F = \bigwedge_{i,j \in [n]} (x_{ij} \vee x_{i+1,j}) \wedge (x_{ij} \vee x_{i,j+1})$ has the $n \times n$-grid as a minor and thus treewidth n, see e.g. [12, Chapt. 12]. But the incidence graph of F has no modules and also has the $n \times n$-grid as a minor, so F has modular incidence treewidth at least n as well. So we gain nothing by going from primal treewidth to modular treewidth without changing the encoding. What Theorem 3 tells us is that there is another formula F' that encodes the function of F, potentially with some additional variables, such that the treewidth of F' is at most $O(n/\log(n))$.

Let us note that encodings with dependent auxiliary variables are often useful, e.g. when considering counting problems. In fact, for such clausal encodings the number of models is the same as for the function they encode. It is thus interesting to see that dependence of the auxiliary variables can be maintained by the construction of Theorem 3. We will see that this is also the case for most other constructions we make.

We now show that the reverse of Theorem 3 is also true: upper bounds for many width measures imply also bounds for the primal treewidth of clausal encodings. Note that this is at first sight surprising since without auxiliary variables many of those width measures are known to be far stronger than primal treewidth.

Theorem 4. *Let f be a Boolean function of n variables.*

(a) *If F has a clausal encoding of modular treewidth, cliquewidth or mim-width k then f also has a clausal encoding F' of primal treewidth $O(k \log(n))$ with $O(kn \log(n))$ auxiliary variables and $n^{O(k)}$ clauses.*

(b) *If F has a clausal encoding of incidence treewidth, dual treewidth, or signed incidence cliquewidth k, then f also has a clausal encoding F' of primal treewidth $O(k)$ with $O(nk)$ auxiliary variables and $2^{O(k)}n$ clauses.*

Theorem 4 is a direct consequence of the following result:

Lemma 3. *Let f be a Boolean function in n variables that is computed by a complete structured DNNF of width k. Then f has a clausal encoding F of primal treewidth $9 \log(k)$ with $O(n \log(k))$ variables and $O(nk^3)$ clauses. Moreover, if D is deterministic then F has dependent auxiliary variables.*

The proof of Lemma 3 will rely on so-called proof trees in DNNF, a concept that has found wide application in circuit complexity and in particular also in knowledge compilation. To this end, we make the following definition: a *proof tree* $\mathcal{T}$ of a complete structured DNNF D is a circuit constructed as follows:

1. The output gate of D belongs to $\mathcal{T}$.
2. Whenever $\mathcal{T}$ contains an $\lor$-gate, we add exactly one of its inputs.
3. Whenever $\mathcal{T}$ contains an $\land$-gate, we add both of its inputs.
4. No other gates are added to $\mathcal{T}$.

Note that the choice in Step 2 is non-deterministic, so there are in general many proof trees for D. Observe also that due to the structure of D given by its v-tree, every proof tree is in fact a tree which justifies the name. Moreover, letting T be the v-tree of D, every proof tree of D has exactly one $\lor$-gate and one $\land$-gate in the set $\mu(t)$ for every non-leaf node t of T. For every leaf t, every proof tree contains an input gate x or $\neg x$ where x is the label of t in T.

The following simple observation that can easily be shown by using distributivity is the main reason for the usefulness of proof trees.

Observation 1. *Let D be a complete structured DNNF and a an assignment to its variables. Then a satisfies D if and only if it satisfies one of its proof trees. Moreover, if D is deterministic, then every assignment a that satisfies D satisfies exactly one proof tree of D.*

Proof (of Lemma 3). Let D be the complete structured DNNF computing f and let T be the v-tree of D. The idea of the proof is to use auxiliary variables to guess for every t an $\lor$-gate and an $\land$-gate. Then we use clauses along the v-tree T to verify that the guessed gates in fact form a proof tree and check in the leaves of T if the assignment to the variables of f satisfies the encoded proof tree. We now give the details of the construction.

We first note that it was shown in [9] that in complete structured DNNF of width k one may assume that every set $\mu(t)$ contains at most k^2 $\land$-gates so we assume this to be the case for D. For every node t of T, we introduce a set X_t of $3 \log(k)$ auxiliary variables to encode one $\lor$-gate and one $\land$-gate of $\mu(t)$ if t is an internal node. If t is a leaf, X_t encodes one of the at most 2 input gates in $\mu(t)$. We now add clauses that verify that the gates chosen by the variables X_t encode a proof tree by doing the following for every t that is not a leaf: first, add clauses in X_t that check if the chosen $\land$-gate is in fact an input of the chosen $\lor$-gate. Since X_t has at most $3 \log(k)$ variables, this introduces at most k^3 clauses. Let t_1 and t_2 be the children of t in T. Then we add clauses that verify if the $\land$-gate chosen in t has as input either the $\lor$-gate chosen in t_1 if t_1 is not a leaf, or the input gate chosen in t_1 if t_1 is a leaf. Finally, we add analogous clauses for t_2.

Each of these clause sets is again in $3\log(k)$ variables, so there are at most $2k^3$ clauses in them overall. The result is a CNF-formula that accepts an assignment if and only if it encodes a proof tree of D.

We now show how to verify if the chosen proof tree is satisfied by an assignment to f. To this end, for every leaf t of T labeled by a variable x, add clauses that check if an assignment to x satisfies the corresponding input gate of D. Since $\mu(t)$ contains at most 2 gates, this only requires at most 4 clauses. This completes the construction of the clausal encoding. Overall, since T has n internal nodes, the CNF has $n(3\log(k)+1)$ variables and $3nk^3 + 4n$ clauses.

It remains to show the bound on the primal treewidth. To this end, we construct a tree decomposition $(T, (B_t)_{t \in V(T)})$ with the v-tree T as underlying tree as follows: for every internal node $t \in V(T)$, we set $B_t := X_t \cup X_{t_1} \cup X_{t_2}$ where t_1 and t_2 are the children of t. Note that for every clause that is used for checking if the chosen nodes form a proof tree, the variables are thus in a bag B_t. For every leaf t, set $B_t := X_t \cup \{x\}$ where x is the variable that is the label of t. This covers the remaining clauses. It follows that all edges of the primal graph are covered. To check the third condition of the definition of tree decompositions, note that every auxiliary variables in a set X_t appears only in B_t and potentially in $B_{t'}$ where t' is the parent of t in T. Thus $(T, (B_t)_{t \in V(T)})$ constructed in this way is a tree decomposition of the primal graph of C. Obviously, the width is bounded by $9\log(k)$ since every X_t has size $3\log(k)$, which completes the proof.   □

Proof (of Theorem 4). We first show (a). By [4], whenever the function f has a clausal encoding F with one of the width measures from this statement bounded by k, then there is also a complete structured DNNF D of width $n^{O(k)}$ computing F. Now forget all auxiliary variables of F to get a DNNF representation D' of f. Note that since forgetting does not increase the width, see [9], D' also has width at most $n^{O(k)}$. We then simply apply Lemma 3 to get the result.

To see (b), just observe that, following the same construction, the width of D is $2^{O(k)}$ for all considered width measures [4].   □

Remark one slightly unexpected property of Theorem 4: the size and the number of auxiliary variables of the constructed encoding F' does *not* depend on the size of the initial encoding at all. Both depend only on the number of variables in f and the width.

To maintain dependence of the auxiliary variables in the above construction, we have to work some more than for Theorem 4. We give the details in the full version of this paper.

We now can state the main result of this section.

Theorem 5. *Let $A = \{\mathsf{tw_p}, \mathsf{tw_d}, \mathsf{tw_i}, \mathsf{scw}\}$ and $B = \{\mathsf{mtw}, \mathsf{cw}, \mathsf{mimw}\}$. Let f be a Boolean function in n variables.*

(a) Let $w_a \in A$ and $w_b \in B$. Then there are constants c_1 and c_2 such that the following holds: let F_a and F_b be clausal representations for f with minimal w_a-width and w_b-width, respectively. Then

$$w_a(F_a) \le k\log(n) \Rightarrow w_b(F_b) \le c_1 k$$

and

$$w_b(F_b) \leq k \Rightarrow w_a(F_a) \leq c_2 k \log(n).$$

(b) *Let* $w_a \in A$ *and* $w_b \in A$ *or* $w_a \in B$ *and* $w_b \in B$. *Then there are constants* c_1 *and* c_2 *such that the following holds: let* F_a *and* F_b *be clausal representations for* f *of minimal* w_a-width *and* w_b-width, *respectively. Then*

$$w_a(F_a) \leq k \Rightarrow w_b(F_b) \leq c_1 k$$

and

$$w_b(F_b) \leq k \Rightarrow w_a(F_a) \leq c_2 k.$$

Proof. Assume first that $w_a = \mathsf{tw_p}$. For (a) we get the second statement directly from Theorem 4 (a). For cw and mtw we get the first statement by Theorem 3. For mimw it follows by the fact that for every graph $\mathsf{mimw}(G) \leq c \cdot \mathsf{cw}(G)$ for some absolute constant c, see [33, Sect. 4].

For (b), the second statement is Theorem 4 (b). Since for every formula F we have $\mathsf{tw_i}(F) \leq \mathsf{tw_p}(F) + 1$, see e.g. [13], the first statement for $\mathsf{tw_i}$ is immediate. For scw it is shown in the full version of this paper, while for $\mathsf{tw_d}$ it can be found in [30].

All other combinations of w_a and w_b can now be shown by an intermediate step using $\mathsf{tw_p}$. $\qquad\square$

6   Applications

6.1   Cardinality Constraints

In this section, we consider cardinality constraints, i.e., constraints of the form $\sum_{i \in [n]} x_i \leq k$ in the Boolean variables $x_1, \ldots, x_n$. The value k is commonly called the *degree* or the *threshold* of the constraint. Let us denote by C_n^k the cardinality constraint with n variables and degree k. Cardinality constraints have been studied extensively and many encodings are known, see e.g. [31]. Here we add another perspective on cardinality constraint encodings by determining their optimal treewidth. We remark that we could have studied cardinality constraints in which the relation is $\geq$ instead of $\leq$ with essentially the same results.

We start with an easy observation:

Observation 2. C_n^k *has an encoding of primal treewidth* $O(\log(\min(k, n - k)))$

Proof (sketch). First assume that $k < n/2$. We iteratively compute the partial sums of $S_j := \sum_{i \in [j]} x_i$ and code their values in $\log(k) + 1$ bits $Y^j := \{y_1^j, \ldots, y_{\log(k)+1}^j\}$. We cut these sums off at $k+1$ (if we have seen at least $k+1$ variables set to 1, this is sufficient to compute the output). In the end we code a comparator comparing the last sum S_n to k.

Since the computation of S_{j+1} can be done from S_j and x_{j+1}, we can compute the partial sums with clauses containing only the variables in $Y^j \cup Y^{j+1} \cup \{x_{j+1}\}$, so $O(\log(k))$ variables. The resulting CNF-formula can easily be seen to be of treewidth $O(\log(k))$.

If $k > n/2$, we proceed similarly but count variables assigned to 0 instead of those set to 1. □

We now show that Observation 2 is essentially optimal.

Proposition 2. *Let $k < n/2$. Then*

$$\mathsf{cc}_{\mathsf{best}}^{1/3}(C_n^k) = \Omega(\log(\min(k, n/3)))$$

Proof. Let $s = \min(k, \frac{n}{3})$. Consider an arbitrary partition Y, Z with $\frac{n}{3} \le |Y| \le \frac{2n}{3}$. We show that every rectangle cover of C_n^k must have s rectangles. To this end, choose assignments $(a_0, b_0), \dots, (a_s, b_s)$ such that $a_i : Y \to \{0,1\}$ assigns i variables to 1 and $b_i : Z \to \{0,1\}$ assigns $k - i$ variables to 1. Note that every (a_i, b_i) satisfies C_n^k. We claim that no rectangle $r_1(Y) \wedge r_2(Z)$ in a rectangle cover of C_n^k can have models (a_i, b_i) and (a_j, b_j) for $i \ne j$. To see this, assume that such model exists and that $i < j$. Then the assignment (a_j, b_i) is also a model of the rectangle since a_j satisfies $r_1(Y)$ and b_i satisfies $r_2(Z)$. But (a_j, b_i) contains more than k variables assigned to 1, so the rectangle $r_1(Y) \wedge r_2(Z)$ cannot appear in a rectangle cover of C_n^k. Thus, every rectangle cover of C_n^k must have a different rectangle for every model (a_i, b_i) and thus at least s rectangles. This completes the proof for this case. □

A symmetric argument shows that for $k > n/2$ we have the lower bound $\mathsf{cc}_{\mathsf{best}}^{1/3}(C_n^k) = \Omega(\log(\min(n - k, n/3)))$ Observing that $k < n$ for non-trivial cardinality constraints, we get the following from Theorem 1.

Corollary 5. *Clausal encodings of smallest primal treewidth for C_n^k have primal treewidth $\Theta(\log(s))$ for $s = \min(k, n - k)$. The same statement is true for dual and incidence treewidth and signed incidence cliquewidth.*

For incidence cliquewidth, modular treewidth and mim-width, there are clausal encodings of C_n^k of constant width.

6.2   The Permutation Function

We now consider the permutation function $\mathbf{PERM}_n$ which has the n^2 input variables $X_n = \{x_{i,j} \mid i, j \in [n]\}$ thought of as a matrix in these variables. $\mathbf{PERM}_n$ evaluates to 1 on an input a if and only if a is a permutation matrix, i.e., in every row and in every column of a there is exactly one 1. $\mathbf{PERM}_n$ is known to be hard in several versions of branching programs, see [34]. In [8], it was shown that clausal encodings of $\mathbf{PERM}_n$ require treewidth $\Omega(n/\log(n))$. We here give an improvement by a logarithmic factor.

Lemma 4. *For every v-tree T on variables X_n, there is a node t of T such that*

$$\mathsf{cc}(\mathbf{PERM}, Y, Z) = \Omega(n)$$

where $Y = \mathsf{var}(T_t)$ and $Z = X \setminus Y$.

The proof of Lemma 4 uses techniques similar to those in [8,21] and is given in the full version of this paper.

As a consequence of Lemma 4, we get an asymptotically tight treewidth bound for encodings of **PERM**$_n$.

Corollary 6. *Clausal encodings of smallest primal treewidth for C_n^k have primal treewidth $\Theta(n)$.*

Proof (sketch). The lower bound follows by using Lemma 4 and Proposition 1 and then arguing as in the proof of Theorem 2.

For the upper bound, observe that checking if out of n variables exactly one has the value 1 can easily be done with n variables. We apply this for every row in a bag of a tree decomposition. We perform these checks for one row after the other and additionally use variables for the columns that remember if in a column we have seen a variables assigned 1 so far. Overall, to implement this, one needs $O(n^2)$ auxiliary variables and gets a formula of treewidth $O(n)$.   □

From Corollary 6 we get the following bound by applying Theorem 3. This answers an open problem from [8] who showed only conditional lower bounds for the incidence cliquewidth of encodings of **PERM**$_n$.

Corollary 7. *Clausal encodings of smallest incidence cliquewidth for C_n^k have width $\Theta(n/log(n))$.*

7   Conclusion

We have shown several results on the expressivity of clausal encodings with restricted underlying graphs. In particular, we have seen that many graph width measures from the literature put strong restrictions on the expressivity of encodings. We have also seen that, contrary to the case of representations by CNF-formulas, in the case where auxiliary variables are allowed, all width measures we have considered are strongly related to primal treewidth and never differ by more than a logarithmic factor. Moreover, most of our results are also true while maintaining dependence of auxiliary variables.

To close the paper, let us discuss several questions. First, the number of clauses of the encodings guaranteed by Theorem 4 is very high. In particular, it is exponential in the width k. It would be interesting to understand if this can be avoided, i.e., if there are encodings of roughly the same primal treewidth whose size is polynomial in k.

It would also be interesting to see if our results can be extended to other classes of CNF-formulas on which SAT is tractable. Interesting classes to consider would e.g. be the classes in [14]. In this paper, the authors define another graph for CNF-formulas for which bounded treewidth yields tractable model counting. It is not clear if the classes characterized that way allow small complete structured DNNF so our framework does not apply directly. It would still be interesting to see if one can show similar expressivity results to those here. Other interesting classes one could consider are those defined by backdoors, see e.g. [15].

Acknowledgements. The authors are grateful to the anonymous reviewers for their comments, which greatly helped to improve the presentation of the paper. The first author would also like to thank David Mitchell for asking the right question at the right moment. This paper grew largely out of an answer to this question.

References

1. Amarilli, A., Capelli, F., Monet, M., Senellart, P.: Connecting knowledge compilation classes and width parameters CoRR, abs/1811.02944 (2018)
2. Amarilli, A., Monet, M., Senellart, P.: Connecting width and structure in knowledge compilation. In: 21st International Conference on Database Theory, ICDT, Vienna, Austria, 26–29 March 2018 (2018)
3. Biere, A., Heule, M., van Maaren, H., Walsh, T. (eds.): Handbook of Satisfiability, volume 185 of Frontiers in Artificial Intelligence and Applications. IOS Press (2009)
4. Bova, S., Capelli, F., Mengel, S., Slivovsky, F.: On compiling CNFs into structured deterministic DNNFs. In: Heule, M., Weaver, S. (eds.) SAT 2015. LNCS, vol. 9340, pp. 199–214. Springer, Cham (2015). https://doi.org/10.1007/978-3-319-24318-4_15
5. Bova, S., Capelli, F., Mengel, S., Slivovsky, F.: Knowledge compilation meets communication complexity. In: Proceedings of the Twenty-Fifth International Joint Conference on Artificial Intelligence, IJCAI 2016 (2016)
6. Bova, S., Szeider, S.: Circuit treewidth, sentential decision, and query compilation (2017)
7. Brault-Baron, J., Capelli, F., Mengel, S.: Understanding model counting for β-acyclic CNF-formulas CoRR, abs/1405.6043 (2014)
8. Briquel, I., Koiran, P., Meer, K.: On the expressive power of CNF formulas of bounded tree- and clique-width. Discrete Appl. Math. **159**(1), 1–14 (2011)
9. Capelli, F., Mengel, S.: Tractable QBF by knowledge compilation. In: 36th International Symposium on Theoretical Aspects of Computer Science, STACS 2019, vol. 126, pp. 18:1–18:16 (2019)
10. Chen, H.: Quantified constraint satisfaction and bounded treewidth. In: Proceedings of the 16th Eureopean Conference on Artificial Intelligence, ECAI 2004 (2004)
11. Darwiche, A., Marquis, P.: A knowledge compilation map. J. Artif. Intell. Res. **17**, 229–264 (2002)
12. Diestel, R.: Graph Theory. Volume 173 of Graduate texts in Mathematics, 4th edn. Springer, Heidelberg (2012)
13. Fischer, E., Makowsky, J., Ravve, E.: Counting truth assignments of formulas of bounded tree-width or clique-width. Discrete Appl. Math. **156**(4), 511–529 (2008)
14. Ganian, R., Szeider, S.: New width parameters for model counting. In: Gaspers, S., Walsh, T. (eds.) SAT 2017. LNCS, vol. 10491, pp. 38–52. Springer, Cham (2017). https://doi.org/10.1007/978-3-319-66263-3_3
15. Gaspers, S., Szeider, S.: Strong backdoors to bounded treewidth SAT. In: 54th Annual IEEE Symposium on Foundations of Computer Science, FOCS 2013 (2013)
16. Gent, I.P., Nightingale, P.: A new encoding of AllDifferent into SAT. In: International Workshop on Modelling and Reformulating Constraint Satisfaction Problems (2004)
17. Giunchiglia, E., Maratea, E., Tacchella, A.: Dependent and independent variables in propositional satisfiability. In: Logics in Artificial Intelligence. JELIA 2002 (2002)

18. Golumbic, M.C., Gurvich, V.: Read-once functions. In: Boolean Functions: Theory, Algorithms and Applications, pp. 519–560 (2011)
19. Gottlob, G., Pichler, R., Wei, F.: Bounded treewidth as a key to tractability of knowledge representation and reasoning. Artif. Intell., **174**(1), 105–132 (2010)
20. Jakl, M., Pichler, R., Woltran, S.: Answer-set programming with bounded treewidth. In: Proceedings of the 21st International Joint Conference on Artificial Intelligence. IJCAI 2009 (2009)
21. Krause, M.: Exponential lower bounds on the complexity of local and real-time branching programs. Elektronische Informationsverarbeitung und Kybernetik **24**(3), 99–110 (1988)
22. Kushilevitz, E., Nisan, N.: Communication Complexity. Cambridge University Press, Cambridge (1997)
23. Lampis, M., Mengel, S., Mitsou, V.: QBF as an alternative to Courcelle's theorem. In: Beyersdorff, O., Wintersteiger, C.M. (eds.) SAT 2018. LNCS, vol. 10929, pp. 235–252. Springer, Cham (2018). https://doi.org/10.1007/978-3-319-94144-8_15
24. Paulusma, D., Slivovsky, F., Szeider, S.: Model counting for CNF formulas of bounded modular treewidth. In: 30th International Symposium on Theoretical Aspects of Computer Science, STACS 2013 (2013)
25. Paulusma, D., Slivovsky, F., Szeider, S.: Model counting for CNF formulas of bounded modular treewidth. Algorithmica **76**(1), 168–194 (2016)
26. Pipatsrisawat, K., Darwiche, A.: New compilation languages based on structured decomposability. In: Proceedings of the Twenty-Third AAAI Conference on Artificial Intelligence, AAAI 2008 (2008)
27. Pipatsrisawat, T., Darwiche, A.: A lower bound on the size of decomposable negation normal form. In: Proceedings of the Twenty-Fourth AAAI Conference on Artificial Intelligence, AAAI 2010 (2010)
28. Sæther, S.H., Telle, J.A., Vatshelle, M.: Solving #SAT and MAXSAT by dynamic programming. J. Artif. Intell. Res. **54**, 59–82 (2015)
29. Samer, M., Szeider, S.: Algorithms for propositional model counting. J. Discrete Algorithms **8**(1), 50–64 (2010)
30. Samer, M., Szeider, S.: Constraint satisfaction with bounded treewidth revisited. J. Comput. Syst. Sci. **76**(2), 103–114 (2010)
31. Sinz, C.: Towards an optimal CNF encoding of boolean cardinality constraints. In: 11th International Conference Principles and Practice of Constraint Programming - CP 2005. CP 2005 (2005)
32. Slivovsky, F., Szeider, S.: Model counting for formulas of bounded clique-width. In: 24th International Symposium on Algorithms and Computation. ISAAC 2013 (2013)
33. Vatshelle, M.: New width parameters of graphs. Ph.D. Thesis, University of Bergen (2012)
34. Wegener, I.: Branching Programs and Binary Decision Diagrams. SIAM (2000)

DRMaxSAT with MaxHS: First Contact

Antonio Morgado[1](✉), Alexey Ignatiev[1,4], Maria Luisa Bonet[2],
Joao Marques-Silva[1], and Sam Buss[3]

[1] Faculty of Science, University of Lisbon, Lisbon, Portugal
{ajmorgado,aignatiev,jpms}@ciencias.ulisboa.pt
[2] Computer Science, Universidad Politécnica de Cataluña, Barcelona, Spain
bonet@cs.upc.edu
[3] Department of Mathematics, University of California, San Diego, USA
sbuss@ucsd.edu
[4] ISDCT SB RAS, Irkutsk, Russia

Abstract. The proof system of Dual-Rail MaxSAT (DRMaxSAT) was recently shown to be capable of efficiently refuting families of formulas that are well-known to be hard for resolution, concretely when the MaxSAT solving approach is either MaxSAT resolution or core-guided algorithms. Moreover, DRMaxSAT based on MaxSAT resolution was shown to be stronger than general resolution. Nevertheless, existing experimental evidence indicates that the use of MaxSAT algorithms based on the computation of minimum hitting sets (MHSes), i.e. Max-HS-like algorithms, are as effective, and often more effective, than core-guided algorithms and algorithms based on MaxSAT resolution. This paper investigates the use of MaxHS-like algorithms in the DRMaxSAT proof system. Concretely, the paper proves that the propositional encoding of the pigenonhole and doubled pigenonhole principles have polynomial time refutations when the DRMaxSAT proof system uses a MaxHS-like algorithm.

1 Introduction

The practical success of Conflict-Driven Clause Learning (CDCL) Boolean Satisfiability (SAT) solvers demonstrates the reach of the (general) resolution proof system. Nevertheless, from a proof complexity point of view, resolution is generally viewed as a rather weak proof system. As a result, there have been recent efforts towards developing practically efficient implementations of stronger proof systems. Concrete examples include extended resolution [2,19,21] (ExtRes), the cutting planes (CP) proof system [4,17] and, more recently, the dual rail (DR) maximum satisfiability (MaxSAT) (i.e. DRMaxSAT) proof sys-

---

This work was supported by FCT grants ABSOLV (PTDC/CCI-COM/28986/2017), FaultLocker (PTDC/CCI-COM/29300/2017), SAFETY (SFRH/BPD/120315/2016), and SAMPLE (CEECIND/04549/2017); grant TIN2016-76573-C2-2-P (TASSAT 3); and Simons Foundation grant 578919.

M. Janota and I. Lynce (Eds.): SAT 2019, LNCS 11628, pp. 239–249, 2019.
https://doi.org/10.1007/978-3-030-24258-9_17

tem [9,20][1]. Although expected to be weaker than CP, and so weaker than ExtRes, DRMaxSAT can in practice build on practically efficient MaxSAT solvers, and so indirectly tap on the practical efficiency of CDCL SAT solvers.

The DRMaxSAT proof system translates a proposition formula using the dual-rail encoding [11,20], and then uses a MaxSAT algorithm. Initial results for DRMaxSAT focused on MaxSAT resolution and on core-guided MaxSAT solvers [20]. Nevertheless, the empirical evidence from earlier work also indicated the MaxSAT solvers based on the iterative computation of minimum hitting sets (MHSes), concretely MaxHS [13] and LMHS [34], performed comparably, if not better, than the use of core-guided MaxSAT solvers, on different families of problems known to be hard for resolution. The reason for this good performance was left as open research.

This paper investigates MHS-based MaxSAT algorithms (also referred to MaxHS-like) on two dual-rail encoded families of problems, encoding respectively the pigeonhole and the doubled pigeonhole problems. These two problems are known to be hard to general resolution, and were studied in earlier work [9, 20]. Moreover, the paper proves that these two families of problems can be refuted in polynomial time when DRMaxSAT proof system uses a MaxHS-like MaxSAT solver. These results thus provide a theoretical justification for the good performance observed in practice for MaxHS-like algorithms (in both families of problems).

2   Motivation

Experimental results presented in [20] and later in [9] confirmed the ability of the DRMaxSAT proof system to refute pigeonhole principle formulas and also doubled pigeonhole formulas in polynomial time. (These families of formulas are defined later.) Figure 1 summarizes the results from both works.[2] Here, formulas encoding the pigeonhole and doubled pigeonhole principles were generated with up to 100 holes[3]. Core-guided MaxSAT solvers are represented by MSCG [28,30] and OpenWBO16 [27], while Eva500a [31] acts for *core-guided* MaxSAT resolution. For comparison purposes, Fig. 1 also depicts the performance of lingeling [5,6] as best performing CDCL SAT solver (see [9,20]). Additionally, earlier work [9,20] assessed the performance of hitting set based MaxSAT solvers MaxHS [3,13–15] and LMHS [34]. Note that the efficiency of the default versions of MaxHS and LMHS is not surprising and can be attributed to a number of powerful heuristics used [3,14,15,34] for improving the *basic MaxHS algorithm*

---

[1] Despite ongoing efforts, and with the exception of families of problem instances known to be hard for resolution, the performance of implementations of proof systems stronger than resolution is still far from what CDCL SAT solvers achieve in practice.

[2] The plots show the performance of the competitors for a specifically constructed PHP and 2PHP formulas, with $\mathcal{P}$ clauses being disabled. (The notation and rationale will be explained below. A reader can also refer to [9,20] for details.)

[3] Some of the (2)PHP instances are skipped in the plots. Please, refer to [9,20] for details.

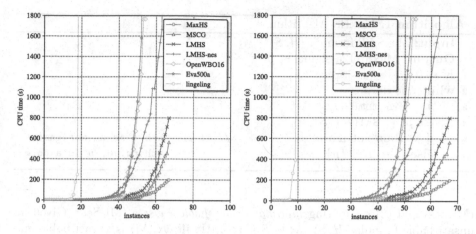

Fig. 1. Performance of MaxSAT solvers and lingeling on PHP and 2PHP formulas.

of MaxHS and LMHS [13]. However, the remarkable efficiency of this basic algorithm (referred to as LMHS-nes in Fig. 1) is yet to be explained. The following sections represent a first attempt to understand the power of the *basic* hitting set based MaxSAT with respect to two families of pigeonhole formulas, namely PHP and 2PHP.

3   Preliminaries

The paper assumes definitions and notation standard in propositional satisfiability (SAT) and maximum satisfiability (MaxSAT) [7].

SAT. A conjunctive normal form (CNF) formula $\mathcal{F}$ is a conjunction of clauses, a clause is a disjunction of literals, and a literal is a variable or its complement. A truth assignment $\mathcal{A}$ is a mapping from variables to $\{0, 1\}$. Given $\mathcal{A}$, a clause is satisfied iff at least one of its literals is assigned value 1. A formula is satisfied iff all of its clauses are satisfied. If there is no satisfying assignment for $\mathcal{F}$, then $\mathcal{F}$ is referred to as *unsatisfiable*.

MaxSAT. For unsatisfiable formulas, the maximum satisfiability (MaxSAT) problem is to find an assignment that maximizes the number of satisfied clauses [24]. This paper considers the *partial MaxSAT* problem $\langle \mathcal{H}, \mathcal{S} \rangle$, which allows for a set of *hard* clauses $\mathcal{H}$ (that must be satisfied), and a set of *soft* clauses $\mathcal{S}$ (represent a preference to satisfy these clauses). In the paper, a MaxSAT *solution* represents either a maximum cardinality set of satisfied soft clauses or an assignment that satisfies all hard clauses and also maximizes (minimizes) the number of satisfied (falsified, resp.) soft clauses. The number of clauses falsified by a MaxSAT solution is referred to as its *cost*. A few other optimization problems exist with respect to MaxSAT: (1) computing *minimal correction subsets*

Algorithm 1. MaxHS Algorithm

Input: MaxSAT formula $\langle \mathcal{H}, \mathcal{S} \rangle$
Output: MaxSAT assignment μ

1 $K \leftarrow \emptyset$
2 **while** true **do**
3 $\quad$ $h \leftarrow \texttt{MinimumHS}(K)$
4 $\quad$ $(st, \mu) \leftarrow \texttt{SAT}(\mathcal{H} \cup \mathcal{S} \setminus h)$
5 $\quad$ **if** st **then return** μ $\qquad$ // If st, then μ is an assignment
6 $\quad$ **else** $K \leftarrow K \cup \{\mu\}$ $\qquad$ // Otherwise, μ is a core

---

(MCSes) and (2) computing *minimal unsatisfiable subsets* (MUSes). Given an unsatisfiable formula $\langle \mathcal{H}, \mathcal{S} \rangle$, $\mathcal{M} \subseteq \mathcal{S}$ is an MUS iff $\langle \mathcal{H}, \mathcal{M} \rangle$ is unsatisfiable and $\forall_{\mathcal{M}' \subsetneq \mathcal{M}} \langle \mathcal{H}, \mathcal{M}' \rangle$ is satisfiable. Given an unsatisfiable formula $\langle \mathcal{H}, \mathcal{S} \rangle$, $\mathcal{C} \subseteq \mathcal{S}$ is an MCS iff $\langle \mathcal{H}, \mathcal{S} \setminus \mathcal{C} \rangle$ is satisfiable and $\forall_{\mathcal{C}' \subsetneq \mathcal{C}} \langle \mathcal{H}, \mathcal{S} \setminus \mathcal{C}' \rangle$ is unsatisfiable. MCSes and MUSes of an unsatisfiable formula are known to be connected through the *hitting set duality* [8,33], i.e. MCSes are minimal hitting sets of MUSes, and vice versa.

MaxHS Algorithm. Many algorithms for MaxSAT have been proposed over the years [24]. The most widely investigated ones can be broadly organized into branch and bound [24], iterative-search [4,18,22], core-guided [1,18,25,26,28,29, 31], and minimum hitting sets [13,34]. This paper focuses solely on the minimum hitting set MaxSAT algorithm referred to as basic MaxHS [13]. Its setup is shown in Algorithm 1. MaxHS builds on the minimal hitting set duality between MCSes and MUSes. Every iteration of the algorithm computes a minimum size hitting set (MHS) h of a set K of unsatisfiable cores found so far and checks if h is an MCS of the formula, i.e. it tests satisfiability of $\mathcal{H} \cup \mathcal{S} \setminus h$. If it is, the algorithm returns a model of $\mathcal{H} \cup \mathcal{S} \setminus h$. Otherwise, a new unsatisfiable core is extracted, added to K, and the algorithm proceeds.

Dual-Rail MaxSAT. The proof system of *dual-rail MaxSAT* (DRMaxSAT) was proposed in [20] and heavily relies on the variant of the *dual-rail encoding* (DRE) [11,32]. Let $\mathcal{F}$ be a CNF formula on the set of N variables $X = \{x_1 \ldots, x_N\}$. Given $\mathcal{F}$, the dual-rail MaxSAT encoding [9,20] creates a (Horn) MaxSAT problem $\langle \mathcal{H}, \mathcal{S} \rangle$, where $\mathcal{H}$ is the set of hard clauses and $\mathcal{S}$ is the set of soft clauses s.t. $|\mathcal{S}| = 2N$. For each variable $x_i \in X$, DRE associates two new variables p_i and n_i, where $p_i = 1$ iff $x_i = 1$, and $n_i = 1$ iff $x_i = 0$. It also adds the soft clauses (p_i) and (n_i) to $\mathcal{S}$ while adding a hard clause $(\neg p_i \vee \neg n_i)$ to $\mathcal{H}$, to ensure that $x_i = 1$ and $x_i = 0$ are not set simultaneously. (Hard clauses of this form are referred to as $\mathcal{P}$ clauses.) For each clause $c \in \mathcal{F}$, we obtain a hard clause $c' \in \mathcal{H}$, as follows: if $x_i \in c$, then $\neg n_i \in c'$, and if $\neg x_i \in c$ then $\neg p_i \in c'$. The formula encoded by DRE is then $\texttt{DREnc}(\mathcal{F}) \triangleq \langle \mathcal{H}, \mathcal{S} \rangle$. One of the major results of [20] is the following.

Theorem 1. *CNF formula $\mathcal{F}$ is satisfiable iff there exists a truth assignment that satisfies $\mathcal{H}$ and at least N clauses in $\mathcal{S}$.*

Applying DRE followed either by MaxSAT resolution [10,23] or core-guided MaxSAT solving [29] consitutes the DRMaxSAT proof system [9], which is able to refute pigeonhole principle [20] and doubled pigeonhole principle [9] in polynomial time.

4 Polynomial Time Refutations with DRMaxHS

4.1 Pigeonhole Principle

Definition 1 (PHP [12]). *The pigeonhole principle states that if $m+1$ pigeons are distributed by m holes, then at least one hole contains more than one pigeon.*

The propositional encoding of the PHP_m^{m+1} problem is as follows. Let the variables be x_{ij}, with $i \in [m+1]$, $j \in [m]$, with $x_{ij} = 1$ iff the ith pigeon is placed in the jth hole. The constraints state that each pigeon must be placed in at least one hole, and each hole must not have more than one pigeon: $\bigwedge_{i=1}^{m+1} \mathsf{AtLeast1}(x_{i1}, \ldots, x_{im}) \wedge \bigwedge_{j=1}^{m} \mathsf{AtMost1}(x_{1j}, \ldots, x_{m+1,j})$. Each $\mathsf{AtLeast1}$ constraint can be encoded with a single clause. For the $\mathsf{AtMost1}$ constraints there are different encodings, including [7,16,35]. In this work, we will consider the pairwise cardinality constraint encoding [7], which encodes $\mathsf{AtMost1}(x_{1j}, \ldots, x_{m+1,j})$ as $\bigwedge_{i_1=1}^{m} \bigwedge_{i_2=i_1+1}^{m+1} (\neg x_{i_1 j} \vee \neg x_{i_2 j})$.

The reduction of the PHP_m^{m+1} problem into MaxSAT using the Dual-Rail encoding is as follows [20]. With each variable x_{ij}, $i \in [m+1], j \in [m]$, we associate two new variables n_{ij} and p_{ij}. The set of clauses $\mathcal{P}$ is $\{(\neg n_{ij} \vee \neg p_{ij}) \mid i \in [m+1], j \in [m]\}$. Let $\mathcal{L}_i$ represent the encoding of each $\mathsf{AtLeast1}$ constraint; $\mathcal{L}_i = (\neg n_{i1} \vee \ldots \vee \neg n_{im})$, and $\mathcal{M}_j$ represent the encoding of each $\mathsf{AtMost1}$ constraint; $\mathcal{M}_j = \bigwedge_{i_1=1}^{m} \bigwedge_{i_2=i_1+1}^{m+1} (\neg p_{i_1 j} \vee \neg p_{i_2 j})$. The soft clauses $\mathcal{S}$ are given by $\{(n_{11}), \ldots, (n_{(m+1)m}), (p_{11}), \ldots, (p_{(m+1)m})\}$, with $|\mathcal{S}| = 2m(m+1)$. The complete encoding of PHP is:

$$\mathsf{DREnc}\left(\mathrm{PHP}_m^{m+1}\right) \triangleq \langle \mathcal{H}, \mathcal{S} \rangle = \langle \wedge_{i=1}^{m+1} \mathcal{L}_i \wedge \wedge_{j=1}^{m} \mathcal{M}_j \wedge \mathcal{P}, \mathcal{S} \rangle \tag{1}$$

From Theorem 1 [20], unsatisfiability of PHP_m^{m+1} implies that DREnc $\left(\mathrm{PHP}_m^{m+1}\right)$ has a MaxSAT cost of at least $m(m+1)+1$. The following shows that the basic MaxHS algorithm can derive this MaxSAT cost for $\mathsf{DREnc}\left(\mathrm{PHP}_m^{m+1}\right)$ in polynomial time. Observe that if the $\mathcal{P}$ clauses from $\mathsf{DREnc}\left(\mathrm{PHP}_m^{m+1}\right)$ are ignored, then the formula can be partitioned into the disjoint formulas $\mathcal{L}_i$, $i \in [m+1]$, and the disjoint formulas $\mathcal{M}_j$, $j \in [m]$. Thus, one can compute a solution for each of these formulas separately and obtain a lower bound on the MaxSAT solution for the complete formula $\mathsf{DREnc}\left(\mathrm{PHP}_m^{m+1}\right)$. We show that the contribution of each $\mathcal{L}_i$ to the total cost is 1. Since there are $m+1$ such formulas, the contribution of all $\mathcal{L}_i$ formulas is $m+1$. Then, we show that each $\mathcal{M}_j$ contributes with a cost of m, and since there are m such formulas, the contribution of all $\mathcal{M}_j$ formulas is m^2. Therefore, we have a lower bound on the total cost of $m(m+1)+1$, proving the original formula to be unsatisfiable.

Proposition 1. *Given a formula* $\langle \mathcal{L}_i, \mathcal{S} \rangle$, *where* $\mathcal{L}_i$ *and* $\mathcal{S}$ *are from* $DREnc\left(PHP_m^{m+1}\right)$, *there is an execution of the basic MaxHS algorithm that computes a MaxSAT solution of cost 1 in polynomial time.*

Proof. (Sketch) Consider Algorithm 1. In the first iteration, an empty MHS is computed in line 3. The SAT solver (line 4) tests the satisfiability of the only clause $(\neg n_{i1} \vee \ldots \vee \neg n_{im})$ with the complement of the MHS as unit clauses, that is, $(n_{i1}), \ldots, (n_{im})$. The formula is unsatisfiable and a new set to hit is added to K, corresponding to the complete set of variables $\{n_{i1}, \ldots, n_{im}\}$ (line 6). Observe that the SAT solver proves the formula to be unsatisfiable by unit propagation.

In the second iteration, K contains only 1 set to hit, in which any of its elements can be selected as a minimum hitting set. The SAT solver tests for the satisfiability of $(\neg n_{i1} \vee \ldots \vee \neg n_{im})$ with the complement of the (unit size) MHS, reporting the formula to be satisfiable. The reported cost of the solution is 1. □

Before presenting the result for the $\mathcal{M}_j$ formulas, we make a few observations.

Observation 1. *Consider a complete graph G, i.e. a clique, of $m+1$ vertices. A vertex cover of a G can be computed in polynomial time and has size m. Simply randomly pick one of the vertices to be out of the cover.*

Observation 2. *Let graph G be composed of a clique of size $m - 1$ plus one extra vertex that is connected to at least one of the vertices of the clique. Then a vertex cover of G has size $m - 2$, and can be computed in polynomial time by including all vertices, except for the two of them that have the smallest degree, i.e. number of adjacents.*

Observation 3. *Assume two possible cases: (1) graph G is a clique, or (2) G is composed of a clique plus one extra vertex connected to some of the vertices of the clique. Then checking which of the cases holds can be done by checking if all vertices of G have the same degree (in this case G is a clique).*

Proposition 2. *Given a formula* $\langle \mathcal{M}_j, \mathcal{S} \rangle$ *s.t.* $\mathcal{M}_j$ *and* $\mathcal{S}$ *are from* $DREnc\left(PHP_m^{m+1}\right)$, *there is an execution of the basic MaxHS algorithm that computes a MaxSAT solution of cost m in polynomial time.*

Proof. (Sketch) The idea of the proof is to show that there is a possible ordering of the set of cores returned by the SAT solver that will make the sets in K to induce a graph that is either a clique or composed of a clique plus one extra vertex connected to some of the other vertices. Then from the previous observations a MHS can be computed in polynomial time. In the final iteration, the graph induced by the sets in K will correspond to a clique of size $m + 1$. As such, the final MHS will have size m, thus reporting a solution with a cost of m.

Consider an order of the clauses to consider in $\mathcal{M}_j$, induced by the following choice of variables. First, consider the clauses that contain $\{p_{1j}, p_{2j}\}$ (only 1 clause); then the clauses that contain $\{p_{1j}, p_{2j}, p_{3j}\}$ (2 more clauses); then $\{p_{1j}, p_{2j}, p_{3j}, p_{4j}\}$ (3 more clauses); and so on until all variables/clauses are considered. Observe that due to the structure of $\mathcal{M}_j$, every unsatisfiable core returned by the SAT solver is of size 2 (obtained by unit propagation). Consequently, the chosen order of variables implies that all pairs of the current set of variables are added to K before considering a new variable. The first set added to K this way is $\{p_{1j}, p_{2j}\}$, followed by $\{p_{1j}, p_{3j}\}$, $\{p_{2j}, p_{3j}\}$, etc.

Since sets in K are pairs, then each set can be regarded as an edge of an induced graph. Given the previous ordering of the variables (and consequently of the sets in K), the induced graph forms a "growing" clique, that is, it is either a clique with all the variables considered so far, or it is a clique with the previous variables plus a new variable connected to some of the previous variables.

Finally, since each clause in $\mathcal{M}_j$ produces an unsatisfiable core returned by the SAT solver (corresponding to a new set to hit in K), then the total number of iterations is equal to the number of clauses in $\mathcal{M}_j$ plus 1, which is $C_2^{m+1} + 1 = \frac{(m+1)m}{2} + 1$. $\qquad\qquad\square$

4.2   Doubled Pigeonhole Principle

Here we consider an extension of the pigeonhole principle, which targets m holes, $2m + 1$ pigeons, and each hole has a maximum capacity of 2 pigeons.

Definition 2 (2PHP). *The doubled pigeonhole principle states that if $2m + 1$ pigeons are distributed evenly by m holes, then at least one hole contains more than 2 pigeons.*

The propositional encoding of the 2PHP_m^{2m+1} problem is as follows. Let the variables be x_{ij}, with $i \in [2m + 1]$, $j \in [m]$, with $x_{ij} = 1$ iff pigeon i is placed in the hole j. The constraints state that each pigeon must be placed in at least one hole, and each hole must not have more than 2 pigeons, $\bigwedge_{i=1}^{2m+1} \text{AtLeast1}(x_{i1}, \ldots, x_{im}) \wedge \bigwedge_{j=1}^{m} \text{AtMost2}(x_{1j}, \ldots, x_{(2m+1)j})$. Similar to the PHP case, each AtLeast1 constraint is encoded with a single clause. The AtMost2 constraints are encoded into clauses using the pairwise encoding [7], as $\bigwedge_{i_1=1}^{2m-1} \bigwedge_{i_2=i_1+1}^{2m} \bigwedge_{i_3=i_2+1}^{2m+1} (\neg x_{i_1 j} \vee \neg x_{i_2 j} \vee \neg x_{i_3 j})$.

The reduction of the 2PHP_m^{2m+1} problem into MaxSAT using the Dual-Rail encoding is similar to the PHP case as follows. With each variable x_{ij}, $i \in [2m + 1]$, $j \in [m]$, we associate two new variables: n_{ij} and p_{ij}. The $\mathcal{P}$ clauses are encoded as $\{(\neg n_{ij} \vee \neg p_{ij}) \mid i \in [2m + 1], j \in [m]\}$. $\mathcal{L}_i$ represents the encoding of an AtLeast1 constraint for pigeon i: $\mathcal{L}_i = (\neg n_{i1} \vee \ldots \vee \neg n_{im})$. $\mathcal{M}_j$ represents the encoding of an AtMost2 constraint for hole j: $\mathcal{M}_j = \bigwedge_{i_1=1}^{2m-1} \bigwedge_{i_2=i_1+1}^{2m} \bigwedge_{i_3=i_2+1}^{2m+1} (\neg p_{i_1 j} \vee \neg p_{i_2 j} \vee \neg p_{i_3 j})$. The soft clauses $\mathcal{S}$ are given by $\{(n_{11}), \ldots, (n_{(2m+1)m}), (p_{11}), \ldots, (p_{(2m+1)m})\}$ with $|\mathcal{S}| = 2m(2m + 1)$. Thus, the complete encoding of 2PHP is:

$$\text{DREnc}\left(2\text{PHP}_m^{2m+1}\right) \triangleq \langle \mathcal{H}, \mathcal{S} \rangle = \langle \bigwedge_{i=1}^{2m+1} \mathcal{L}_i \wedge \bigwedge_{j=1}^{m} \mathcal{M}_j \wedge \mathcal{P}, \mathcal{S} \rangle \qquad (2)$$

2PHP_m^{2m+1} is unsatisfiable if and only if the cost of $\text{DREnc}\left(2\text{PHP}_m^{2m+1}\right)$ is at least $m(2m + 1) + 1$ (from Theorem 1 [20]). Similar to the PHP case, if the $\mathcal{P}$

clauses from $\mathsf{DREnc}\left(2\mathrm{PHP}_m^{2m+1}\right)$ are ignored, then the resulting formula can be partitioned into the disjoint formulas $\mathcal{L}_i$ $(i \in [2m+1])$ and the disjoint formulas $\mathcal{M}_j$ $(j \in [m])$. One can compute a MaxSAT solution for each of $\mathcal{L}_i$ and $\mathcal{M}_j$ separately and obtain a lower bound on the cost of the MaxSAT solution for the complete formula $\mathsf{DREnc}\left(2\mathrm{PHP}_m^{2m+1}\right)$. Processing each formula $\mathcal{L}_i$ can be done as in the PHP case (see Proposition 1).

As shown below, the contribution of each $\mathcal{M}_j$ to the MaxSAT cost is $2m-1$, and since there are m such formulas, then the contribution of all $\mathcal{M}_j$ formulas is $m(2m-1)$. As a result, the lower bound on the total cost for $\mathsf{DREnc}\left(2\mathrm{PHP}_m^{2m+1}\right)$ is $m(2m-1)+2m+1 = m(2m+1)+1$, thus, proving formula $2\mathrm{PHP}_m^{2m+1}$ to be unsatisfiable. We also show that the basic MaxHS algorithm is able to derive the MaxSAT cost for each $\mathcal{M}_j$ in polynomial time. To proceed, let us first show that the following holds.

Proposition 3. *Let X be a set of elements of size $|X| = s+2$. Let K be a set of all possible triples $\{x_i, x_j, x_r\}$ of elements of X, $1 \leq i < j < r \leq s+2$. Then any set of s different elements from X is a minimum hitting set for K.*

Proof. (Sketch) Proof by induction on s. Base case, $s = 1$. X contains 3 elements and, thus, $|K| = 1$. Randomly selecting 1 element from X is an MHS of the only set to hit.

Step case $s = n+1$. Suppose that K contains all the combinations of size 3 of elements from X, with $|X| = n+3$. Select randomly 1 element from X, let it be p. Partition K into the sets that contain p and the sets to hit that do not contain p. The sets to hit that dot not contain p form a minimum hitting set subproblem, with an initial set of elements of size $n+2$ (i.e. $s' = n$), and all sets to hit of size 3. By induction hypotheses, this minimum hitting set subproblem has a solution of size n (and is minimum). The solution to the subproblem is not a an MHS of K. This can be seen by considering w.l.o.g. the solution to be $\{x_1, \ldots, x_n\}$, then in K there is the set to hit $\{x_{n+1}, x_{n+2}, p\}$, which is not covered by $\{x_1, \ldots, x_n\}$. Nevertheless, we can extend the solution of the subproblem into a MHS of K by including p in the solution. The size of the new solution is $n + 1 = s$. Note that a smaller solution is not possible; otherwise, we could disregard p from the solution, and that would correspond to a solution to the subproblem of the induction hypothesis with fewer than $n = s'$ elements, thus, contradicting the induction hypothesis. □

Proposition 4. *Let X be a set of elements of size $|X| = s+2$, and an additional element p not in X. Let K be a set of all possible triples $\{x_i, x_j, x_r\}$ of elements of X, $1 \leq i < j < r \leq s+2$, together with some (not all) triples $\{x_i, x_j, p\}$, $x_i, x_j \in X$, $1 \leq i < j \leq s+2$. A minimum hitting set of K has size s and does not contain p.*

Proof. (Sketch) Consider by contradiction that the size of the solution is smaller than s. Since the sets to hit contain all the possible combinations of 3 elements from X, then these sets would have a solution smaller than s, which contradicts Proposition 3.

Consider now by contradiction that the solution has size s and includes p, then we could disregard p from the solution, and hit all the possible combinations of 3 elements from X with fewer than s elements, again contradicting Proposition 3.

Consider now that h is a set with s elements from X not including p. Suppose w.l.o.g. that $\{x_{s+1}, x_{s+2}, p\}$ is missing from the set of elements to hit, and that $h = \{x_1, \cdots, x_s\}$. Since h has size s, then by Proposition 3, h covers the sets to hit that do not contain p. On the other hand, if a set to hit contains p, then it doesn't contain both x_{s+1} and x_{s+2}. So it contains an element from h. Thus h is a MHS of K.   □

Proposition 5. *Given a formula $\langle \mathcal{M}_j, \mathcal{S} \rangle$ s.t. $\mathcal{M}_j$ and $\mathcal{S}$ are from DREnc$\left(2PHP_m^{2m+1}\right)$, there is execution of the basic MaxHS algorithm that computes a MaxSAT solution of cost $2m - 1$ in polynomial time.*

Proof. (Sketch) The proof illustrates a possible setup of the MHS-algorithm that does a polynomial number of iterations s.t. each minimum hitting set is computed in polynomial time. This setup is achieved by ordering the cores computed by the SAT solver (line 4 of Algorithm 1). Similar to the PHP case, we can order the clauses in the SAT solver, by considering an order on the variables. We consider the clauses that contain $\{p_{1j}, p_{2j}, p_{3j}\}$ (only 1 clause), then the clauses that contain $\{p_{1j}, p_{2j}, p_{3j}, p_{4j}\}$ (the 3 clauses $\neg p_{1j} \lor \neg p_{2j} \lor \neg p_{4j}$, $\neg p_{1j} \lor \neg p_{3j} \lor \neg p_{4j}$ and $\neg p_{2j} \lor \neg p_{3j} \lor \neg p_{4j}$), and so on until all variables/clauses are considered. In contrast to the PHP case, when considering the clauses with a new element, these are also ordered. For example, after considering all clauses with $\{p_{1j}, p_{2j}, p_{3j}, p_{4j}\}$, we will consider the clauses with new element p_{5j}. We order these clauses by first considering the clauses that contain $\{p_{1j}, p_{2j}, p_{5j}\}$ (1 clause), then the clauses that contain $\{p_{1j}, p_{2j}, p_{3j}, p_{5j}\}$ (2 more clauses), and finally the clauses that contain $\{p_{1j}, p_{2j}, p_{3j}, p_{4j}, p_{5j}\}$ (3 more clauses). Note that, the new sets to hit include the new element being added (in the example above the element p_{5j}). On the other hand, by Proposition 4, the minimum hitting set solution does not include the element being added. As such, if we disregard the new element being added in the new sets to hit, then we have pairs which can be regarded as edges of a graph. The graph induced by the pairs in the hitting sets will be a growing clique, as in the PHP case Proposition 2. The orderings of the variables guarantee that the sets to hit in K either contain all the possible combinations of size 3 of the variables we are considering, or they induce a "growing" clique. In the first case a minimum hitting set is obtained in polynomial time using the result of Proposition 3. For the second case, we obtain a minimum hitting set in polynomial time similarly to Proposition 2, using Proposition 4. The process of creating a core (and the corresponding set to hit in K) is repeated for each clause in $\mathcal{M}_j$, thus the total number of iterations is equal to the number of clauses plus 1, which is $C_3^{2m+1} + 1 = \frac{(2m+1)(2m)(2m-1)}{6} + 1$. Additionally, the reported cost corresponds to the size of the MHS found in the last iteration, i.e., when all variables are considered. Thus, by Proposition 3, the reported cost is $2m - 1$.   □

5 Conclusions

This paper is motivated by the unexpected good performance of MaxHS-like MaxSAT algorithms on dual-rail encoded families of instances that are hard for general resolution [9,20]. We prove that for the PHP and 2PHP principles, the DRMaxSAT proof system has polynomial time refutations when a MaxHS-like algorithm is used. Future research will seek to understand how MaxHS-like algorithms compare with core-guided algorithms in the DRMaxSAT proof system.

References

1. Ansótegui, C., Bonet, M.L., Levy, J.: SAT-based MaxSAT algorithms. Artif. Intell. **196**, 77–105 (2013)
2. Audemard, G., Katsirelos, G., Simon, L.: A restriction of extended resolution for clause learning SAT solvers. In: AAAI (2010)
3. Bacchus, F., Hyttinen, A., Järvisalo, M., Saikko, P.: Reduced cost fixing in MaxSAT. In: Beck, J.C. (ed.) CP 2017. LNCS, vol. 10416, pp. 641–651. Springer, Cham (2017). https://doi.org/10.1007/978-3-319-66158-2_41
4. Berre, D.L., Parrain, A.: The Sat4j library, release 2.2. JSAT **7**(2–3), 59–6 (2010)
5. Biere, A.: Lingeling, plingeling and treengeling entering the SAT competition 2013. In: Balint, A., Belov, A., Heule, M., Järvisalo, M. (eds.) Proceedings of SAT Competition 2013, Department of Computer Science Series of Publications B, vol. B-2013-1, pp. 51–52. University of Helsinki (2013)
6. Biere, A.: Lingeling essentials, a tutorial on design and implementation aspects of the SAT solver lingeling. In: Pragmatics of SAT workshop, p. 88 (2014)
7. Biere, A., Heule, M., van Maaren, H., Walsh, T. (eds.): Handbook of Satisfiability, Frontiers in Artificial Intelligence and Applications, vol. 185. IOS Press (2009)
8. Birnbaum, E., Lozinskii, E.L.: Consistent subsets of inconsistent systems: structure and behaviour. J. Exp. Theor. Artif. Intell. **15**(1), 25–46 (2003)
9. Bonet, M.L., Buss, S., Ignatiev, A., Marques-Silva, J., Morgado, A.: Maxsat resolution with the dual rail encoding. In: AAAI. pp. 6565–6572 (2018)
10. Bonet, M.L., Levy, J., Manyà, F.: Resolution for Max-SAT. Artif. Intell. **171**(8–9), 606–618 (2007)
11. Bryant, R.E., Beatty, D.L., Brace, K.S., Cho, K., Sheffler, T.J.: COSMOS: a compiled simulator for MOS circuits. In: DAC, pp. 9–16 (1987)
12. Cook, S.A., Reckhow, R.A.: The relative efficiency of propositional proof systems. J. Symb. Log. **44**(1), 36–50 (1979)
13. Davies, J., Bacchus, F.: Solving MAXSAT by solving a sequence of simpler SAT instances. In: Lee, J. (ed.) CP 2011. LNCS, vol. 6876, pp. 225–239. Springer, Heidelberg (2011). https://doi.org/10.1007/978-3-642-23786-7_19
14. Davies, J., Bacchus, F.: Exploiting the power of MIP solvers in MAXSAT. In: Järvisalo, M., Van Gelder, A. (eds.) SAT 2013. LNCS, vol. 7962, pp. 166–181. Springer, Heidelberg (2013). https://doi.org/10.1007/978-3-642-39071-5_13
15. Davies, J., Bacchus, F.: Postponing optimization to speed Up MAXSAT solving. In: Schulte, C. (ed.) CP 2013. LNCS, vol. 8124, pp. 247–262. Springer, Heidelberg (2013). https://doi.org/10.1007/978-3-642-40627-0_21
16. Eén, N., Sörensson, N.: Translating pseudo-boolean constraints into SAT. JSAT **2**(1–4), 1–26 (2006)

17. Elffers, J., Nordström, J.: Divide and conquer: towards faster pseudo-boolean solving. In: IJCAI, pp. 1291–1299 (2018). www.ijcai.org
18. Fu, Z., Malik, S.: On solving the partial MAX-SAT problem. In: Biere, A., Gomes, C.P. (eds.) SAT 2006. LNCS, vol. 4121, pp. 252–265. Springer, Heidelberg (2006). https://doi.org/10.1007/11814948_25
19. Huang, J.: Extended clause learning. Artif. Intell. **174**(15), 1277–1284 (2010)
20. Ignatiev, A., Morgado, A., Marques-Silva, J.: On Tackling the limits of resolution in SAT solving. In: Gaspers, S., Walsh, T. (eds.) SAT 2017. LNCS, vol. 10491, pp. 164–183. Springer, Cham (2017). https://doi.org/10.1007/978-3-319-66263-3_11
21. Kiesl, B., Rebola-Pardo, A., Heule, M.J.H.: Extended resolution simulates DRAT. In: Galmiche, D., Schulz, S., Sebastiani, R. (eds.) IJCAR 2018. LNCS (LNAI), vol. 10900, pp. 516–531. Springer, Cham (2018). https://doi.org/10.1007/978-3-319-94205-6_34
22. Koshimura, M., Zhang, T., Fujita, H., Hasegawa, R.: QMaxSAT: a partial Max-SAT solver. JSAT **8**(1/2), 95–100 (2012)
23. Larrosa, J., Heras, F., de Givry, S.: A logical approach to efficient Max-SAT solving. Artif. Intell. **172**(2–3), 204–233 (2008)
24. Li, C.M., Manyà, F.: MaxSAT. In: Biere et al. [7], pp. 613–631
25. Marques-Silva, J., Planes, J.: On using unsatisfiability for solving maximum satisfiability CoRR abs/0712.1097 (2007)
26. Martins, R., Joshi, S., Manquinho, V., Lynce, I.: Incremental cardinality constraints for MaxSAT. In: O'Sullivan, B. (ed.) CP 2014. LNCS, vol. 8656, pp. 531–548. Springer, Cham (2014). https://doi.org/10.1007/978-3-319-10428-7_39
27. Martins, R., Manquinho, V., Lynce, I.: Open-WBO: a modular MaxSAT solver'. In: Sinz, C., Egly, U. (eds.) SAT 2014. LNCS, vol. 8561, pp. 438–445. Springer, Cham (2014). https://doi.org/10.1007/978-3-319-09284-3_33
28. Morgado, A., Dodaro, C., Marques-Silva, J.: Core-guided MaxSAT with soft cardinality constraints. In: O'Sullivan, B. (ed.) CP 2014. LNCS, vol. 8656, pp. 564–573. Springer, Cham (2014). https://doi.org/10.1007/978-3-319-10428-7_41
29. Morgado, A., Heras, F., Liffiton, M.H., Planes, J., Marques-Silva, J.: Iterative and core-guided MaxSAT solving: a survey and assessment. Constraints **18**(4), 478–534 (2013)
30. Morgado, A., Ignatiev, A., Marques-Silva, J.: MSCG: robust core-guided MaxSAT solving. JSAT **9**, 129–134 (2015)
31. Narodytska, N., Bacchus, F.: Maximum satisfiability using core-guided MaxSAT resolution. In: AAAI, pp. 2717–2723 (2014)
32. Palopoli, L., Pirri, F., Pizzuti, C.: Algorithms for selective enumeration of prime implicants. Artif. Intell. **111**(1–2), 41–72 (1999)
33. Reiter, R.: A theory of diagnosis from first principles. Artif. Intell. **32**(1), 57–95 (1987)
34. Saikko, P., Berg, J., Järvisalo, M.: LMHS: a SAT-IP hybrid MaxSAT solver. In: Creignou, N., Le Berre, D. (eds.) SAT 2016. LNCS, vol. 9710, pp. 539–546. Springer, Cham (2016). https://doi.org/10.1007/978-3-319-40970-2_34
35. Sinz, C.: Towards an optimal CNF encoding of boolean cardinality constraints. In: van Beek, P. (ed.) CP 2005. LNCS, vol. 3709, pp. 827–831. Springer, Heidelberg (2005). https://doi.org/10.1007/11564751_73

Backing Backtracking

Sibylle Möhle[(✉)] and Armin Biere

Johannes Kepler University Linz, Linz, Austria
{sibylle.moehle-rotondi,biere}@jku.at

Abstract. Non-chronological backtracking was considered an important and necessary feature of conflict-driven clause learning (CDCL). However, a SAT solver combining CDCL with chronological backtracking succeeded in the main track of the SAT Competition 2018. In that solver, multiple invariants considered crucial for CDCL were violated. In particular, decision levels of literals on the trail were not necessarily increasing anymore. The corresponding paper presented at SAT 2018 described the algorithm and provided empirical evidence of its correctness, but a formalization and proofs were missing. Our contribution is to fill this gap. We further generalize the approach, discuss implementation details, and empirically confirm its effectiveness in an independent implementation.

1 Introduction

Most state-of-the-art SAT solvers are based on the CDCL framework [8,9]. The performance gain of SAT solvers achieved in the last two decades is to some extent attributed to combining conflict-driven backjumping and learning. It enables the solver to escape regions of the search space with no solution.

Non-chronological backtracking during learning enforces the lowest decision level at which the learned clause becomes unit and then is used as a reason. While backtracking to a higher level still enables propagation of a literal in the learned clause, the resulting propagations might conflict with previous assignments. Resolving these conflicts introduces additional work which is prevented by backtracking non-chronologically to the lowest level [15].

However, in some cases a significant amount of the assignments undone is repeated later in the search [10,16], and a need for methods to save redundant work has been identified. Chronological backtracking avoids redundant work by keeping assignments which otherwise would be repeated at a later stage of the search. As our experiments show, satisfiable instances benefit most from chronological backtracking. Thus this technique should probably also be seen as a method to optimize SAT solving for satisfiable instances similar to [2,14].

The combination of chronological backtracking with CDCL is challenging since invariants classically considered crucial to CDCL cease to hold. Nonetheless, taking appropriate measures preserves the solver's correctness, and the

Supported by Austrian Science Fund (FWF) grant S11408-N23 (RiSE) and by the LIT Secure and Correct Systems Lab funded by the State of Upper Austria.

© Springer Nature Switzerland AG 2019
M. Janota and I. Lynce (Eds.): SAT 2019, LNCS 11628, pp. 250–266, 2019.
https://doi.org/10.1007/978-3-030-24258-9_18

Chronological and non-chronological CDCL:

Trail:	The assignment trail contains neither complementary pairs of literals nor duplicates.
ConflictLower:	The assignment trail preceding the current decision level does not falsify the formula.

Non-chronological CDCL only:

Propagation:	On every decision level preceding the current decision level all unit clauses are propagated until completion.
LevelOrder:	The literals are ordered on the assignment trail in ascending order with respect to their decision level.
ConflictingClause:	At decision levels greater than zero the conflicting clause contains at least two literals with the current decision level.

Fig. 1. The CDCL invariants listed in the box are usually considered crucial to CDCL. By combining CDCL with chronological backtracking, the last three are violated.

combination of chronological backtracking and CDCL appeared to be a winning strategy: The SAT solver MAPLE_LCM_DIST_CHRONOBT [11] was ranked first in the main track of the SAT Competition 2018.

In Fig. 1 we give invariants classically considered crucial to CDCL which are relevant for the further discussion. Our aim is to demonstrate that although some of them do not hold in [10], the solving procedure remains correct.

Clearly, if upon conflict analysis the solver jumps to a decision level higher than the asserting level, invariant Propagation is violated. Measures to fix potential conflicting assignments were proposed in [10] which in addition violated invariants LevelOrder and ConflictingClause. The algorithm's correctness as well as its efficiency were empirically demonstrated. However, a formal treatment with proofs was not provided.

Our Contribution. Our main contribution consists in providing a generalization of the method presented in [10] together with a formalization. We prove that despite violating some of the invariants given above, the approach is correct. Our experiments confirm the effectiveness of chronological backtracking with an independent implementation in our SAT solver CaDiCaL [4].

2 Preliminaries

Let F be a formula over a set of variables V. A *literal* ℓ is either a variable $v \in V$ or its negation $\neg v$. The variable of ℓ is obtained by $V(\ell)$. We denote by $\bar{\ell}$ the *complement* of ℓ, i.e., $\bar{\ell} = \neg\ell$, and assume $\neg\neg\ell = \ell$. We consider formulae in conjunctive normal form (CNF) defined as conjunctions of clauses which are disjunctions of literals. We write $C \in F$ if C is a clause in F and $\ell \in C$ for a literal ℓ occurring in C interpreting F as a set of clauses and C as a set of literals. We use set notation for formulae and clauses where convenient.

We call *trail* a sequence of literals with no duplicated variables and write $I = \ell_1 \ldots \ell_n$. We refer to an element ℓ of I by writing $\ell \in I$ interpreting I as a set of literals and denote the set of its variables by $V(I)$. Trails can be concatenated, $I = JK$, assuming $V(J) \cap V(K) = \emptyset$. We denote by $\tau(I, \ell)$ the position of the literal ℓ on the trail I. A *total assignment* is a mapping from V to the truth values 1 and 0. A trail may be interpreted as a *partial assignment* where $I(\ell) = 1$ iff $\ell \in I$. Similarly, $I(C)$ and $I(F)$ are defined.

The *residual* of the formula F under the trail I, denoted by $F|_I$, is obtained by replacing in F the literals ℓ where $V(\ell) \in I$ with their truth value. We define the residual of a clause in an analogous manner. The empty clause and the literal assigned truth value 0 are denoted by $\bot$, the empty formula by $\top$. If $I(F) = \top$, i.e., $F|_I = \top$, we say that I *satisfies* F and call I a *model* of F. If $I(C) = \bot$ for a clause $C \in F$, i.e., $C|_I = \bot$ and hence $F|_I = \bot$, we say that I *falsifies* C (and therefore F) and call C the *conflicting clause*.

We call *unit clause* a clause $\{\ell\}$ containing one single literal ℓ which we refer to as *unit literal*. We denote by $\mathsf{units}(F)$ the sequence of unit literals in F and extend this notion to the residual of F under I by writing $\mathsf{units}(F|_I)$. We write $\ell \in \mathsf{units}(F|_I)$ for referring to the unit literal ℓ in the residual of F under I.

3   Generalizing CDCL with Chronological Backtracking

In classical CDCL SAT solvers based on **non-chronological** backtracking [8] the trail reflects the order in which literals are assigned. The trail is used during conflict analysis to simplify traversal of the implication graph in reverse assignment order and in general during backtracking to undo assignments in the proper reverse assignment order.

In CDCL with non-chronological backtracking, the trail is partitioned into subsequences of literals between decisions in which all literals have the same decision level. Each subsequence starts with a decision literal and extends until the last literal before the next decision. Literals assigned before any decision may form an additional subsequence at decision level zero.

After adding **chronological** backtracking to CDCL as described in [10], the trail is not partitioned in the same way but subsequences of the same decision level are interleaved, while still respecting the assignment order.

Let $\delta \colon V \mapsto \mathbb{N} \cup \{\infty\}$ return the decision level of a variable v in the set of variables V, with $\delta(v) = \infty$ if v is unassigned. This function is updated whenever a variable is either assigned or unassigned. The function δ is extended to literals ℓ, clauses C and trails I by defining $\delta(\ell) = \delta(V(\ell))$, $\delta(C) = \max\{\delta(\ell) \mid \ell \in C\}$ for $C \neq \bot$, and $\delta(I) = \max\{\delta(\ell) \mid \ell \in I\}$. We further define $\delta(\bot) = 0$.

Given a set of literals L, we denote by $\delta(L) = \{\delta(\ell) \mid \ell \in L\}$ the set containing the decision levels of its elements. The function δ updated with decision level d

assigned to $V(\ell)$ is denoted by $\delta[\ell \mapsto d]$. Similarly, $\delta[I \mapsto \infty]$ represents the function δ where all literals on the trail I are unassigned. In the same manner, $\delta[V \mapsto \infty]$ assigns all variables in V to decision level ∞. We may write $\delta \equiv \infty$ as a shortcut. The function δ is left-associative. We write $\delta[L \mapsto \infty][\ell \mapsto b]$ to express that the function δ is updated by first unassigning all literals in a sequence of literals L and then assigning literal ℓ to decision level b.

For the sake of readability, we write $J \leqslant I$ where J is a subsequence of I and the elements in J have the same order as in I and $J < I$ when furthermore $J \neq I$. We denote by $I_{\leqslant b}$ the subsequence of I containing exactly the literals ℓ where $\delta(\ell) \leqslant b$.

Due to the interleaved trail structure we need to define decision literals differently than in CDCL. We refer to the set consisting of all decision literals on I by writing $\mathsf{decs}(I)$ and define a *decision literal* ℓ as

$$\ell \in \mathsf{decs}(I) \quad \text{iff} \quad \ell \in I, \; \delta(\ell) > 0, \; \forall k \in I \,.\, \tau(I,k) < \tau(I,\ell) \Rightarrow \delta(k) < \delta(\ell) \quad (1)$$

Thus, the decision level of a decision literal $\ell \in I$ is strictly higher than the decision level of any literal preceding it on I. If $C|_I = \{\ell\}$ for a literal ℓ, then ℓ is not a decision literal. The set $\mathsf{decs}(I)$ can be restricted to decision literals with decision level lower or equal to i by writing $\mathsf{decs}_{\leqslant i}(I) = \mathsf{decs}(I_{\leqslant i})$.

As in [13] we use an abstract representation of the assignment trail I by writing $I = I_0\ell_1 I_1 \ldots \ell_n I_n$ where $\{\ell_1, \ldots, \ell_n\} = \mathsf{decs}(I)$. We denote by $\mathsf{slice}(I, i)$ the i-th *slice* of I, i.e., the subsequence of I containing all literals ℓ with the same decision level $\delta(\ell) = i$. The i-th *block*, denoted by $\mathsf{block}(I, i)$, is defined as the subsequence of I starting with the decision literal with decision level i and extending until the last literal before the next decision:

$$\mathsf{slice}(I, i) = I_{=i}$$
$$\mathsf{block}(I, i) = \ell_i I_i$$

Note that in general $I_{=i} \neq I_i$, since I_i (due to the interleaved structure of the trail I) may contain literals with different decision levels, while this is not the case in $I_{=i}$. In particular, there might be literals with a lower decision level than some literal preceding them on the trail. We call these literals *out-of-order literals*. Contrarily to classical CDCL with non-chronological backtracking, upon *backtracking* to a decision level b, blocks must not be discarded as a whole, but only the literals in $\mathsf{slice}(I, i)$ where $i > b$ need to be unassigned.

Consider the trail I on the left hand side of Fig. 2 over variables $\{1, \ldots, 5\}$ (in DIMACS format) where τ represents the position of a literal on I and δ represents its decision level:

Literals 1 and 3 were propagated at decision level zero, literal 5 was propagated at decision level one. The literals 3 and 5 are out-of-order literals: We have $\delta(2) = 1 > 0 = \delta(3)$, whereas $\tau(I, 2) = 1 < 2 = \tau(I, 3)$. In a similar manner, $\delta(4) = 2 > 1 = \delta(5)$, and $\tau(I, 4) = 3 < 4 = \tau(I, 5)$. Moreover, $I_{\leqslant 1} = 1\,2\,3\,5$, $\mathsf{decs}(I) = 2\,4$, $\mathsf{decs}_{\leqslant 1}(I) = 2$, $\mathsf{slice}(I, 1) = 2\,5$, and $\mathsf{block}(I, 1) = 2\,3$.

Upon backtracking to decision level one, the literals in $\mathsf{slice}(I, 2)$ are unassigned. The resulting trail is visualized in the middle of Fig. 2. Note that since

τ	0	1	2	3	4
I	1	2	3	4	5
δ	0	1	0	2	1

τ	0	1	2	3
I	1	2	3	5
δ	0	1	0	1

τ	0	1
I	1	3
δ	0	0

Fig. 2. In the trail I on the left, from the three trails shown, literals 3 and 5 are placed out of order. In fact, their decision level δ is lower than the decision level of a literal preceding them on the trail, i.e., with lower position τ. The trails in the middle and on the right hand side show the results of backtracking to decision levels 1 and 0. When backtracking to the *backtrack level* b, only literals ℓ with $\delta(\ell) > b$ are removed from the trail, while the assignment order is preserved.

the assignment order is preserved, the trail still contains one out-of-order literal, namely 3. Backtracking to decision level zero unassigns all literals in $\mathsf{slice}(I, 2)$ and $\mathsf{slice}(I, 1)$ resulting in the trail in which all literals are placed in order depicted on the right hand side.

4   Calculus

We devise our calculus as a transition system over a set of states S, a transition relation $\leadsto \, \subseteq S \times S$ and an initial state s_0. Non-terminal states are described by (F, I, δ) where F denotes a formula over variables V, I denotes the current trail and δ refers to the decision level function.

The *initial* state is given by $s_0 = (F, \varepsilon, \delta_0)$. In this context, F is the original formula, ε denotes the empty trail and $\delta_0 \equiv \infty$. The *terminal* state is either SAT or UNSAT expressing satisfiability or unsatisfiability of F. The transition relation $\leadsto$ is defined as the union of transition relations $\leadsto_{\mathsf{R}}$ where R is either True, False, Unit, Jump or Decide. These rules are listed in Fig. 3. We first explain the intuition behind these rules before proving correctness in Sect. 5:

True / False. If $F|_I = \top$, F is satisfiable and the search terminates in the state SAT (rule True). If $F|_I = \bot$, a clause $C \in F$ exists where $I(C) = \bot$. The *conflict level* is $\delta(C) = 0$. Obviously $I_{\leqslant 0}(F) = \bot$ and consequently F is unsatisfiable. Then the procedure terminates in state UNSAT (False).

Unit. Propagated unit literals are assigned the maximum decision level of their reason which may be lower than the current decision level. Requiring that the residual of F under I is conflict-free ensures that invariant ConflictLower holds.

Jump. We have $F|_I = \bot$, i.e., there exists a clause $C \in F$ for which we have $I(C) = \bot$. Since the conflict level is $\delta(C) = c > 0$, there is a decision left on I. We assume to obtain a clause D implied by F (usually through conflict analysis) with $\delta(D) = c > 0$ whose residual is unit, e.g., $\{\ell\}$, at *jump level* $j = \delta(D \setminus \{\ell\})$, the second highest decision level in D. In fact, the residual of D under the trail is unit at any backtrack level b where $j \leqslant b < c \leqslant d$, with $d = \delta(I)$ denoting the current decision level. Using D as a reason, we may backtrack to any of these decision levels. Remember that the decision levels on the trail do not have

True:	$(F, I, \delta) \rightsquigarrow_{\text{True}} \text{SAT}$ if $F\|_I = \top$

False:	$(F, I, \delta) \rightsquigarrow_{\text{False}} \text{UNSAT}$ if exists $C \in F$ with $C\|_I = \bot$ and $\delta(C) = 0$

Unit: $(F, I, \delta) \rightsquigarrow_{\text{Unit}} (F, I\ell, \delta[\ell \mapsto a])$ if $F\|_I \neq \top$ and $\bot \notin F\|_I$ and exists $C \in F$ with $\{\ell\} = C\|_I$ and $a = \delta(C \setminus \{\ell\})$

Jump: $(F, I, \delta) \rightsquigarrow_{\text{Jump}} (F \wedge D, PK\ell, \delta[L \mapsto \infty][\ell \mapsto j])$ if exists $C \in F$ with $PQ = I$ and $C\|_I = \bot$ such that $c = \delta(C) = \delta(D) > 0$ and $\ell \in D$ and $\ell|_Q = \bot$ and $F \models D$ and $j = \delta(D \setminus \{\ell\})$ and $b = \delta(P)$ and $j \leqslant b < c$ and $K = Q_{\leqslant b}$ and $L = Q_{>b}$

Decide: $(F, I, \delta) \rightsquigarrow_{\text{Decide}} (F, I\ell, \delta[\ell \mapsto d])$ if $F\|_I \neq \top$ and $\bot \notin F\|_I$ and units$(F\|_I) = \emptyset$ and $V(\ell) \in V$ and $\delta(\ell) = \infty$ and $d = \delta(I) + 1$

Fig. 3. In the transition system of our framework non-terminal states (F, I, δ) consist of a CNF formula F, the current trail I and the decision level function δ. The rules formalize termination (True and False), backtracking (Jump), unit propagation (Unit) and picking decisions (Decide).

to be sorted in ascending order and that upon backtracking only the literals in the i-th slice with $i > b$ need to be unassigned as discussed in Sect. 3. After backtracking we propagate ℓ and assign it decision level j to obtain $\delta(PK\ell) = b$.

Note. If the conflicting clause C contains exactly one literal ℓ assigned at conflict level c, its residual is unit at decision level $c - 1$. The solver therefore could backtrack to decision level $c-1$ and propagate ℓ. An optimization is to use $D = C$ as reason saving the computational effort of conflict analysis. This corresponds to learning C instead of a new clause D and is a special case of rule Jump. It is also explicitly included in the pseudocode in [10].

Decide. If F is neither satisfied nor falsified and there are no unit literals in $F\|_I$, an unassigned variable is assigned. Invariants ConflictLower and Propagation hold.

Example. As pointed out above, the conflicting clause C may contain one single literal ℓ assigned at decision level c. While according to the pseudocode in [10] backtracking is executed to the second highest decision level j in C, the implementation MAPLE_LCM_DIST_CHRONOBT backtracks chronologically to decision level $c - 1$, which may be higher than j. We adopt this strategy in our own solver, but unlike MAPLE_LCM_DIST_CHRONOBT we eagerly propagate ℓ and assign it decision level j, as described in the explanation of rule Jump above.

The authors of [10] focused on unit propagation and backtracking, and an in-depth discussion of the case in which the conflicting clause contains exactly one literal at conflict level is missing. We fill this gap and explain our calculus in detail by means of an example for this case. We generated this example with our model-based tester Mobical for CaDiCaL based on ideas in [1,12]. Our example is larger and provides a good intuition regarding the occurrence of multiple nested decision levels on the trail as well as its effect on backtracking.

We represent variables by natural numbers and consider a formula F over variables $\{1,\ldots,48\}$ as above where negative numbers encode negated variables. We further use set notation for representing clauses. Consider the following assignment trail excerpt where the trail I is represented as a sequence of literals, τ denotes the position of a literal on I and δ its decision level:

τ	$\cdots$	4	5	6	7	8	9	10	11	12	13	14	15	16	17	18	19
I	$\cdots$	4	5	30	47	15	18	6	-7	-8	45	9	38	-23	17	44	-16
δ	$\cdots$	3	4	4	4	4	4	5	5	5	5	6	6	6	6	6	6

| C | $\{$ | | | -47, | | | | | | | | | | | -17,-44 | | $\}$ |
| D | $\{$ | | | -30, -47, | | | -18, | | | | | | | | 23 | | $\}$ |

Initially, the literals are placed in order on I, i.e., they are sorted in ascending order with respect to their decision level. At this point, a conflict occurs. The conflicting clause is $C = \{-47, -17, -44\}$ depicted below the trail containing two literals at conflict level $c = \delta(C) = 6$, i.e., -17 and -44 depicted in boldface in the above outline. Conflict analysis in our implementation learned the clause $D = \{-30, -47, -18, 23\}$ where $\delta(-30) = \delta(-47) = \delta(-18) = 4$ and $\delta(23) = 6$. Since $\delta(D) = c = 6$ and $j = \delta(D \setminus \{23\}) = 4$, the solver in principle would be free to backtrack to either decision level 4 or 5.

Let the solver backtrack *chronologically* to decision level 5 where D becomes unit, specifically $\{23\}$. The position on the trail the solver backtracks to is marked with a vertical dotted line. Accordingly all literals with decision level higher than 5 are removed from I (literals at positions higher than 13). Then literal 23 is propagated. The jump level is $j = 4$, hence literal 23 is assigned decision level 4 out of order. Literal -38 is propagated due to reason$(-38) = \{-15, -23, -38\}$ (not shown). Since $\delta(-15) = \delta(-23) = 4$, literal -38 is assigned decision level 4. Then literal -9 is propagated with reason$(-9) = \{-45, 38, -9\}$ with $\delta(-45) = 5$ and $\delta(38) = 4$. Thus, -9 is assigned decision level 5. The resulting trail is

τ	$\cdots$	4	5	6	7	8	9	10	11	12	13	14	15	16
I	$\cdots$	4	5	30	47	15	18	6	-7	-8	45	23	-38	-9
δ	$\cdots$	3	4	4	4	4	4	5	5	5	5	4	4	5

where the literals 23 and -38 (depicted in boldface) are placed out of order on I. Later in the search we might have the following situation:

τ ···	9	10	11	12	13	14	15	16	17	18	19	20	21	22	23	24	25
I ···	18	6	-7	-8	45	23	-38	-9	10	-11	13	16	-17	-25	42	12	-41
δ ···	4	5	5	5	5	4	4	5	6	7	5	4	4	4	4	5	5
C	{													17,		-42,	**-12** }

The first assignment after analyzing the last conflict is placed right after the dashed vertical line. Again, a conflict occurs. Let $C = \{17, -42, -12\}$ be the conflicting clause. The conflict level is $\delta(C) = 5$ and the decision level of I is $\delta(I) = 7$. Clause C contains exactly one literal at conflict level, namely -12 depicted in boldface. The solver backtracks to decision level $c - 1 = 4$ marked with a thick solid line. After removing from I all literals with decision level higher than 4 and propagating literal -12, the resulting trail is

τ ···	9	10	11	12	13	14	15	16
I ···	18	23	-38	16	-17	-25	42	-12
δ ···	4	4	4	4	4	4	4	4

Note that as discussed above we use $D = C \in F$ without actually adding it.

5 Proofs

For proving the correctness of our method, we first show that the system terminates in the correct state which can be done in a straightforward manner. Proving that the system always makes progress is more involved. Last we prove that our procedure terminates by showing that no infinite sequence of states is generated. By $\delta(\mathsf{decs}(I))$ we denote the set consisting of the decision levels of the set of decision literals on I. We start by proving the following invariants:

(1) $\forall k, \ell \in \mathsf{decs}(I) . \tau(I, k) < \tau(I, \ell) \implies \delta(k) < \delta(\ell)$
(2) $\delta(\mathsf{decs}(I)) = \{1, \dots, \delta(I)\}$
(3) $\forall n \in \mathbb{N} . F \wedge \mathsf{decs}_{\leqslant n}(I) \models I_{\leqslant n}$

Lemma 1 (Invariants). *Invariants* (1) – (3) *hold in non-terminal states.*

Proof. The proof is carried out by induction over the number of rule applications. We assume Inv. (1) – (3) hold in a non-terminal state (F, I, δ) and show that they are still met after the transition to another non-terminal state for all rules.

Unit: The trail I is extended by a literal ℓ. We need to show that ℓ is not a decision literal. To this end it is sufficient to consider the case where $a > 0$. There exists a clause $C \in F$ with $\{\ell\} = C|_I$. Since $a = \delta(C \setminus \{\ell\})$, there exists a literal $k \in C$ where $k \neq \ell$ and such that $\delta(k) = a$. Obviously, k was assigned prior to ℓ and $\tau(I, k) < \tau(I, \ell)$. Since $\delta(k) = \delta(\ell)$ and by the definition of decision literal in Eq. (1), ℓ is not a decision literal. The decisions remain unchanged, and Inv. (1) and (2) hold after executing rule Unit.

We have $F \wedge \mathsf{decs}_{\leqslant n}(I) \models \overline{C \setminus \{\ell\}}$ and $F \wedge \mathsf{decs}_{\leqslant n}(I) \models C$, therefore, by modus ponens we get $F \wedge \mathsf{decs}_{\leqslant n}(I) \models \ell$. Since ℓ is not a decision literal, as shown above, $F \wedge \mathsf{decs}_{\leqslant n}(I\ell) \equiv F \wedge \mathsf{decs}_{\leqslant n}(I) \models I_{\leqslant n}$. Hence, $F \wedge \mathsf{decs}_{\leqslant n}(I\ell) \models (I\ell)_{\leqslant n}$, and Inv. (3) holds after executing rule Unit.

Jump: We first show that K contains no decision literal. In fact, the trail I is of the form $I = PQ$, and K is obtained from Q by removing all literals with decision level greater than b. In particular, the order of the remaining (decision) literals remains unaffected, and Inv. (1) still holds. We further have $\delta(K) \leqslant \delta(P) = b$. Since $\forall p \in P, k \in K . \tau(PK, p) < \tau(PK, k)$ and by the definition of decision literals in Eq. (1), the decision literal with decision level b is contained in P. Therefore, since K contains no (decision) literal with decision level greater than b, it contains no decision literal.

Now we show that ℓ is not a decision literal either. As in the proof for rule Unit, it is sufficient to consider the case where $j > 0$. There exists a clause D where $F \models D$ such that $\delta(D) = c$ and a literal $\ell \in D$ for which $\ell|_Q = \bot$ and $\ell \in Q$. Since $j = \delta(D \setminus \{\ell\})$, $\delta(\ell) = \delta(D) = c > b$, and $\ell \notin K$. Instead, $\ell \in L$, and ℓ is unassigned during backtracking to any decision level smaller than c, i.e., $\ell \notin PK$. Furthermore, there exists a literal $k \in D$ where $k \neq \ell$ and such that $\delta(k) = j$ which precedes ℓ on the trail $PK\ell$. Therefore, following the argument in rule Unit, literal ℓ is not a decision literal, and since the decisions remain unchanged, Inv. (1) and (2) hold after applying rule Jump.

Invariant (3) holds prior to applying rule Jump, i.e., $F \wedge \mathsf{decs}_{\leqslant n}(I) \models I_{\leqslant n}$. We have that $F \models D$, and therefore $F \wedge D \equiv F$. Since $I = PQ$, $PK < I$ and obviously $F \wedge \mathsf{decs}_{\leqslant n}(PK) \implies (PK)_{\leqslant n}$. From $j = \delta(D \setminus \{\ell\})$ we get $D|_{PK} = \{\ell\}$. Repeating the argument in the proof for rule Unit by replacing I by PK and C by D, we have that $F \wedge \mathsf{decs}_{\leqslant n}(PQ\ell) \models (PQ\ell)_{\leqslant n}$, and Inv. (3) is met after executing rule Jump.

Decide: Literal ℓ is a decision literal by the definition of a decision literal in Eq. 1: It is assigned decision level $d = \delta(I) + 1$, and $\forall k \in I . \delta(k) < \delta(\ell)$. Further, $\forall k \in I\ell . k \neq \ell \implies \tau(I\ell, k) < \tau(I\ell, \ell)$. Since $\ell \in \mathsf{decs}(I\ell)$, we have $\delta(\mathsf{decs}(I\ell)) = \{1, \ldots, d\}$, and Inv. (1) and (2) hold after applying rule Decide.

Since ℓ is a decision, $F \wedge \mathsf{decs}_{\leqslant n}(I\ell) \equiv F \wedge \mathsf{decs}_{\leqslant n}(I) \wedge \ell_{\leqslant n}$ and since Inv. (3) holds prior to applying Decide, obviously $F \wedge \mathsf{decs}_{\leqslant}(I\ell) \models I_{\leqslant n} \wedge \ell_{\leqslant n} \equiv (I\ell)_{\leqslant n}$, and Inv. (3) is met after applying rule Decide. $\qquad \square$

Proposition 1 (Correctness of Terminal State). SEARCH *terminates in the correct state, i.e., if the terminal state is* SAT, *then F is satisfiable, and if the terminal state is* UNSAT, *then F is unsatisfiable.*

Proof. We show that the terminal state is correct for all terminal states.

SAT: We must show that an unsatisfiable formula can not be turned into a satisfiable one by any of the transition rules. Only equivalence-preserving transformations are executed: Rules Unit and Decide do not affect F, and in rule Jump a clause implied by F is added. Therefore, if the system terminates in state SAT, F is indeed satisfiable.

<u>UNSAT:</u> It must be proven that a satisfiable formula can not be made unsatisfiable. Only equivalence-preserving transformations are executed. Rules Unit and Decide do not affect F, and in rule Jump a clause implied by F is added. We need to show that if rule False is applied, the formula F is unsatisfiable. We have to consider Inv. (3) for $n = 0$. There exists a clause $C \in F$ such that $I_{\leqslant 0}(C) = \bot$, which leads to $F \wedge \mathsf{decs}_{\leqslant 0}(F) \equiv F \models I_{\leqslant 0}(C) = \bot$. $\qquad \square$

Proposition 2 (Progress and Termination). SEARCH *makes progress in non-terminal states (a rule is applicable) and always reaches a terminal state.*

Proof. We first prove progress by showing that in every non-terminal state a transition rule is applicable. Then we prove termination by showing that no infinite state sequence is generated.

<u>Progress:</u> We show that in every non-terminal state a transition rule is applicable. The proof is by induction over the number of rule applications. Assume we reached a non-terminal state (F, I, δ). We show that one rule is applicable.

If $F|_I = \top$, rule True can be applied. If $F|_I = \bot$, there exists a clause $C \in F$ such that $C|_I = \bot$. The conflict level $\delta(C) = c$ may be either zero or positive. If $c = 0$, rule False is applicable. Now assuming $c > 0$ we obtain with Inv. (3):

$$F \wedge \mathsf{decs}_{\leqslant c}(I) \quad \equiv \quad F \wedge \mathsf{decs}_{\leqslant c}(I) \wedge I_{\leqslant c} \quad \models \quad I_{\leqslant c}.$$

Due to $I_{\leqslant c}(F) \equiv \bot$ we further have $F \wedge I_{\leqslant c} \equiv F \wedge \mathsf{decs}_{\leqslant c}(I) \equiv \bot$. By simply picking $\neg D = \mathsf{decs}_{\leqslant c}(I)$ we obtain $F \wedge \neg D \equiv F \wedge \neg D \wedge I_{\leqslant c} \equiv \bot$, thus $F \models D$. Clause D contains only decision literals and $\delta(D) = c$. From Inv. (1) and (2) we know that D contains exactly one decision literal for each decision level in $\{1, \ldots, c\}$. We choose $\ell \in D$ such that $\delta(\ell) = c$. Then the asserting level is given by $j = \delta(D \setminus \{\ell\})$ and we pick some backtrack level b where $j \leqslant b < c$. Without loss of generalization we assume the trail to be of the form $I = PQ$ where $\delta(P) = b$. After backtracking to decision level b, the trail is equal to $I_{\leqslant b} = PK$ where $K = Q_{\leqslant b}$. Since $D|_{PK} = \{\ell\}$, all conditions of rule Jump hold.

If $F|_I \notin \{\top, \bot\}$, there are still unassigned variables in V. If there exists a clause $C \in F$ where $C|_I = \{\ell\}$, the preconditions of rule Unit are met. If instead $\mathsf{units}(F|_I) = \emptyset$, there exists a literal ℓ with $V(\ell) \in V$ and $\delta(\ell) = \infty$, and the preconditions of rule Decide are satisfied.

In this argument, all possible cases are covered and thus in any non-terminal state a transition rule can be executed, i.e., the system never gets stuck.

<u>Termination:</u> To show termination we follow the arguments in [7,13] or more precisely the one in [5], except that our blocks (as formalized above with the block notion) might contain literals with different decision levels, i.e., subsequences of literals of the same decision level are interleaved as discussed in Sect. 3. This has an impact on the backtracking procedure adopted in rule Jump, where after backtracking to the end of $\mathsf{block}(I, b)$, trail P is extended by $K = Q_{\leqslant b}$. As discussed in the proof of Lemma 1, K contains no decision literals. Apart from that, the same argument applies as in [5], and SEARCH always terminates. $\qquad \square$

6   Algorithm

The transition system presented in Sect. 4 can be turned into an algorithm described in Fig. 4 providing a foundation for our implementation. Unlike in [10], we refrain from giving implementation details but provide pseudocode on a higher abstraction level covering exclusively chronological backtracking.

Search: The main function Search takes as input a formula F, a set of variables V, a trail I and a decision level function δ. Initially, I is equal to the empty trail and all variables are assigned decision level ∞.

   If all variables are assigned and no conflict occurred, it terminates and returns SAT. Otherwise, unit propagation by means of Propagate is executed until either a conflict occurs or all units are propagated.

   If a conflict at decision level zero occurs, Search returns UNSAT, since conflict analysis would yield the empty clause even if the trail contains literals with decision level higher than zero. These literals are irrelevant for conflict analysis (line 7), and they may be removed from I prior to conflict analysis without affecting the computation of the learned clause. The resulting trail contains only propagation literals, and the new (current) decision level is zero upon which the empty clause is learned.

   If a conflict at a decision level higher than zero occurs, conflict analysis (function Analyze) is executed. If no conflict occurs and there are still unassigned variables, a decision is taken and a new block started.

Propagate: Unit propagation is carried out until completion. Unlike in CDCL with non-chronological backtracking, the propagated literals may be assigned a decision level lower than the current one (line 3). In this case invariant LevelOrder presented in Sect. 1 does not hold anymore. Propagate returns the empty clause if no conflict occurs and the conflicting clause otherwise.

Analyze: If the conflict level is higher than zero and the conflicting clause C contains exactly one literal ℓ at conflict level c, then C can be used as reason instead of performing conflict analysis (lines 1–3). The idea is to save the computational effort of executing conflict analysis and adding redundant clauses.

   Otherwise, a clause D is learned as in CDCL, e.g., the first unique implication point (1st-UIP) containing exactly one literal ℓ at conflict level. Let j be the lowest decision level at which D (or C, if it contains exactly one literal at conflict level) becomes unit. Then according to some heuristics the solver backtracks to a decision level $b \in [j, c-1]$.

   This for instance, can be used to retain part of the trail, to avoid redundant work which would repeat the same assignments after backtracking. Remember that the decision levels on the trail may not be in ascending order. When backtracking to b, the solver removes all literals with decision level higher than b from I, i.e., all i-th slices with $i > b$.

Input: formula F, set of variables V, trail I, decision level function δ
Output: SAT iff F is satisfiable, UNSAT otherwise

Search (F)
1 $V := V(F)$
2 $I := \varepsilon$
3 $\delta := \infty$
4 **while** there are unassigned variables in V **do**
5     $C := Propagate\,(F, I, \delta)$
6   **if** $C \neq \bot$ **then**
7        $c := \delta(C)$
8   **if** $c = 0$ **then return** UNSAT
9        $Analyze\,(F, I, C, c)$
10   **else**
11       $Decide\,(I, \delta)$
12 **return** SAT

Propagate (F, I, δ)
1 **while** some $C \in F$ is unit $\{\ell\}$ under I **do**
2     $I := I\ell$
3     $\delta(\ell) := \delta(C \setminus \{\ell\})$
4   **for all** clauses $D \in F$ containing $\neg\ell$ **do**
5   **if** $I(D) = \bot$ **then return** D
6 **return** $\bot$

Analyze (F, I, C, c)
1 **if** C contains exactly one literal at decision level c **then**
2     $\ell :=$ literal in C at decision level c
3     $j := \delta(C \setminus \{\ell\})$
4 **else**
5     $D := Learn\,(I, C)$
6     $F := F \wedge D$
7     $\ell :=$ literal in D at decision level c
8     $j := \delta(D \setminus \{\ell\})$
9 pick $b \in [\,j,\ c - 1\,]$
10 **for all** literals $k \in I$ with decision level $> b$ **do**
11   assign k decision level ∞
12   remove k from I
13 $I := I\ell$
14 assign ℓ decision level j

Fig. 4. This is the algorithm for CDCL with chronological backtracking, which differs from its non-chronological backtracking version as follows: Propagated literals ℓ are assigned a decision level which may be lower than the current one (line 3 in Propagate). The conflict level may be lower than the current decision level (line 7 in Search). If the conflicting clause contains only one literal at conflict level, it is used as reason and no conflict analysis is performed (lines 1–3 in Analyze). Picking the backtracking level is usually non-deterministic (line 9 in Analyze). Backtracking involves removing from the trail I all (literals in) $\mathsf{slice}(I, i)$ with $i > b$ (line 12 in Analyze).

7   Implementation

We added chronological backtracking to our SAT solver CADiCaL [4] based on the rules presented in Sect. 4, in essence implementing the algorithm presented in Fig. 4, on top of a classical CDCL solver. This required the following four changes, similar to those described in [10] and implemented in the source code which was submitted by the authors to the SAT 2018 competition [11]. This list is meant to be comprehensive and confirms that the changes are indeed local.

Asserting Level. During unit propagation the decision level $a = \delta(C \backslash \{\ell\})$, also called asserting level, of propagated literals ℓ needs to be computed based on the decision level of all the falsified literals in the reason clause C. This is factored out in a new function called `assignment_level`[1], which needs to be called during assignment of a variable if chronological backtracking is enabled.

Conflict Level. At the beginning of the conflict analysis the current conflict level c is computed in a function called `find_conflict_level` (see footnote 1). This function also determines if the conflicting clause has one or more falsified literals on the conflict level. In the former case we can simply backtrack to backtrack level $b = c - 1$ and use the conflicting clause as reason for assigning that single literal on the conflict level. Even though not described in [10] but implemented in their code, it is also necessary to update watched literals of the conflict. Otherwise important low-level invariants are violated and propagation might miss falsified clauses later. In order to restrict the changes to the conflict analysis code to a minimum, it is then best to backtrack to the conflict level, if it happens to be smaller than the current decision level. The procedure for deriving the learned clause D can then remain unchanged (minimizing the 1st-UIP clause).

Backtrack Level. Then we select the backtrack level b with $j \leqslant b < c$, where j is the minimum backjump level j (the second largest decision level in D) in the function `determine_actual_backtrack_level` (see footnote 1). By default we adopted the heuristic from the original paper [10] to always force chronological backtracking ($b = c - 1$) if $c - j > 100$ (T in [10]) but in our implementation we do not prohibit chronological backtracking initially (C in [10]). Beside that we adopted a variant of reusing the trail [16] as follows. Among the literals on the trail assigned after and including the decision at level $j + 1$ we find the literal k with the largest variable score and backtrack to b with $b + 1 = \delta(k)$.

Flushing. Finally, the last required change was to flush literals from the trail with decision level larger than b but keep those smaller or equal than b. Instead of using an extra data structure (queue) as proposed in [10] we simply traverse the trail starting from block $b + 1$, flushing out literals with decision level larger than b. It is important to make sure that all the kept literals are propagated again (resetting the `propagated` (see footnote 1) level).

---

[1] Please refer to the source code of CADiCaL provided at http://fmv.jku.at/chrono.

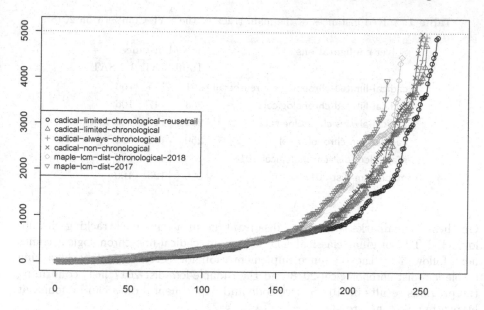

Fig. 5. Cactus plot for benchmarks of the main track of the SAT Competition 2018.

8 Experiments

We evaluated our implementation on the benchmarks from the main track of the SAT Competition 2018 and compare four configurations of CADICAL [4]. We also consider maple-lcm-dist-2017 [17], also called MAPLE_LCM_DIST, which won the main track of the SAT Competition 2017, on which maple-lcm-dist-chronological-2018 [11], also called MAPLE_LCM_DIST_CHRONOBT, is based. We consider the latter as reference implementation for [10]. It won the main track of the SAT Competition 2018 on the considered benchmark set.

The experiments were executed on our cluster where each compute node has two Intel Xeon E5-2620 v4 CPUs running at 2.10 GHz with turbo-mode disabled. Time limit was set to 3600 seconds and memory limit to 7 GB. We used version "0nd" of CADICAL. Compared to the SAT Competition 2018 version [4] it incorporates new phase saving heuristics and cleaner separation between stabilizing and non-stabilizing phases [14]. This gave substantial performance improvements on satisfiable formulae [3]. Nevertheless adding chronological backtracking improves performance even further as the cactus plot in Fig. 5 and Table 1 show.

The default version cadical-limited-chronological-reusetrail is best (preliminary experiments with CADICAL optimized for SAT Race 2019 did not confirm this result though). It uses the limit $C = 100$ to chronologically backtrack if $c - j > 100$ and further reuses the trail as explained in the previous section. For cadical-limited-chronological reusing the trail is disabled and less instances are solved. Quite remarkable is that configuration cadical-always-chronological ranks third, even though it always enforces chronological backtracking ($b = c-1$).

Table 1. Solved instances of the main track of the SAT Competition 2018.

Solver configurations	Solved instances		
	Total	SAT	UNSAT
cadical-limited-chronological-reusetrail	261	155	106
cadical-limited-chronological	253	147	106
cadical-always-chronological	253	148	105
cadical-non-chronological	250	144	106
maple-lcm-dist-chronological-2018	236	134	102
maple-lcm-dist-2017	226	126	100

On these benchmarks there is no disadvantage in always backtracking chronologically! The original classical CDCL variant cadical-non-chronological comes next followed by the reference implementation for chronological backtracking maple-lcm-dist-chronological-2018 and then maple-lcm-dist-2017 last, confirming the previous results in [10]. Source code and experimental data can be found at http://fmv.jku.at/chrono.

9   Conclusion

The success of MAPLE_LCM_DIST_CHRONOBT [11] is quite remarkable in the SAT Competition 2018, since the solver violates various invariants previously considered crucial for CDCL solvers (summarized in Fig. 1). The corresponding paper [10] however was lacking proofs. In this paper we described and formalized a framework for combining CDCL with chronological backtracking. Understanding precisely which invariants are crucial and which are redundant was the main motivation for this paper. Another goal was to empirically confirm the effectiveness of chronological backtracking within an independendent implementation.

Our main contribution is to precisely define the concepts introduced in [10]. The rules of our framework simplify and generalize chronological backtracking. We may relax even more CDCL invariants without compromising the procedure's correctness. For instance first experiments show that during the application of the Unit rule it is not necessary to require that the formula is not falsified by the trail. Similarly, requiring the formula not to be falsified appears to be sufficient for rule Decide (no need to require that there are no units).

Our experiments confirm that combining chronological backtracking and CDCL has a positive impact on solver performance. We have further explored reusing the trail [16] during backjumping, which requires a limited form of chronological backtracking, too. Our experiments also show that performing chronological backtracking exclusively does not degrade performance much and thus for instance has potential to be used in propositional model counting. Furthermore, besides counting, possible applications may be found in SMT and QBF. We further plan to investigate the combination of these ideas with total assignments following [6].

References

1. Artho, C., Biere, A., Seidl, M.: Model-based testing for verification back-ends. In: Veanes, M., Viganò, L. (eds.) TAP 2013. LNCS, vol. 7942, pp. 39–55. Springer, Heidelberg (2013). https://doi.org/10.1007/978-3-642-38916-0_3
2. Audemard, G., Simon, L.: Refining restarts strategies for SAT and UNSAT. In: Milano, M. (ed.) CP 2012. LNCS, vol. 7514, pp. 118–126. Springer, Heidelberg (2012). https://doi.org/10.1007/978-3-642-33558-7_11
3. Biere, A.: CaDiCaL at the SAT Race 2019. In: Proceedings of SAT Race (2019, Submitted)
4. Biere, A.: CaDiCaL, Lingeling, Plingeling, Treengeling and YalSAT entering the SAT competition 2018. In: Proceedings of the SAT Competition 2018 - Solver and Benchmark Descriptions. Department of Computer Science Series of Publications B, vol. B-2018-1, pp. 13–14. University of Helsinki (2018)
5. Blanchette, J.C., Fleury, M., Weidenbach, C.: A verified SAT solver framework with learn, forget, restart, and incrementality. In: Olivetti, N., Tiwari, A. (eds.) IJCAR 2016. LNCS (LNAI), vol. 9706, pp. 25–44. Springer, Cham (2016). https://doi.org/10.1007/978-3-319-40229-1_4
6. Goultiaeva, A., Bacchus, F.: Off the trail: re-examining the CDCL algorithm. In: Cimatti, A., Sebastiani, R. (eds.) SAT 2012. LNCS, vol. 7317, pp. 30–43. Springer, Heidelberg (2012). https://doi.org/10.1007/978-3-642-31612-8_4
7. Marić, F., Janičić, P.: Formalization of abstract state transition systems for SAT. Logical Methods Comput. Sci. **7**(3), 1–37 (2011)
8. Marques-Silva, J.P., Lynce, I., Malik, S.: Conflict-driven clause learning SAT solvers. In: Handbook of Satisfiability, Frontiers in Artificial Intelligence and Applications, vol. 185, pp. 131–153. IOS Press (2009)
9. Marques-Silva, J.P., Sakallah, K.A.: GRASP - a new search algorithm for satisfiability. In: Proceedings of ICCAD 1996, pp. 220–227 (1996)
10. Nadel, A., Ryvchin, V.: Chronological backtracking. In: Beyersdorff, O., Wintersteiger, C.M. (eds.) SAT 2018. LNCS, vol. 10929, pp. 111–121. Springer, Cham (2018). https://doi.org/10.1007/978-3-319-94144-8_7
11. Nadel, A., Ryvchin, V.: Maple LCM dist ChronoBT: featuring chronological backtracking. In: Proceedings of SAT Competition 2018 - Solver and Benchmark Descriptions. Department of Computer Science Series of Publications B, vol. B-2018-1, p. 29. University of Helsinki (2018)
12. Niemetz, A., Preiner, M., Biere, A.: Model-based API testing for SMT solvers. In: Proceedings of SMT 2017, Affiliated with CAV 2017, p. 10 (2017)
13. Nieuwenhuis, R., Oliveras, A., Tinelli, C.: Solving SAT and SAT modulo theories: from an abstract Davis-Putnam-Logemann-Loveland procedure to DPLL(T). J. ACM **53**(6), 937–977 (2006)
14. Oh, C.: Between SAT and UNSAT: the fundamental difference in CDCL SAT. In: Heule, M., Weaver, S. (eds.) SAT 2015. LNCS, vol. 9340, pp. 307–323. Springer, Cham (2015). https://doi.org/10.1007/978-3-319-24318-4_23
15. Oh, C.: Improving SAT solvers by exploiting empirical characteristics of CDCL. Ph.D. Thesis, New York University, Department of Computer Science (2016)

16. van der Tak, P., Ramos, A., Heule, M.: Reusing the assignment trail in CDCL solvers. JSAT **7**(4), 133–138 (2011)
17. Xiao, F., Luo, M., Li, C.M., Manyà, F., Lü, Z.: MapleLRB_LCM, Maple_LCM, Maple_LCM_Dist, MapleLRB_LCMoccRestart, and Glucose-3.0+width in SAT Competition 2017. In: Proceedings of SAT Competition 2017 - Solver and Benchmark Descriptions. Department of Computer Science Series of Publications B, vol. B-2017-1, pp. 22–23. University of Helsinki (2017)

Assessing Heuristic Machine Learning Explanations with Model Counting

Nina Narodytska[1]([✉]), Aditya Shrotri[2], Kuldeep S. Meel[3], Alexey Ignatiev[4,5], and Joao Marques-Silva[4]

[1] VMware Research, Palo Alto, CA, USA
nnarodytska@vmware.com
[2] Rice University, Houston, USA
as128@rice.edu
[3] National University of Singapore, Singapore, Singapore
meel@comp.nus.edu.sg
[4] Faculty of Science, University of Lisbon, Lisbon, Portugal
{aignatiev,jpms}@ciencias.ulisboa.pt
[5] ISDCT SB RAS, Irkutsk, Russia

Abstract. Machine Learning (ML) models are widely used in decision making procedures in finance, medicine, education, etc. In these areas, ML outcomes can directly affect humans, e.g. by deciding whether a person should get a loan or be released from prison. Therefore, we cannot blindly rely on black box ML models and need to explain the decisions made by them. This motivated the development of a variety of ML-explainer systems, including LIME and its successor ANCHOR. Due to the heuristic nature of explanations produced by existing tools, it is necessary to validate them. We propose a SAT-based method to assess the quality of explanations produced by ANCHOR. We encode a trained ML model and an explanation for a given prediction as a propositional formula. Then, by using a state-of-the-art approximate model counter, we estimate the quality of the provided explanation as the number of solutions supporting it.

1 Introduction

The advances in Machine Learning (ML) in recent years explain, to a large extent, the impact and societal significance commonly attributed to Artificial Intelligence (AI). As an indirect consequence, the fast growing range of ML applications includes settings where safety is paramount (e.g. self-driving vehicles), but also settings were humans are directly affected (e.g. finance, medicine, education, and judiciary). Work on improving confidence in ML models includes

This work was supported by FCT grants ABSOLV (PTDC/CCI-COM/28986/2017), FaultLocker (PTDC/CCI-COM/29300/2017), SAFETY (SFRH/BPD/120315/2016), SAMPLE (CEECIND/04549/2017), National Research Foundation Singapore under its AI Singapore Programme AISG-RP-2018-005 and NUS ODPRT Grant R-252-000-685-133.

© Springer Nature Switzerland AG 2019
M. Janota and I. Lynce (Eds.): SAT 2019, LNCS 11628, pp. 267–278, 2019.
https://doi.org/10.1007/978-3-030-24258-9_19

both solutions for verifying properties of these models, but also approaches for explaining predictions, namely in situations where the operation of the ML model is not readily interpretable by a human decision maker. The general field of eXplainable AI (XAI) targets both the development of naturally interpretable ML models (e.g. decision trees or sets), but also the computation of explanations in settings where the ML model is viewed as a black-box, e.g. neural networks, ensembles of classifiers, among others.

The best-known XAI approaches are heuristic-based, and offer so-called *local* explanations, in the sense that the space of feature values is *not* analyzed exhaustively. Among a number of recently proposed approaches for computing local explanations, LIME [24] and its successor ANCHOR [25] represent two successful instantiations. However, and since both LIME and ANCHOR are heuristic-based, a natural question is: *how accurate in practice are heuristic-based explanations?* This paper focuses on ANCHOR [25] (since it improves upon LIME [24]) and proposes a novel approach for assessing the quality of heuristic approaches for computing (local) explanations. For each computed explanation, ANCHOR reports a measure of quality, namely the estimated precision of the explanation, i.e. the percentage of instances where the explanation applies and the prediction matches. Starting from an encoding of the ML model, the explanation computed by ANCHOR, and the target prediction, this paper proposes to use (approximate) model counting for assessing the actual precision of the computed explanation. Concretely, the paper considers Binarized Neural Networks (BNNs) [14], exploits a recently proposed propositional encoding for BNNs [23], and assesses the quality of ANCHOR on well-known datasets. As we demonstrate, the quality of ANCHOR's explanations can vary wildly, indicating that there are datasets where the explanations of ANCHOR are fairly accurate, but also that there are datasets where the explanations of ANCHOR can be rather inaccurate. The somewhat unexpected conclusions of our experimental evaluation offer further evidence to the need of formal techniques in XAI.

2   Preliminaries

Boolean Satisfiability. We assume notation and definitions standard in the area of Boolean Satisfiability (SAT), i.e. the decision problem for propositional logic [4]. Formulas are represented in Conjunctive Normal Form (CNF) and defined over a set of variables $Y = \{y_1, \ldots, y_n\}$. A CNF formula $\mathcal{F}$ is a conjunction of clauses, a clause is a disjunction of literals, and a literal l_i is a variable y_i or its complement $\neg y_i$. A truth assignment is a map from variables to {FALSE, TRUE}. Given a truth assignment, a clause is satisfied iff at least one of its literals is assigned value TRUE. A formula is *satisfied* iff all of its clauses are satisfied. If there is an assignment μ that satisfies formula $\mathcal{F}$, then $\mathcal{F}$ is said to be *satifiable*, and assignment μ is called a *model* of formula $\mathcal{F}$.

CNF encodings of *cardinality constraints*, i.e. constraints of the form $\sum_{i=0}^{n} l_i \circ k$ where $\circ \in \{<, \leq, =, \neq, \geq, >\}$, have been studied extensively, and will be assumed throughout [4]. In this work we employ *reified* cardinality constraints:

$y \Leftrightarrow \sum_{i=0}^{n} l_i \geq k$. We use the full sequential counters encoding [32] to model this constraint in SAT. However, other encodings can be used.

Model Counting. Given a CNF formula, the problem of model counting is to calculate the number satisfying assignments or models. For a formula $\mathcal{F}$, we denote its count by $\#\mathcal{F}$. This problem is complete for the complexity class $\#\mathcal{P}$ [37], which contains the entire polynomial hierarchy [36]. Despite the high complexity, a number of tools for exact model counting have been developed [18,22,28,35], which were shown to work well on certain benchmarks arising in practice. However, for many applications, obtaining exact counts is not necessary and a good approximation often suffices, especially when it leads to better scalability. These requirements have led to the emergence of *approximate* counting approaches [6,12,33] which employ universal hash functions [5] along with specialized SAT solvers [34] for balancing accuracy and scalability. In our experiments, we use a state-of-the-art tool called ApproxMC3 [33] which takes as input a CNF formula $\mathcal{F}$ along with a tolerance value ε and confidence δ, and outputs an approximate count C such that $\Pr[\frac{1}{(1+\varepsilon)}\#\mathcal{F} \leq C \leq (1+\varepsilon)\#\mathcal{F}] \geq 1 - \delta$, where the probability is defined over the random choice of hash functions. A key advantage of ApproxMC3 is that it supports *projected* model counting, i.e. for a formula $\mathcal{F}$ over variable set $Y = Y_1 \cup Y_2$, ApproxMC3 can approximately count the number of assignments to the variables in Y_1, called *sampling set*, such that the formula $\exists Y_2(\mathcal{F})$ evaluates to true. Since we use parsimonious encodings of cardinality constraints in the current work, projection is not strictly required from a correctness perspective. Nevertheless, it greatly speeds up computation as our encoding scheme provides us access to independent support [16], which is specified as sampling set.

Classification Problems. We define the supervised classification problem. We are given a dataset of training samples (i.e. a training set) $\mathcal{E} = \{e_1, \ldots, e_M\}$ over a set of categorical features $\mathcal{F} = \{f_1, \ldots, f_K\}$. Each sample e_i is a pair $\{(v_1^i, \ldots, v_K^i), Q^i\}$, where $(v_1^i, \ldots, v_K^i)$ are values of the corresponding features, $v_i^j \in \mathbb{Z}$, and Q^i determines the class of the sample e_i, $Q^i \in Q$, where Q is a set of possible classes. Note that we assume that all features are categorical, so v_i^js take only discrete values.

Solving a classification problem consists of building a classifier function that maps a sample to its class label, $\mathcal{G} : \{0,1\}^K \rightarrow Q$. Given a training set, a classifier is learned during the training phase so that it captures the relationship between samples and class labels. After the training phase is complete, the classifier is *fixed*. W.l.o.g. we work with a binary classification problem, so that $Q = \{0,1\}$. However, our approach also works for the classification with multiple class labels case, without additional modifications.

Binarized Neural Networks. A binarized neural network (BNN) is a feed-forward network where weights and activations are binary and take values $\{-1,1\}$ [14]. A BNN is built from blocks of layers. Each block performs a number of linear and non-linear transformations such as batch normalization, hyperbolic tangent, and binarization. The output of the last layer is a real-valued vector of size $|Q|$ determines the winner.

3   Explanations for Machine Learning Models

There is a large body of work on generating explanations for ML models [1,2,15,20,21,26,27,30,38]. The main motivation for this line of research is the practical need to interpret, analyze and understand decisions of ML models, as these decisions can affect humans. For example, ML models can be used to decide whether a person should get a loan or be released from prison [17,29]. There are multiple ways to attack the explainability problem depending on our assumptions about the model. One approach is to treat the model as a white box. For example, we can analyze an ML model and extract a (nearly) equivalent interpretable model, e.g. one can distill a neural network into a decision tree model [13,39]. However, this approach has several drawbacks, e.g. a converter needs to be developed for each pair of model classes, the extraction of an equivalent model can be computationally hard, etc. There are also heuristic gradient-based white-box methods [31], but they are mostly designed for computer vision tasks. An alternative approach is to treat the ML model as a black box. Methods that make no assumptions about the underlying ML model are known as model-agnostic explanation generators. Since working with black-box models is challenging, these methods are restricted to finding explanations that hold *locally* in a subspace of the input space. Prominent examples include ML-explainer LIME and its successor ANCHOR [24,25].

In this work, we consider ANCHOR, proposed by Ribeiro *et al.* [25], which is the only system that generates explanations, but also provides quality metrics for them. However, these metrics are obtained purely heuristically and are not guaranteed to be accurate. We take this approach a step further, proposing a rigorous method with guaranteed error bounds. Since this problem is intractable in the general case, we focus on types of ML models that allow a succinct representation as a Boolean formula. For such models, we reformulate the computation of quality metrics as a Boolean logic problem, specifically the problem of determining the number of solutions of a Boolean formula.

3.1   ANCHOR's Heuristic Explanations

We start by describing the concepts behind ANCHOR. We use notations and definitions from [25]. ANCHOR introduces the notion of an anchor explanation, which is *"a rule [over inputs] that sufficiently anchors the prediction locally so that changes to the rest of inputs does not affect the prediction"*. Let $e = ((v_1, \ldots, v_K), e_q)$ be a sample from the dataset $\mathcal{E}$ and let us denote vector $(v_1, \ldots, v_K)$ of input feature values by e_v.

Example 1. Consider a sample e from the *adult* dataset describing characteristics of a person [17]. A sample contains twelve features such as age, sex, race, occupation, marriage status, etc. We consider a binary classifier that predicts whether the income of a person is above or below \$50K. Suppose we have a sample:

$$e = ((\text{Age is } (37, 48], \text{Private, Husband, Sales, White, Male, Married}), > \$50K).$$

Hence, $e_v = (\text{Age is } (37, 48], \text{Private, Husband, Sales, White, Male, Married})$ is the vector of features (some features are undefined) and the class label e_q is $> \$50K$. □

Let A be a set of predicates over input features e_v such that e_v satisfies all predicates in A, denoted $A(e_v) = 1$. An anchor is defined as follows.

Definition 1. *A set of unary predicates A is an anchor for prediction e_q of model $\mathcal{G}$ if*

$$\mathbb{E}_{\mathcal{D}(e'_v | A)}[\mathcal{G}(e'_v) = e_q] \geq \tau, A(e_v) = 1, \tag{1}$$

where $\mathcal{D}(e'_v \mid A)$ denotes samples that satisfy A and τ is a parameter close to 1. In other words, A is an anchor if all samples that match A are most likely classified as e_q.

Example 2. Continuing with our example, the ANCHOR algorithm starts from a sample e and a trained ML model. Assume that it produces an anchor $A = (\text{White, Male})$. The interpretation of this anchor is that if we consider a population of white males then the ML model mostly likely predicts the class $> \$50K$. □

To estimate the quality of the explanation, the precision metric was introduced:

$$prec(A) = \mathbb{E}_{\mathcal{D}(e'_v | A)}[\mathcal{G}(e'_v) = e_q], \tag{2}$$

In other words, the precision metric measures the fraction of samples that are classified as e_q among those that match A. As pointed out in [25], for an arbitrary dataset and a black-box model, *it is intractable to compute this precision directly*, prompting a probabilistic definition of precision $P(prec(A) \geq \tau) \geq 1 - \delta$, where τ and δ are parameters of the algorithm. The problem of finding a high-precision anchor was formulated as a pure-exploration multi-armed bandit problem and solved using a dedicated algorithm. In our example, the precision of A is 0.99%.

Indeed, for an arbitrary model computing the exact precision value is intractable. However, we argue that for a rich class of models we can compute the precision metric exactly or get a good approximation of this value.

3.2 Model-Counting Based Assessment of Explanations

In this section, we discuss how to evaluate the quality of explanations produced by ANCHOR for a subclass of ML models. Our main idea is to compute the precision metric directly using an logical representation of the ML model, following the ideas of [15]. Here we focus on ML models that admit an equivalent CNF encoding. For such models, we can formulate the problem of computing the precision metric as a model counting problem. To obtain such formulation, we need to encode the following three components as a CNF: (a) the ML model, (b) the anchor and (c) the set of valid inputs.

First, we define a subclass of ML models suitable for our approach. Consider an ML model $\mathcal{G}$ that maps an input vector x to an output vector o, $o = \mathcal{G}(x)$. The prediction of the model is given by ARGMAX(o). For simplicity, we consider a binary classification problem, so we have two classes, hence, o is a 2-dimensional vector. We require that there exists a CNF representation of the model, denoted BIN$\mathcal{G}(x, s)$, that simulates the model in the following sense: all models of BIN$\mathcal{G}(x, s)$ such that $s = 0$ are exactly the samples that are classified as class 0 by $\mathcal{G}$. Likewise, all models of BIN$\mathcal{G}(x, s)$ such that $s = 1$ are samples that are classified as class 1 by $\mathcal{G}$. For now, we assume that such models exist and consider a concrete classifier in the next session. Second, we consider a CNF encoding of an anchor A. We recall that A is a set of unary predicates over input categorical features, so it can be easily translated to CNF. We denote a CNF encoding of A as BINA. Third, we need a declarative representation of the space of samples. We require that a set of valid samples can be defined using a propositional formula that we denote by BIN$\mathcal{D}(x)$. The assumption that the input space can be represented as a logical formula is standard in the line of work on verification of neural networks [11,19]. However, it might not hold for some datasets, like images. An encoding of a valid instance space is an interesting research direction, which is outside the scope of this work. There are a number of natural cases that allow such representation. First, we can assume that all possible combinations of input features are valid. For instance, in Example 1, any combination of features is a plausible sample. Second, we may want to restrict the considered inputs, e.g. we would like to consider all inputs that are close to a given sample e w.r.t. a given distance measure. Finally, a user might have a set of preferences over samples that are expressible as a Boolean formula.

Now we put all three components together into the following formula:

$$\mathcal{P}_A(x, s) = \text{BIN}\mathcal{G}(x, s) \wedge \text{BINA}(x) \wedge \text{BIN}\mathcal{D}(x). \tag{3}$$

The precision of an anchor A for $\mathcal{G}$ and e_q defined by (2) can be written as:

$$M = \frac{\#(\mathcal{P}_A(x, s) \wedge s = e_q)}{\#(\mathcal{P}_A(x, s))}. \tag{4}$$

In other words, M measures the fraction of models that are classified as e_q among those that match A and satisfy $\mathcal{D}$. Hence, this is exactly the fraction of models that defines the precision of the anchor A in (2). In practice, even for small ML models exact model counting is not feasible. Therefore, we use an efficient approximate model counting algorithm called ApproxMC3.

4   Encoding of Binarized Neural Networks

Let us discuss BNNs, which is our underlying machine learning model. A BNN can be described in terms of blocks of layers that map binary vectors to binary vectors. Hence, we define a *block* of binarized neural network (referred to as BLOCK) as a function that maps an input x to an output x', i.e. BLOCK : $\{-1,1\}^n \rightarrow \{-1,1\}^m$. The last block has a different structure OUTPUT : $\{-1,1\}^n \rightarrow \mathbb{R}^2$. Each BLOCK takes an input vector x and applies three transformations: LIN, BN and BIN[1]. Table 1 shows transformations of internal and output blocks in detail. It was shown in [23], that BLOCK can be encoded as a system of reified cardinality constraints that can be translated into a Boolean formula efficiently [32]. We recall the main idea of the encoding.

Encoding of BLOCK. BLOCK applies three transformations: LIN, BN and BIN to an input vector x. The main insight here is that instead of applying these transformations sequentially, we can consider a composition of these functions. Hence, we relate input and output of a block as follows.

$$x' = \begin{cases} 1, & \text{if } \alpha \left(\frac{(Ax+b)-\mu}{\sigma} \right) + \gamma \geq 0 \\ -1, & \text{otherwise} \end{cases}$$

Next we note that we can re-group the inequality condition in such a way that the left side of the inequality must take the integer value and the right side is a real value. Namely, assuming $\alpha > 0$ ($\alpha < 0$ is similar), we rewrite inequality condition as $Ax \geq \frac{-\gamma\sigma}{\alpha} - b + \mu$. Note that Ax is an integer vector and $\frac{-\gamma\sigma}{\alpha} - b + \mu$ is a real-valued vector. Hence, we can perform rounding of the right hand side safely. This gives a relation that contains only integers, so it can be encoded into CNF efficiently.

Table 1. Structure of internal and output blocks which, stacked together, form a BNN.

Structure of kth internal block, BLOCK$_k$: $\{-1,1\}^{n_k} \rightarrow \{-1,1\}^{n_{k+1}}$ on input $x_k \in \{-1,1\}^{n_k}$
LIN $\quad\quad\quad y = A_k x_k + b_k$, where $A_k \in \{-1,1\}^{n_{k+1} \times n_k}$ and $b_k \in \mathbb{R}^{n_{k+1}}$
BN $\quad z_i = \alpha_{k_i} \left(\frac{y_i - \mu_{k_i}}{\sigma_{k_i}} \right) + \gamma_{k_i}$, where $y = (y_1, \ldots, y_{n_{k+1}})$, and $\alpha_{k_i}, \gamma_{k_i}, \mu_{k_i}, \sigma_{k_i} \in \mathbb{R}^{n_{k+1}}$
BIN $\quad\quad x_{k+1} = \text{sign}(z)$ where $z = (z_1, \ldots, z_{n_{k+1}}) \in \mathbb{R}^{n_{k+1}}$ and $x_{k+1} \in \{-1,1\}^{n_{k+1}}$

Structure of output block, OUTPUT : $\{-1,1\}^{n_d} \rightarrow \{0,1\}$ on input $x_d \in \{-1,1\}^{n_d}$
LIN $\quad\quad\quad\quad w = A_d x_d + b_d$, where $A_d \in \{-1,1\}^{2 \times n_d}$ and $b_d \in \mathbb{R}^2$
BN $\quad\quad o = \alpha_o \left(\frac{w - \mu_o}{\sigma_o} \right) + \gamma_o$, where $w = (w_1, w_2)$, and $\alpha_o, \gamma_o, \mu_o, \sigma_o \in \mathbb{R}^2$

---

[1] In the training phase, there is an additional *hard tanh* layer after batch normalization but it is redundant in the inference phase.

Encoding of OUTPUT. For our purpose, we only need to know whether $o_1 \geq o_2$. We introduce a Boolean variable s that captures this relation. Namely, $s = 0$ if $\alpha_{o_1} \left(\frac{w_1 - \mu_{o_1}}{\sigma_{o_1}} \right) + \gamma_{o_1} \geq \alpha_{o_2} \left(\frac{w_2 - \mu_{o_2}}{\sigma_{o_2}} \right) + \gamma_{o_2}$ and $s = 1$ otherwise. As $w = A_d x_d + b_d$, we can re-group the inequality condition the same way as for the BLOCK to obtain a reified constraint over integers.

CNF Encoding. Based on the encoding of blocks BLOCK and OUTPUT, the network can be represented as a Boolean formula: $\text{BINBNN}(x, s) \equiv \bigwedge_{i=1}^{D} \text{BINBLOCK}_d(x^{d-1}, x^d) \wedge \text{BINO}(x_D, s)$, where $\text{BINBLOCK}_d(x^{d-1}, x^d)$ encodes the dth block BLOCK with an input x_d and an output x_{d+1}, $d \in [1, D]$, BINO is a Boolean encoding of the last set of layers. Note that $\text{BINBNN}(x, s)$ satisfies the requirements for ML model encoding that we stated in the previous section.

5   Experimental Results

We present our evaluation to assess the quality of ANCHOR's precision metric.

Datasets. We consider three well-known publicly available datasets that were used in [25]. These datasets were processed the same way[2] as in [25]. The *adult* dataset [17] was collected by the Census bureau, where it was used to predict whether a given adult person earns more than \$50K a year depending on various attributes. The *lending* dataset was used to predict whether a loan on the Lending Club website will be granted. The *recidivism* dataset was used to predict recidivism for individuals released from North Carolina prisons in 1978 and 1980 [29].

ML Model. We trained a BNN on each benchmark with three internal BLOCKs and an OUTPUT block. There are 25 neurons per layer on average. We use a standard onehot encoding to convert categorical features to Boolean inputs for BNN. We split the input data into training (80% of data) and test (20%) sets. The accuracy of BNN is 0.82 for the *adult*, 0.83 for *lending* and 0.63 for *recidivism*, which matches XGBoost [8] used in [25]. To generate anchors, we used the native implementation of ANCHOR. The sizes of CNF encodings are (a) $50K$ variables and $202K$ clauses for *adult*, (b) $21K$ variables and $80K$ clauses for *lending*, (c) $75K$ variables and $290K$ clauses for *recidivism*.

Quality Assessment. We performed two sets of experiments depending on constraints on the sample space. First, we consider the case of an unrestricted set of samples. Second, we restrict samples to a local neighborhood of a sample we started with to generate an anchor. For the first case, the rejection sampling based algorithm by Dagum *et al.* [9] can be used for measuring accuracy up to desired tolerance and confidence, very efficiently. However, in general there can be additional constraints over the input, such as for encoding some notion of neighborhood of a sample, or for incorporating a set of user preferences. In such cases, rejection sampling based approaches fail. To ensure wider applicability, we

---

[2] https://github.com/marcotcr/anchor-experiments.

use the tool ApproxMC3, which can perform projected counting over the input variables of arbitrary constraints encoded as CNF formulas.

We invoke ApproxMC3 with the default tolerance and confidence ($\varepsilon = 0.8$ and $\delta = 0.2$), which has been the standard in earlier works [6,7,33]. Studies have reported that the error observed in practice ($\varepsilon_{obs} = 0.037$) is an order of magnitude better than the theoretical guarantee of 0.8 [33], which is similar to what we found in our preliminary experiments even when considering the quotient of two approximate counts as in (4). This obviates the need for stronger (ε, δ) as input, as the default settings are sufficiently accurate in practice for our purposes.

Unrestricted Set of Samples. We consider the case of an unrestricted set of samples. We compute high-precision anchors for 300 randomly selected inputs of each test dataset with a default value of $\tau = 0.95$. On average, the precision metric reported by ANCHOR is high, over 0.99. For each anchor, we perform approximate model counting of solutions according to (4). Then, we compute the discrepancy between the precision metric returned by ANCHOR and the estimate computed by our method.

Figure 1a shows our results. Each cactus plot shows the precision that is returned by ANCHOR and is computed with ApproxMC3 for the corresponding dataset. ANCHOR's precision estimates are around 0.99 for the three datasets and so the corresponding lines in Fig. 1 merge. Figure 1 shows that ANCHOR's estimates of the precision metric are good for the *lending* dataset. On average the discrepancy is 0.13 in this set. In contrast, the discrepancy was high in the *adult* dataset, 0.34 on average. For the *recidivism* dataset the mean discrepancy is 0.25. Overall, we can conclude that the metric produced by ANCHOR is more on the optimistic side, as we cannot confirm 0.99 precision.

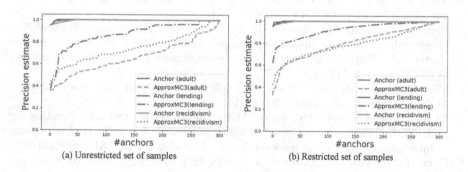

(a) Unrestricted set of samples   (b) Restricted set of samples

Fig. 1. The precision metric estimates for three datasets.

Constrained Set of Samples. Second, we consider the case when we want to restrict the space of samples. One interesting case is to consider how good the anchor are among the samples that are close to the original sample e we started with. We define a neighborhood of e given an anchor A as all samples that match A and differ from e in at most 50% of the remaining features. We expect that ANCHOR performs better in the local neighborhood of e. Figure 1b shows our results. We obtain that on average the discrepancy is 0.08 for the *lending* dataset, 0.2 for the *adult* dataset and 0.21 for *recidivism*. So, we see a significant improvement for the *adult* dataset (the discrepancy dropped from 0.34 to 0.2) and minor improvements for the other two sets.

6   Conclusions and Future Work

This paper investigates the quality of heuristic (or local) explanations computed by a recent XAI tool, namely ANCHOR [25]. Although for some datasets the precisions claimed by Anchor can be confirmed, it is also the case that for several other datasets Anchor estimates precisions that are unrealistically high. There are a number of possible directions for future work. For example, it is interesting to consider powerful model compilation techniques [10] that can compute the exact number of solutions and, therefore, the precision metrics exactly. The main challenge here is to build an effective compiler from BNNs to BDDs [3]. Another direction is to investigate the application of the ideas in this paper to other XAI tools and consider other ML models that can be translated to SAT.

References

1. Adebayo, J., Gilmer, J., Muelly, M., Goodfellow, I.J., Hardt, M., Kim, B.: Sanity checks for saliency maps. In: NeurIPS, pp. 9525–9536 (2018)
2. Alvarez-Melis, D., Jaakkola, T.S.: Towards robust interpretability with self-explaining neural networks. In: NeurIPS, pp. 7786–7795 (2018)
3. Shih, A., Darwiche, A., Choi, A.: Verifying binarized neural networks by local automaton learning. In: VNN (2019)
4. Biere, A., Heule, M., van Maaren, H., Walsh, T. (eds.): Handbook of Satisfiability, Frontiers in Artificial Intelligence and Applications, vol. 185. IOS Press (2009)
5. Carter, J.L., Wegman, M.N.: Universal classes of hash functions. In: Proceedings of STOC, pp. 106–112. ACM (1977)
6. Chakraborty, S., Meel, K.S., Vardi, M.Y.: A scalable approximate model counter. In: Proceedings of CP, pp. 200–216 (2013)
7. Chakraborty, S., Meel, K.S., Vardi, M.Y.: Improving approximate counting for probabilistic inference: from linear to logarithmic sat solver calls. In: Proceedings of International Joint Conference on Artificial Intelligence (IJCAI), July 2016
8. Chen, T., Guestrin, C.: XGBoost: a scalable tree boosting system. In: KDD, pp. 785–794. ACM (2016)
9. Dagum, P., Karp, R., Luby, M., Ross, S.: An optimal algorithm for Monte Carlo estimation. SIAM J. Comput. **29**(5), 1484–1496 (2000)
10. Darwiche, A., Marquis, P.: A knowledge compilation map. J. Artif. Intell. Res. **17**, 229–264 (2002). https://doi.org/10.1613/jair.989
11. Dreossi, T., Ghosh, S., Sangiovanni-Vincentelli, A.L., Seshia, S.A.: A formalization of robustness for deep neural networks. CoRR abs/1903.10033 (2019). http://arxiv.org/abs/1903.10033

12. Ermon, S., Gomes, C.P., Sabharwal, A., Selman, B.: Taming the curse of dimensionality: discrete integration by hashing and optimization. In: Proceedings of ICML, pp. 334–342 (2013)
13. Frosst, N., Hinton, G.E.: Distilling a neural network into a soft decision tree. In: Besold, T.R., Kutz, O. (eds.) Proceedings of the First International Workshop on Comprehensibility and Explanation in AI and ML 2017 co-located with 16th International Conference of the Italian Association for Artificial Intelligence (AI*IA 2017), Bari, Italy, 16–17 November 2017. CEUR Workshop Proceedings, vol. 2071. CEUR-WS.org (2017)
14. Hubara, I., Courbariaux, M., Soudry, D., El-Yaniv, R., Bengio, Y.: Binarized neural networks. In: NIPS, pp. 4107–4115 (2016)
15. Ignatiev, A., Narodytska, N., Marques-Silva, J.: Abduction-based explanations for machine learning models. In: AAAI (2019)
16. Ivrii, A., Malik, S., Meel, K.S., Vardi, M.Y.: On computing minimal independent support and its applications to sampling and counting. Constraints **21**(1), 41–58 (2016). https://doi.org/10.1007/s10601-015-9204-z
17. Kohavi, R.: Scaling up the accuracy of naive-bayes classifiers: a decision-tree hybrid. In: KDD, pp. 202–207 (1996)
18. Lagniez, J.M., Marquis, P.: An improved decision-DNNF compiler. In: IJCAI, pp. 667–673 (2017)
19. Leofante, F., Narodytska, N., Pulina, L., Tacchella, A.: Automated verification of neural networks: advances, challenges and perspectives. CoRR abs/1805.09938 (2018). http://arxiv.org/abs/1805.09938
20. Li, O., Liu, H., Chen, C., Rudin, C.: Deep learning for case-based reasoning through prototypes: a neural network that explains its predictions. In: AAAI, pp. 3530–3537 (2018)
21. Montavon, G., Samek, W., Müller, K.: Methods for interpreting and understanding deep neural networks. Digital Sig. Process. **73**, 1–15 (2018)
22. Muise, C., McIlraith, S.A., Beck, J.C., Hsu, E.: DSHARP: Fast d-DNNF Compilation with sharpSAT. In: Canadian Conference on Artificial Intelligence (2012)
23. Narodytska, N., Kasiviswanathan, S.P., Ryzhyk, L., Sagiv, M., Walsh, T.: Verifying properties of binarized deep neural networks. In: AAAI, pp. 6615–6624 (2018)
24. Ribeiro, M.T., Singh, S., Guestrin, C.: Why should I trust you?: explaining the predictions of any classifier. In: KDD, pp. 1135–1144 (2016)
25. Ribeiro, M.T., Singh, S., Guestrin, C.: Anchors: high-precision model-agnostic explanations. In: AAAI (2018)
26. Ross, A.S., Doshi-Velez, F.: Improving the adversarial robustness and interpretability of deep neural networks by regularizing their input gradients. In: AAAI, pp. 1660–1669 (2018)
27. Ross, A.S., Hughes, M.C., Doshi-Velez, F.: Right for the right reasons: training differentiable models by constraining their explanations. In: IJCAI, pp. 2662–2670 (2017)
28. Sang, T., Beame, P., Kautz, H.: Performing Bayesian inference by weighted model counting. In: Proceedings of AAAI, pp. 475–481 (2005)
29. Schmidt, P., Witte, A.D.: Predicting recidivism in north carolina, 1978 and 1980. Inter-University Consortium for Political and Social Research (1988). https://www.ncjrs.gov/App/Publications/abstract.aspx?ID=115306
30. Shih, A., Choi, A., Darwiche, A.: A symbolic approach to explaining Bayesian network classifiers. In: IJCAI, pp. 5103–5111 (2018)

31. Simonyan, K., Vedaldi, A., Zisserman, A.: Deep inside convolutional networks: visualising image classification models and saliency maps. CoRR abs/1312.6034 (2013). http://arxiv.org/abs/1312.6034
32. Sinz, C.: Towards an optimal CNF encoding of Boolean cardinality constraints. In: CP, pp. 827–831 (2005)
33. Soos, M., Meel, K.S.: Bird: Engineering an efficient CNF-XOR SAT solver and its applications to approximate model counting. In: Proceedings of AAAI Conference on Artificial Intelligence (AAAI), Jan 2019
34. Soos, M., Nohl, K., Castelluccia, C.: Extending SAT solvers to cryptographic problems. In: Kullmann, O. (ed.) SAT 2009. LNCS, vol. 5584, pp. 244–257. Springer, Heidelberg (2009). https://doi.org/10.1007/978-3-642-02777-2_24
35. Thurley, M.: SharpSAT: counting models with advanced component caching and implicit BCP. In: Proceedings of SAT, pp. 424–429 (2006)
36. Toda, S.: PP is as hard as the polynomial-time hierarchy. SIAM J. Comput. **20**(5), 865–877 (1991)
37. Valiant, L.: The complexity of enumeration and reliability problems. SIAM J. Comput. **8**(3), 410–421 (1979)
38. Wu, M., Hughes, M.C., Parbhoo, S., Zazzi, M., Roth, V., Doshi-Velez, F.: Beyond sparsity: tree regularization of deep models for interpretability. In: AAAI, pp. 1670–1678 (2018)
39. Zhang, Q., Yang, Y., Wu, Y.N., Zhu, S.: Interpreting CNNs via decision trees. CoRR abs/1802.00121 (2018), http://arxiv.org/abs/1802.00121

Syntax-Guided Rewrite Rule Enumeration for SMT Solvers

Andres Nötzli[1]([✉]) [iD], Andrew Reynolds[2] [iD], Haniel Barbosa[2] [iD],
Aina Niemetz[1] [iD], Mathias Preiner[1] [iD], Clark Barrett[1] [iD], and Cesare Tinelli[2] [iD]

[1] Stanford University, Stanford, USA
noetzli@cs.stanford.edu
[2] University of Iowa, Iowa City, USA

Abstract. The performance of modern Satisfiability Modulo Theories (SMT) solvers relies crucially on efficient decision procedures as well as static simplification techniques, which include large sets of rewrite rules. Manually discovering and implementing rewrite rules is challenging. In this work, we propose a framework that uses enumerative syntax-guided synthesis (SyGuS) to propose rewrite rules that are not implemented in a given SMT solver. We implement this framework in CVC4, a state-of-the-art SMT and SyGuS solver, and evaluate several use cases. We show that some SMT solvers miss rewriting opportunities, or worse, have bugs in their rewriters. We also show that a variation of our approach can be used to test the correctness of a rewriter. Finally, we show that rewrites discovered with this technique lead to significant improvements in CVC4 on both SMT and SyGuS problems over bit-vectors and strings.

1 Introduction

Developing state-of-the-art Satisfiability Modulo Theories (SMT) solvers is challenging. Implementing only basic decision procedures is usually not enough, since in practice, many problems can only be solved after they have been simplified to a form for which the SMT solver is effective. Typically, such simplifications are implemented as a set of *rewrite rules* applied by a solver component that we will call the *rewriter*. Depending on the theory, optimizing the rewriter can be as important as optimizing the decision procedure. Designing an effective and correct rewriter requires extensive domain knowledge and the analysis of many specific problem instances. New rewrite rules are often only introduced when a new problem requires a particular simplification.

SMT rewriters also have other applications such as accelerating enumerative syntax-guided synthesis (SyGuS). The SyGuS Interchangeable Format [6] uses a subset of the theories defined in SMT-LIB [11] to assign meaning to pre-defined function symbols. Because a term t is equivalent to its *rewritten form* $t\downarrow$, an enumerative SyGuS solver can limit its search space to rewritten terms only. This significantly prunes the search.

Rewriters are typically developed manually by SMT experts. Because this process is difficult and error-prone, automating parts of this process is extremely

© Springer Nature Switzerland AG 2019
M. Janota and I. Lynce (Eds.): SAT 2019, LNCS 11628, pp. 279–297, 2019.
https://doi.org/10.1007/978-3-030-24258-9_20

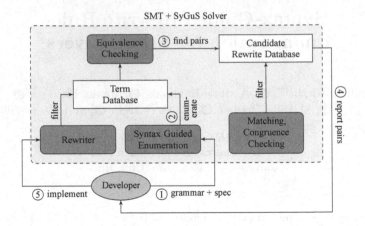

Fig. 1. Overview of our workflow for the development of rewriters.

useful. In this paper, we propose a partially automated workflow that increases the productivity of SMT solver developers by systematically identifying rewriting opportunities that the solver misses. We leverage the common foundations of SyGuS and SMT solving to guide developers in the analysis and implementation of the rewriter, independently of the theory under consideration.

Figure 1 shows an overview of our proposed workflow. In Step 1, the developer provides a *grammar* and a *specification* as inputs (see Sect. 5 for examples). This input describes the class of terms that the developer is interested in targeting in the rewriter. In Step 2, we use previous techniques (implemented in the SyGuS module of the SMT solver CVC4 [9]) to efficiently enumerate target terms into a *term database*. In Steps 3 and 4, we pair those terms together to form a *candidate rewrite database*. A subset of this database is then reported to the developer as a set of *unoriented* pairs $t_1 \approx s_1, \ldots, t_n \approx s_n$ with the following key properties:

1. Terms t_i and s_i were inferred to be equivalent based on some criteria in Step 3.
2. $t_i\downarrow$ is not equal to $s_i\downarrow$; i.e., the rewriter does not treat t_i and s_i as equivalent.

We can interpret these pairs as rewrites by picking an orientation ($t_i \rightsquigarrow s_i$ or $s_i \rightsquigarrow t_i$) and interpreting the free variables on the left-hand side as place-holders that match arbitrary subterms. For example, $x + 0 \approx x$ can be interpreted as a rewrite $x + 0 \rightsquigarrow x$ that rewrites matching terms such as $y \cdot z + 0$ to $y \cdot z$.

The set of pairs can be understood as a *to do* list of rewrites that are currently missing in the rewriter. In Step 5, based on this list, the developer can extend the rewriter to incorporate rules for these unhandled equivalences. We have found that our workflow is most effective when candidate rewrite rules act as hints to inspire developers, who, through creativity and careful engineering, are subsequently able to improve the rewriter. In our experience, this workflow leads to a positive feedback loop. As the developer improves the rewriter, term enumeration produces fewer redundant terms, thus yielding more complex terms on the next iteration. As a result, the candidate rewrite rules become more complex as well. Since our workflow is *partially* automated, there are no restrictions on the complexity of the rewrites implemented. As we illustrate in Sect. 5.1, some rewrites can only be defined with an expressive language, and fully automated approaches tend to restrict themselves to a relatively simple form of rewrites.

Contributions

- We present a novel theory-independent workflow for generating candidate rewrite rules, based on a user-provided grammar, that uses the solver under development as an oracle for determining the missing rules.
- We propose techniques for efficient equivalence checking between terms and for filtering to reduce redundancy in the candidate rewrite rules.
- We introduce several metrics for measuring coverage of the rewrite rules implemented by SMT solvers and demonstrate their impact in practice.
- We show that our workflow is highly effective both for discovering shortcomings in state-of-the-art SMT solvers, and as means of improving the performance of our own solver CVC4 on a wide range of SyGuS and SMT benchmarks.

We discuss our approach for choosing suitable grammars (Step 1) in Sect. 5.1. We describe Steps 2–4 of Fig. 1 in Sects. 2, 3, and 4 respectively. In Sect. 5, we report on our experience with the proposed workflow and discuss using the workflow to test other SMT solvers and to improve our confidence in our implementation. Finally, we evaluate the impact of the rewrites on solving performance. We discuss related work in Sect. 6 and future work in Sect. 7.

Preliminaries. We use standard notions of typed (higher-order) logic, including formula, term, quantifier, etc. We recall a few definitions here. A *substitution* $\sigma = \{\bar{x} \mapsto \bar{t}\}$ is a finite map from variables to terms of the same type, and $s\sigma$ denotes the result of replacing all occurrences of $\bar{x}$ in s by $\bar{t}$. A *theory* T is a pair (Σ, I), where Σ is a signature (a set of types and function symbols), and I is a set of Σ-interpretations, called the *models* of T. We assume that Σ contains the equality relation $\approx$ for all types τ from Σ, interpreted as the identity relation. A formula is T-*satisfiable* (resp., T-*valid*) if it is satisfied by some (resp., all) models of T. Terms t_1 and t_2 are T-*equivalent*, written $t_1 \approx_T t_2$, if the formula $t_1 \approx t_2$ is T-valid.

2   Syntax-Guided Synthesis for Term Enumeration

Step 2 of the workflow in Fig. 1 is to enumerate terms from a given grammar. For this task, we leverage previous work on enumerative approaches to the SyGuS problem [25].

Syntax-Guided Synthesis. A *SyGuS problem* for an n-ary (first-order) function f in a background theory T consists of: *(i)* a set of *syntactic restrictions*, given by a grammar $\mathcal{R}$; and *(ii)* a set of *semantic restrictions*, a *specification*, given by a T-formula of the form $\exists f. \forall \bar{x}. \varphi[f, \bar{x}]$, where φ is typically a quantifier-free formula over the (second-order) variable f and the first-order variables $\bar{x} = (x_1, ..., x_n)$. A grammar $\mathcal{R}$ consists of an initial symbol s_0, a set S of *non-terminal symbols*, where $s_0 \in S$, and a set R of rules $s \to t$, with $s \in S$, and where t is a term built from symbols of S, free variables, and symbols in the signature of T—with the latter two acting as *terminal symbols*. The set of rules R defines a rewrite relation over terms s and t, denoted by $\to$. A term t is *generated* by $\mathcal{R}$ if $s_0 \to^* t$, where $\to^*$ is the reflexive-transitive closure of $\to$ and t contains no symbols from S. A *solution* for f is a closed lambda term $\lambda \bar{y}.t$ of the same type as f such that t is generated by $\mathcal{R}$ and $\forall \bar{x}. \varphi[\lambda \bar{y}.t, \bar{x}]$ is T-valid.

An *enumerative* SyGuS solver consists of a *candidate solution generator*, which produces a stream of terms t_i in the language generated by grammar $\mathcal{R}$, and a *verifier*, which, given a candidate solution t_i, checks whether $\forall \bar{x}.\varphi[\lambda \bar{y}.t_i, \bar{x}]$ is T-valid. The solver terminates as soon as it generates a candidate solution t_i that the verifier accepts. In practice, most state-of-the-art enumerative SyGuS solvers [7,33] are implemented on top of an SMT solver with support for quantifier-free T-formulas, which can be used as a verifier in this approach. Our solver CVC4 acts as both the candidate solution generator and the verifier [26].

Generating Rewrite Rules With SyGuS. In the context of our proposed workflow, an enumerative SyGuS solver can be used as a term generator in Step 2 from Fig. 1. In particular, it can be used to produce *multiple* solutions to the SyGuS problem specified in Step 1, where each solution is added to the term database. Recall that free variables in the solutions are interpreted as place-holders for arbitrary subterms. Thus, the arguments of f determine the number and types of the place-holders in the generated rewrites. A SyGuS specification $\exists f. \forall \bar{x}. \varphi[f, \bar{x}]$ acts as a filtering mechanism to discard terms that should not be included in rewrite rules. If we do not wish to impose any semantic restrictions, the common case in our workflow, we can use the specification $\exists f. \top$, which causes the enumeration of all terms that meet the syntactic restrictions.

To achieve high performance, SyGuS solvers implement techniques that limit the enumeration of equivalent candidate solutions in order to prune the search space. The rationale for this is that if $\forall \bar{x}.\varphi[\lambda \bar{y}.t_1, \bar{x}]$ is not T-valid for a candidate t_1, then it is fruitless to consider any candidate t_2 if $t_1 \approx_T t_2$ since $\forall \bar{x}. \varphi[\lambda \bar{y}.t_2, \bar{x}]$ will not be T-valid either. CVC4 uses its own rewriter as an (incomplete but fast) T-equivalence oracle, where the syntactic equality of $t_1{\downarrow}$ and $t_2{\downarrow}$ implies, by the soundness of the rewriter, that $t_1 \approx_T t_2$ [27]. When the SyGuS solver of CVC4 discovers the equivalence of two terms t_1 and t_2, it generates constraints to ensure that subsequent solutions only include either t_1 or t_2 as a subterm. As a consequence, since CVC4 never generates two candidate solutions that are identical up to rewriting, a better rewriter leads to a more efficient term enumeration. This also ensures that the term database generated in Step 2 in Fig. 1 contains no distinct terms s and t such that $s{\downarrow} = t{\downarrow}$, which, in turn, also ensures that no existing rewrites are considered as candidates.

3   Equivalence Checking Techniques for Rewrite Rules

In Step 3 in Fig. 1, we are given a database of terms, which may contain free variables $\bar{y}$ corresponding to the arguments of the function to be synthesized. The term database, a set of terms D, is generated by syntax-guided enumeration. From this set, we generate a set of (unoriented) pairs of the form $t_1 \approx s_1, \ldots, t_n \approx s_n$ where for each $i = 1, \ldots, n$, terms t_i and s_i are in D and meet some criterion for equivalence. We call such pairs *candidate rewrite rules*.

Our techniques apply to any background theory with a distinguished set of *values* for each of its types, i.e., variable-free terms for denoting the elements of that type, e.g. (negated) numerals $(-)n$ for an integer type. We assume that the initial rewriter reduces any variable-free term to a value, e.g., $4 - 5$ to -1. These assumptions hold for the initial rewriter and the theories in our evaluation.

A naïve way to find candidate rewrite rules is to consider each pair of distinct terms $s[\bar{y}], t[\bar{y}] \in D$, check the satisfiability of $\exists \bar{y}. t \not\approx s$, and, if unsatisfiable, include $t \approx s$ as a candidate. However, this can be inefficient and may even be infeasible for some theories.[1] To mitigate these issues, we have developed techniques based on evaluating terms on a set of *sample points*, i.e. tuples of values $\bar{c}$ of the same type as $\bar{y}$ above. We describe those techniques below.

We compute an equivalence relation E over the terms in our (evolving) term database D. Two terms t_i, t_j are related in E if for every sample point $\bar{c}$ they have the same evaluation, i.e. $(t_i\{\bar{y} \mapsto \bar{c}\})\!\downarrow = (t_j\{\bar{y} \mapsto \bar{c}\})\!\downarrow$. While equivalence in E does not entail T-equivalence, terms disequivalent in E are guaranteed to be T-disequivalent. To see how E evolves, let $\{r_1, \ldots, r_n\} \subseteq D$ be a set of *representative* terms from the equivalence relation E containing exactly one term from each equivalence class. For each new term t added to D, we either *(i)* determine t is equivalent to some r_i, output $t \approx r_i$ as a candidate rewrite rule, and add t to the equivalence class of r_i; or *(ii)* determine t is not equivalent to any of $r_1, \ldots, r_n$, i.e. for each r_i there is at least one sample point on which the evaluations differ, and add $\{t\}$ as an equivalence class. Thus, each equivalence class $\{t_1, \ldots, t_n\}$ of E is such that, for each $i = 1, \ldots, n$, a pair of the form $t_i \approx t_j$ has been generated for some $j \neq i$. In other words, E is the transitive closure of the set of pairs generated so far.

To optimize the evaluation of terms on sample points we rely on a *lazy evaluation trie*. This data structure maintains a list $P = [\bar{c}_1, \ldots, \bar{c}_n]$ of sample points, all of the same type as $\bar{y}$. It indexes a term t by its *evaluation sequence* on the set of sample points P, i.e., term t is indexed by a sequence of the form $[(t\{\bar{y} \mapsto \bar{c}_1\})\!\downarrow, \ldots, (t\{\bar{y} \mapsto \bar{c}_n\})\!\downarrow]$. Due to our assumptions, each term in this list is a value. For example, if $\bar{y} = (y_1, y_2)$ and $P = [(0, 1), (3, 2), (5, 5)]$, then the term $y_1 + 1$ is indexed by the list $[1, 4, 6]$. When a new term t is added to D, it is evaluated on each of the points in P. If the resulting sequence is already in the trie, t is added to the equivalence class of the term indexed by that sequence. If not, the new sequence is added to the trie and t becomes a singleton equivalence class. This guarantees that each representative from E is indexed to a different

---

[1] E.g. for checks in the theory of strings with length whose decidability is unknown [17].

location in this trie. The technique can be made more efficient by performing certain evaluations *lazily*. In particular, it is sufficient to only use a *prefix* of the above sequence, provided that the prefix suffices to show that t is distinct from all other terms in the trie. For the previous example, if $y_1 + 1$ and $y_1 + y_2$ were the only two terms in the trie, they would be indexed by $[1, 4]$ and $[1, 5]$ respectively, since the second sample point $(3, 2)$ shows their disequality. We now discuss our different equivalence checking criteria by the way the sample points P are constructed.

Random Sampling. A naïve method for constructing P is to choose n points at random. For that, we have implemented a random value generator for each type we are interested in. For Booleans and fixed-width bit-vectors, it simply returns a uniformly random value in the (finite) range. For integers, we first pick a sign and then iteratively concatenate digits $0-9$, with a fixed probability to terminate after each iteration. For rationals, we pick a fraction c_1/c_2 with integer c_1 and non-zero integer c_2 chosen at random. For strings, we concatenate a random number of characters over a fixed alphabet that includes all characters occurring in a rule of $\mathcal{R}$ and dummy character(s) for ensuring that the cardinality of this alphabet is at least two. The latter is needed because, for instance, if the alphabet contained only one letter, it would be impossible to generate witnesses that disprove equivalences such as `contains("A" ++ x ++ "A", "AA")` $\approx$ `true` where `++` denotes string concatenation and `contains`(t, s) is true if s is a substring of t.

Grammar-Based Sampling. While random sampling is easy to implement, it is not effective at generating points that witness the disequivalence of certain term pairs. Thus, we have developed an alternative method that constructs points based on the user-provided grammar. In this method, each sample point in P (of arity n) is generated by choosing random points $\bar{c}_1, \ldots, \bar{c}_n$, random terms $t_1, \ldots, t_n$ generated by the input grammar $\mathcal{R}$, and then computing the result of $((t_1\{\bar{y} \mapsto \bar{c}_1\})\downarrow, \ldots, (t_n\{\bar{y} \mapsto \bar{c}_n\})\downarrow)$. The intuition is that sample points of this form are biased towards interesting values. In particular, they are likely to include non-trivial combinations of the user-provided constants that occur in the input grammar. For example, if the grammar contains `++`, `"A"`, and `"B"`, grammar-based sampling may return samples such as `"BA"` or `"AAB"`.

Exact Checking with Model-based Sampling. In contrast to the previous methods, this method makes two terms equivalent only if they are T-equivalent. It is based on satisfiability checking and *dynamic, model-based* generation of a set of sample points P, which is initially empty. When a term t is generated, we check if it evaluates to the same values on all sample points in P as any previously generated term s. If so, we separately check the T-satisfiability of $t \not\approx s$. If $t \not\approx s$ is unsatisfiable, we put t and s in the same equivalence class. Otherwise, $t \not\approx s$ is satisfied by some model $\mathcal{M}$, and we add $\mathcal{M}(\bar{y})$ as a new sample point to P, guaranteeing that s and t evaluate differently on the updated set. The new sample point remains in P for use as other terms are generated. For example, $x + 1 \not\approx x$ is satisfied by $x = 1$, so $x = 1$ is added to P.

4   Filtering Techniques for Rewrite Rules

For developers, it is desirable for the workflow in Fig. 1 to identify *useful* rewrites. For instance, it is desirable for the set of candidate rewrite rules to omit trivial consequences of other rules in the set. A rewrite rule $t \approx s$ is *redundant* with respect to a set $\{t_1 \approx s_1, \ldots, t_n \approx s_n\}$ of equations with free variables from $\bar{y}$ if $\forall \bar{y}. (t_1 \approx s_1 \wedge \ldots \wedge t_n \approx s_n)$ entails $\forall \bar{y}. t \approx s$ in the theory of equality. Redundant rules are not useful to the user, since they typically provide no new information. For example, if the framework generates $x \approx x + 0$ as a candidate rewrite rule, then it should not also generate the redundant rule $x \cdot y \approx (x+0) \cdot y$, which is entailed by the former equality. Checking this entailment with a solver, however, is expensive, since it involves first-order quantification. Instead, we use several incomplete but sound and efficient filtering techniques. These techniques significantly reduce the number of rewrite rules printed (as we show empirically in Sect. 5.2).

Filtering Based on Matching. One simple way to detect whether a rewrite rule is redundant is to check if it is an instance of a previously generated rule. For example, $y_1 + 1 \approx 1 + y_1$ is redundant with respect to any set that includes $y_1 + y_2 \approx y_2 + y_1$. Our implementation caches in a database all representative terms generated as a result of our equivalence checking. For each new candidate rewrite rule $t \approx s$, we query this database for a set of *matches* of the form $t_1 \sigma_1, \ldots, t_n \sigma_n$ where for each $i = 1, \ldots, n$, we have that t_i is a previous term added to the structure, and $t_i \sigma_i = t$. If for any such i, we have that $t_i \approx s_i$ was a previously generated candidate rewrite rule and $s_i \sigma_i = s$, we discard $t \approx s$.

Filtering Based on Variable Ordering. This technique discards rewrite rules whose variables are not in a given canonical order. For example, assume a grammar $\mathcal{R}$ with free variables y_1, y_2 ordered as $y_1 \prec y_2$. Furthermore, assume that y_1 and y_2 are indistinguishable in $\mathcal{R}$ in the sense that if $\mathcal{R}$ has a rule $s \rightarrow t$, then it also has rules $s \rightarrow t\{y_1 \mapsto y_2\}$ and $s \rightarrow t\{y_2 \mapsto y_1\}$. In this case, we can discard candidate rewrite rules $t_1 \approx t_2$ where y_2 appears before y_1 in (say, left-to-right, depth-first) traversals of both t_1 and t_2. For example, we can pre-emptively discard the rewrite $y_2 + 0 \approx y_2$, without needing to apply the above filtering based on matching, by reasoning that we will eventually generate $y_1 + 0 \approx y_1$.

Filtering Based on Congruence. Another inexpensive way to discover that a rewrite rule is redundant is to verify that it can be deduced from previous ones by congruence. For example, for any function f, it is easy to check that $f(y_1 + 0) \approx f(y_1)$ is redundant with respect to a set of rules containing $y_1 + 0 \approx y_1$. Our implementation maintains a data structure representing the congruence closure $\mathcal{C}(S)$ of the current set S of rewrite rules, i.e., the smallest superset of S closed under entailment in the theory of equality. Then, it discards any new candidate rewrite rule if it is already in $\mathcal{C}(S)$.

5   Evaluation

We now discuss our experience with the proposed workflow, evaluate different configurations, and show their impact on benchmarks. We ran experiments on a cluster equipped with Intel E5-2637 v4 CPUs running Ubuntu 16.04. All jobs used one core and 8 GB RAM unless otherwise indicated.

5.1   Experience

We implemented our framework for enumerating rewrite rules in CVC4, a state-of-the-art SMT solver. We used four grammars (Sect. 5.2), over strings (using the semantics of the latest draft of the SMT-LIB standard [32]), bit-vectors, and Booleans, as inputs to our framework. The grammar determines the operators and the number and types of different subterms that can be matched by the generated rewrites. To generate rewrites to accelerate the search for a particular SyGuS problem, we can simply use the grammar provided by the problem. To generate more general rewrites, it is helpful to divide the problem into rewrites for function symbols and rewrites for predicate symbols, as each of these formulations will use a different return type for f, the function to synthesize. We typically picked a small number of arguments for f, using types that we are interested in. In practice, we found two arguments to be sufficient to generate a large number of interesting rewrites for complex types. Rewrites with more variables on the left side are rarer than rewrites with fewer because they are more general. Thus, increasing the number of arguments is often not helpful and results in extra overhead.

Guided by the candidates generated using these grammars, we wrote approximately 5,300 lines of source code to implement the new rewrite rules. We refer to this implementation as the *extended rewriter* (ext), which can be optionally enabled. The rewrites implemented in ext are a superset of the rewrites in the default rewriter (std). Our implementation is public [3]. We implemented approximately 80 classes of string rewrites with a focus on eliminating expensive operators, e.g. contains(replace($x, y, z), z$) $\rightsquigarrow$ contains(x, y) $\lor$ contains(x, z), where replace(x, y, z) denotes the string obtained by replacing the first occurrence (if any) of the string y in x by z. A less intuitive example is replace(x, replace($x, y, x), x$) $\rightsquigarrow x$. To see why this holds, assume that the inner replace returns a string other than x (otherwise it holds trivially). In that case, the returned string must be longer than x, so the outer replacement does nothing. We implemented roughly 30 classes of bit-vector rewrites, including $x + 1 \rightsquigarrow -\sim x$, $x - (x \mathbin{\&} y) \rightsquigarrow x \mathbin{\&} \sim y$, and $x \mathbin{\&} \sim x \rightsquigarrow 0$ where $\sim$ and $\&$ are respectively bitwise negation and conjunction. Note that our workflow suggests bit-vector rewrites for fixed bit-widths, so the developer has to establish which rewrites hold *for all* bit-widths. For Booleans, we implemented several classes of rewrite rules for negation normal form, commutative argument sorting, equality chain normalization, and constraint propagation.

By design, our framework does not generalize the candidate rewrite rules. We found instead that the framework naturally suggests candidates that can be seen as instances of the same generalized rule. This allows the developer to devise rules over conditions that are not easily expressible as logical formulas. For example, consider the candidates: x ++ "A" $\approx$ "AA" $\leadsto x \approx$ "A", x ++ "A" $\approx$ "A" $\leadsto x \approx$ "", and "BBB" $\approx x$ ++ "B" ++ $y \leadsto$ "BB" $\approx x$ ++ y. These suggest that if one side of an equality is just a repetition of a single character, one can drop any number of occurrences of it from both sides. The condition and conclusion of such rules are more easily understood operationally than as logical formulas.

```
(synth-fun f ((x String)                        (synth-fun f ((s (BitVec 4))
 (y String) (z Int)) String (                    (t (BitVec 4))) (BitVec 4) (
 (Start String (x y "A" "B" ""                   (Start (BitVec 4) (
  (str.++ Start Start)                             s t #x0
  (str.replace Start Start Start)                  (bvneg  Start) (bvnot  Start)
  (str.at Start ie) (int.to.str ie)               (bvadd  Start Start) (bvmul Start Start)
  (str.substr Start ie ie)))                      (bvand  Start Start) (bvor  Start Start)
 (ie Int (0 1 z (+ ie ie) (- ie ie)              (bvlshr Start Start)
  (str.len Start) (str.to.int Start)              (bvshl  kStart Start)))))
  (str.indexof Start Start ie)))))
```

Fig. 2. Examples of grammars used in our workflow: strterm (left), bvterm$_4$ (right).

5.2 Evaluating Internal Metrics of Our Workflow

We now address the following questions about the effectiveness of our workflow:

- How does the number of unique terms scale with the number of grammar terms?
- How do rewriters affect term redundancy and enumeration performance?
- What is the accuracy and performance of different equivalence checks?
- How many candidate rewrites do our filtering techniques eliminate?

We consider four grammars: strterm and strpred for the theory of strings, bvterm for the theory of bit-vectors, and crci for Booleans. To show the impact of bit-width, we consider 4-bit (bvterm$_4$) and a 32-bit (bvterm$_{32}$) variant of bvterm. Figure 2 shows strterm and bvterm$_4$ in SyGuS syntax. A SyGuS problem specifies a function f with a set of parameters (e.g. two strings x and y and an integer z for strterm) and a return type (e.g. a string for strterm) to synthesize. The initial symbol of the grammar is Start and each rule of the grammar is given by a symbol (e.g. ie in strterm) and a list of terms (e.g. the constants 0, 1 and the functions len, str.to.int, indexof for ie). We omit semantic constraints because we want to enumerate all terms for a given grammar. The number of terms is infinite, so our evaluation restricts the search for candidate rewrites by limiting the *size* of the enumerated terms, using a 24 h timeout and a 32 GB RAM limit. The size of a term refers to the number of non-nullary symbols in it. Our implementation provides three rewriter settings: none disables rewriting; and std and ext are as defined in Sect. 5.1. For equivalence checking, we implemented the three methods from Sect. 3: random sampling (random); grammar-based sampling (grammar); and exact equivalence checking (exact).

Table 1. Impact of different rewriters on term redundancy using grammar equivalence checking. * may be inaccurate because Z3's rewriter is incorrect for strings (see Sect. 5.3 for details). † estimate is based on the unique terms for bvterm$_4$.

Grammar	Size	Terms	T-Unique	none Red. %	none Time [s]	std Red. %	std Time [s]	ext Red. %	ext Time [s]	z3 Red. %
strterm	1	218	86	60.6%	0.18	17.3%	0.09	0.0%	0.09	21.1%*
	2	24587	4204	82.9%	22.64	49.4%	8.78	20.0%	3.57	52.5%*
strpred	1	104	31	70.2%	0.13	34.0%	0.13	0.0%	0.07	32.6%*
	2	8726	1057	87.9%	9.32	66.5%	7.66	26.2%	2.16	68.1%*
	3	1100144	≥53671	—	t/o	≤82.5%	1154.02	≤57.0%	376.61	—
bvterm$_4$	1	63	22	65.1%	0.16	8.3%	0.12	0.0%	0.13	29.0%
	2	2343	288	87.7%	1.03	22.0%	0.24	0.7%	0.14	58.3%
	3	110583	4744	95.7%	89.84	39.3%	11.03	9.9%	3.55	76.9%
	4	5865303	84048	—	t/o	—	t/o	23.9%	242.15	—
bvterm$_{32}$	1	63	22	65.1%	0.09	8.3%	0.05	0.0%	0.05	29.0%
	2	2343	290	87.6%	4.53	21.4%	1.45	0.0%	0.85	58.0%
	3	110583	4925	95.5%	462.47	37.0%	85.62	6.5%	51.79	—
	4	5865303	≥84229†	—	t/o	—	t/o	≤23.8%	1955.97	—
crci	1	4	3	25.0%	0.11	25.0%	0.13	0.0%	0.11	0.0%
	2	32	12	62.5%	0.11	52.0%	0.12	0.0%	0.12	33.3%
	3	276	44	84.1%	0.15	74.4%	0.13	0.0%	0.12	62.1%
	4	2656	176	93.4%	0.38	87.5%	0.28	0.0%	0.13	81.1%
	5	17920	228	98.7%	2.11	96.9%	1.05	0.0%	0.15	93.1%
	6	107632	348	99.7%	15.97	99.0%	6.33	36.7%	0.24	97.8%
	7	596128	396	99.9%	112.71	99.8%	31.62	68.3%	0.43	99.3%
	8	2902432	396	—	t/o	99.9%	124.28	71.4%	0.45	—

T-Unique Solutions. In Table 1, we show the number of terms (Terms) and the number of unique terms modulo T-equivalence (T-Unique) at different sizes for each grammar. We established the number of unique terms for the Boolean and bvterm$_4$ grammars using the exact equivalence checking technique. For bvterm$_{32}$ and the string grammars, some equivalence checks are challenging despite their small size, as we discuss in Sect. 5.3. For terms of size 2 from the string grammars and terms of size 3 from the bvterm$_{32}$ grammar, we resolved the challenging equivalence checks in a semi-automated fashion in some cases and manually in others. For larger term sizes, we approximated the number of unique terms using grammar with 10,000 samples.[2] Our equivalence checks do not produce false negatives, i.e. they will not declare two terms to be different when they are actually equivalent. Thus, grammar can be used to compute a lower bound on the actual number of unique terms. For all grammars, the number of terms grows rapidly with increasing size while the number of unique terms grows much slower. Thus, enumerating terms without rewriting is increasingly inefficient as terms grow in size, indicating the utility of aggressive rewriting in these domains. The number of unique terms differs between the 4-bit and the 32-bit versions of bvterm at size 2, showing that some rewrite rules are valid for smaller bit-widths only.

---

[2] For a better estimate for bvterm$_{32}$, we approximate the number as $u_{4,n} + u_{32,n-1} - u_{4,n-1}$ where $u_{m,n}$ is the number of unique terms for bit-width m and term size n.

Rewriter Comparison. To measure the impact of different rewriters, we measured the redundancy of our rewriter configurations at different sizes for each grammar, and the wall-clock time to enumerate and check all the solutions. We define the *redundancy* of a rewriter for a set of terms S to be $(n - u)/n$, where n is the cardinality of $\{t\downarrow \mid t \in S\}$ and u is the number of T-unique terms in S. We used grammar with 1,000 samples for the equivalence checks. As a point of reference, we include the redundancy of z3's `simplify` command by counting the number of unique terms after applying the command to terms generated by none. Table 1 summarizes the results. With none, the redundancy is very high at larger sizes, whereas std keeps the redundancy much lower, except for crci. This is because std only performs basic rewriting for Boolean terms as CVC4 relies on a SAT solver for Boolean reasoning. Overall, std is competitive with z3's rewriter, indicating that it is a decent representative of a state-of-the-art rewriter. As expected, ext fares much better in all cases, lowering the percentage of redundant terms by over 95% in the case of crci at size 5. This has a significant effect on the time it takes to enumerate all solutions: ext consistently and significantly outperforms std, in some cases by almost two orders of magnitude (crci at size 7), especially at larger sizes. Compared to none, both std and ext perform much better.

Equivalence Check Comparison. To compare the different equivalence checks, we measured their error with respect to the set of terms enumerated by the ext rewriter, and the wall-clock time to enumerate all the solutions. For a set of terms S, we define the error of an equivalence check as $(u - n)/u$, where n is number of equivalence classes of S induced by the check and u is the number of T-unique terms in S, where $u \geq n$. For both random and grammar, we used 1,000 samples. We summarize the results in Table 2. For crci and bvterm$_4$, 1,000 samples are enough to cover all possible inputs, so there is no error in those cases and we do not report grammar. While sampling performs similarly to exact for crci and bvterm$_4$, for bvterm$_{32}$, exact is more precise and slightly faster for size 2. At sizes 3 and 4, exact ran out of memory. For strterm, we found that grammar was much more effective than random, having an error that was around 2.7 times smaller for term size 2. Similarly, grammar-based sampling was more effective on strpred with an error that was 1.5 times smaller for term size 2. Recall that we determined the numbers of unique terms at this size manually and grammar is good enough to discard a majority of spurious rewrites at those small sizes. As expected, exact gets stuck and times out for strterm and strpred at sizes 2 and 3.

Impact of Filtering. Table 2 also lists how many candidate rewrites the filtering techniques in Sect. 4 eliminate. We used ext and exact if available and grammar otherwise. Filtering eliminates up to 69.7% of the rules, which significantly lowers the burden on the developer when analyzing the proposed rewrite rules.

5.3 Evaluating SMT Solvers for Equivalence Checking

In this section, we demonstrate the use of our workflow to generate small queries that correspond to checking the equivalence of two enumerated terms to evaluate other SMT solvers. The motivation for this is twofold. First, we are interested in

Table 2. Comparison of different equivalence checks, the number of candidates filtered and the overhead of checking rewrites for soundness (Sect. 5.4), using the ext rewriter. For bvterm$_4$ and crci random and grammar, are the same.

Grammar	Size	no − eqc Time	random Error	Time	grammar Error	Time	exact Error	Time	Rewrites Filtered	Confidence Overhead
strterm	1	0.04	0.0%	0.03	0.0%	0.09	0.0%	0.13	0.0%	111.1%
	2	0.60	2.5%	3.75	0.9%	3.57	—	t/o	63.8%	2.0%
strpred	1	0.03	0.0%	0.03	0.0%	0.07	0.0%	0.12	0.0%	85.7%
	2	0.49	6.8%	2.09	4.4%	2.16	—	t/o	59.8%	6.9%
	3	59.54	≤16.1%	380.23	≤13.6%	376.61	—	t/o	66.5%	0.6%
bvterm$_4$	1	0.02	0.0%	0.04			0.0%	0.02	0.0%	92.3%
	2	0.03	0.0%	0.06			0.0%	0.12	50.0%	85.7%
	3	0.35	0.0%	3.56			0.0%	2.10	45.0%	3.1%
	4	9.71	0.0%	266.09			0.0%	215.93	60.8%	2.3%
bvterm$_{32}$	1	0.01	27.3%	0.04	0.0%	0.05	0.0%	0.12	0.0%	80.0%
	2	0.03	62.8%	1.54	15.9%	0.85	0.0%	0.47	0.0%	16.5%
	3	0.35	79.9%	69.40	40.9%	51.79	—	t/o	57.8%	7.5%
	4	9.06	≤87.3%	2502.11	≤57.3%	1955.97	—	t/o	69.7%	2.9%
crci	1	0.02	0.0%	0.03			0.0%	0.02	0.0%	163.6%
	2	0.02	0.0%	0.05			0.0%	0.02	0.0%	150.0%
	3	0.04	0.0%	0.03			0.0%	0.03	0.0%	191.7%
	4	0.05	0.0%	0.06			0.0%	0.05	0.0%	146.2%
	5	0.08	0.0%	0.08			0.0%	0.12	0.0%	113.3%
	6	0.10	0.0%	0.17			0.0%	0.43	0.0%	87.5%
	7	0.22	0.0%	0.37			0.0%	1.58	0.0%	48.8%
	8	0.21	0.0%	0.38			0.0%	1.70	0.0%	44.4%

how other SMT solvers perform as equivalence checkers in our workflow. Second, we are interested in finding queries that uncover issues in SMT solvers or serve as small but challenging benchmarks. In contrast to random testing for SMT solvers [13,14,23], our approach can be seen as a form of *exhaustive* testing, where all relevant queries up to a given term size are considered.

First, we logged the candidate rewrites generated by our workflow up to a fixed size using a basic rewriter that only evaluates operators with constant arguments. We considered the grammars strterm, strpred, and bvterm$_{32}$, for which equivalence checking is challenging. We considered a size of bound 2 for the string grammars and 3 for bvterm$_{32}$. Then, we used grammar-based sampling with 1,000 samples to compute candidate rewrite rules, outputting for each candidate $t \approx s$ the (quantifier-free) satisfiability query $t \not\approx s$. Finally, we tested CVC4 with the ext and std rewriters as well as state-of-the-art SMT solvers for string and bit-vector domains. Specifically, we tested Z3 [20] 4.8.1 for all grammars, Z3STR3 [12] for the string grammars and BOOLECTOR [22] for the bit-vector grammar. This set of solvers includes the winners of the QF_BV (BOOLECTOR) and BV (CVC4) divisions of SMT-COMP 2018 [1].

Table 3 summarizes the results. For each grammar and solver, we give the number of unsatisfiable, satisfiable, and unsolved responses. We additionally provide the number of unsatisfiable responses for which a solver did not require any SAT decisions, i.e., it solved the benchmark with rewriting only.

Table 3. Results for equivalence checking with a 300 s timeout. The number of benchmarks for each grammar is below its name. The number of responses x (y) indicates that the solver solved x benchmarks, of which it solved y by simplification only. Incorrect responses from solvers are given in the columns #w.

Grammar	Result	CVC4ext	CVC4std	z3 Solved	#w	z3STR3 Solved	#w	BOOLECTOR
strterm (1045)	unsat	1030 (666)	991 (254)	888 (348)		953	3	
	sat	10	9	5	93	4	28	
	unsolved	5	45	59		57		
strpred (835)	unsat	807 (569)	775 (297)	716 (287)		779		
	sat	13	13	6	32	11	17	
	unsolved	15	47	81		28		
bvterm$_{32}$ (1575)	unsat	1484 (1271)	1406 (641)	1426 (743)				1399 (766)
	sat	85	85	89				89
	unsolved	6	84	60				87

For benchmarks from the string grammars, we found that z3 and z3STR3 generated 125 and 45 incorrect "sat" responses respectively.[3] For 122 and 44 of these cases respectively, CVC4ext answered "unsat". For the other 3 cases, CVC4ext produced a different model that was accepted by all solvers. Additionally, we found that z3STR3 gave 3 confirmable incorrect "unsat" responses.[4] We filed these cases as bug reports and the developers confirmed that reasons for the incorrect responses include the existence of unsound rewrites. As expected, CVC4ext significantly outperforms the other solvers because the ext rewriter is highly specialized for this domain. CVC4ext solves a total of 1,235 instances using rewriting alone, which is 684 more than CVC4std. Even on instances that CVC4ext's rewriter does not solve directly, it aids solving. This is illustrated by CVC4ext solving 96.8% whereas CVC4std solving only 92.2% of the string benchmarks that neither of them solved by rewriting alone.

For bvterm$_{32}$, we found no incorrect answers. Again, CVC4ext outperforms the other solvers due to the fact that its rewriter was trained for this grammar. Among the other solvers, BOOLECTOR solved the most benchmarks using simplification alone. Surprisingly, it solved 27 fewer unsatisfiable benchmarks than z3, the second best performer. This can be primarily attributed to the fact that BOOLECTOR currently does not rewrite $x \cdot -y$ and $-(x \cdot y)$ to the same term. As a result, it could not prove the disequality $x \cdot -y \not\approx -(x \cdot y)$ within 300s. Two variants of $(s \cdot s) \gg (s \ll s) \not\approx s \cdot (s \gg (s \ll s))$ were challenging for all solvers, which confirms that our workflow can be used to find small, challenging queries.

5.4   Improving Confidence in the Rewriter

The soundness of rewriters is of utmost importance because an unsound rewriter often implies that the overall SMT solver is unsound. A rewriter is unsound if

---

[3] The solver answered "sat", but produced a model that did not satisfy the constraints.
[4] The solver answered "unsat", but accepted a model generated by CVC4ext.

there exists a pair of T-disequivalent terms t and s such that $t{\downarrow} = s{\downarrow}$. To accommodate the rapid development of rewrite rules in our workflow in Fig. 1, CVC4 supports an optional mode that attempts to detect unsoundness in its rewriter. When this mode is enabled, for each term t enumerated in Step 2, we use grammar to test the equivalence of t and $t{\downarrow}$. In particular, this technique discovers points where t and $t{\downarrow}$ have different values. This functionality has been critical for discovering subtle bugs in the implementation of new rules. It even caught *previously existing* bugs in CVC4's rewriter. For instance, an older version of CVC4 implemented replace$(x, x, y) \rightsquigarrow y$ which was incompatible with the (now outdated) semantics of replace (if the second argument was empty, the operation would leave the first argument unchanged). Table 2, shows the overhead of running these checks for each grammar and term size, where grammar-based sampling is used both for computing the rewrite rules and for checking the soundness of the ext rewriter. For example, adding checks that ensure that no unsound rewrites are produced for terms up to size 3 in the bvterm$_{32}$ grammar has a 7.5% overhead. As term size increases and the enumeration rate decreases, the relative overhead tends to becomes smaller. Overall, this option leads to a noticeable, but not prohibitively large, slowdown in the workflow.

5.5　Impact of Rewrites on Solving

Finally, we evaluate the impact of the extended rewrites on CVC4's solving performance on SyGuS and SMT problems. Figure 3 summarizes the results.

Impact on SyGuS Problems. The Boolean and the string problems in our evaluation are from SyGuS-COMP 2018 [2]. Their grammars are similar to the ones we used in our workflow. We distinguish between two types of bit-vector (BV) benchmarks: programming-by-examples (PBE) benchmarks where the constraints are input-output pairs and benchmarks with arbitrary non-PBE constraints. We make this distinction because for PBE problems, CVC4 can rely on input examples to decide term equivalence (the function to synthesize can return arbitrary values for the domain outside of the input examples). For the non-PBE benchmarks, we use the benchmarks from the general track of SyGuS-COMP, from work on synthesizing invertibility conditions (IC) for bit-vector operators [24], and invariant synthesis [4]. The PBE bit-vector benchmarks [5,19] are from the 2013 ICFP programming competition. Their grammars vary significantly.

The top table in Fig. 3 lists the number of solved instances within a 3600 s timeout for ext and std and the speedup of ext over std on commonly solved benchmarks. With ext, CVC4 solves more benchmarks on all benchmark sets except IC and the PBE bit-vector problems. We experienced a significant speedup on the commonly solved instances from the Boolean benchmarks, the SyGuS-COMP general track bit-vector benchmarks, and the string benchmarks. The IC grammars are focused on bit-vector comparisons, which our grammar bvterm lacks. Thus, the new rewrites do not significantly reduce the redundancy for that set. On the PBE bit-vector problems, ext is slightly faster overall and solves a problem that std does not solve but times out on two hard problems that std solves, which is likely due to the rewrites affecting CVC4's search heuristic.

Benchmark Set		#	std	ext	S%
Strings		108	73	**88**	1.81×
BV (non-PBE)		361	236	**238**	1.02×
▷ IC		160	**131**	130	0.76×
▷ CegisT		79	**41**	**41**	1.09×
▷ General		122	64	**67**	1.73×
BV (PBE)		803	**774**	773	1.11×
Boolean		214	153	**159**	1.83×

Logic	Result	std	ext	S%
QF_SLIA	unsat	3803	**3823**	0.46×
(25421)	sat	20887	**20950**	1.87×
BV	unsat	4969	**4974**	1.12×
(5751)	sat	536	**542**	1.41×

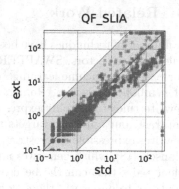

Fig. 3. Impact of ext on SyGuS (top left) and SMT (bottom left and right) solving. They ran with a 3600 s and 300 s timeouts respectively. Best results are in **bold**. The scatter plot is logarithmic. "S%" is the speedup of ext over std on commonly solved problems.

Impact on SMT Problems. We evaluated the impact of ext on quantifier-free string benchmarks (QF_SLIA) from the symbolic execution of Python programs [28] and found that ext has a significant positive impact. While ext is faster than std on commonly solved satisfiable benchmarks, it is slower on unsatisfiable ones due to three outliers. The scatter plot in Fig. 3 shows that for benchmarks that take more than one second, there is a trend towards shorter solving times with ext.

For benchmarks in the quantified bit-vector logic (BV) of SMT-LIB [10], the extended rewriter improves the overall performance by solving 11 additional instances as shown in the bottom table in Fig. 3.

For quantifier-free bit-vector (QF_BV) problems, naively using all new bit-vector rules from ext resulted in fewer solved instances. This is due to the fact that ext performs aggressive rewriting to eliminate as much redundancy as possible, which is helpful in enumerative SyGuS. For QF_BV solving, however, it is important to consider the effects of a rewrite at the word-level *and* the bit-level. Rewrites at the word-level can destroy common structures between different terms at the bit-level, which may increase the size of the formula at the bit-level while decreasing it at the word level. Rewrites that eliminate redundancy at the cost of introducing expensive operators (e.g., multipliers) may also harm performance. We evaluated each bit-vector rewrite in ext w.r.t. solving QF_BV problems. We generalized one family involving bitwise operators over concatenations that e.g. rewrites $x \ \& \ (0 \circ y)$ to $0 \circ (x[m-1:0] \ \& \ y)$, where x has bit-width n and y has bit-width m for some $m < n$. These resulted in a net gain of 41 solved instances over 40,102 QF_BV benchmarks with 1,089 timeouts, a 3.8% improvement in the success rate of CVC4 on this set. This suggests that developing a library of rewrite rules and selectively enabling some can be beneficial in this domain.

6   Related Work

A number of techniques have been proposed for automatically generating rewrite rules for bit-vectors. SWAPPER [31] is an automatic formula simplifier generator based on machine learning and program synthesis techniques. In the context of symbolic execution, Romanoe et al. [29] propose an approach that learns rewrite rules to simplify expressions before sending them to an SMT solver. In contrast, our approach targets the SMT solver developer and is not limited to rewrites expressible in a restricted language. A related approach was explored by Hansen [18], which generates all the terms that fit a grammar and finds equivalent pairs that can the be used by the developer to implement new rules. In contrast to our work, the candidate rules are not filtered, and the grammar is hard-coded and only considers bit-vector operations. Nadel [21] proposed generating bit-vector rewrite rules in SMT solvers *at runtime* for a given problem. Syntax-guided synthesis was used by Niemetz et al. [24] to synthesize conditions that characterize when bit-vector constraints have solutions. Rewrite rules in SMT solvers—especially the ones for the theories of bit-vectors and Booleans— bear similarities with local optimizations in compilers [8,15]. Finally, caching counterexamples as we do in our exact equivalence check is similar to techniques used in symbolic execution engines, e.g. KLEE [16], and superoptimizers, e.g. STOKE [30].

7   Conclusion

We have presented a syntax-guided paradigm for developing rewriters for SMT solvers. In ongoing work, we are exploring an automated analysis of how likely particular rewrites are to help solving constraints, which we found was the most tedious aspect of our current workflow. Furthermore, we plan to adapt existing techniques for automatically constructing grammars from a set of problems to use them as inputs to our approach. In the shorter term, we plan to use the existing framework to identify useful rewrite rules for emerging SMT domains, notably the theory of floating-point arithmetic, for which developing a rewriter is notoriously difficult.

Acknowledgements. This material is based upon work partially supported by the National Science Foundation (Award No. 1656926), the Office of Naval Research (Contract No. 68335-17-C-0558), and DARPA (N66001-18-C-4012, FA8650-18-2-7854 and FA8650-18-2-7861).

References

1. SMT-COMP 2018 (2018) http://smtcomp.sourceforge.net/2018/
2. SyGuS-COMP 2018 (2018). http://sygus.seas.upenn.edu/SyGuS-COMP2018.html
3. CVC4 sat2019 branch (2019). https://github.com/4tXJ7f/CVC4/tree/sat2019

4. Abate, A., David, C., Kesseli, P., Kroening, D., Polgreen, E.: Counterexample guided inductive synthesis modulo theories. In: Chockler, H., Weissenbacher, G. (eds.) CAV 2018. LNCS, vol. 10981, pp. 270–288. Springer, Cham (2018). https://doi.org/10.1007/978-3-319-96145-3_15

5. Akiba, T., et al.: Calibrating research in program synthesis using 72,000 hours of programmer time. Technical Report, MSR, Redmond, WA, USA (2013)

6. Alur, R., et al.: Syntax-guided synthesis. In: Formal Methods in Computer-Aided Design, FMCAD 2013, Portland, OR, USA, 20–23 October 2013. pp. 1–8. IEEE (2013) http://ieeexplore.ieee.org/document/6679385/

7. Alur, R., Radhakrishna, A., Udupa, A.: Scaling enumerative program synthesis via divide and conquer. In: Legay, A., Margaria, T. (eds.) TACAS 2017. LNCS, vol. 10205, pp. 319–336. Springer, Heidelberg (2017). https://doi.org/10.1007/978-3-662-54577-5_18

8. Bansal, S., Aiken, A.: Automatic generation of peephole superoptimizers. In: Shen, J.P., Martonosi, M. (eds.) Proceedings of the 12th International Conference on Architectural Support for Programming Languages and Operating Systems, ASP-LOS 2006, San Jose, CA, USA, 21–25 October 2006. pp. 394–403. ACM (2006), https://doi.org/10.1145/1168857.1168906

9. Barrett, C., et al.: CVC4. In: Gopalakrishnan, G., Qadeer, S. (eds.) CAV 2011. LNCS, vol. 6806, pp. 171–177. Springer, Heidelberg (2011). https://doi.org/10.1007/978-3-642-22110-1_14

10. Barrett, C., Fontaine, P., Tinelli, C.: The Satisfiability Modulo Theories Library (SMT-LIB) (2016). www.SMT-LIB.org

11. Barrett, C., Fontaine, P., Tinelli, C.: The SMT-LIB Standard: Version 2.6. Technical report, Department of Computer Science, The University of Iowa (2017). www.SMT-LIB.org

12. Berzish, M., Ganesh, V., Zheng, Y.: Z3str3: A string solver with theory-aware heuristics. In: 2017 Formal Methods in Computer Aided Design, FMCAD 2017, Vienna, Austria, 2–6 October 2017. pp. 55–59 (2017). https://doi.org/10.23919/FMCAD.2017.8102241

13. Blotsky, D., Mora, F., Berzish, M., Zheng, Y., Kabir, I., Ganesh, V.: StringFuzz: a fuzzer for string solvers. In: Chockler, H., Weissenbacher, G. (eds.) CAV 2018. LNCS, vol. 10982, pp. 45–51. Springer, Cham (2018). https://doi.org/10.1007/978-3-319-96142-2_6

14. Brummayer, R., Biere, A.: Fuzzing and delta-debugging SMT solvers. In: Proceedings of the 7th International Workshop on Satisfiability Modulo Theories, SMT 2009, p. 5. ACM (2009)

15. Buchwald, S.: OPTGEN: a generator for local optimizations. In: Franke, B. (ed.) CC 2015. LNCS, vol. 9031, pp. 171–189. Springer, Heidelberg (2015). https://doi.org/10.1007/978-3-662-46663-6_9

16. Cadar, C., Dunbar, D., Engler, D.R.: KLEE: unassisted and automatic generation of high-coverage tests for complex systems programs. In: Draves, R., van Renesse, R. (eds.) Proceedings of 8th USENIX Symposium on Operating Systems Design and Implementation, OSDI 2008, 8–10 December 2008, San Diego, California, USA, pp. 209–224. USENIX Association (2008). http://www.usenix.org/events/osdi08/tech/full_papers/cadar/cadar.pdf

17. Ganesh, V., Minnes, M., Solar-Lezama, A., Rinard, M.: Word equations with length constraints: what's decidable? In: Biere, A., Nahir, A., Vos, T. (eds.) HVC 2012. LNCS, vol. 7857, pp. 209–226. Springer, Heidelberg (2013). https://doi.org/10.1007/978-3-642-39611-3_21
18. Hansen, T.: A constraint solver and its application to machine code test generation. Ph.D. thesis, University of Melbourne, Australia (2012). http://hdl.handle.net/11343/37952
19. Warren Jr., H.S.: Hacker's Delight, 2nd edn. Pearson Education, London (2013). http://www.hackersdelight.org/
20. de Moura, L.M., Bjørner, N.: Z3: an efficient SMT solver. In: Tools and Algorithms for the Construction and Analysis of Systems, 14th International Conference, TACAS 2008, Held as Part of the Joint European Conferences on Theory and Practice of Software, ETAPS 2008, Budapest, Hungary, March 29-April 6, 2008. Proceedings, pp. 337–340 (2008). https://doi.org/10.1007/978-3-540-78800-3_24
21. Nadel, A.: Bit-vector rewriting with automatic rule generation. In: Biere, A., Bloem, R. (eds.) CAV 2014. LNCS, vol. 8559, pp. 663–679. Springer, Cham (2014). https://doi.org/10.1007/978-3-319-08867-9_44
22. Niemetz, A., Preiner, M., Biere, A.: Boolector 2.0 system description. J. Satis. Boolean Model. Comput. 9, 53–58 (2014, published 2015)
23. Niemetz, A., Preiner, M., Biere, A.: Model-Based API Testing for SMT Solvers. In: Brain, M., Hadarean, L. (eds.) Proceedings of the 15th International Workshop on Satisfiability Modulo Theories, SMT 2017), affiliated with the 29th International Conference on Computer Aided Verification, CAV 2017, Heidelberg, Germany, 24–28 July 2017, p. 10 (2017)
24. Niemetz, A., Preiner, M., Reynolds, A., Barrett, C., Tinelli, C.: Solving quantified bit-vectors using invertibility conditions. In: Chockler, H., Weissenbacher, G. (eds.) CAV 2018. LNCS, vol. 10982, pp. 236–255. Springer, Cham (2018). https://doi.org/10.1007/978-3-319-96142-2_16
25. Reynolds, A., Barbosa, H., Nötzil, A., Barrett, C., Tinelli, C.: CVC4Sy: Smart and fast term enumeration for syntax-guided synthesis. In: Dilig, I., Tasiran, S. (eds.) Computer Aided Verification (CAV) - 31st International Conference. Lecture Notes in Computer Science, Springer (2019, Accepted for publication)
26. Reynolds, A., Deters, M., Kuncak, V., Tinelli, C., Barrett, C.: Counterexample-guided quantifier instantiation for synthesis in SMT. In: Kroening, D., Păsăreanu, C.S. (eds.) CAV 2015. LNCS, vol. 9207, pp. 198–216. Springer, Cham (2015). https://doi.org/10.1007/978-3-319-21668-3_12
27. Reynolds, A., Tinelli, C.: SyGuS techniques in the core of an SMT solver. In: Proceedings Sixth Workshop on Synthesis, SYNT@CAV 2017, Heidelberg, Germany, 22nd July 2017, pp. 81–96 (2017). https://doi.org/10.4204/EPTCS.260.8
28. Reynolds, A., Woo, M., Barrett, C., Brumley, D., Liang, T., Tinelli, C.: Scaling Up DPLL(T) string solvers using context-dependent simplification. In: Majumdar, R., Kunčak, V. (eds.) CAV 2017. LNCS, vol. 10427, pp. 453–474. Springer, Cham (2017). https://doi.org/10.1007/978-3-319-63390-9_24
29. Romano, A., Engler, D.: Expression reduction from programs in a symbolic binary executor. In: Bartocci, E., Ramakrishnan, C.R. (eds.) SPIN 2013. LNCS, vol. 7976, pp. 301–319. Springer, Heidelberg (2013). https://doi.org/10.1007/978-3-642-39176-7_19
30. Schkufza, E., Sharma, R., Aiken, A.: Stochastic superoptimization. In: Sarkar, V., Bodík, R. (eds.) Architectural Support for Programming Languages and Operating Systems, ASPLOS 2013, Houston, TX, USA - 16–20 March 2013, pp. 305–316. ACM (2013). https://doi.org/10.1145/2451116.2451150

31. Singh, R., Solar-Lezama, A.: SWAPPER: A framework for automatic generation of formula simplifiers based on conditional rewrite rules. In: Piskac, R., Talupur, M. (eds.) 2016 Formal Methods in Computer-Aided Design, FMCAD 2016, Mountain View, CA, USA, 3–6 October 2016, pp. 185–192. IEEE (2016). https://doi.org/10.1109/FMCAD.2016.7886678

32. Tinelli, C., Barrett, C., Fontaine, P.: Unicode Strings (Draft 1.0) (2018). http://smtlib.cs.uiowa.edu/theories-UnicodeStrings.shtml

33. Udupa, A., Raghavan, A., Deshmukh, J.V., Mador-Haim, S., Martin, M.M.K., Alur, R.: TRANSIT: specifying protocols with concolic snippets. In: ACM SIGPLAN Conference on Programming Language Design and Implementation, PLDI 2013, Seattle, WA, USA, 16–19 June 2013, pp. 287–296 (2013). https://doi.org/10.1145/2462156.2462174

DRAT-based Bit-Vector Proofs in CVC4

Alex Ozdemir, Aina Niemetz, Mathias Preiner, Yoni Zohar$^{(\boxtimes)}$,
and Clark Barrett

Stanford University, Stanford, CA, USA
{aozdemir,niemetz,preiner,yoniz,barrett}@cs.stanford.edu

Abstract. Many state-of-the-art Satisfiability Modulo Theories (SMT) solvers for the theory of fixed-size bit-vectors employ an approach called bit-blasting, where a given formula is translated into a Boolean satisfiability (SAT) problem and delegated to a SAT solver. Consequently, producing bit-vector proofs in an SMT solver requires incorporating SAT proofs into its proof infrastructure. In this paper, we describe three approaches for integrating DRAT proofs generated by an off-the-shelf SAT solver into the proof infrastructure of the SMT solver CVC4 and explore their strengths and weaknesses. We implemented all three approaches using CryptoMiniSat as the SAT back-end for its bit-blasting engine and evaluated performance in terms of proof-production and proof-checking.

1 Introduction

The majority of Satisfiability Modulo Theories (SMT) solvers for the theory of fixed-size bit-vectors employ an approach called bit-blasting. That is, an input formula is first simplified, and then eagerly translated into propositional logic and handed to a Boolean satisfiability (SAT) solver. Thus, when producing a proof of unsatisfiability for a given bit-vector input, it is crucial to obtain the unsatisfiability proof from the SAT solver back-end and incorporate it into a possibly larger SMT proof. The bit-blasting engine of the SMT solver CVC4 [1] currently supports several SAT solvers as back-ends. Producing proofs, however, is only supported with a modified version of MiniSat [5], which was extended to record resolution proofs that can be embedded into CVC4 proofs [8]. This custom MiniSat implementation requires extra maintenance and is less competitive than more recent off-the-shelf SAT solvers.

In recent years, the *Delete Resolution Asymmetric Tautologies* (DRAT) proof system [16], a generalization of *extended resolution* (ER) [15], has become the de facto standard for validating unsatisfiability in SAT solvers. Using a state-of-the-art SAT solver with support for DRAT inside CVC4 would allow CVC4 to use the latest, best SAT techniques while being able to produce bit-vector proofs without additional customization of the SAT solver code. However, in

This work was supported in part by DARPA (N66001-18-C-4012 and FA8650-18-2-7861, NSF (1814369), and the Stanford Center for Blockchain Research.

M. Janota and I. Lynce (Eds.): SAT 2019, LNCS 11628, pp. 298–305, 2019.
https://doi.org/10.1007/978-3-030-24258-9_21

order to support this, CVC4 must be able to incorporate DRAT proofs into its proof infrastructure, which is based on LFSC, an extension of *Edinburgh's Logical Framework* [9] (LF) with functional programs called *side conditions* (see [13] for more details on LFSC and [2] for a more general survey of proofs in SMT-solvers). In this paper, we examine three approaches for translating DRAT proofs to LFSC: (i) a direct translation from DRAT to LFSC proofs, (ii) an intermediate translation from DRAT to Linear RAT (LRAT) proofs [4], and (iii) an intermediate translation from DRAT to ER proofs [10], which are then translated to LFSC. The produced proofs can be independently checked by any proof checker for LFSC. We describe the implementation of these three approaches for generating bit-vector proofs in CVC4, discuss their strengths and weaknesses, and evaluate their performance in terms of proof production and proof checking.

2   From DRAT to LFSC

We briefly review the definitions relevant to the proof systems DRAT, LRAT, and ER. More details can be found in [4,10,16].

A *literal* is either a propositional variable or its negation. A *clause* is a disjunction of literals, sometimes interpreted as a set of literals. A clause is *unit* if it is a singleton. A *formula* in conjunctive normal form (CNF) is a conjunction of clauses, sometimes interpreted as a set of clauses.

A proof for formula F in CNF is a sequence $\pi = C_1, \ldots, C_m, I_{m+1}, \ldots, I_n$ with clauses $C_1, \ldots, C_m \in F$ and pairs I_i of the form $\langle \diamond, X \rangle$, where either $\diamond \in \{a, d\}$ and X is a clause, or $\diamond = e$ and X is a CNF formula. Letters a, d, and e indicate addition, deletion, and extension, respectively. Sequence π induces a sequence of CNF formulas $F_0, \ldots, F_n$ such that $F_i = \{C_1, \ldots, C_i\}$ for $1 \le i \le m$, and for $i > m$, $F_i = F_{i-1} \cup \{C\}$ if $I_i = \langle a, C \rangle$, $F_i = F_{i-1} \setminus \{C\}$ if $I_i = \langle d, C \rangle$, and $F_i = F_{i-1} \cup G$ if $I_i = \langle e, G \rangle$. It is a proof of unsatisfiability of F if $\emptyset \in F_n$.

A proof π of unsatisfiability of F is a valid ER proof if every I_i is either: (i) $\langle a, C \cup D \rangle$, where $C \cup \{p\}$ and $D \cup \{\overline{p}\} \in F_{i-1}$ for some p; or (ii) $\langle e, G \rangle$, where G is the CNF translation of $x \leftrightarrow \varphi$ with x a fresh variable and φ some formula over variables occurring in F_{i-1}. Proof π is a valid DRAT proof if every I_i is either $\langle d, C \rangle$ or $\langle a, C \rangle$ and for the latter, one of the following holds:

- C is a *reverse unit propagation* (RUP) [7] in F_{i-1}, i.e., the empty clause is derivable from F_{i-1} and the negations of literals in C using unit propagation.
- C is a *resolution asymmetric tautology* (RAT) in F_{i-1}, i.e., there is some $p \in C$ such that for every $D \cup \{\overline{p}\} \in F_{i-1}$, $C \cup D$ is a RUP in F_{i-1}. If C is a RAT but not a RUP, we call it a *proper* RAT.

LRAT proofs are obtained from DRAT proofs by allowing a third element in each I_i that includes hints regarding the clauses and literals that are relevant for verifying the corresponding proof step.

```
(program is_specified_drat_proof ((f cnf) (proof DRATProof)) bool
  (match proof
    (DRATProofn (cnf_has_bottom f))
    ((DRATProofa c p) (
        match (is_rat f c) (tt (is_specified_drat_proof (cnfc c f) p)) (ff ff)))
    ((DRATProofd c p) (is_specified_drat_proof (cnf_remove_clause c f) p))))
```

Fig. 1. Side condition for checking a specified DRAT proof. The side conditions cnf_has_bottom, is_rat, and cnf_remove_clause are defined in the same signature. Type cnfc is a constructor for CNF formulas, and is defined in a separate signature.

2.1 Integration Methods

The *Logical Framework with Side Conditions* (LFSC) [13] is a statically and dependently typed Lisp-style meta language based on the Edinburgh Logical Framework (LF) [9]. It can be used to define logical systems and check proofs written within them by way of the Curry-Howard correspondence. Like LF, LFSC is a framework in which axioms and derivation rules can be defined for multiple theories and their combination. LFSC additionally adds the notion of *side conditions* as functional programs, which can restrict the application of derivation rules. This is convenient for expressing proof-checking rules that are computational in nature. In order to use DRAT proofs in CVC4, the proofs need to be representable in LFSC. We consider the following three approaches for integrating DRAT proofs into LFSC.

Checking DRAT Proofs in LFSC. This approach directly translates DRAT proofs into LFSC. It requires creating a signature for DRAT in LFSC, which essentially is an LFSC implementation of a DRAT checker.

Checking LRAT Proofs in LFSC. LRAT proofs include hints to accelerate unit propagation while proof checking. We use the tool DRAT-trim [16] to translate DRAT proofs into the LRAT format and then check the resulting proof with an LRAT LFSC signature.

Checking ER Proofs in LFSC. This approach aims at further reducing computational overhead during proof checking by translating a DRAT proof into an ER proof with the tool drat2er [10]. The ER proof is then translated to LFSC and checked with an ER LFSC signature.

3   LFSC Signatures

In this section, we describe the main characteristics of the LFSC signatures[1] that we have defined for checking DRAT, LRAT, and ER proofs.

The *LFSC DRAT signature* makes extensive use of side conditions to express processes such as unit propagation and the search for the resolvents of a proper RAT. Because of the divergence between operational and specified DRAT and

---

[1] https://github.com/CVC4/CVC4/blob/master/proofs/signatures/.

```
(declare definition (! x var (! p lit (! ls lit_list type))))
(declare decl_definition
   (! p lit (! ls lit_list (! pf_continuation
            (! x var (! def (definition x p ls) (holds empty_clause)))
      (holds empty_clause)))))
```

Fig. 2. Derivation rules for checking that a clause constitutes an extension.

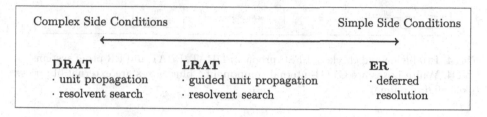

Fig. 3. Side conditions across our signatures.

the resulting ambiguity (see [11] for further details), our signature accepts both kinds of proofs. Figure 1 shows the main side condition that is used to check a DRAT proof. Though we do not explain the LFSC syntax in detail here due to lack of space, the general idea can be easily understood. Given a proof candidate **proof**, it covers three cases: (i) the proof is empty and the working formula includes a contradiction; (ii) the proof begins with an addition of a (proper or improper) RAT; or (iii) the proof begins with a deletion of some clause. In (ii) and (iii), the same side condition is recursively called on the rest of **proof**, with an updated working formula. In (ii), side condition **is_rat** checks whether the added clause is indeed a RAT via resolvent search and unit propagation.

The *LFSC LRAT signature* is similar in nature, and also makes extensive use of side conditions—albeit less computationally expensive ones. In particular, this signature uses hints provided in the LRAT proofs to accelerate unit propagation.

The *LFSC ER signature* is an extension of the LFSC signature for resolution proofs that is currently employed by CVC4. It implements *deferred resolution* to quickly check large resolution proofs using only a single side condition [13]. The signature extends resolution in order to check the ER proofs produced by the **drat2er** tool. These proofs feature extensions of the form $x \leftrightarrow (p \vee (l_1 \wedge l_2 \wedge \cdots \wedge l_k))$, where x is fresh and p and l_i are not. Our signature includes side-condition-free rules for introducing such extensions and translating them to CNFs of the form $\left\{ \{x, \overline{p}\}, \{x, \overline{l_1}, \ldots, \overline{l_k}\}, \{\overline{x}, p, l_1\}, \ldots, \{\overline{x}, p, l_k\} \right\}$. The **decl_definition** rule in Fig. 2 is used to introduce these extensions. Its first two arguments are literal p and the list of literals l_i (denoted as **ls** of type **lit_list**) from the definition. The third argument is a function that receives a fresh variable **x** and connects the introduced definition to the rest of the proof. Figure 3 illustrates the difference in side conditions between the three signatures.

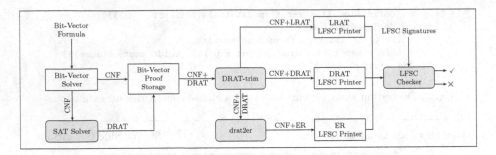

Fig. 4. Producing and checking LFSC proofs in DRAT, LRAT and ER proof systems in CVC4. White boxes are CVC4-internal components; blue boxes are external libraries. (Color figure online)

4   Workflow: From CVC4 to LFSC

Figure 4 shows the general workflow for incorporating DRAT proofs in the LFSC proof infrastructure of CVC4 after bit-blasting. LFSC proofs for the bit-blasting step are described in [8]. A given bit-vector formula is bit-blasted to SAT, and the resulting CNF is then sent to the underlying SAT solver. We use DRAT-trim to trim the original formula, optimize the proof produced by the SAT solver, and optionally produce an LRAT proof that is forwarded to the LRAT LFSC pipeline. In case of DRAT LFSC proofs, we can directly use the optimized proof and formula emitted by DRAT-trim. For ER LFSC proofs, we first use drat2er to translate the optimized DRAT proof into an ER proof, which is then sent to the ER LFSC pipeline. The result of each pipeline is an LFSC proof in the corresponding proof system, which can be checked with the corresponding signature (see Sect. 3) using the LFSC proof checker. Note that prior to bit-blasting, the input is usually simplified via rewriting and other preprocessing techniques, for which CVC4 currently does not produce proofs. The addition of such proofs is left as future work and orthogonal to incorporating DRAT proofs from the SAT solver back-end, which is the focus of this paper.

5   Experiments

We implemented the three approaches described in Sect. 2.1 in CVC4 using CryptoMiniSat 5.6 [12] as the SAT back-end. We compared them against the resolution-based proof machinery currently employed in CVC4 and evaluated our techniques on all 21125 benchmarks from the quantifier-free bit-vector logic QF_BV of SMT-LIB [3] with status *unsat* or *unknown*. All experiments were performed on a cluster with Intel Xeon E5-2620v4 CPUs with 2.1 GHz and 128GB of memory. We used a time limit of 600 s (CPU time) and a memory limit of 32GB for each solver/benchmark pair. For each error or memory-out, we added a penalty of 600 s.

Table 1. Impact of proof logging, production, and checking on # solved problems.

Proof system	solve		log		log,prod		log,prod,check	
	#	[s]	#	[s]	#	[s]	#	[s]
Resolution	20480	464k	20396	524k	**14217**	**4400k**	13510	4973k
DRAT	**20767**	**283k**	**20736**	**319k**	14098	4492k	12563	5616k
LRAT					14088	4500k	12877	5370k
ER					14035	4565k	**13782**	**4886k**

Table 1 shows the results for the Resolution approach with MiniSat, and the DRAT, LRAT and ER approaches with CryptoMiniSat. For each, we ran the following four configurations: proofs disabled (*solve*), proof logging enabled (*log*), proof production enabled (*prod*), and proof checking enabled (*check*). Proof logging records proof-related information but does not produce the actual proof, e.g., when producing DRAT proofs, proof logging stores the DRAT proof from the SAT-solver, which is only translated to LFSC during proof production. In the *solve* configuration, the DRAT-based approaches (using CryptoMiniSat) solve 287 more problems than the Resolution approach (which uses CVC4's custom version of MiniSat). This indicates that the custom version of MiniSat was a bottleneck for solving. In the *log* configuration, the DRAT-based approaches solve 31 fewer problems than in the *solve* configuration; and in the *prod* configuration the DRAT-based approaches produce proofs for ∼6600 fewer problems. This indicates that the bottleneck in the DRAT-based approaches is the translation of DRAT to LFSC. For all approaches, about 30% of the solved problems require more than 8GB of memory to produce a proof, showing that proof production can in general be very memory-intensive. Finally, with proof checking enabled, the ER-based approach outperforms all other approaches. Note that in ∼270 cases, CryptoMiniSat produced a DRAT proof that was rejected by DRAT-trim, which we counted as error. Further, for each *check* configuration, our LFSC checker reported ∼200 errors, which are not related to our new approach. Both issues need further investigation.

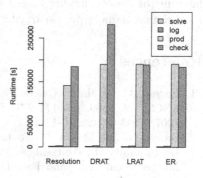

Fig. 5. Runtime distribution on 12539 commonly proved problems.

Figure 5 shows the runtime distribution for all approaches and configurations over the commonly proved problems (12539 in total). The runtime overhead of proof production for the DRAT-based approaches is 1.35 times higher compared to resolution. This is due to the fact that we post-process the DRAT-proof prior to translating it to LFSC, which involves writing temporary files and calling external libraries. The proof checking time correlates with the complexity of the side conditions (see Fig. 3), where ER clearly outperforms DRAT.

6   Conclusion

We have described three approaches for integrating DRAT proofs in LFSC, which enable us to use off-the-shelf SAT solvers as the SAT back-end for the bit-blasting engine of CVC4 while supporting bit-vector proofs. For future work, we plan to reduce the complexity of the side conditions in the DRAT and LRAT signatures and the proof production overhead in the translation workflows. We also plan to add support for the new signatures in SMTCoq [6], a tool that increases automation in Coq [14] using proofs generated by CVC4. In a more applicative direction, we plan to explore the potential DRAT proofs in SMT-solvers may have in the proof-carrying code paradigm [17], as well as its recent variant in blockchains, namely proof-carrying smart contracts [18].

References

1. Barrett, C., et al.: CVC4. In: Gopalakrishnan, G., Qadeer, S. (eds.) CAV 2011. LNCS, vol. 6806, pp. 171–177. Springer, Heidelberg (2011). https://doi.org/10.1007/978-3-642-22110-1_14

2. Barrett, C., de Moura, L., Fontaine, P.: Proofs in satisfiability modulo theories. In: Delahaye, D., Woltzenlogel Paleo, B. (eds.) All about Proofs, Proofs for All, Mathematical Logic and Foundations, vol. 55, pp. 23–44. College Publications, London, UK (2015)

3. Barrett, C., Stump, A., Tinelli, C.: The SMT-LIB standard: version 2.0. In: Gupta, A., Kroening, D. (eds.) Proceedings of the 8th International Workshop on Satisfiability Modulo Theories, Edinburgh, UK (2010)

4. Cruz-Filipe, L., Heule, M.J.H., Hunt, W.A., Kaufmann, M., Schneider-Kamp, P.: Efficient certified RAT verification. In: de Moura, L. (ed.) CADE 2017. LNCS (LNAI), vol. 10395, pp. 220–236. Springer, Cham (2017). https://doi.org/10.1007/978-3-319-63046-5_14

5. Eén, N., Sörensson, N.: An extensible SAT-solver. In: Giunchiglia, E., Tacchella, A. (eds.) SAT 2003. LNCS, vol. 2919, pp. 502–518. Springer, Heidelberg (2004). https://doi.org/10.1007/978-3-540-24605-3_37

6. Ekici, B., et al.: SMTCoq: a plug-in for integrating SMT solvers into Coq. In: Majumdar, R., Kunčak, V. (eds.) CAV 2017. LNCS, vol. 10427, pp. 126–133. Springer, Cham (2017). https://doi.org/10.1007/978-3-319-63390-9_7

7. Gelder, A.V.: Verifying RUP proofs of propositional unsatisfiability. In: International Symposium on Artificial Intelligence and Mathematics (ISAIM). Springer (2008)

8. Hadarean, L., Barrett, C., Reynolds, A., Tinelli, C., Deters, M.: Fine grained SMT proofs for the theory of fixed-width bit-vectors. In: Davis, M., Fehnker, A., McIver, A., Voronkov, A. (eds.) LPAR 2015. LNCS, vol. 9450, pp. 340–355. Springer, Heidelberg (2015). https://doi.org/10.1007/978-3-662-48899-7_24

9. Harper, R., Honsell, F., Plotkin, G.: A framework for defining logics. J. ACM 40(1), 143–184 (1993)

10. Kiesl, B., Rebola-Pardo, A., Heule, M.J.H.: Extended resolution simulates DRAT. In: Galmiche, D., Schulz, S., Sebastiani, R. (eds.) IJCAR 2018. LNCS (LNAI), vol. 10900, pp. 516–531. Springer, Cham (2018). https://doi.org/10.1007/978-3-319-94205-6_34

11. Pardo, A.R., Biere, A.: Two flavors of drat. In: Berre, D.L., Järvisalo, M. (eds.) Proceedings of Pragmatics of SAT 2015 and 2018. EPiC Series in Computing, vol. 59, pp. 94–110. EasyChair (2019)

12. Soos, M., Nohl, K., Castelluccia, C.: Extending SAT Solvers to cryptographic problems. In: Kullmann, O. (ed.) SAT 2009. LNCS, vol. 5584, pp. 244–257. Springer, Heidelberg (2009). https://doi.org/10.1007/978-3-642-02777-2_24

13. Stump, A., Oe, D., Reynolds, A., Hadarean, L., Tinelli, C.: SMT proof checking using a logical framework. Form. Methods Syst. Des. **42**(1), 91–118 (2013)

14. Development team, T.C.: The Coq proof assistant reference manual version 8.9 (2019). https://coq.inria.fr/distrib/current/refman/

15. Tseitin, G.S.: On the complexity of derivation in propositional calculus. In: Siekmann, J.H., Wrightson, G. (eds.) Automation of Reasoning. Symbolic Computation (Artificial Intelligence). Springer, Heidelberg (1983). https://doi.org/10.1007/978-3-642-81955-1_28

16. Wetzler, N., Heule, M.J.H., Hunt, W.A.: DRAT-trim: efficient checking and trimming using expressive clausal proofs. In: Sinz, C., Egly, U. (eds.) SAT 2014. LNCS, vol. 8561, pp. 422–429. Springer, Cham (2014). https://doi.org/10.1007/978-3-319-09284-3_31

17. Necula, G.C.: Proof-carrying code. In: POPL, pp. 106–119. ACM Press (1997)

18. Dickerson, T., Gazzillo, P., Herlihy, M., Saraph, V., Koskinen, E.: Proof-carrying smart contracts. In: Zohar, A., Eyal, I., Teague, V., Clark, J., Bracciali, A., Pintore, F., Sala, M. (eds.) FC 2018. LNCS, vol. 10958, pp. 325–338. Springer, Heidelberg (2019). https://doi.org/10.1007/978-3-662-58820-8_22

Combining Resolution-Path Dependencies
with Dependency Learning

Tomáš Peitl[(✉)], Friedrich Slivovsky, and Stefan Szeider

Algorithms and Complexity Group, TU Wien, Vienna, Austria
{peitl,fs,sz}@ac.tuwien.ac.at

Abstract. We present the first practical implementation of the reflexive resolution-path dependency scheme in a QBF solver. Unlike in DepQBF, which uses the less general standard dependency scheme, we do not compute the dependency relation upfront, but instead query relevant dependencies on demand during dependency conflicts, when the solver is about to learn a missing dependency. Thus, our approach is fundamentally tied to dependency learning, and shows that the two techniques for dependency analysis can be fruitfully combined. As a byproduct, we propose a quasilinear-time algorithm to compute all resolution-path dependencies of a given variable. Experimental results on the QBF library confirm the viability of our technique and identify families of formulas where the speedup is particularly promising.

1 Introduction

Dependency analysis is a state-of-the-art technique in QBF solving, in which the QBF solver attempts to identify spurious syntactic dependencies between variables and by doing so simplify the quantifier prefix. The historically older approach to dependency analysis are *dependency schemes* [14], first implemented in the QCDCL (Quantified Conflict-Driven Clause/Cube/Constraint Learning) solver DepQBF [2]. A dependency scheme is a mapping that, given a formula, identifies pairs of variables that are syntactic dependencies according to the quantifier prefix, but can in fact be safely ignored. DepQBF employs the standard dependency scheme in order to identify pairs of variables that are guaranteed to be independent. A more recent idea, implemented in the QCDCL solver Qute [11], is *dependency learning*, where the solver speculatively assumes all pairs of variables to be independent, and updates the information whenever it proves wrong during search.

Since the dawn of DepQBF and its use of the standard dependency scheme, one of the main open questions in QBF dependency analysis has been whether stronger dependency schemes can be utilized as well. Since DepQBF uses tailor-made data structures to efficiently compute the standard dependency scheme [3], one cannot answer this question by simply substituting a different dependency

This research was partially supported by FWF grants P27721 and W1255-N23.

M. Janota and I. Lynce (Eds.): SAT 2019, LNCS 11628, pp. 306–318, 2019.
https://doi.org/10.1007/978-3-030-24258-9_22

scheme into DepQBF. Of particular interest would be an efficient implementation of the *reflexive resolution-path dependency scheme* [16,17], the strongest known tractable sound one.

The main issue with implementing dependency schemes is that the number of pairs of variables in the dependency relation can in the worst case be quadratic, which turns out to be impractical for a large number of relevant formulas. We therefore take a different approach, and only compute parts of the dependency relation on demand. We do this during *dependency conflicts*, a state of the solver unique to QCDCL with dependency learning, in which the solver attempts to perform a resolution step, but fails due to universal literals (in the case of clause resolution) left of the pivot variable appearing in different polarities in the two clauses and thus blocking the resolution step. When the solver encounters a dependency conflict, it would normally have to learn a new dependency. Instead, we compute the dependencies of the pivot variable and filter out any blocking variables that are actually independent. If it turns out that no blocking variables remain, the resolution step can be carried out, otherwise a dependency on one or more of the remaining blocking variables is learned.

While it is known that resolution paths are equivalent to directed paths in the implication graph of the formula [15], using this result directly would require us to perform one search for each blocking variable, resulting in an overall quadratic running time. Instead, we show that all dependencies of a given variable can be found by searching for *widest paths* in a weighted variant of the implication graph. This is a well-studied problem that can be solved efficiently, for instance in overall quasilinear time using a variant of Dijkstra's algorithm [9].

We implemented the dependency scheme in the QBF solver Qute, and evaluated our implementation on the entire QBF Library [5], preprocessed by the preprocessor HQSpre [18]. We observed a modest increase in the total number of solved instances. We also identified families of formulas on which the use of the dependency scheme appears to be particularly beneficial.

2   Preliminaries

A *CNF formula* is a finite conjunction $C_1 \wedge \cdots \wedge C_m$ of clauses, a *clause* is a finite disjunction $(\ell_1 \vee \cdots \vee \ell_k)$ of literals, and a *literal* is a variable x or a negated variable $\bar{x}$. We will also refer to *terms* (also known as *cubes*), which are finite conjunctions of literals. Whenever convenient, we consider clauses and terms as sets of literals, and CNF formulas as sets of clauses. The length of a CNF formula $\varphi = C_1 \wedge \cdots \wedge C_m$ is defined as $\|\varphi\| = \sum_{i=1}^{m} |C_m|$.

We consider QBFs in Prenex Conjunctive Normal Form (PCNF), i.e., formulas $\mathcal{F} = \mathcal{Q}.\varphi$ consisting of a (quantifier) prefix $\mathcal{Q}$ and a propositional CNF formula φ, called the *matrix* of $\mathcal{F}$. The *prefix* is a sequence $\mathcal{Q} = Q_1 x_1 \ldots Q_n x_n$, where $Q_i \in \{\forall, \exists\}$ is a universal or existential quantifier and the x_i are (universal or existential) variables. The *depth* of a variable x_i is defined as $\delta(x_i) = i$, the depth of a literal ℓ is $\delta(\ell) = \delta(\text{var}(\ell))$. If $\delta(x) < \delta(y)$ we say that x is left of y, y is right of x, and we write $R_{\mathcal{F}}(x) = \{ v \in \text{var}(\mathcal{F}) \mid \delta(x) < \delta(v) \}$, and

$x_i \prec_{\mathcal{F}} x_j$ if $1 \leq i < j \leq n$ and $Q_i \neq Q_j$, dropping the subscript if the formula $\mathcal{F}$ is understood. The length of a PCNF formula $\mathcal{F} = \mathcal{Q}.\varphi$ is defined as $\|\mathcal{F}\| = \|\varphi\|$.

We assume that PCNF formulas are *closed*, so that every variable occurring in the matrix appears in the prefix, and that each variable appearing in the prefix occurs in the matrix. We write $\text{var}(x) = \text{var}(\overline{x}) = x$ to denote the variable associated with a literal and let $\text{var}(C) = \{\, \text{var}(\ell) \mid \ell \in C \,\}$ if C is a clause (term), $\text{var}(\varphi) = \bigcup_{C \in \varphi} \text{var}(C)$ if φ is a CNF formula, and $\text{var}(\mathcal{F}) = \text{var}(\varphi)$ if $\mathcal{F} = \mathcal{Q}.\varphi$ is a PCNF formula.

The semantics of a PCNF formula Φ are defined as follows. If Φ does not contain any variables then Φ is true if its matrix is empty and false if its matrix contains the empty clause $\emptyset$. Otherwise, let $\Phi = Qx\mathcal{Q}.\varphi$. If $Q = \exists$ then Φ is true if $\Phi[(x)]$ is true or $\Phi[(\neg x)]$ is true, and if $Q = \forall$ then Φ is true if both $\Phi[(x)]$ and $\Phi[(\neg x)]$ are true.

2.1   QCDCL and Q-Resolution

We briefly review QCDCL and Q-resolution [10], its underlying proof system. More specifically, we consider *long-distance Q-resolution*, a version of Q-resolution that admits the derivation of tautological clauses in certain cases. Although this proof system was already used in early QCDCL solvers [19], the formal definition shown in Fig. 1 was given only recently [1]. A dual proof system called *(long-distance) Q-consensus*, which operates on terms instead of clauses, is obtained by swapping the roles of existential and universal variables (the analogue of universal reduction for terms is called *existential reduction*).

$$\frac{C_1 \vee e \qquad \neg e \vee C_2}{C_1 \vee C_2} \text{ (resolution)}$$

The *resolution* rule allows the derivation of $C_1 \vee C_2$ from clauses $C_1 \vee e$ and $\neg e \vee C_2$, provided that the *pivot* variable e is existential and that $e \prec \text{var}(\ell_u)$ for each universal literal $\ell_u \in C_1$ such that $\overline{\ell_u} \in C_2$. The clause $C_1 \vee C_2$ is called the resolvent of $C_1 \vee e$ and $\neg e \vee C_2$. Each variable u for which $\ell_u \in C_1$ and $\overline{\ell_u} \in C_2$ is said to be *merged over* e in this resolution step.

$$\frac{C}{C \setminus \{u, \neg u\}} \text{ (universal reduction)}$$

The *universal reduction* rule admits the deletion of a universal variable u from a clause C under the condition that $e \prec u$ for each existential variable e in C.

Fig. 1. Long-distance Q-resolution.

A (long-distance) Q-resolution *derivation* from a PCNF formula Φ is a sequence of clauses such that each clause appears in the matrix of Φ or can be derived from clauses appearing earlier in the sequence using resolution or universal reduction. A derivation of the empty clause is called a *refutation*, and one can show that a PCNF formula is false, if, and only if, it has a long-distance Q-resolution refutation [1]. Dually, a PCNF formula is true, if, and only if, the empty term can be derived from a DNF representation of its matrix by Q-consensus.

Starting from an input PCNF formula, QCDCL generates ("learns") *constraints*—clauses and terms—until it produces an empty constraint. Every clause learned by QCDCL can be derived from the input formula by Q-resolution, and every term learned by QCDCL can be derived by Q-consensus [4,6]. Accordingly, the solver outputs TRUE if the empty term is learned, and FALSE if the empty clause is learned.

3 Resolution-Path Dependency Scheme

The reflexive resolution path dependency scheme detects spurious dependencies of PCNF formulas based on *resolution paths* [16,17].

Definition 1 (Resolution Path). *Let $\mathcal{F} = Q.\varphi$ be a PCNF formula and let X be a set of variables. A* resolution path *from ℓ_1 to ℓ_{2k} in $\mathcal{F}$ is a sequence $\pi = \ell_1, \ldots, \ell_{2k}$ of literals satisfying the following properties:*

1. *for all $i \in \{1, \ldots, k\}$, there is a $C_i \in \varphi$ such that $\ell_{2i-1}, \ell_{2i} \in C_i$,*
2. *for all $i \in \{1, \ldots, k\}$, $var(\ell_{2i-1}) \neq var(\ell_{2i})$,*
3. *for all $i \in \{1, \ldots, k-1\}$, $\overline{\ell_{2i}} = \ell_{2i+1}$.*

If additionally

4. *for all $i \in \{1, \ldots, k-1\}$, $\{\ell_{2i}, \ell_{2i+1}\} \subseteq X \cup \overline{X}$,*

then we say that π is a resolution path via X. If $\pi = \ell_1, \ldots, \ell_{2k}$ is a resolution path in $\mathcal{F}$ (via X), we say that ℓ_1 and ℓ_{2k} are connected *in $\mathcal{F}$ (with respect to X). For every $i \in \{1, \ldots, k-1\}$ we say that π* goes through $var(\ell_{2i})$ *and $var(\ell_{2i}), 1 \leq i < k$ are the* connecting variables *of π.*

Definition 2 (Proper Resolution Path). *Let ℓ, ℓ' be two literals of a PCNF formula $\mathcal{F}$ such that $\delta(\ell') < \delta(\ell)$. A resolution path from ℓ to ℓ' is called* proper, *if it is a resolution path via $R_{\mathcal{F}}(var(\ell')) \cap var_\exists(\mathcal{F})$. If there is a proper resolution path from ℓ to ℓ', we say that ℓ and ℓ' are* properly connected *(in $\mathcal{F}$).*

Resolution paths can be understood in terms of walks in the *implication graph* of a formula [15].

Definition 3 (Implication graph). *Let $\mathcal{F} = Q.\varphi$ be a PCNF formula. The* implication graph *of $\mathcal{F}$, denoted by $IG(\mathcal{F})$ is the directed graph with vertex set $var(\mathcal{F}) \cup \overline{var(\mathcal{F})}$ and edge set $\{(\overline{\ell}, \ell') \mid$ there is a $C \in \varphi$ such that $\ell, \ell' \in C$ and $\ell \neq \ell'\}$.*

Lemma 1 ([15]). *Let $\mathcal{F}$ be a PCNF formula and let $\ell, \ell' \in var(\mathcal{F}) \cup \overline{var(\mathcal{F})}$ be distinct literals. The following statements are equivalent:*

1. $\ell, \ell_1, \overline{\ell_1}, \ldots, \ell_k, \overline{\ell_k}, \ell'$ is a resolution path from ℓ to ℓ',
2. $\overline{\ell}, \ell_1, \ldots, \ell_k, \ell'$ is a path in $IG(\mathcal{F})$.

The resolution path dependency scheme identifies variables connected by a pair of resolution paths as potentially dependent on each other. We call a pair of variables connected in this way a *dependency pair*.

Definition 4 (Dependency pair). *Let $\mathcal{F}$ be a PCNF formula and $x, y \in var(\mathcal{F})$. We say $\{x, y\}$ is a resolution-path dependency pair of $\mathcal{F}$ with respect to $X \subseteq var_\exists(\mathcal{F})$ if at least one of the following conditions holds:*

- *x and y, as well as $\neg x$ and $\neg y$, are connected in $\mathcal{F}$ with respect to X.*
- *x and $\neg y$, as well as $\neg x$ and y, are connected in $\mathcal{F}$ with respect to X.*

It remains to determine the set X of variables with respect to which a pair x, y of variables needs to be connected to induce a dependency. For $x \prec_\mathcal{F} y$, the original *resolution-path dependency scheme* only included dependency pairs $\{x, y\}$ connected with respect to existential variables to the right of x, excluding x and y. It turns out that this dependency scheme can be used for reordering the quantifier prefix [15] but does not lead to a sound generalization of Q-resolution as required for use within a QCDCL-solver [16]. By dropping the restriction that x and y must not appear on the resolution paths inducing a dependency pair, we obtain the *reflexive resolution-path dependency scheme*, which yields a sound generalization of Q-resolution [16].

Definition 5 (Proper dependency pair). *Let $\mathcal{F}$ be a PCNF formula and $x, y \in var(\mathcal{F})$, $\delta(x) < \delta(y)$. We say $\{x, y\}$ is a proper resolution-path dependency pair of $\mathcal{F}$ if at least one of the following conditions holds:*

- *x and y, as well as $\neg x$ and $\neg y$, are properly connected in $\mathcal{F}$.*
- *x and $\neg y$, as well as $\neg x$ and y, are properly connected in $\mathcal{F}$.*

Definition 6. *The reflexive resolution-path dependency scheme is the mapping $\mathcal{D}^{rrs}$ that assigns to each PCNF formula $\mathcal{F} = \mathcal{Q}.\varphi$ the relation*

$$\mathcal{D}_\mathcal{F}^{rrs} = \{ x \prec_\mathcal{F} y \mid \{x, y\} \text{ is a proper resolution-path dependency pair of } \mathcal{F} \}.$$

When $\mathcal{D}^{rrs}$ is used in QCDCL solving, the solver learns clauses in a generalization of long-distance Q-resolution called LDQ($\mathcal{D}^{rrs}$)-resolution. Figure 3 shows the proof rules of LDQ($\mathcal{D}^{rrs}$)-resolution. Soundness of LDQ($\mathcal{D}^{rrs}$)-resolution has been established by [12].

Theorem 1 (Corollary 3, [12]). *LDQ($\mathcal{D}^{rrs}$)-resolution is sound.*

We note that the soundness of the corresponding LDQ($\mathcal{D}^{rrs}$)-consensus for terms still remains as an open problem. In our experiments with the proof system, we have been able to independently verify the truth value of all formulas by a different QBF solver.

$$\frac{}{C} \text{ (input clause)} \qquad \frac{C_1 \vee e \qquad \neg e \vee C_2}{C_1 \vee C_2} \text{ (resolution)}$$

An *input clause* $C \in \varphi$ can be used as an axiom. From two clauses $C_1 \vee e$ and $\neg e \vee C_1$, where e is an existential variable, the *(long-distance) resolution* rule can derive the clause $C_1 \vee C_2$, provided that $(u, e) \notin \mathcal{D}_{\mathcal{F}}^{\mathrm{rrs}}$ for each universal variable u with $u \in C_1$ and $\overline{u} \in C_2$ (or vice versa), and that $C_1 \vee C_2$ does not contain an existential variable in both polarities.

$$\frac{C}{C \setminus \{u, \neg u\}} \text{ (generalized } \forall\text{-reduction)}$$

The $\forall$-*reduction* rule derives the clause $C \setminus \{u, \neg u\}$ from C, where $u \in \mathrm{var}(C)$ is a universal variable such that $(u, e) \notin \mathcal{D}_{\mathcal{F}}^{\mathrm{rrs}}$ for every existential variable $e \in \mathrm{var}(C)$.

Fig. 2. Derivation rules of $\mathrm{LDQ}(\mathcal{D}^{\mathrm{rrs}})$-resolution for a PCNF formula $\mathcal{F} = \mathcal{Q}.\varphi$.

4 Using Resolution-Path Dependencies in Practice

The major issue with implementing any dependency scheme for use in a QBF solver is the fact that the size of the dependency relation is inherently worst-case quadratic in the number of variables—all pairs of variables of opposite quantifier type potentially need to be stored. QBFs of interest often contain hundreds of thousands of variables, and therefore any procedure with quadratic complexity is infeasible. DepQBF overcomes this by identifying equivalence classes of variables with identical dependency information, and storing only one chunk of data per equivalence class [3]. This compressed form, however, is specifically tailored to the standard dependency scheme, and cannot directly be transferred to other dependency schemes.

4.1 Dynamically Applying $\mathcal{D}^{\mathrm{rrs}}$

In order to avoid the quadratic blowup, we take a different approach. We do not aim at computing the entire dependency relation, but instead compute parts of it on demand, when a *dependency conflict* occurs.

Dependency conflicts in clause learning in QCDCL with dependency learning take place in the following way (in this entire section we focus on the case of clauses, but the case of term learning is dual): the solver attempts to resolve two clauses, C_1 and C_2, over a pivot variable e, but there is a non-empty set of universal variables U, such that

$$\forall u \in U \ u \prec e, (u \in C_1 \wedge \overline{u} \in C_2) \vee (\overline{u} \in C_1 \wedge u \in C_2).$$

These variables are blocking the resolution step, as is shown in the pseudocode snippet in Algorithm 1 (for a more thorough treatment of QCDCL with dependency learning we refer to [11]). The reason why this occurs is that the solver mistakenly assumed e not to depend on any $u \in U$, and this erroneous assumption is now to be rectified by learning the dependency of e on at least one variable from U.

Algorithm 1. Conflict Analysis with Dependency Learning

```
 1: procedure ANALYZECONFLICT(conflict)
 2:     constraint = GETCONFLICTCONSTRAINT(conflict)
 3:     while NOT ASSERTING(constraint) do
 4:         pivot = GETPIVOT(constraint)
 5:         reason = GETANTECEDENT(pivot)
 6:         if EXISTSRESOLVENT(constraint, reason, pivot) then
 7:             constraint = RESOLVE(constraint, reason, pivot)
 8:             constraint = REDUCE(constraint)
 9:         else // dependency conflict
10:             U = ILLEGALMERGES(constraint, reason, pivot)
11:             D = D ∪ { (v, pivot) | v ∈ U }
12:             return NONE, DECISIONLEVEL(pivot)
13:         end if
14:     end while
15:     btlevel = GETBACKTRACKLEVEL(constraint)
16:     return constraint, btlevel
17: end procedure
```

We can conveniently insert a dynamically computed dependency scheme at this moment. Before any dependency of e is learned, the dependencies of e according to the dependency scheme are computed. Any $u \in U$ that turns out to be independent of e can be removed from the set of blocking variables. If everything in U is independent, no dependency needs to be learned, and conflict analysis can proceed by performing a resolution step in $\mathrm{LDQ}(\mathcal{D}^{\mathrm{rrs}})$-resolution, in which all $u \in U$ are merged over e. If some variables in U turn out to be actual dependencies of e, at least one of them has to be learned as usual. The modification to the conflict analysis process is shown in Algorithm 2.

The computed dependencies of e are then stored and re-used in any future dependency conflicts featuring e as the pivot variable, as well as in strengthening the reduction rule.

Soundness of QCDCL with dependency learning *and* the reflexive resolution-path dependency scheme follows from the soundness of long-distance $\mathrm{Q}(\mathcal{D}^{\mathrm{rrs}})$-resolution, the underlying proof system used by the algorithm.

4.2 Dynamically Computing $\mathcal{D}^{\mathrm{rrs}}$

When computing resolution-path connections, it is natural to start with a variable v, and compute all variables which depend on v. This is because in this case, the set of connecting variables that can form proper resolution paths is

---

Algorithm 2. Conflict Analysis with DL and a Dependency Scheme

```
 1: procedure ANALYZECONFLICT(conflict)
 2:     constraint = GETCONFLICTCONSTRAINT(conflict)
 3:     while NOT ASSERTING(constraint) do
 4:         pivot = GETPIVOT(constraint)
 5:         reason = GETANTECEDENT(pivot)
 6:         if EXISTSRESOLVENT(constraint, reason, pivot) then
 7:             constraint = RESOLVE(constraint, reason, pivot)
 8:             constraint = REDUCE(constraint)
 9:         else // dependency conflict
10:             U = ILLEGALMERGES(constraint, reason, pivot)
11:             rrs_deps[pivot] = getDependencies(pivot)
12:             U = U ∩ rrs_deps[pivot]
13:             if U = ∅ then
14:                 goto 7
15:             else
16:                 D = D ∪ { (v, pivot) | v ∈ U }
17:                 return NONE, DECISIONLEVEL(pivot)
18:             end if
19:         end if
20:     end while
21:     btlevel = GETBACKTRACKLEVEL(constraint)
22:     return constraint, btlevel
23: end procedure
```

---

fixed—all existential variables right of v are permitted—and the task of finding everything that depends on v is reducible to reachability in a single directed graph. However, since a dependency conflict may feature any number of blocking variables, we would potentially need to perform the search many times in order to check each dependency. It would be preferable to compute all dependencies of the pivot variable instead. However, since for every blocking variable $u \in U$, the set of allowed connecting variables may be different, we cannot reduce the task of finding all dependencies of the pivot e to just reachability in a single directed graph, and we need a different approach.[1]

Definition 7. *Let $\mathcal{F}$ be a PCNF formula, ℓ a literal of $\mathcal{F}$, and $w_\ell : var(\mathcal{F}) \cup var(\mathcal{F}) \to \mathbb{R} \cup \{\pm\infty\}$ the mapping defined by*

$$w_\ell(l) = \begin{cases} \infty & \text{if } l = \overline{\ell}, \\ \delta(\ell) & \text{if } l \neq \overline{\ell} \text{ and } var(l) \text{ is existential}, \\ -\infty & \text{otherwise.} \end{cases}$$

The depth-implication graph for $\mathcal{F}$ at ℓ, denoted $DIG(\mathcal{F}, \ell)$ is the weighted version of $IG(\mathcal{F})$ where the weight of an edge (ℓ_1, ℓ_2) is defined as $w(\ell_1, \ell_2) = w_\ell(\ell_1)$.

---

[1] This is the case regardless of the quantifier type of the pivot, the issue is that different targets in the set of blocking variables can be reached using different connecting variables.

For a path π in a weighted directed graph G, the *width* of π is defined as the minimum weight over all edges of π. The following theorem relates resolution paths in a formula with *widest paths* in its depth-implication graph.

Theorem 2. *Let ℓ, ℓ' be two literals of a PCNF formula $\mathcal{F}$ such that $\delta(\ell') < \delta(\ell)$. There is a proper resolution path from ℓ to ℓ' if, and only if, the widest path from $\overline{\ell}$ to ℓ' in $DIG(\mathcal{F}, \ell)$ has width larger than $\delta(\ell')$.*

Proof. Let $\pi = \ell, \ell_2, \ldots, \ell_{2k-1}, \ell'$ be a proper resolution path, and let $\pi' = \overline{\ell}, \ell_2, \ldots, \ell_{2k-2}, \ell'$ be the corresponding path in $DIG(\mathcal{F}, \ell)$ (by Lemma 1). The width of π' is defined as

$$w(\pi') = \min \left\{ w(\overline{\ell}, \ell_2), \ldots, w(\ell_{2k-2}, \ell') \right\}$$
$$= \min \left\{ w_\ell(\overline{\ell}), \ldots, w_\ell(\ell_{2k-2}) \right\}.$$

Since $w_\ell(\overline{\ell}) = \infty$ and π is proper and hence none of its connecting variables are universal, we have that $w(\pi') = \min \left\{ \delta(\ell_2), \ldots, \delta(\ell_{2k-2}) \right\} > \delta(\ell')$, where the inequality follows from π being proper.

Conversely, let $\pi' = \overline{\ell_1}, \ell_2, \ldots, \ell_k, \ell'$ be a path of width greater than $\delta(\ell')$, and let $\pi = \ell_1, \ell_2, \overline{\ell_2}, \ldots, \ell_k, \overline{\ell_k}, \ell'$ be the corresponding resolution path. Since $w(\pi') > \delta(\ell')$, no connecting variables in π can be universal, and they all have to be right of ℓ', hence π is proper. $\qquad\square$

Naively applying the algorithm from [15] would result in an overall quadratic running time needed to determine all dependencies of a given variable v. Using Theorem 2 we can reduce the task to two searches for widest paths, and obtain a much more favourable time bound.

Theorem 3. *Given a variable v of a PCNF formula $\mathcal{F}$, all resolution-path dependencies, i.e., the set $\left\{ x \in var(\mathcal{F}) \mid (x, v) \in \mathcal{D}_{\mathcal{F}}^{rrs} \right\}$, can be computed in time $O\left(\|\mathcal{F}\| \log \|\mathcal{F}\|\right)$.*

Proof. In order to find out whether a given candidate variable x is a dependency of v, one has to determine whether there is a pair of proper resolution paths, either from v to x and from $\overline{v}$ to $\overline{x}$, or from v to $\overline{x}$ and from $\overline{v}$ to x. Theorem 2 tells us that the existence of proper resolution paths is equivalent to existence of wide paths. A generalization of Dijkstra's algorithm can compute widest paths from a single source to all destinations in a given graph in quasilinear time [9]. The key observation is that the entire computation is performed within two graphs, namely $DIG(\mathcal{F}, v)$ and $DIG(\mathcal{F}, \overline{v})$. By computing all widest paths from both v and $\overline{v}$, and then subsequently checking for which candidate variables x both polarities of x are reached by a wide enough path, we can find all dependencies of v.

By using the clause-splitting trick like in [15] we can, in linear time, obtain an equisatisfiable formula $\mathcal{F}'$ with $var(\mathcal{F}) \subseteq var(\mathcal{F}')$ such that the resolution-path connections between variables of $\mathcal{F}$ are the same. Since $\mathcal{F}'$ has bounded clause size, we get that the number of edges in $IG(\mathcal{F}')$ is $O\left(\|\mathcal{F}'\|\right) = O\left(\|\mathcal{F}\|\right)$, and the stated running time is then simply the running time of Dijkstra's algorithm. $\quad\square$

5   Experiments

We modified the dependency-learning solver Qute so as to perform the procedure described above—when a dependency is about to be learned, resolution-path dependencies of the pivot variable are computed, and all blocking variables that turned out to be spurious dependencies are eliminated. Furthermore, the computed dependencies are kept for re-use in future dependency conflicts featuring the same pivot variable, as well as to be used in generalized ∀-reduction.

We evaluated our solver on a cluster of 16 machines each having two 10-core Intel Xeon E5-2640 v4, 2.40 GHz processors and 160 GB of RAM, running Ubuntu 16.04. We set the time limit to 900 s and the memory limit to 4 GB. As our benchmark set, we selected the QDIMACS instances available in the QBF Library[2] [5]. We first preprocessed them using the preprocessor HQSpre[3] [18] with a time limit of 400 seconds, resulting in a set of 14893 instances not solved by HQSpre. Out of these instances, we further identified the set of easy instances as those solved within 10 seconds by each of the following solvers: CaQE[4] 3.0.0 [13], DepQBF[5] 6.03 [2], QESTO[6] 1.0 [8], Qute[7] 1.1 [11], and RaReQS[8] 1.1 [7]. We decided to focus only on instances not solved by at least one of these solvers in under 10 s, as it arguably makes little sense to try and push state of the art for formulas that can already be solved in almost no time regardless of the choice of the solver. That left us with a set of 11262 instances.

Table 1 and Fig. 3 show the comparison between plain Qute and the version which implements the dependency scheme (Qute-$\mathcal{D}^{rrs}$). The version with the dependency scheme solved 176 (roughly 4.5%) more instances than the version without. The scatter plot in Fig. 3 deserves further attention. While the overall number of solved instances is higher for Qute-$\mathcal{D}^{rrs}$, the plot is skewed towards Qute-$\mathcal{D}^{rrs}$. We attribute this to a small overhead associated with the use of the dependency scheme, which is most apparent for the easiest formulas. The plot also shows that there are a few formulas solved by the plain version, but not by Qute-$\mathcal{D}^{rrs}$. This is only partly due to the additional time spent computing resolution paths, and is, in our opinion, in much larger part due to the heuristics being led off the right track towards a proof of the formula.

We found two families of instances where the increase in number of solved instances is even more significant, as is documented in Table 1. Particularly on the *matrix multiplication* and *reduction finding* benchmarks the dependency scheme provides a tremendous boost of performance, resulting in almost four times as many solved instances.

---

[2] http://www.qbflib.org/.

[3] https://projects.informatik.uni-freiburg.de/users/4.

[4] https://www.react.uni-saarland.de/tools/caqe.

[5] https://github.com/lonsing/depqbf.

[6] http://sat.inesc-id.pt/~mikolas/sw/qesto.

[7] https://github.com/perebor/qute.

[8] http://sat.inesc-id.pt/~mikolas/sw/areqs.

Table 1. Number of instances solved by plain Qute vs Qute using the reflexive resolution-path dependency scheme on the 'matrix multiplication' and 'reduction finding' families of formulas, as well as on all instances.

	MM-family	RF-family	all instances
# of instances	334	2269	11262
solved by Qute (SAT / UNSAT)	34 (4/30)	423 (140/283)	3959 (1467/2492)
solved by Qute-$\mathcal{D}^{\mathrm{rrs}}$ (SAT / UNSAT)	123 (4/119)	484 (144/340)	4135 (1489/2646)

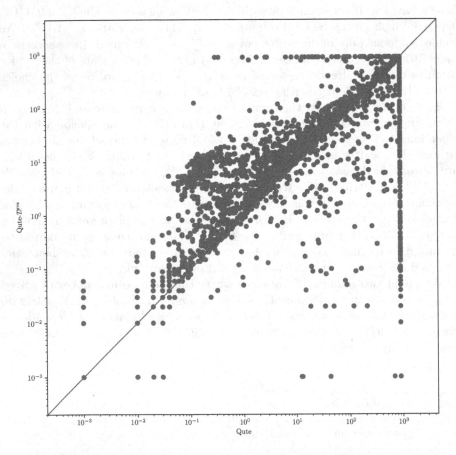

Fig. 3. Runtimes of Qute with and without $\mathcal{D}^{\mathrm{rrs}}$ on all instances.

6   Conclusion and Future Work

We presented the first practical implementation of $\mathcal{D}^{rrs}$ in a QBF solver. Thus, we have demonstrated that the strongest known tractable sound dependency scheme can be efficiently used in QBF solving. Our approach shows that dependency schemes can be fruitfully combined with dependency learning. Our algorithm for the computation of all resolution-path dependencies of a given variable may also be of independent interest.

While the additional prefix relaxation that comes from $\mathcal{D}^{rrs}$ is no cure-all for the hardness of QBF, we have found families of formulas where it provides a significant speedup. In particular, the use of the dependency scheme turned out very beneficial on the 'matrix multiplication' and 'reduction finding' classes, which are both practically relevant applications and further improvement using QBF would be valuable.

A possible direction for future work is to try to further improve the time bound of our algorithm for computing the resolution-path dependencies of a variable either by using data structures more suitable for this concrete scenario, or by preprocessing the formula. A succinct, possibly implicit, representation of $\mathcal{D}^{rrs}$ for use in other solver architectures would also be very interesting.

References

1. Balabanov, V., Jiang, J.H.R.: Unified QBF certification and its applications. Formal Methods Syst. Des. **41**(1), 45–65 (2012)
2. Lonsing, F., Biere, A.: Integrating Dependency schemes in search-based QBF solvers. In: Strichman, O., Szeider, S. (eds.) SAT 2010. LNCS, vol. 6175, pp. 158–171. Springer, Heidelberg (2010). https://doi.org/10.1007/978-3-642-14186-7_14
3. Lonsing, F., Biere, A.: A compact representation for syntactic dependencies in QBFs. In: Kullmann, O. (ed.) SAT 2009. LNCS, vol. 5584, pp. 398–411. Springer, Heidelberg (2009). https://doi.org/10.1007/978-3-642-02777-2_37
4. Egly, U., Lonsing, F., Widl, M.: Long-distance resolution: proof generation and strategy extraction in search-based QBF solving. In: McMillan, K., Middeldorp, A., Voronkov, A. (eds.) LPAR 2013. LNCS, vol. 8312, pp. 291–308. Springer, Heidelberg (2013). https://doi.org/10.1007/978-3-642-45221-5_21
5. Giunchiglia, E., Narizzano, M., Pulina, L., Tacchella, A.: Quantified Boolean Formulas satisfiability library (QBFLIB) (2005). www.qbflib.org
6. Giunchiglia, E., Narizzano, M., Tacchella, A.: Clause/term resolution and learning in the evaluation of quantified Boolean formulas. J. Artif. Intell. Res. **26**, 371–416 (2006)
7. Janota, M., Klieber, W., Marques-Silva, J., Clarke, E.: Solving QBF with counterexample guided refinement. In: Cimatti, A., Sebastiani, R. (eds.) SAT 2012. LNCS, vol. 7317, pp. 114–128. Springer, Heidelberg (2012). https://doi.org/10.1007/978-3-642-31612-8_10
8. Janota, M., Marques-Silva, J.: Solving QBF by clause selection. In: Yang, Q., Wooldridge, M. (eds.) Proceedings of the Twenty-Fourth International Joint Conference on Artificial Intelligence, IJCAI 2015, pp. 325–331. AAAI Press (2015)
9. Kaibel, V., Peinhardt, M.: On the bottleneck shortest path problem. Zib-report 06–22, Zuse Institute Berlin (2006)

10. Büning, H.K., Karpinski, M., Flögel, A.: Resolution for quantified Boolean formulas. Information and Computation **17**(1), 12–18 (1995)
11. Peitl, T., Slivovsky, F., Szeider, S.: Dependency learning for QBF. J. Artif. Intell. Res., 65 (2019)
12. Peitl, T., Slivovsky, F., Szeider, S.: Long-distance Q-resolution with dependency schemes. J. Autom. Reason. **63**(1), 127–155 (2019)
13. Rabe, M.N., Tentrup, L.: CAQE: A certifying QBF solver. In: Kaivola, R., Wahl, T. (eds.) Formal Methods in Computer-Aided Design - FMCAD 2015, pp. 136–143. IEEE Computer Society (2015)
14. Samer, M., Szeider, S.: Backdoor sets of quantified Boolean formulas. J. Autom. Reason. **42**(1), 77–97 (2009)
15. Slivovsky, F., Szeider, S.: Quantifier reordering for QBF. J. Autom. Reason. **56**(4), 459–477 (2016)
16. Slivovsky, F., Szeider, S.: Soundness of Q-resolution with dependency schemes. Theor. Comput. Sci. **612**, 83–101 (2016)
17. Gelder, A.: Variable Independence and resolution paths for quantified Boolean formulas. In: Lee, J. (ed.) CP 2011. LNCS, vol. 6876, pp. 789–803. Springer, Heidelberg (2011). https://doi.org/10.1007/978-3-642-23786-7_59
18. Wimmer, R., Reimer, S., Marin, P., Becker, B.: HQSpre – an effective preprocessor for QBF and DQBF. In: Legay, A., Margaria, T. (eds.) TACAS 2017. LNCS, vol. 10205, pp. 373–390. Springer, Heidelberg (2017). https://doi.org/10.1007/978-3-662-54577-5_21
19. Zhang, L., Malik, S.: Conflict driven learning in a quantified Boolean satisfiability solver. In: Pileggi, L.T., Kuehlmann, A. (eds.) Proceedings of the 2002 IEEE/ACM International Conference on Computer-aided Design, ICCAD 2002, San Jose, California, USA, 10–14 November 2002, pp. 442–449. ACM/IEEE Computer Society (2002)

Proof Complexity of Fragments of Long-Distance Q-Resolution

Tomáš Peitl, Friedrich Slivovsky[✉], and Stefan Szeider

Algorithms and Complexity Group, TU Wien, Vienna, Austria
{peitl,fs,sz}@ac.tuwien.ac.at

Abstract. Q-resolution is perhaps the most well-studied proof system for Quantified Boolean Formulas (QBFs). Its proof complexity is by now well understood, and several general proof size lower bound techniques have been developed. The situation is quite different for long-distance Q-resolution (LDQ-resolution). While lower bounds on LDQ-resolution proof size have been established for specific families of formulas, we lack semantically grounded lower bound techniques for LDQ-resolution.

In this work, we study restrictions of LDQ-resolution. We show that a specific lower bound technique based on bounded-depth strategy extraction does not work even for *reductionless Q-resolution* by presenting short proofs of the QPARITY formulas. Reductionless Q-resolution is a variant of LDQ-resolution that admits merging but no universal reduction. We also prove a lower bound on the proof size of the *completion principle* formulas in reductionless Q-resolution. This shows that two natural fragments of LDQ-resolution are incomparable: Q-resolution, which allows universal reductions but no merging, and reductionless Q-resolution, which allows merging but no universal reductions. Finally, we develop semantically grounded lower bound techniques for fragments of LDQ-resolution, specifically tree-like LDQ-resolution and regular reductionless Q-resolution.

1 Introduction

The effectiveness of modern satisfiability (SAT) solvers has established propositional logic as the language of choice for encoding hard combinatorial problems from areas such as formal verification [8,33] and AI planning [27]. However, since the computational complexity of these problems usually exceeds the complexity of SAT, propositional encodings of such problems can be exponentially larger than their original descriptions. This imposes a limit on the problem instances that can be feasibly solved even with extremely efficient SAT solvers, and has prompted research on decision procedures for more succinct logical formalisms such as Quantified Boolean Formulas (QBFs).

QBFs augment propositional formulas with existential and universal quantification over truth values and can be exponentially more succinct. The down-

---

This research was partially supported by FWF grants P27721 and W1255-N23.

M. Janota and I. Lynce (Eds.): SAT 2019, LNCS 11628, pp. 319–335, 2019.
https://doi.org/10.1007/978-3-030-24258-9_23

side of this conciseness is that the satisfiability problem of QBFs is PSPACE-complete [29], and in spite of substantial progress in solver technology, practically relevant instances remain hard to solve. Unlike in the case of SAT, where Conflict-Driven Clause Learning (CDCL) [15] has emerged as the single dominant solving paradigm, there is a variety of competing solver architectures for QBF, most of which are either based on a generalization of CDCL (QCDCL) [11,35], quantifier expansion [7,17], or clausal abstraction [18,26,30].

Research in proof complexity has provided valuable insights into the theoretical limits of different solving approaches and their relative strengths and weaknesses. *Q-resolution* [21] is perhaps the most well-studied QBF proof system, largely due to the fact that it is used by QCDCL solvers for proof generation (see Sect. 3). Early proof size lower bounds for Q-resolution relied on propositional hardness or ad-hoc arguments [21]. Semantically grounded lower bound techniques based on strategy extraction have been developed only recently [4,6]. These techniques identify properties of winning strategies extracted from proofs and use them to derive proof size lower bounds. They not only help us prove lower bounds for new classes of formulas but afford a better understanding of what *kinds* of problems certain proof systems can solve efficiently.

Long-distance Q-resolution is a variant of Q-resolution which allows the derivation of syntactically tautological clauses in certain cases [1] and which is arguably the most natural proof system for use in a QCDCL solver [13,35]. Although lower bounds for long-distance Q-resolution have been proved [3,6], we lack semantically grounded lower bound techniques for this proof system. In this paper, we present results on the proof complexity of restricted versions of long-distance Q-resolution:

1. We prove an exponential lower bound on the *reductionless* Q-resolution [9] proof size of a class of QBFs with short Q-resolution refutations [20].
2. We observe that the QPARITY formulas [6] have short proofs in reductionless Q-resolution. It has already been shown that these formulas have short (linear) proofs in long-distance Q-resolution [12], and in fact these proofs are reductionless. In combination with the first result, this proves the incomparability of Q-resolution and reductionless Q-resolution. It also marks the breakdown of a semantically grounded lower bound technique for Q-resolution [6]—strategies corresponding to reductionless Q-resolution proofs cannot be efficiently represented by bounded-depth circuits.
3. Finally, we develop semantically grounded lower bound techniques for restricted subsystems of long-distance Q-resolution. Specifically, we show that the strategy functions computed by proofs in *regular* reductionless Q-resolution are read-once branching programs, and that *tree-like* long-distance Q-resolution proofs correspond to bounded depth circuits.

2   Preliminaries

We assume a countably infinite set V of propositional *variables* and consider *propositional formulas* constructed from V using the connectives ¬ (negation),

$\wedge$ (conjunction), $\vee$ (disjunction), $\rightarrow$ (implication), and $\leftrightarrow$ (the biconditional). The *size* $|\varphi|$ of a propositional formula φ is number of variable occurrences in φ plus the number of connectives. Given a propositional formula φ, we write $var(\varphi)$ to denote the set of variables occurring in φ. A *literal* is a variable v or a negated variable $\neg v$. A *clause* is a finite disjunction of literals. A clause is *tautological* if it contains both v and $\neg v$ for some variable v. A propositional formula is in *conjunctive normal form (CNF)* if it is a finite conjunction of non-tautological clauses. An *assignment* (or *variable assignment*) is a function that maps a subset $X \subseteq V$ of variables to the set $\{0, 1\}$ of truth values. Given a propositional formula φ and an assignment $\tau : X \rightarrow \{0,1\}$ with $var(\varphi) \subseteq X$, we let $\varphi[\tau]$ denote the truth value obtained by evaluating φ under τ. The formula φ is *satisfied* by τ if $\varphi[\tau] = 1$, otherwise it is *falsified* by τ.

We consider Quantified Boolean Formulas in *Prenex Conjunctive Normal Form (PCNF)*. A PCNF formula $\mathcal{F} = \mathcal{Q}.\varphi$ consists of a *quantifier prefix* $\mathcal{Q}$ and a propositional formula φ in conjunctive normal form, called the *matrix* of $\mathcal{F}$. The quantifier prefix is a sequence $Q_1 x_1 \ldots Q_n x_n$ where $Q_i \in \{\exists, \forall\}$ and the x_i are pairwise distinct variables for $1 \le i \le n$. The quantifier prefix defines an ordering $<_{\mathcal{F}}$ on its variables as $x_i <_{\mathcal{F}} x_j$ for $1 \le i < j \le n$. We assume that $\{x_1, \ldots, x_n\} = var(\varphi)$ and write $var(\mathcal{F}) = var(\varphi)$. The set of *existential* variables of $\mathcal{F}$ is $var_\exists(\mathcal{F}) = \{ x_i \mid 1 \le i \le n, Q_i = \exists \}$, and the set of *universal* variables of $\mathcal{F}$ is $var_\forall(\mathcal{F}) = \{ x_i \mid 1 \le i \le n, Q_i = \forall \}$.

A *strategy function* for a universal variable $u \in var_\forall(\mathcal{F})$ is a Boolean function $f_u : 2^{D_{\mathcal{F}}^u} \rightarrow \{0, 1\}$. Here, $D_{\mathcal{F}}^u = \{ v \in var_\exists(\mathcal{F}) \mid v <_{\mathcal{F}} u \}$ is the set of existential variables to the left of u in the quantifier prefix and $2^{D_{\mathcal{F}}^u}$ denotes the set of assignments of $D_{\mathcal{F}}^u$. A *universal winning strategy* for a PCNF formula $\mathcal{F}$ is a family $\{ f_u \mid u \in var_\forall(\mathcal{F}) \}$ of strategy functions with the following property. Let $\tau : var(\mathcal{F}) \rightarrow \{0, 1\}$ be an assignment satisfying $\tau(u) = f_u(\tau|_{D_{\mathcal{F}}^u})$ for each universal variable u, where $\tau|_{D_{\mathcal{F}}^u}$ denotes the restriction of τ to $D_{\mathcal{F}}^u$. Then the matrix of $\mathcal{F}$ is falsified by τ. A PCNF formula $\mathcal{F}$ is *false* if there exists a universal winning strategy for $\mathcal{F}$ and *true* otherwise.

3   Q-Resolution Proof Systems

In this section, we are going to introduce several clausal proof systems for PCNF formulas. The original *Q-resolution* proof system consists of propositional resolution and the *universal reduction* rule for dealing with universally quantified variables. This system—which is displayed in Fig. 1—was shown to be sound and complete for false PCNF formulas [21]. Different variants of Q-resolution can be identified as the proof systems underlying search-based QBF solvers. For instance, the traces of certain DPLL-style algorithms can be mapped to Q-resolution proofs [14]. However, since ordinary Q-resolution explicitly forbids tautological resolvents, this requires that literals that would result in a tautology be resolved away in a recursive manner, a process that was shown to require exponential time in the worst case [32].

The exponential overhead can be avoided by a more intricate analysis of the implication graph [23], but arguably the more natural solution is to allow

$$\frac{}{C}\ \text{(input clause)} \qquad \frac{C_1 \vee e \qquad \neg e \vee C_2}{C_1 \vee C_2}\ \text{(resolution)}$$

An *input clause* $C \in \varphi$ can be used as an axiom. From two clauses $C_1 \vee e$ and $\neg e \vee C_1$, the *resolution* rule can derive the clause $C_1 \vee C_2$ provided that it is non-tautological. Here, e is an existential variable called the *pivot*.

$$\frac{C \vee \ell}{C}\ \text{(universal reduction)}$$

The *universal reduction* rule can derive the clause C from $C \vee \ell$ if $var(\ell)$ is universal and there is no existential variable $e \in var(C)$ with $\ell <_{\mathcal{F}} e$.

Fig. 1. Derivation rules of Q-resolution for a PCNF formula $\mathcal{F} = \mathcal{Q}.\varphi$.

tautological clauses during learning. This was the approach taken by an early version of CDCL for QBF [35], but the resulting proof system was only studied and proven sound under the name of *long-distance Q-resolution* much later [1]. Long-distance Q-resolution involves a generalized resolution rule that allows for the derivation of tautologies, or equivalently, of *merged literals*. A merged literal u^* is generated for a universal variable u upon resolving a clause $C_1 \vee e \vee u$ with a clause $C_2 \vee \neg e \vee \neg u$. Here, it is required that $e < u$ in the quantifier prefix. Since u^* is essentially a shorthand for $u \vee \neg u$, we let $\overline{u^*} = u^*$ and $var(u^*) = u$. The resolution rule of long-distance Q-resolution is shown in Fig. 2. The long-distance Q-resolution (LDQ-resolution) proof system is comprised of the input clause rule, long-distance resolution, and universal reduction.

$$\frac{C_1 \vee e \qquad \neg e \vee C_2}{(C_1 \setminus \overline{C_2}) \vee (C_2 \setminus \overline{C_1}) \vee \{\, u^* \mid u \in var(C_1 \cap \overline{C_2})\,\}}\ \text{(long-distance resolution)}$$

The *long-distance resolution* rule can derive clauses containing merged literals u^* for universal variables u. We require that $e <_{\mathcal{F}} u$ for each such variable $u \in var(C_1 \cap \overline{C_2})$, and that $C_1 \vee C_2$ does not contain an existential variable and its negation.

Fig. 2. The long-distance resolution rule for a PCNF formula $\mathcal{F}$.

The QBF solver GHOSTQ [22] uses a restricted version of long-distance Q-resolution without universal reduction. Indeed, in a search-based solver that assigns variables in the order of the quantifier prefix, universal reduction is not required to derive a learned clause. One only needs to identify purely universal clauses, which are treated as if they were empty. The corresponding traces can be construed as derivations in the proof system shown in Fig. 3. We refer to this system—which was first studied under the name of Q^w-resolution [9]—as *reductionless Q-resolution*.

$$\frac{}{C} \text{ (input clause)}$$

Every input clause $C \in \varphi$ can be used as an axiom.

$$\frac{C_1 \vee e \qquad \neg e \vee C_2}{(C_1 \setminus \overline{C_2}) \vee (C_2 \setminus \overline{C_1}) \vee \{\, u^* \mid u \in var(C_1 \cap \overline{C_2}) \,\}} \text{ (long-distance resolution)}$$

It is required that $e <_{\mathcal{F}} u$ for each variable $u \in var(C_1 \cap \overline{C_2})$ and that $C_1 \vee C_2$ does not contain an existential variable and its negation.

Fig. 3. Derivation rules of reductionless Q-resolution for a PCNF formula $\mathcal{F} = \mathcal{Q}.\varphi$.

As usual, we consider *derivations* in these proof systems which are sequences $C_1, \ldots, C_k$ of clauses such that each clause C_i is an axiom or derived from clauses appearing earlier in the sequence using one of the proof rules. The *size* of a derivation is the number k of clauses in the sequence. A *refutation* is a derivation of the empty clause or, in the case of reductionless Q-resolution, a derivation of a purely universal clause.

In the following sections, we will sometimes write $C_1 \odot_x C_2$ to denote the resolvent of C_1 and C_2 on pivot x. If the pivot is understood we may simply write $C_1 \odot C_2$.

4   A Lower Bound for Reductionless Q-Resolution

We generalize an exponential lower bound for *level-ordered* Q-resolution [19] for the *completion principle* formulas CR_n defined below. A Q-resolution derivation is level-ordered if the order of pivot variables encountered on any path in the derivation follows the order in the quantifier prefix. A level-ordered Q-resolution refutation can be turned into a reductionless Q-resolution refutation simply by omitting the reduction steps.

Definition 1 ([19]). *Let*

$$CR_n = \mathop{\exists}_{1 \leq i,j \leq n} x_{ij} \, \forall z \, \mathop{\exists}_{i=1}^{n} a_i \, \mathop{\exists}_{j=1}^{n} b_j \left(\bigwedge_{1 \leq i,j \leq n} A_{ij} \wedge B_{ij} \right) \wedge A \wedge B,$$

where

$$A_{ij} = x_{ij} \vee z \vee a_i, \qquad\qquad A = \overline{a_1} \vee \cdots \vee \overline{a_n},$$
$$B_{ij} = \overline{x_{ij}} \vee \overline{z} \vee b_j, \qquad\qquad B = \overline{b_1} \vee \cdots \vee \overline{b_n}.$$

We will prove the following result.

Theorem 1. *Any reductionless Q-resolution refutation of the formula* CR_n *has size at least* 2^n.

In the following we assume $n \geq 2$ (Theorem 1 obviously holds for $n = 1$), Π is a reductionless Q-resolution refutation of CR_n, C is a clause of Π, and $C_\perp$ the conclusion of Π (recall that a reductionless Q-resolution refutation ends in a purely universal clause). For the purposes of this subsection, we will consider a merged literal u^* as a shorthand for $u \vee \neg u$.

Claim 1. For all $1 \leq i \leq n$ and $1 \leq j \leq n$,

- if $a_i \in C$ or $x_{ij} \in C$ then $z \in C$,
- if $b_j \in C$ or $\overline{x_{ij}} \in C$ then $\overline{z} \in C$.

Proof. The statement holds for input clauses and universal literals are never removed from clauses.   □

Claim 2. For all $1 \leq i \leq n$ and $1 \leq j \leq n$,

- if $\overline{a_i} \in C$ then $z \in C$ or $C = A$,
- if $\overline{b_j} \in C$ then $\overline{z} \in C$ or $C = B$.

Proof. The statement holds for input clauses. Let C be the resolvent of C_1 and C_2 and assume without loss of generality that $\overline{a_i} \in C_1$. By induction hypothesis either $z \in C_1$, in which case $z \in C$, or $C_1 = A$, in which case the resolution step is over some pivot a_j. That means $a_j \in C_2$, and by Claim 1 we have $z \in C_2$ and so $z \in C$.   □

Claim 3.

- For all $1 \leq i \leq n$, if $\overline{z} \notin C \wedge \overline{a_i} \notin C \wedge C \neq B$, then there is j, such that $x_{ij} \in C$,
- For all $1 \leq j \leq n$, if $z \notin C \wedge \overline{b_j} \notin C \wedge C \neq A$, then there is i, such that $\overline{x_{ij}} \in C$.

Proof. By induction on the proof size. The statement clearly holds for input clauses. Let C be the resolvent of C_1 and C_2. Since $\overline{z} \notin C$, both $\overline{z} \notin C_1$ and $\overline{z} \notin C_2$. Since $\overline{a_i} \notin C$, either $\overline{a_i} \notin C_1$ or $\overline{a_i} \notin C_2$, assume the first. If $C_1 = B$, then the resolution step can only resolve away one of the literals, and so there is j such that $\overline{b_j} \in C$. By Claim 2 (since $C \neq B$) we have $\overline{z} \in C$, a contradiction. Therefore $C_1 \neq B$ and by induction hypothesis there is j such that $x_{ij} \in C_1$. Unless the resolution step is on x_{ij}, we have $x_{ij} \in C$, so it remains to prove that this is indeed not the case. If x_{ij} were the pivot, then $\overline{x_{ij}} \in C_2$, hence $\overline{z} \in C_2$, a contradiction.   □

Claim 4. The conclusion $C_\perp$ contains z^*.

Proof. $C_\perp$ contains no existential literals and is distinct from A and B, so in order not to be in contradiction with Claim 3, the statement has to hold.   □

Claim 5. If $z, \overline{z} \in C$, then for all $1 \leq i, j \leq n$ we have $a_i, \overline{a_i}, b_j, \overline{b_j} \notin C$.

Proof. Consider the last C that violates this implication. Clearly $C \neq C_\perp$. Therefore, there is C_0 and C_1 such that C_1 is the resolvent of C and C_0. Clearly $z, \overline{z} \in C_1$. Since C_1 no longer violates the condition, there is no literal right of z in C_1. Hence C_0 is neither A nor B. Therefore by Claims 1 and 2, a literal on z is in C_0. Since C violates the implication, there is a literal right of z in C, but since there is none in C_1, the pivot variable must be right of z. That means the resolution step is illegal, a contradiction.   □

Claim 6. If C is the resolvent of C_1 and C_2 and $z, \overline{z} \in C$, then neither C_1 nor C_2 contains any of the literals $a_i, \overline{a_i}, b_j, \overline{b_j}$.

Proof. If $z, \overline{z} \in C_1$ or $z, \overline{z} \in C_2$, then the statement follows from Claim 5. Otherwise, the resolution step merges z and $\overline{z}$. That means the pivot must be left of z, and so any of the literals $a_i, \overline{a_i}, b_j, \overline{b_j}$ would end up in C if contained in any of the premises, a contradiction with Claim 5.   □

For the next claims, we need to introduce sets M and S as follows. We let

$$M = \{ C \in \Pi \mid z^* \in C \}$$

be the set of clauses which contain a merged literal. We define S as the "boundary" of M, i.e., the set of clauses that do not contain a merged literal but have a direct descendant that does, formally

$$S = \{ C \in \Pi \mid C \notin M \text{ and there are } C_0, C_1 \in \Pi \text{ s.t. } C_1 = C \odot C_0 \text{ and } C_1 \in M \}.$$

Claim 7. If $C \in S$, then $a_i, \overline{a_i}, b_j, \overline{b_j} \notin C$.

Proof. Follows from Claim 6 as $C \in S$ has a direct descendant with z^*.   □

Claim 8. If $C \in S$ then $|C \setminus \{z, \overline{z}\}| \geq n$.

Proof. By Claim 7 and the fact that $C \notin M$ we get that all preconditions of one of the rows of Claim 3 are satisfied for C, which means that either for all i there is a j such that $x_{ij} \in C$, or for all j there is an i such that $\overline{x_{ij}} \in C$, in both cases a total of n distinct literals, none of which is a literal on z.   □

Proof (of Theorem 1). Consider $\Pi' = S \cup M$. Clauses in M have direct ancestors only in M or S (any direct ancestor that is not in M is by definition in S). Since $C_\perp \in M$, Π' is a reductionless Q-resolution refutation of S. If we disregard literals on z, it is in fact a resolution refutation of the propositional formula S, which means that S is unsatisfiable. By Claim 8, every clause in S has at least n literals, and so it excludes at most 2^{n^2-n} of the assignments to the variables of S. Therefore, S must have at least 2^n clauses in order to exclude all assignments and be unsatisfiable.   □

Corollary 1. *Reductionless Q-resolution does not p-simulate tree-like Q-resolution.*

Proof. Since CR_n have short proofs in tree-like Q-resolution [20], the separation follows from Theorem 1. □

Remark 1. Theorem 1 has another interesting consequence. Since QCDCL with dependency learning can solve CR_n in polynomial time [25], Theorem 1 implies that in order to harness the full power of QCDCL with dependency learning, one has to perform universal reduction during clause learning. This is in contrast with "ordinary" QCDCL where universal reduction is not required to derive a learned clause [22].

5   Short Proofs of QParity in Reductionless Q-Resolution

In this section, we prove that the QPARITY formulas, which require exponentially long proofs in Q-resolution [6], have short proofs in reductionless Q-resolution. It has already been shown that these formulas have short proofs in long-distance Q-resolution [12, Theorem 9]. We simply observe that these proofs—which we reproduce below for the sake of completeness—are in fact reductionless.

Definition 2 ([6]). *Let* $\text{QPARITY}_n = \exists x_1, \ldots, x_n \, \forall z \, \exists t_2, \ldots, t_n. \, \phi_n,$ *where*

$$\phi_n = T_2^1 \wedge T_2^2 \wedge T_2^3 \wedge T_2^4 \wedge \left(\bigwedge_{i=3}^{n} T_i^1 \wedge T_i^2 \wedge T_i^3 \wedge T_i^4 \right) \wedge Z^1 \wedge Z^2, \text{ and}$$

$$
\begin{aligned}
T_2^1 &= x_1 \vee x_2 \vee \overline{t_2}, & T_i^1 &= t_{i-1} \vee x_i \vee \overline{t_i}, & Z^1 &= t_n \vee \overline{z}, \\
T_2^2 &= x_1 \vee \overline{x_2} \vee t_2, & T_i^2 &= t_{i-1} \vee \overline{x_i} \vee t_i, & Z^2 &= \overline{t_n} \vee z. \\
T_2^3 &= \overline{x_1} \vee x_2 \vee t_2, & T_i^3 &= \overline{t_{i-1}} \vee x_i \vee t_i, \\
T_2^4 &= \overline{x_1} \vee \overline{x_2} \vee \overline{t_2}, & T_i^4 &= \overline{t_{i-1}} \vee \overline{x_i} \vee \overline{t_i},
\end{aligned}
$$

Theorem 2. *There is a reductionless Q-resolution refutation of* QPARITY_n *of size* $6n - 5$.

Proof. For $2 \le i \le n - 1$ and $1 \le j \le 2$, we let $Z_i^1 = t_i \vee z^*$, $Z_i^2 = \overline{t_i} \vee z^*$, and $Z_n^j = Z^j$, and it is easy to verify that

$$Z_{i-1}^j = \left(T_i^{3j-2} \odot_{t_i} Z_i^1 \right) \odot_{x_i} \left(T_i^{j+1} \odot_{t_i} Z_i^2 \right).$$

Hence, we derive Z_2^1 and Z_2^2 in a total of $6(n-2)$ steps. Next, we have

$$(z^*) = \left(\left(T_2^1 \odot_{t_2} Z_2^1 \right) \odot_{x_2} \left(T_2^2 \odot_{t_2} Z_2^2 \right) \right) \odot_{x_1} \left(\left(T_2^4 \odot_{t_2} Z_2^1 \right) \odot_{x_2} \left(T_2^3 \odot_{t_2} Z_2^2 \right) \right),$$

and so the formula is refuted. The resolution steps on the x_i are sound, because $x_i < z$ for all $1 \le i \le n$. The total number of resolution steps is $6n - 5$. □

Corollary 2. *Q-resolution does not p-simulate reductionless Q-resolution.*

Remark 2. While strategies extracted from Q-resolution refutations (of PCNF formulas containing a single universal variable) correspond to bounded-depth circuits [6], Theorem 2 implies that reductionless Q-resolution proofs cannot be efficiently transformed into bounded-depth circuits.

6 Lower Bounds from Strategy Extraction

In this section, we will show how to extend the scope of lower bound techniques based on strategy extraction [6] to fragments of long-distance Q-resolution. We begin by observing that strategies extracted from reductionless Q-resolution proofs correspond to branching programs [5].

We briefly review the definition of a branching program and refer to the book by Wegener for more details [34]. A *branching program* or *binary decision diagram (BDD)* on a set X of variables is a directed acyclic graph with a unique source node and at most two sink nodes. Each node v is labelled with a variable $\lambda(v) \in X$, except for the sinks, which are labelled with 0 or 1. If there are two sink nodes, their labels must be distinct. Moreover, every node has exactly two outgoing edges labelled with 0 and 1, respectively. A path $v_1, \ldots, v_n$ from the source of a branching program to its sink is *consistent* if the label of edge (v_i, v_{i+1}) agrees with the label of edge (v_j, v_{j+1}) whenever v_i and v_j are labelled with the same variable. A consistent path corresponds to an assignment in the obvious way. A branching program B on X computes a Boolean function $f(B)$ in the following way. Let $\tau : X \to \{0, 1\}$ be an assignment. We follow the (consistent) path induced by τ to a sink node v, and set $f(B)(\tau) = \lambda(v)$. The *size* of a branching program is the number of nodes in it.

Let $\pi = C_1, \ldots, C_k$ be a reductionless Q-resolution derivation from a PCNF formula $\mathcal{F}$. For each universal variable $u \in var_\forall(\mathcal{F})$, we construct a branching program $B_\mathcal{F}^u(\pi)$ in the following way [5,9]. We first introduce two nodes v_0 and v_1 with $\lambda(v_0) = 0$ and $\lambda(v_1) = 1$. We now consider the clauses C_i in the order of their derivation and associate a node v_i with each one. Depending on how clause C_i was derived, we distinguish two cases:

1. If C_i is an input clause, we let

$$v_i = \begin{cases} v_0, & \text{if } u \in C_i; \\ v_1, & \text{otherwise.} \end{cases}$$

2. If C_i is the resolvent of clauses $C_j = C_j' \vee e$ and $C_l = \neg e \vee C_l'$, there are two possibilities depending on the order of variables e and u in the prefix:
   - If $e < u$, we introduce a fresh node v_i to $B_\mathcal{F}^u(\pi)$ and label it $\lambda(v) = e$. Moreover, we add a 0-labelled edge from v_i to v_j and a 1-labelled edge from v_i to v_l.
   - Otherwise, $u < e$ and we cannot have $u \in var(C_j \cap \overline{C_l})$ by the rules of reductionless Q-resolution (see Fig. 3). If $u \in var(C_j)$, we let $v_i = v_j$. Otherwise, we let $v_i = v_l$.

Finally, we remove all nodes that cannot be reached from v_k. The following statement is immediate from the construction.

Lemma 1. *If* $\pi = C_1, \ldots, C_k$ *is a reductionless Q-resolution derivation from a PCNF formula* $\mathcal{F}$ *and* $u \in var_\forall(\mathcal{F})$ *is a universal variable, then* $B^u_{\mathcal{F}}(\pi)$ *is a branching program on* $D^u_{\mathcal{F}}$ *of size at most* k.

Moreover, if the derivation π is a refutation, these branching programs compute a universal winning strategy. To show this, we first prove the following statement.

Lemma 2. *Let* $\pi = C_1, \ldots, C_k$ *be a reductionless Q-resolution derivation from a PCNF formula* $\mathcal{F}$. *Let* $\tau : var_\exists(\mathcal{F}) \to \{0,1\}$ *be an assignment that does not satisfy* C_k, *and let*

$$\sigma^\pi_{\mathcal{F}} = \{ u \mapsto f(B^u_{\mathcal{F}}(\pi))(\tau|_{D^u_{\mathcal{F}}}) \mid u \in var_\forall(\mathcal{F}) \}$$

be the assignment computed by the branching programs $B^u_{\mathcal{F}}(\pi)$ *in response. Then* $C_i[\sigma^\pi_{\mathcal{F}} \cup \tau] = 0$ *for some input clause* $C_i \in \pi$.

Proof. We proceed by induction on the size k of the derivation. If $\pi = C_1$ then C_1 must be an input clause, and $B^u_{\mathcal{F}}(\pi)$ consists of a single node labelled 0 if $u \in C_1$ and labelled 1 if $\neg u \in C_1$. Accordingly, the function $f(B^u_{\mathcal{F}}(\pi))$ constantly returns an assignment that falsifies any universal literal on variable u. This proves the base case.

Suppose the statement of the lemma holds for derivations of size up to $k - 1$. If C_k is an input clause, the same reasoning as in the base case applies, so suppose C_k is derived by resolution from clauses $C_i = C'_i \vee e$ and $C_j = \neg e \vee C'_j$ with $1 \leq i, j < k$. Let $\pi_i = C_1, \ldots, C_i$ and let $\pi_j = C_1, \ldots, C_j$ be the derivations of the corresponding clauses. We claim that $\sigma^\pi_{\mathcal{F}}(u) = \sigma^{\pi_i}_{\mathcal{F}}(u)$ for each universal variable $u \in var(C_i)$ if $\tau(e) = 0$, and $\sigma^\pi_{\mathcal{F}}(u) = \sigma^{\pi_j}_{\mathcal{F}}(u)$ for each universal variable $u \in var(C_j)$ if $\tau(e) = 1$.

Choose a universal variable u and let $\tau' = \tau|_{D^u_{\mathcal{F}}}$ be the corresponding restriction of τ. We consider two cases. If $e < u$ it is not difficult to see that

$$f(B^u_{\mathcal{F}}(\pi))(\tau') = \begin{cases} f(B^u_{\mathcal{F}}(\pi_i))(\tau') \text{ if } \tau(e) = 0, \text{ and} \\ f(B^u_{\mathcal{F}}(\pi_j))(\tau') \text{ otherwise.} \end{cases}$$

Accordingly, we have $\sigma^\pi_{\mathcal{F}}(u) = \sigma^{\pi_i}_{\mathcal{F}}(u)$ if $\tau(e) = 0$ and otherwise $\sigma^\pi_{\mathcal{F}}(u) = \sigma^{\pi_j}_{\mathcal{F}}(u)$. On the other hand, if $u < e$ by construction of the branching program we have

$$f(B^u_{\mathcal{F}}(\pi)) = \begin{cases} f(B^u_{\mathcal{F}}(\pi_i)) \text{ if } u \in var(C_i), \text{ and} \\ f(B^u_{\mathcal{F}}(\pi_j)) \text{ otherwise.} \end{cases}$$

If $u \in var(C_i)$ then $\sigma^\pi_{\mathcal{F}}(u) = \sigma^{\pi_i}_{\mathcal{F}}(u)$. If $u \in var(C_j)$ as well then we must have $u \in C_i \cap C_j$ or $\neg u \in C_i \cap C_j$ since $u \notin var(C_i \cap \overline{C_j})$ by definition of reductionless Q-resolution. It follows that $f(B^u_{\mathcal{F}}(\pi_i))$ and $f(B^u_{\mathcal{F}}(\pi_j))$ compute the same constant function. Otherwise, if $u \notin var(C_i)$ then $\sigma_{\mathcal{F}}^{\pi_i}(u) = \sigma_{\mathcal{F}}^{\pi_j}(u)$. This proves the claim.

If $\tau(e) = 0$ then by induction hypothesis $C_l[\sigma_{\mathcal{F}}^{\pi_i} \cup \tau] = 0$ for an input clause $C_l \in \pi_i$. We can assume without loss of generality that π_i does not contain any universal variable besides those in the clause C_i. It follows from the claim that the assignment $\sigma_{\mathcal{F}}^{\pi} \cup \tau$ falsifies C_l as well. The case $\tau(e) = 1$ is symmetric. $\square$

Lemma 3. *Let* $\pi = C_1, \ldots, C_k$ *be a reductionless Q-resolution refutation of a PCNF formula* $\mathcal{F}$. *The set* $\{ f(B_{\mathcal{F}}^u(\pi)) \mid u \in var_\forall(\mathcal{F}) \}$ *is a universal winning strategy.*

Proof. Because π is a refutation the clause C_k must not contain existential variables. Thus every assignment $\tau : var_\exists(\mathcal{F}) \to \{0, 1\}$ is an assignment that does not satisfy C_k, and by Lemma 2 the universal response $\sigma_{\mathcal{F}}^{\pi}$ (defined as in the statement of that lemma), in conjunction with the assignment τ, must falsify an input clause. $\square$

These results allow us to translate lower bounds for branching programs to lower bounds on the size of reductionless Q-resolution refutations. Let $f : X \to \{0, 1\}$ be a Boolean function, let $\varphi(X)$ be a Boolean circuit encoding f, and let u be a variable not occurring in φ. Using Tseitin transformation [31], we can construct a CNF formula $\psi(X, u, Y)$ such that $\exists Y.\psi(X, u, Y)$ is logically equivalent to $\varphi(X) \neq u$. The PCNF formula $\mathcal{F} = \exists X \forall u \exists Y.\psi(X, u, Y)$ is a false PCNF formula with f as a unique universal winning strategy (cf. the lower bounds from strategy extraction for Q-resolution [6]). We call such a formula $\mathcal{F}$ a *PCNF encoding of* f.

Proposition 1. *Let* $\mathcal{F}$ *be a PCNF encoding of a Boolean function* f *such that any branching program computing* f *has size at least* m. *Then any reductionless Q-resolution refutation of* $\mathcal{F}$ *requires at least* m *clauses.*

Proof. Since f is the unique universal winning strategy for $\mathcal{F}$, the statement follows immediately from Lemmas 1 and 3. $\square$

To the best of our knowledge, the only lower bounds on the size of general branching programs for explicitly defined Boolean functions currently known are polynomial [24]. Accordingly, Proposition 1 does not yield strong lower bounds for reductionless Q-resolution. However, we can lift lower bounds for restricted classes of branching programs to lower bounds on the proof size in restricted versions of reductionless Q-resolution.

6.1   Regular Reductionless Q-Resolution

Every reductionless Q-resolution derivation $\pi = C_1, \ldots, C_k$ can be represented by a directed acyclic graph $G(\pi)$ on vertices $v_1, \ldots, v_k$ where v_i is labelled with C_i and there is an edge from v_j to v_i if $i < j$ and C_i is one of the clauses C_j was derived from (that is, edges are oriented from conclusions to premises). Each edge is labelled with the corresponding pivot variable.

A reductionless Q-resolution refutation $\pi = C_1, \ldots, C_k$ is *regular* if each variable occurs at most once as a label on any directed path starting from the vertex labelled with clause C_k. Each strategy function computed by such a proof corresponds to a so-called *read-once branching programs* or *free binary decision diagram (FBDD)*. A read-once branching program is a branching program where each variable is encountered at most once on any path from the source to a sink [34].

Lemma 4. *Let $\pi = C_1, \ldots, C_k$ be a regular reductionless Q-resolution refutation of a PCNF formula $\mathcal{F}$. Then $B^u_{\mathcal{F}}(\pi)$ is a read-once branching program of size at most k for each universal variable $u \in var_\forall(\mathcal{F})$.*

Proof. Consider $B^u_{\mathcal{F}}(\pi)$ for any universal variable $u \in var_\forall(\mathcal{F})$. By construction, the sequence of variables encountered on any path starting from the source of $B^u_{\mathcal{F}}(\pi)$ is a subsequence of the pivot variables seen as edge labels on any path starting from the source of $G(\pi)$. In particular, every variable occurs at most once. Since $B^u_{\mathcal{F}}(\pi)$ is a branching program of size at most k by Lemma 1, it is in fact a read-once branching program of size at most k. $\qquad\square$

The *FBDD size* of a Boolean function f is the size of the smallest read-once branching program representing f. We can transfer lower bounds on the FBDD size of Boolean functions into lower bounds on the regular reductionless Q-resolution proof size of certain PCNF formulas, as stated in the next result.

Proposition 2. *Let $\mathcal{F}$ be a PCNF encoding of a Boolean function f with FBDD size m. Any regular reductionless Q-resolution refutation of $\mathcal{F}$ has size at least m.*

Proof. The statement follows from Lemmas 4 and 3. $\qquad\square$

Unlike in the case of general branching programs, strong lower bounds on the FBDD size of many explicitly defined Boolean functions are known [34]. For instance, we can use the following result due to Bollig and Wegener [10].

Theorem 3 ([10]). *There is a Boolean function g in n variables that can be computed by a Boolean circuit of size $O(n^{3/2})$ but has FBDD size $\Omega(2^{\sqrt{n}})$.*

Corollary 3. *There is a Boolean function g in n variables with a PCNF encoding $\mathcal{F}$ of size polynomial in n such that any regular reductionless Q-resolution refutation of $\mathcal{F}$ has size $\Omega(2^{\sqrt{n}})$.*

6.2   Tree-Like Long-Distance Q-Resolution

In this subsection, we are going to prove lower bounds on the size of *tree-like* long-distance Q-resolution. As in the case of reductionless Q-resolution, a long-distance Q-resolution derivation $\pi = C_1, \ldots, C_k$ can be represented by a labelled DAG $G(\pi)$. A derivation π is called *tree-like* if the DAG $G(\pi)$ is a tree.

We want to show that every tree-like long-distance Q-resolution refutation of a PCNF encoding of a Boolean function f can be efficiently turned into a bounded-depth circuit computing f. First, we generalize the construction of the branching programs $B_{\mathcal{F}}^u$ described at the beginning of this section to long-distance Q-resolution derivations. Let $\pi = C_1, \ldots, C_k$ be a long-distance Q-resolution derivation from a PCNF formula $\mathcal{F}$. For each universal variable $u \in var_\forall(\mathcal{F})$, we construct a labelled DAG $B_{\mathcal{F}}^u(\pi)$ in the same way as for a reduction-less Q-resolution derivation, except for the following modification: if clause C_i is derived from a clause C_j by universal reduction and $u \in var(C_j) \setminus var(C_i)$, we set $v_i = v_0$, where $\lambda(v_0) = 0$. It is readily verified that we obtain a branching program of size at most k, as stated in the following lemma.

Lemma 5. *If $\pi = C_1, \ldots, C_k$ is a long-distance Q-resolution derivation from a PCNF formula $\mathcal{F}$ and $u \in var_\forall(\mathcal{F})$ is a universal variable, then $B_{\mathcal{F}}^u(\pi)$ is a branching program on $D_{\mathcal{F}}^u$ of size at most k.*

A universal winning strategy can be computed from a long-distance Q-resolution refutation as follows [2]. We maintain a kind of *decision list* [28] for each universal variable that is intended to encode a strategy function. Specifically, we consider sequences $L = (\varphi_1 \to \psi_1), \ldots, (\varphi_k \to \psi_k)$ where each of the φ_i and ψ_i are propositional formulas. Such a list, which we call a *generalized decision list*, represents a Boolean function f_L in the following way. Consider an assignment $\tau : \bigcup_{i=1}^k var(\varphi_i) \cup var(\psi_i) \to \{0,1\}$ to all the variables appearing in formulas on the list. If there is no index i with $1 \leq i \leq k$ such that τ satisfies φ_i, we define $f_L(\tau) = 1$. Otherwise, let i be the smallest index such that τ satisfies φ_i. Then $f_L(\tau) = \psi_i[\tau]$. The *size* of a decision list $L = (\varphi_1 \to \psi_1), \ldots, (\varphi_k \to \psi_k)$ is $|L| = \sum_{i=1}^k (|\varphi_i| + |\psi_i|)$.

Given a long-distance Q-resolution refutation $\pi = C_1, \ldots, C_k$ of a PCNF formula $\mathcal{F}$, we construct a family $\mathcal{L}_{\mathcal{F}}(\pi) = \{ L_u \mid u \in var_\forall(\mathcal{F}) \}$ of generalized decision lists representing a universal winning strategy for $\mathcal{F}$. For $Q \in \{\exists, \forall\}$, let $C_i^Q = \{ \ell \in C_i \mid var(\ell) \in var_Q(\mathcal{F}) \}$ denote the restriction of C_i to existential or universal literals. Moreover, for a Boolean function f, let $\phi(f)$ denote an encoding of f as a propositional formula. We consider applications of universal reduction in the same order as they appear in the proof. Let C_i be a clause derived by universal reduction from a clause C_j, and let $u \in var(C_j) \setminus var(C_i)$. Let $\pi_i = C_1, \ldots, C_i$ and $\pi_j = C_1, \ldots, C_j$ denote the subderivations ending in clauses C_i and C_j, respectively. We add an entry

$$\left(\overline{C_i^\exists} \wedge \bigwedge_{v \in var(C_i^\forall)} v \leftrightarrow \phi(f(B_{\mathcal{F}}^v(\pi_i))) \right) \to \phi(f(B_{\mathcal{F}}^u(\pi_j))) \tag{1}$$

at the end of the decision list L_u. Observing that the functions $f(B_{\mathcal{F}}^u(\pi_j))$ correspond to the (negated) *phase functions* introduced for the purpose of efficiently extracting universal winning strategies from long-distance Q-resolution refutations [2], it can be verified that the strategy functions computed by the corresponding algorithm coincide with the functions computed by the decision lists defined according to (1).

Proposition 3 ([2]).   *Let π be a long-distance Q-resolution refutation of a PCNF formula $\mathcal{F}$. The set $\{\, f_{L_u} \mid L_u \in \mathcal{L}_{\mathcal{F}}(\pi)\,\}$ is a universal winning strategy.*

We now argue that this winning strategy can be represented by a bounded-depth circuit for certain proofs in *tree-like* long-distance Q-resolution. Specifically, we will show that this is the case for every tree-like refutation of a PCNF encoding of a Boolean function. We first observe that the branching programs $B_{\mathcal{F}}^u$ for tree-like proofs are decision trees. A *decision tree* is a branching program that can be turned into a tree by deleting the sink nodes.

Lemma 6. *If $\pi = C_1, \ldots, C_k$ is a tree-like long-distance Q-resolution derivation from a PCNF formula $\mathcal{F}$, then $B_{\mathcal{F}}^u(\pi)$ is a decision tree for each universal variable $u \in var_\forall(\mathcal{F})$.*

Proof. It is not difficult to see that after deleting the sink nodes labelled with 0 and 1 from $B_{\mathcal{F}}^u(\pi)$, the corresponding DAG can be obtained from $G(\pi)$ by deleting vertices and edges as well as contracting induced paths. Since $G(\pi)$ is a tree, the result is also a tree.   □

Every decision tree can be efficiently translated into a CNF formula by taking the conjunction over the negations of its consistent paths [28]. Moreover, a generalized decision list $L = (\varphi_1 \to \psi_1), \ldots, (\varphi_k \to \psi_k)$ can be represented by a circuit

$$\phi(L) = \bigvee_{i=1}^{k} \left((\bigwedge_{j=1}^{i-1} \neg\varphi_j \wedge \varphi_i) \to \psi_i \right). \tag{2}$$

Lemma 7. *Let $L = (\varphi_1 \to \psi_1), \ldots, (\varphi_k \to \psi_k)$ be a generalized decision list such that d is the maximum depth of any formula φ_i and ψ_i, for $1 \le i \le k$. Then $\phi(L)$ is equivalent to f_L. Moreover, $\phi(L)$ has depth at most $d + 4$ and $|\phi(L)| = O(|L|^2)$.*

Let $\mathcal{F}$ be the PCNF encoding of a Boolean function f and consider a tree-like long-distance Q-resolution refutation π of $\mathcal{F}$. Because $\mathcal{F}$ contains only a single universal variable, each entry in a decision list of $\mathcal{L}_{\mathcal{F}}(\pi)$ given by (1) simplifies to $\overline{C} \to \phi(f(B_{\mathcal{F}}^u(\pi_j)))$, and the right hand side of this implication can be efficiently transformed into a CNF because $B_{\mathcal{F}}^u$ is a decision tree by Lemma 6. We thus obtain the following result.

Proposition 4. *There is a polynomial $p(\cdot)$ and a constant d such that, for any tree-like long-distance Q-resolution refutation π of the PCNF encoding of a function f, there exists a Boolean circuit of size at most $p(|\pi|)$ and depth at most d computing f.*

Proof. By Lemmas 5 and 6, each labelled DAG $B_{\mathcal{F}}^u(\pi')$ is a decision tree of size at most $|\pi|$ for the universal variable u of $\mathcal{F}$ and each subproof π' of π. Each such decision tree can be efficiently encoded as a CNF formula and the decision list has no more than $|\pi|$ entries of size polynomial in $|\pi|$, so it follows from

Lemma 7 that there is a polynomial $p(\cdot)$ such that $\phi(L_u)$ has size at most $p(|\pi|)$ and depth at most 6. Finally, $\{f_{L_u}\}$ is a universal winning strategy for $\mathcal{F}$ by Proposition 3, so f_{L_u} must coincide with f.   □

Since any bounded-depth circuit computing the n-bit parity function has size exponential in n [16], Proposition 4 allows us to obtain the following exponential lower bound on the size of refutations of QPARITY in tree-like long-distance Q-resolution.

Theorem 4. *Any refutation of* QPARITY$_n$ *in tree-like long-distance Q-resolution requires size exponential in* n.

7   Conclusion

We studied the proof complexity of fragments of long-distance Q-resolution. We proved that reductionless Q-resolution cannot p-simulate even tree-like Q-resolution. Since reductionless Q-resolution can be used to derive learned clauses in QCDCL solvers [22], this is another indication that QCDCL[1] proofs correspond to a fairly weak fragment of (long-distance) Q-resolution [20]. The QPARITY formulas, on the other hand, have short refutations in reductionless Q-resolution. These formulas require Q-resolution refutations of exponential size [6], so Q-resolution and reductionless Q-resolution turn out to be incomparable.

The existence of short proofs of QPARITY also marks the breakdown of an elegant technique for obtaining lower bounds on the size of Q-resolution refutations through strategy extraction [6]. Evidently, strategies corresponding to reductionless Q-resolution proofs do not correspond to bounded-depth circuits.

We proved that arguments based on strategy extraction can nevertheless be used to obtain lower bounds for restricted versions of long-distance Q-resolution. Specifically, we showed that *regular* reductionless Q-resolution proofs correspond to *read-once branching programs*, and that *tree-like* long-distance Q-resolution proofs correspond to *bounded-depth* circuits, allowing us to transfer known lower bounds.

Obtaining a characterization of the strategies corresponding to (even reductionless) long-distance Q-resolution refutations that could be used in obtaining lower bounds remains as an intriguing open problem.

References

1. Balabanov, V., Roland Jiang, J.-H.: Unified QBF certification and its applications. Formal Methods Syst. Des. **41**(1), 45–65 (2012)
2. Balabanov, V., Jiang, J.H.R., Janota, M., Widl, M.: Efficient extraction of QBF (counter) models from long-distance resolution proofs. In: Bonet, B., Koenig, S., (eds.) Proceedings of the Twenty-Ninth AAAI Conference on Artificial Intelligence, 25–30 January 2015, Austin, Texas, USA., pp. 3694–3701. AAAI Press (2015)

---

[1] At least without techniques like dependency learning [25].

3. Balabanov, V., Widl, M., Jiang, J.-H.R.: QBF resolution systems and their proof complexities. In: Sinz, C., Egly, U. (eds.) SAT 2014. LNCS, vol. 8561, pp. 154–169. Springer, Cham (2014). https://doi.org/10.1007/978-3-319-09284-3_12
4. Beyersdorff, O., Blinkhorn, J., Hinde, L.: Size, cost, and capacity: a semantic technique for hard random QBFs. Log. Methods Comput. Sci. **15**(1), 13:1–13:39 (2019)
5. Beyersdorff, O., Blinkhorn, J., Mahajan, M.: Building strategies into QBF proofs. In: 36th International Symposium on Theoretical Aspects of Computer Science, STACS 2019, 13–16 March (2019), Berlin, Germany, pp. 14:1–14:18 (2019)
6. Beyersdorff, O., Chew, L., Janota, M.: Proof complexity of resolution-based QBF calculi. In: Mayr, E.W., Ollinger, N. (eds.) 32nd International Symposium on Theoretical Aspects of Computer Science, STACS 2015, 4–7 March 2015, Garching, Germany, vol. 30 of LIPIcs, pp. 76–89. Schloss Dagstuhl - Leibniz-Zentrum fuer Informatik (2015)
7. Biere, A.: Resolve and expand. In: Proceedings of SAT 2004 Seventh International Conference on Theory and Applications of Satisfiability Testing, 10–13 May 2004, Vancouver, BC, Canada, pp. 59–70 (2004)
8. Biere, A.: Bounded model checking. In: Biere, A., Heule, M., van Maaren, H., Walsh, T. (eds) Handbook of Satisfiability, volume 185 of Frontiers in Artificial Intelligence and Applications, pp. 457–481. IOS Press (2009)
9. Bjørner, N., Janota, M., Klieber, W.: On conflicts and strategies in QBF. In: Fehnker, A., McIver, A., Sutcliffe, G., Voronkov, A., (eds.) 20th International Conferences on Logic for Programming, Artificial Intelligence and Reasoning - Short Presentations, EPiC Series in Computing, LPAR 2015, Suva, Fiji, 24–28 November 2015, vol. 35 pp. 28–41. EasyChair (2015)
10. Bollig, B., Wegener, I.: A very simple function that requires exponential size read-once branching programs. Inf. Process. Lett. **66**(2), 53–57 (1998)
11. Cadoli, M., Schaerf, M., Giovanardi, A., Giovanardi, M.: An algorithm to evaluate Quantified Boolean Formulae and its experimental evaluation. J. Automat. Reason. **28**(2), 101–142 (2002)
12. Chew, L.N.: QBF proof complexity. Ph.D. thesis, University of Leeds, UK (2017)
13. Egly, U., Lonsing, F., Widl, M.: Long-distance resolution: proof generation and strategy extraction in search-based QBF solving. In: McMillan, K., Middeldorp, A., Voronkov, A. (eds.) LPAR 2013. LNCS, vol. 8312, pp. 291–308. Springer, Heidelberg (2013). https://doi.org/10.1007/978-3-642-45221-5_21
14. Giunchiglia, E., Narizzano, M., Tacchella, A.: Clause/term resolution and learning in the evaluation of quantified Boolean formulas. J. Artif. Intell. Res. **26**, 371–416 (2006)
15. Gomes, C.P., Kautz, H., Sabharwal, A., Selman, B.: Satisfiability solvers. In: Handbook of Knowledge Representation, volume 3 of Foundations of Artificial Intelligence, pp. 89–134. Elsevier (2008)
16. Håstad, J.: Computational Limitations of Small-depth Circuits. MIT Press, Cambridge, MA, USA (1987)
17. Janota, M., Klieber, W., Marques-Silva, J., Clarke, E.: Solving QBF with counterexample guided refinement. In: Cimatti, A., Sebastiani, R. (eds.) SAT 2012. LNCS, vol. 7317, pp. 114–128. Springer, Heidelberg (2012). https://doi.org/10.1007/978-3-642-31612-8_10
18. Janota, M., Marques-Silva, J.: Solving QBF by clause selection. In: Yang, Q., Wooldridge, M. (eds.) Proceedings of the Twenty-Fourth International Joint Conference on Artificial Intelligence, IJCAI 2015, pp. 325–331. AAAI Press (2015)
19. Janota, M., Marques-Silva, J.: Expansion-based QBF solving versus Q-resolution. Theor. Comput. Sci. **577**, 25–42 (2015)

20. Janota, M.: On Q-resolution and CDCL QBF solving. In: Creignou, N., Le Berre, D. (eds.) SAT 2016. LNCS, vol. 9710, pp. 402–418. Springer, Cham (2016). https://doi.org/10.1007/978-3-319-40970-2_25

21. Büning, H.K., Karpinski, M., Flögel, A.: Resolution for quantified Boolean formulas. Inf. Comput. **117**(1), 12–18 (1995)

22. Klieber, W., Sapra, S., Gao, S., Clarke, E.: A non-prenex, non-clausal QBF solver with game-state learning. In: Strichman, O., Szeider, S. (eds.) SAT 2010. LNCS, vol. 6175, pp. 128–142. Springer, Heidelberg (2010). https://doi.org/10.1007/978-3-642-14186-7_12

23. Lonsing, F., Egly, U., Van Gelder, A.: Efficient clause learning for quantified boolean formulas via QBF pseudo unit propagation. In: Järvisalo, M., Van Gelder, A. (eds.) SAT 2013. LNCS, vol. 7962, pp. 100–115. Springer, Heidelberg (2013). https://doi.org/10.1007/978-3-642-39071-5_9

24. Nechiporuk, I.: A Boolean function. Dokl. Akad. Nauk SSSR **169**(4), 765–766 (1966)

25. Peitl, T., Slivovsky, F., Szeider, S.: Dependency learning for QBF. J. Artif. Intell. Res. **65**, 181–208 (2019)

26. Rabe, M.N., Tentrup, L.: CAQE: a certifying QBF solver. In: Kaivola, R., Wahl, R. (eds.) Formal Methods in Computer-Aided Design, FMCAD 2015, pp. 136–143. IEEE Computer Society (2015)

27. Rintanen, J.: Planning and SAT. In: Biere, A., Heule, M., van Maaren, H., Walsh, T. (eds.) Handbook of Satisfiability, volume 185 of Frontiers in Artificial Intelligence and Applications, pp. 483–504. IOS Press (2009)

28. Ronald, L.: Rivest. Learning decision lists. Mach. Learn. **2**(3), 229–246 (1987)

29. Stockmeyer, L.J., Meyer, A.R.: Word problems requiring exponential time. In: Proceedings of Theory of Computing, pp. 1–9. ACM (1973)

30. Tentrup, L.: Non-prenex QBF solving using abstraction. In: Creignou, N., Le Berre, D. (eds.) SAT 2016. LNCS, vol. 9710, pp. 393–401. Springer, Cham (2016). https://doi.org/10.1007/978-3-319-40970-2_24

31. Tseitin, G.S.: On the complexity of derivation in propositional calculus. In: Siekmann, J., Wrightson, G. (eds.) Automation of Reasoning. Classical Papers on Computer Science, pp. 466–483. Springer, Heidelberg (1983). https://doi.org/10.1007/978-3-642-81955-1_28. Zap. Nauchn. Sem. Leningrad Otd. Mat. Inst. Akad. Nauk SSSR, 8:23–41, 1968. Russian. English translation

32. van Gelder, A.: Contributions to the theory of practical quantified Boolean formula solving. In: Milano, M. (ed.) CP 2012. LNCS, pp. 647–663. Springer, Heidelberg (2012). https://doi.org/10.1007/978-3-642-33558-7_47

33. Vizel, Y., Weissenbacher, G., Malik, S.: Boolean satisfiability solvers and their applications in model checking. Proc. IEEE **103**(11), 2021–2035 (2015)

34. Wegener, I.: Branching Programs and Binary Decision Diagrams. SIAM (2000)

35. Zhang, L., Malik, S.: Conflict driven learning in a quantified Boolean satisfiability solver. In: Pileggi, L.T., Kuehlmann, A. (eds.) Proceedings of the 2002 IEEE/ACM International Conference on Computer-aided Design, ICCAD 2002, San Jose, California, USA, 10–14 November 2002, pp. 442–449. ACM/IEEE Computer Society (2002)

Guiding High-Performance SAT Solvers
with Unsat-Core Predictions

Daniel Selsam[1(⊠)] and Nikolaj Bjørner[2]

[1] Stanford University, Stanford, CA 94305, USA
dselsam@cs.stanford.edu
[2] Microsoft Research, Redmond, WA 98052, USA

Abstract. The *NeuroSAT* neural network architecture was introduced in [37] for predicting properties of propositional formulae. When trained to predict the satisfiability of toy problems, it was shown to find solutions and unsatisfiable cores on its own. However, the authors saw "no obvious path" to using the architecture to improve the state-of-the-art. In this work, we train a simplified NeuroSAT architecture to directly predict the unsatisfiable cores of real problems. We modify several state-of-the-art SAT solvers to periodically replace their variable activity scores with NeuroSAT's prediction of how likely the variables are to appear in an unsatisfiable core. The modified MiniSat solves 10% more problems on SATCOMP-2018 within the standard 5,000 second timeout than the original does. The modified Glucose solves 11% more problems than the original, while the modified Z3 solves 6% more. The gains are even greater when the training is specialized for a specific distribution of problems; on a benchmark of hard problems from a scheduling domain, the modified Glucose solves 20% more problems than the original does within a one-hour timeout. Our results demonstrate that NeuroSAT can provide effective guidance to high-performance SAT solvers on real problems.

1 Introduction

Over the past decade, neural networks have dramatically advanced the state of the art on many important problems, most notably object recognition [22], speech recognition [13], and machine translation [45]. There have also been several attempts to apply neural networks to problems in discrete search, such as program synthesis [7,33], first-order theorem proving [17,27] and higher-order theorem proving [16,18,42,44]. More recently, [37] introduce a neural network architecture designed for satisfiability problems, called *NeuroSAT*, and show that when trained to predict satisfiability on toy problems, it learns to find solutions and unsatisfiable cores on its own. Moreover, the neural network is iterative, and the authors show that by running for many more iterations at test time, it can solve problems that are bigger and even from completely different domains

---

D. Selsam—This paper describes work performed while the first author was at Microsoft Research.

M. Janota and I. Lynce (Eds.): SAT 2019, LNCS 11628, pp. 336–353, 2019.
https://doi.org/10.1007/978-3-030-24258-9_24

than the problems it was trained on. While these results may be intriguing, the authors' motivation was to study the capabilities of neural networks rather than to solve real SAT problems, and they admit to seeing "no obvious path" to beating existing SAT solvers.

In this work, we make use of the NeuroSAT architecture, but whereas it was originally used as an end-to-end solver on toy problems, here we use it to help inform variable branching decisions within high-performance SAT solvers on real problems. Given this goal, the main design decision becomes how to produce data to train the network. Our approach is to generate a supervised dataset mapping unsatisfiable problems to the variables in their unsatisfiable cores. Note that perfect predictions would not always yield a useful variable branching heuristic; for some problems, the smallest core may include every variable, and of course for satisfiable problems, there are no cores at all. Thus, our approach is pragmatic; we rely on NeuroSAT predicting *imperfectly*, and hope that the probability NeuroSAT assigns to a given variable being in a core correlates well with that variable being good to branch on.

The next biggest design decision is how to make use of the predictions inside a SAT solver. Even if we wanted to query NeuroSAT for every variable branching decision, doing so would have severe performance implications, particularly for large problems. A SAT solver makes tens of thousands of assignments every second, whereas even with an on-device GPU, querying NeuroSAT on an industrial-sized problem may take hundreds or even thousands of milliseconds. We settle for complementing—rather than trying to replace—the efficient variable branching heuristics used by existing solvers. All three solvers we extend (MiniSat, Glucose, Z3) use the Exponential Variable State-Independent Decaying Sum (EVSIDS) heuristic, which involves maintaining activity scores for every variable and branching on the free variable with the highest score. The only change we make is that we periodically query NeuroSAT on the *entire* problem (i.e. not conditioning on the current trail), and set all variable activity scores at once in proportion to how likely NeuroSAT thinks the variable is to be involved in an unsat core. We refer to our integration strategy as *periodic refocusing*. We remark that the base heuristics are already strong, and they may only need an occasional, globally-informed reprioritization to yield substantial improvements.

We summarize our pipeline:

1. Generate many unsatisfiable problems by decimating existing problems.
2. For each such problem, generate a DRAT proof, and extract the variables that appear in the unsat core.
3. Train NeuroSAT (henceforth NeuroCore) to map unsatisfiable problems to the variables in the core.
4. Instrument state-of-the-art solvers (MiniSat, Glucose, Z3) to query Neuro-Core periodically (using the original and the learnt clauses), and to reset their variable activity scores according to NeuroCore's predictions.

As a result of these modifications, the MiniSat solver solves 10% more problems on SATCOMP-2018 within the standard 5,000 second timeout. The modified Glucose 4.1 solves 11% more problems than the original, while the modified

Z3 solves 6% more. The gains are even greater when the training is specialized for a specific distribution of problems; our training set included (easy) subproblems of a collection of hard scheduling problems, and on that collection of hard problems the modified Glucose solves 20% more problems than the original does within a one-hour timeout. Our results demonstrate that NeuroSAT (and in particular, NeuroCore) can provide effective guidance to high-performance SAT solvers on real problems. All scripts and sources associated with NeuroCore are available from https://github.com/dselsam/neurocore-public.

2   Data Generation

As discussed in Sect. 1, we want to train our neural network architecture to predict which variables will be involved in unsat cores. Unfortunately, there are only roughly one thousand unsatisfiable problems across all SATCOMP competitions, and a network trained on such few examples would be unlikely to generalize well to unseen problems. We overcome this limitation and generate a dataset containing over 150,000 different problems with labeled cores by considering unsatisfiable *subproblems* of existing problems.

Specifically, we generate training data as follows. We use the distributed execution framework ray [30] to coordinate one driver and hundreds of workers distributed over several machines. The driver maintains a queue of (sub)problems, and begins by enqueuing all problems from SATCOMP (through 2017 only) as well as a few hundred hard scheduling problems. It might help to initialize with even more problems, but we did not find it necessary to do so. Whenever a worker becomes free, the driver dequeues a problem and passes it to the worker. The worker tries to solve it using Z3 with a fixed timeout (we used 60 s). If Z3 returns *sat*, it does nothing, but if Z3 returns *unsat*, it passes the generated DRAT proof [43] to DRAT-trim [43] to determine which of the original clauses were used in the proof. It then computes the variables in the core by traversing the clauses in the core, and finally generates a single datapoint in a format suitable for the neural network architecture we will describe in Sect. 3. If Z3 returns *unknown*, the worker uses a relatively expensive, hand-engineered variable branching heuristic (specifically, Z3's implementation of the March heuristic [29]) and returns the two subproblems to the driver to be added to the queue.

This process generates one datapoint roughly every 60 s per worker. Some of the original problems are very difficult, and so the process may not terminate in a reasonable amount of time; thus we stopped it once we had generated 150,000 datapoints.

Note that our data generation process is not guaranteed to generate diverse cores. To the extent that March is successful in selecting variables to branch on that are in the core, the cores of the two subproblems will be different; if it fails to do this, then the cores of the two subproblems may be the same (though the non-core clauses will still be different). We remark that there are many other ways one might augment the dataset, for example by including additional problems from synthetic distributions, or by directly perturbing the signs of the literals in the existing problems. However, our simple approach proved sufficient.

We stress that predicting the (binary) presence of variables in the core is simplistic. As mentioned in Sect. 1, for some problems, the smallest core may include every variable, in which case the datapoint for that problem would contain no information. Even if only a small fraction of variables are in the core, it may still be that only a small fraction of those core variable would make good branches. A more sophisticated approach would analyze the full DRAT proof and calculate a more nuanced score for each variable that reflects its importance in the proof. However, as we will see in Sect. 5, our simplistic approach of predicting the variables in the core proved sufficient to achieve compelling results.

3   Neural Network Architecture

Background on Neural Networks. Before describing our simplified version of the NeuroSAT architecture, we provide elementary background on neural networks. A neural network can be thought of as a computer program that is differentiable with respect to a set of real-valued, unknown parameters. There may be thousands, millions, or even billions of such parameters, and it would be impossible to specify them by hand. Instead, the practitioner specifies a second differentiable program called the *loss function*, which takes a collection of input/output pairs (*i.e.* training data), runs the neural network on the inputs, and computes a scalar score that measures how much the neural network's outputs disagree with the true outputs. Numerical optimization is then used to find values of the unknown parameters that make the loss function as small as possible.

The basic building block of neural networks is the multilayer perceptron (MLP), also called a feed-forward network or a fully-connected network. An MLP takes as input a vector $x \in \mathbb{R}^{d_{in}}$ for a fixed d_{in}, and outputs a vector $y \in \mathbb{R}^{d_{out}}$ for a fixed d_{out}. It computes y from x by applying a sequence of (parameterized) affine transformations, each but the last followed by a component-wise nonlinear function called an *activation function*. The most common activation function (which we use in this work) is the rectified linear unit (ReLU), which is the identity function on positive numbers and sets all negative numbers to zero.

Notation. We use function-call notation to denote the application of MLPs, where the different arguments to the MLP are implicitly concatenated. Thus if $M : \mathbb{R}^{d_1+d_2} \to \mathbb{R}^{d_{out}}$ is an MLP and $x_1 \in \mathbb{R}^{d_1}, x_2 \in \mathbb{R}^{d_2}$ are vectors, we write $M(x_1, x_2) \in \mathbb{R}^{d_{out}}$ to denote the result of applying the MLP M to the concatenation of x_1 and x_2. For performance reasons, one almost never applies an MLP to an individual vector, and instead applies it to a *batch* of vectors of the same dimension concatenated into a matrix. Thus if $X_1 \in \mathbb{R}^{k \times d_1}, X_2 \in \mathbb{R}^{k \times d_2}$, we write $M(X_1, X_2) \in \mathbb{R}^{k \times d_{out}}$ to denote the result of first concatenating X_1 and X_2 into a $\mathbb{R}^{k \times (d_1+d_2)}$ matrix, applying M to the each of the k rows separately and then concatenating the k results back into a matrix.

NeuroCore Asrchitecture. We now describe our simplified version of the NeuroSAT architecture. We represent a Boolean formula in CNF with n_v variables and n_c clauses by an $n_c \times 2n_v$ sparse matrix $\mathcal{G}$, where the (i,j)th element is 1 if and only if the ith clause contains the jth literal. For example, we represent the formula $\underbrace{(x_1 \vee x_2 \vee x_3)}_{c_1} \wedge \underbrace{(x_1 \vee \overline{x_2} \vee \overline{x_3})}_{c_2}$ as the following 2×6 (sparse) matrix:

$$
\mathcal{G}: \quad
\begin{array}{c|cccccc}
 & x_1 & x_2 & x_3 & \overline{x_1} & \overline{x_2} & \overline{x_3} \\
\hline
c_1 & 1 & 1 & 1 & 0 & 0 & 0 \\
c_2 & 1 & 0 & 0 & 0 & 1 & 1
\end{array}
$$

Our neural network itself is made up of three standard MLPs:

$$
\mathbf{C}_{\text{update}} : \mathbb{R}^{2d} \to \mathbb{R}^d, \mathbf{L}_{\text{update}} : \mathbb{R}^{3d} \to \mathbb{R}^d, \mathbf{V}_{\text{proj}} : \mathbb{R}^{2d} \to \mathbb{R}
$$

where d is a fixed hyperparameter (we used $d = 80$). The network computes forward as follows. First, it initializes two matrices $C \in \mathbb{R}^{n_c \times d}$ and $L \in \mathbb{R}^{2n_v \times d}$ to all ones. Each row of C corresponds to a clause, while each row of L corresponds to a literal:

$$
C = \begin{bmatrix} - & c_1 & - \\ & \vdots & \\ - & c_{n_c} & - \end{bmatrix} \in \mathbb{R}^{n_c \times d}, \qquad
L = \begin{bmatrix} - & x_1 & - \\ & \vdots & \\ - & x_{n_v} & - \\ - & \overline{x_1} & - \\ & \vdots & \\ - & \overline{x_{n_v}} & - \end{bmatrix} \in \mathbb{R}^{2n_v \times d}
$$

We refer to the row corresponding to a clause c or a literal ℓ as the *embedding* of that clause or literal. Note that for notational convenience, we conflate clauses and literals with their embeddings, so *e.g.* the symbol c may refer to the actual clause or to the row of C that embeds the clause.

Define the operation Flip to swap the first half of the rows of a matrix with the second half, so that in Flip(L), each literal's row is swapped with its negation's:

$$
\text{Flip}(L) = \begin{bmatrix} - & \overline{x_1} & - \\ & \vdots & \\ - & \overline{x_{n_v}} & - \\ - & x_1 & - \\ & \vdots & \\ - & x_{n_v} & - \end{bmatrix} \in \mathbb{R}^{2n_v \times d}
$$

After initializing C and L, the network performs T iterations of "message passing" (we used $T = 4$), where a single iteration consists of two updates. First, each clause updates its embedding based on the current embeddings of the literals it contains: $\forall c, c \leftarrow \mathbf{C}_{\text{update}}\left(c, \sum_{\ell \in c} \ell\right)$. Next, each literal updates its embedding based on the current embeddings of the clauses it occurs in, as well as the current embedding of its negation: $\forall \ell, \ell \leftarrow \mathbf{L}_{\text{update}}\left(\ell, \sum_{\ell \in c} c, \overline{\ell}\right)$. We can express these updates compactly and implement them efficiently using the matrix $\mathcal{G}$ and the Flip operator:

$$C \leftarrow \mathbf{C}_{\text{update}}\left(C, \mathcal{G}L\right)$$

$$L \leftarrow \mathbf{L}_{\text{update}}\left(L, \mathcal{G}^{\top}C, \text{Flip}(L)\right)$$

Define the operation Flop to concatenate the first half of the rows of a matrix with the second half along the second axis, so that in Flop(L), the two vectors corresponding to the same variable are concatenated:

$$\text{Flop}(L) = \begin{bmatrix} - \ x_1 \ - & - \ \overline{x_1} \ - \\ & \vdots & \\ - \ x_{n_v} \ - & - \ \overline{x_{n_v}} \ - \end{bmatrix} \in \mathbb{R}^{n_v \times 2d}$$

After T iterations, the network flops L to produce the matrix $V \in \mathbb{R}^{n_v \times 2d}$, and then projects V into an n_v-dimensional vector $\hat{v}$ using the third MLP, $\mathbf{V}_{\text{proj}}$:

$$\hat{v} \leftarrow \mathbf{V}_{\text{proj}}(V) \in \mathbb{R}^{n_v}$$

The vector $\hat{v}$ is the output of NeuroCore, and consists of a numerical score for each variable, which can be passed to the softmax function to define a probability distribution $\hat{p}$ over the variables. During training, we turn each labeled bitmask over variables into a probability distribution p^* by assigning uniform probability to each variable in the core and zero probability to the others. We optimize the three MLPs all at once to minimize the Kullback-Leibler divergence [23]:

$$\mathbf{D}_{KL}(p^* \parallel \hat{p}) = \sum_{i=1}^{n_v} p_i^* \log\left(p_i^*/\hat{p}_i\right)$$

Figure 1 summarizes the architecture.

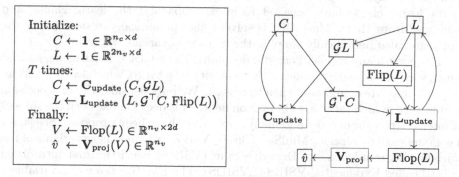

Fig. 1. An overview of the NeuroCore architecture

Comparison to the Original NeuroSAT. While the original NeuroSAT architecture was designed to solve small problems end-to-end, ours is designed to provide cheap, heuristic guidance on (potentially) large problems. Accordingly, our network differs from the original in a few key ways. First, ours only runs for 4 iterations at both train and test time, whereas the original was trained with 26 iterations and ran for upwards of a thousand iterations at test time. Second, our update networks are simple MLPs, whereas the original used Long Short-Term Memories (LSTMs) [14]. Third, as discussed above, ours is trained with supervision at every variable and outputs a vector $\hat{v} \in \mathbb{R}^{n_v}$, whereas the original is trained with only a single bit of supervision and accordingly only outputs a single scalar.

Training NeuroCore. As we discussed in Sect. 1, our goal is not to learn a perfect core predictor, but rather only to learn a coarse heuristic that broadly assigns higher score to more important variables. Thus, fine-tuning the network is relatively unimportant, and we only ever trained with a single set of hyperparameters. We used the ADAM optimizer [19] with a constant learning rate of 10^{-4}, and trained asynchronously with 20 GPUs for under an hour, using distributed TensorFlow [1].

4   Hybrid Solving: Extending CDCL with NeuroCore

Background on CDCL. Modern SAT solvers are based on the Conflict-Driven Clause Learning (CDCL) algorithm [21,28]. Before explaining how we integrate NeuroCore with CDCL solvers , we briefly summarize the parts of CDCL that are relevant to our work. At a high level, a CDCL solver works as follows. It maintains a *trail* of literals that have been given tentative assignments, and continues to assign variables and propagate the implications until reaching a contradiction. It then analyzes the cause of the contradiction and learns a *conflict clause* that is implied by the existing clauses and that would have helped avoid the current conflict. Finally, it pops variables off the trail until all but one of the literals in the learnt clause have been set to false, propagates the learnt clause, and continues from there. Most CDCL solvers also periodically *restart* (clear the trail), and also periodically simplify the clauses in various ways.

There are many crucial, heuristic decisions that a CDCL must make, such as which variable to branch on next, what polarity to set it to, which learned clauses to prune and when, and also when to restart. We only consider the first decision in this work: which variable to branch on next. This decision has been the subject of intense study for decades and many approaches have been proposed. See [4] for a comprehensive overview. MiniSat, Glucose and Z3 all implement variants of the Variable State-Independent Decaying Sum (VSIDS) heuristic (first introduced in [31]) called Exponential-VSIDS (EVSIDS). The EVSIDS score of a variable x after the tth conflict is defined by:

$$\mathbf{InConflict}(x, i) = \begin{cases} 1 & x \text{ was involved in the } i\text{th conflict} \\ 0 & \text{otherwise} \end{cases}$$

$$\mathbf{EVSIDS}(x, t) = \sum_i \mathbf{InConflict}(x, i)\rho^{t-i}$$

where $\rho < 1$ is a hyperparameter. Intuitively, the EVSIDS score of a variable measures how many conflicts the variable has been involved in, with more recent conflicts weighted much more than past conflicts. As we will discuss in Sect. 4, our approach is to periodically reset these EVSIDS scores based on the outputs of NeuroCore.

Integrating NeuroCore. As discussed in Sect. 1, it is too expensive to query NeuroCore for every variable branching decision, and so we settle for querying periodically on the entire problem (*i.e.* not conditioning on the trail) and replacing the variable activity scores with NeuroCore's prediction. We now describe this process in detail.

When we query NeuroCore, we build the sparse clause-literal adjacency matrix $\mathcal{G}$ (see Sect. 3) as follows. First, we collect all non-eliminated variables that are not units at level 0. These are the only variables we tell NeuroCore about. Second, we collect all the clauses that we plan to tell NeuroCore about. We would like to tell NeuroCore about all the clauses, both original and learnt, but the size of the problem can get extremely large as the solver accumulates learnt clauses. At some point the problem would no longer fit in GPU memory, and it might be undesirably expensive even before that point. After collecting the original clauses, we traverse the learned clauses in ascending size order, collecting clauses until the number of literals plus the number of clauses plus the number of cells (*i.e.* literal occurrences in clauses) exceed a fixed cutoff (we used 10 million). If a problem is so big that the original clauses already exceed this cutoff, then for simplicity we do not query NeuroCore at all, although we could have still queried it on random subsets of the clauses. Finally, we traverse the chosen clauses to construct $\mathcal{G}$. Note that because of the learned clauses, the eliminated variables, and the discovered units, NeuroCore is shown a substantially different graph on each query even though we do not condition on the trail.

NeuroCore then returns a vector $\hat{v} \in \mathbb{R}^{n_v}$, where a higher score for a variable indicates that NeuroCore thinks the corresponding variable is more likely to be in the core. We turn $\hat{v}$ into a probability distribution by dividing it by a scalar temperature parameter τ (we used 0.25) and taking the softmax, and then we scale the resulting vector by the number of variables in the problem, and additionally by a fixed constant κ (we used 10^4). Finally, we replace all the EVSIDS scores at once:[1]

$$\forall i, \mathbf{EVSIDS}(x_i, t) \leftarrow \text{Softmax}(\hat{v}/\tau)_i n_v \kappa$$

---

[1] In MiniSat, this involves setting the activity vector to these values, resetting the variable increment to 1.0, and rebuilding the order-heap.

Note that the decay factor ρ is often rather small (MiniSat uses $\rho = 0.95$), and to a first approximation solvers average ten thousand conflicts per second, so these scores decay to 0 in only a fraction of a second. However, such an intervention can still have a powerful effect by refocusing EVSIDS on a more important part of the search space. We refer to our integration strategy as *periodic refocusing* to stress that we are only refocusing EVSIDS rather than trying to replace it. Our hybrid solver based on MiniSat only queries NeuroCore once every 100 s.

5   Solver Experiments

We evaluate the hybrid solver *neuro-minisat* (described in Sect. 4) and the original MiniSat solver *minisat* on the 400 problems from the main track of SATCOMP-2018, with the same 5,000 second timeout used in the competition. For each solver, we solved the 400 problems in 400 different processes in parallel, spread out over 8 identical 64-core machines, with no other compute-intensive processes running on any of the machines. In addition, the hybrid solver also had network access to 5 machines each with 4 GPUs, with the 20 GPUs split evenly and randomly across the 400 processes. We calculate the running time of a solver by adding together its process time with the sum of the wall-clock times of each of the TensorFlow queries it requests on the GPU servers. We ignore the network transmission times since in practice one would often use an on-device hardware accelerator.

Note that although we did not train NeuroCore on any (sub)problems from SATCOMP-2018, we did perform some extremely coarse tuning of hyperparameters (specifically κ, which a-priori might reasonably span 100 orders of magnitude) based on runs of the hybrid solver on problems from SATCOMP-2018. In hindsight we regret not using alternate problems for this, but we strongly suspect that we would have found a similar ballpark by only tuning on problems from other sources.

Results. The main result, alluded to in Sect. 1, is that *neuro-minisat* solves 205 problems within the 5,000 second timeout whereas *minisat* only solves 187. This corresponds to an increase of 10%. Most of the improvement comes from solving more satisfiable problems: *neuro-minisat* solve 125 satisfiable problems compared to *minisat*'s 109, which is a 15% increase. On the other hand, *neuro-minisat* only solved 3% more unsatisfiable problems (80 vs 78). Figure 3 shows a cactus plot of the two solvers, which shows that *neuro-minisat* takes a substantial lead within the first minutes and maintains the lead until the end. Figure 2 shows a scatter plot of the same data, which shows there are quite a few problems that *neuro-minisat* solves within a few minutes that *minisat* times out on. It also shows that there are very few problems on which *neuro-minisat* is substantially worse than *minisat*.

Ablations. The results show that our hybrid approach is effective, but do not tell us much about why it is effective. We do not have a satisfying answer to this

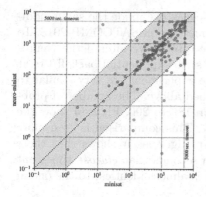

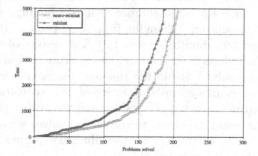

Fig. 2. Scatter plot comparing NeuroCore-assisted MiniSat (*neuro-minisat*) against (*minisat*). Several problems are solved within a few minutes by *neuro-minisat* for which *minisat* times out. The converse scenario is relatively rare.

Fig. 3. Cactus plot comparing NeuroCore-assisted MiniSat (*neuro-minisat*) with the original (*minisat*). It shows that *neuro-minisat* takes a substantial within the first few minutes and maintains the lead until the end.

question yet. As a consolation, we report a few ablations that shed some light on why it may work. For these ablations, we periodically refocus every 10 s instead of every 100 s to make the effect of the quality of the scores more pronounced. When we query NeuroCore every 10 s instead of every 100 s, *neuro-minisat* still solves 205 problems within the timeout.

First, we investigated whether using EVSIDS between queries was necessary, or whether NeuroCore's predictions were sufficient on their own. Simply increasing κ from 10^4 to 10^{40} (which only prevents EVSIDS from taking over for approximately 200ms following each query) already had an substantial negative effect: it solved less than half of the problems that *minisat* solved. Thus NeuroCore is not a replacement to EVSIDS but only a complement to it. Second, we investigated whether NeuroCore's predictions even mattered at all, or if the solver would benefit equally from just periodically setting the EVSIDS scores to random values. When we would otherwise call NeuroCore, we substituted $\hat{v}$ with scores sampled uniformly between $(-1, 1)$, and transformed them to EVSIDS scores using the original τ and κ values. This change had a even more harmful effect: it only solved a handful of problems out of 400. Third, we considered that perhaps NeuroCore's predictions are mostly irrelevant, and that the important part is that they are roughly the same at every query. We tried the same experiment with random scores but with the scores sampled uniformly once at the beginning of search and reused at every query. This did a little better than when the random logits changed each time, but not by much. These experiments do not rule out the possibility that there is a simple, hardcodeable heuristic that could do just as well as NeuroCore, but they do suggest that there is substantial signal in NeuroCore's predictions.

Glucose. As a follow-up experiment and sanity-check, we made the same modifications to Glucose 4.1 and evaluated in the same way on SATCOMP-2018. To provide further assurance that our findings are robust, we altered the Neuro-Core schedule, changing from fixed pauses (100 s) to exponential backoff (5 s at first with multiplier $\gamma = 1.2$). The results of the experiment are very similar to the results from the MiniSat experiment described above. The number of problems solved within the timeout jumps 11% from 186 to 206. Figure 4 show the scatter plot comparing *neuro-glucose* to *glucose*. This comparison is even more favorable to the NeuroCore-assisted solver than Fig. 2, as it shows that there are many problems *neuro-glucose* solves within seconds that *glucose* times out on. The cactus plot for the Glucose experiment is almost identical to the one in Fig. 3 and so is not shown.

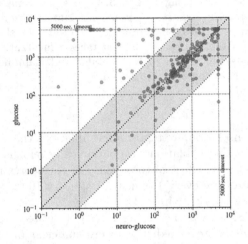

Fig. 4. Scatter plot comparing NeuroCore-assisted glucose (*neuro-glucose*) with the original (*glucose*). It shows that there are quite a few problems that *neuro-glucose* solves within a few seconds that *glucose* times out on, and there are very few problems on which *neuro-glucose* is substantially worse than *glucose*.

Z3. Lastly, we made the same modifications to Z3, except we once again altered the NeuroCore schedule, this time from exponential backoff in terms of user-time to geometric backoff in terms of the number of conflicts. Specifically, we first query NeuroCore after 50,000 conflicts, and then each time wait 50,000 more conflicts than the previous time before querying NeuroCore again. The modified Z3 solves 170 problems within the timeout, up from 161 problems, which is a 6% increase.

Note that for the Z3 experiment, to save on computational costs, we evaluated both solvers simultaneously instead of sequentially. To ensure fairness, we ordered the task queue by problem rather than by solver. The lower absolute scores compared to MiniSat and Glucose are partly the result of the increased contention.

A More Favorable Regime. It is worth remarking that SATCOMP-2018 is an extremely unfavorable regime for machine learning methods. All problems are arbitrarily out of distribution. The 2018 benchmarks include problems arising from a dizzyingly diverse set of domains: proving theorems about bit-vectors, reversing Cellular Automata, verifying floating-point computations, finding efficient polynomial multiplication circuits, mining Bitcoins, allocating time-slots to students with preferences, and finding Hamiltonian cycles as part of a puzzle game, among many others [12].

In practice, one often wants to solve many problems arising from a common source over an extended period of time, in which case it could be worth training a neural network specifically for the problem distribution in question. We approximate this regime by evaluating the same trained network discussed above on the set of 303 (non-public) hard scheduling problems that were included in the data generation process along with SATCOMP 2013–2017. Note that although NeuroCore may have seen the cores of *subproblems* of these problems during training, most of the problems are so hard that many variables need to be set before Z3 can solve them in under a minute. Also, at deployment time we are passing the learned clauses to NeuroCore as well, which may vastly outnumber the original clauses. Thus, although it clearly cannot hurt to train on subproblems of the test problems, NeuroCore is still being queried on problems that are substantially different than those it saw during training.

For this experiment, we compared *glucose* to *neuro-glucose* on the 303 scheduling problems, using a one-hour timeout and the same setting of κ as for the SATCOMP-2018 experiment above. As one might expect, the results are even better than in the SATCOMP regime. The hybrid *neuro-glucose* solver solves 20% more problems than *glucose* within the timeout. Figure 5 shows a cactus plot comparing the two solvers. In contrast to Fig. 3, which showed that on SATCOMP-2018 *neuro-minisat* got off to an early lead and maintained it throughout, here we see that the solvers are roughly tied for the first thirty minutes, at which point *neuro-glucose* begins to pull away, and continues to add to its lead until the one-hour timeout.

6   Related Work

Machine learning in automated deduction has been pursued in several guises. Two established approaches are strategy selection [46] and axiom selection [41].

Strategy selection is to our knowledge mainly applied to setting configuration parameters for SAT, MIP (Mixed-Integer Programming), TSP (the Traveling Salesman Problem), and ATP (First-order Automated Theorem Proving) systems, and enjoy the additional advantage that they can be used in a setting where multiple systems are combined to approach a virtual best solver[2]. ATP systems regularly use strategy selection especially when preparing for competitions, e.g. [36], ever since Gandalf [40] won the 1996 ATP competition (known as CASC [39]) by spending the first few minutes running a suite of different

---

[2] http://ada.liacs.nl/events/sparkle-sat-18/documents/floc-18-sparkle-extended.pdf.

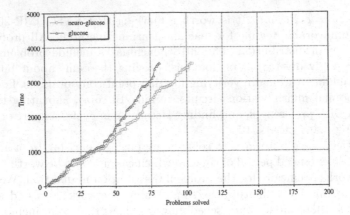

Fig. 5. Cactus plot comparing NeuroCore-assisted Glucose (*neuro-glucose*) with the original (*glucose*) on a benchmark of 303 (non-public) challenging scheduling problems, for which some subproblems were included in the training set. In contrast to Fig. 3, which showed that on SATCOMP-2018 *neuro-minisat* got off to an early lead and maintained it throughout, here we see that the solvers are roughly tied for the first thirty minutes, at which point *neuro-glucose* begins to pull away, and continues to add to its lead until the one-hour timeout.

strategies before selecting one that appeared to make the most progress. Strategy selection as composing tactics [6] was pursued in [3] to speed up performance over baseline tactics.

Axiom selection methods help focus search on a subset of input clauses. The domain-specific Sine [15] method is a prominent example, and selects axioms that share infrequently-appearing symbols with the goal. ATP systems rely on clause selection for driving inferences, and a recent use of machine learning for clause selection [27] was integrated in the E theorem prover [35]. In SAT, CDCL solvers select clauses using unit propagation and conflict analysis, and rely on garbage collection of redundant clauses to balance available inferences from memory and propagation overhead. Carefully crafted methods have been introduced to balance different heuristics within SAT garbage collection, more recently by [32], combining glue levels with activity scores. Ostensibly as a reaction to the opacity and complexity of these heuristics, the CryptoMiniSat solver[3] has recently integrated machine learning to eliminate redundant clauses. Similar to our approach, their approach relies on information from DRAT proofs to relate features of learned clauses (for example, their glue levels) with their usefulness to a derivation. The CryptoMiniSat version of DRAT[4] indeed collects several more features than the original version of DRAT-trim[5] that we used in this work. Their approach then trains a succinct decision tree on this data, that is compiled into a specialized version of CryptoMiniSat.

---

[3] https://github.com/msoos/cryptominisat/.

[4] https://github.com/msoos/drat-trim/.

[5] https://github.com/marijnheule/drat-trim.

Integration of machine learning techniques for branch selection in SAT is to our knowledge relatively unexplored. The VSIDS heuristic (and its descendants such as EVSIDS) presented a breakthrough in SAT solving as it amplified branching on variables that would maximize the conflict-to-branch ratio, thus focusing search within clusters of related clauses. Several refinements of VSIDS are used in newer SAT solvers, including CHB (Conflict History Based) [26] and VMTF (Variable Move-To-Front) [4,34]. Branch selection heuristics within CDCL solvers are finely tuned for performance because they are invoked on every decision.

In contrast, look-ahead solvers [8] afford higher overhead during a look-ahead phase to identify branch literals, also known as so-called *cubes* when look-ahead solving is used in the cube-and-conquer paradigm [10]. Cubes are selected to optimize a carefully crafted metric on clause reduction, such as weighing variable occurrences inversely by the sizes of the clauses they appear in [9,11,24]. Cubing is an appealing target for machine learning because it is used in phases where a global analysis on a problem is feasible.

In MIP solvers, branch-and-bound methods [25] share many similarities to cube-and-conquer methods in SAT. Branch operations split the search into separate parts, and are relatively rare, as the main engine in state-of-the-art MIP solvers remains dual-Simplex, often augmented by interior point methods. Branching is applied when the linear programming optimization is unable to find integral values for integer variables. State-of-the-art branching methods in MIP solvers use heuristics that are related in spirit to look-ahead heuristics: among a set of candidate branch variables they run a limited (cheap) form of linear programming and assemble progress metrics for each candidate variable, and branch on a variable that optimizes a selected metric. As a common trait, these metrics depend on finely tuned parameters, and are therefore ripe targets for machine learning techniques [2].

In the backdrop of the related work, the approach we pursue here is wedged between the fine-grained branching preferences of CDCL solvers and the single-step branch decisions of look-ahead solvers. NeuroCore performs a global analysis to predict a good ordering among *all* unassigned variables, but does this only periodically to allow the fine-grained built-in heuristics to take over during inferences. It provides the capability to rehash a search into a different cluster of clauses where CDCL can perform local tuning.

7   Discussion

There is a vast design space for how to train NeuroSAT and how to use it to guide SAT solvers. In this work we have only considered one tiny point in that design space. We now briefly discuss some of the other possible directions.

For our first experiment predicting unsatisfiable cores, we trained NeuroSAT on a synthetic graph-coloring distribution that happened to have tiny cores. NeuroSAT was able to predict these cores so accurately that we could get almost arbitrarily big speedups over Z3 by only giving it the 0.5% of clauses that NeuroSAT thought most likely to be in the core (and doubling the number of clauses given as necessary until they included the core). Unfortunately, it is much harder to learn a general-purpose core predictor than one on a particular synthetic distribution for which instances may all have similar cores. Real problems also rarely have such tiny cores, so even a perfect core predictor might not be such a silver bullet. With that said, we do think NeuroCore may prove useful in eliminating unhelpful learned clauses, and we plan to pursue this direction in the future.

We have also experimented with training NeuroSAT to imitate the decisions of the March cubing heuristic. Based on preliminary experiments on hard scheduling problems, we found that NeuroSAT trained only to imitate March may actually produce better cubes than March itself, though it remains to be seen if this result holds up to greater scrutiny.

Lastly, inspired by the success of [38], we have experimented with various forms of Monte-Carlo tree search and reinforcement learning, though so far the only competitive heuristic we have been learn de novo is a cubing strategy for uniform random problems. There are two main challenges for learning variable branching heuristics by exploration alone: problems may have a huge number of variables, and it may take substantial time to solve the (sub)problems in order to get feedback about a given branching decision. The former challenge can be mitigated by beginning with imitation learning (e.g. by imitating March). We tried to mitigate the latter by pretraining a value function based on data collected from solving a collection of benchmarks, and then using the value function estimates to make cheap importance-sampling estimates of the size of the search tree under different policies as described in [20]. We found that even in the supervised context, training the value function was difficult; without taking logs it is numerically difficult, and with taking logs, one can get very low loss while ignoring the relatively few hard subproblems towards the roots that make the most difference.

We have only scratched the surface of this design space. We hope that our promising initial results with NeuroCore inspire others to try leveraging NeuroSAT in other, creative ways.

Acknowledgments. We thank Percy Liang, David L. Dill, and Marijn J. H. Heule for helpful discussions.

References

1. Abadi, M., et al.: TensorFlow: a system for large-scale machine learning. In: 12th USENIX Symposium on Operating Systems Design and Implementation OSDI 16, pp. 265–283 (2016)
2. Balcan, M., Dick, T., Sandholm, T., Vitercik, E.: Learning to branch. In: Dy, J.G., Krause, A. (eds.) Proceedings of the 35th International Conference on Machine Learning, ICML 2018, Stockholmsmässan, Stockholm, Sweden, July 10–15, 2018. JMLR Workshop and Conference Proceedings, vol. 80, pp. 353–362. JMLR.org (2018). http://proceedings.mlr.press/v80/balcan18a.html

3. Balunovic, M., Bielik, P., Vechev, M.T.: Learning to solve SMT formulas. In: Bengio, S., Wallach, H.M., Larochelle, H., Grauman, K., Cesa-Bianchi, N., Garnett, R. (eds.) Advances in Neural Information Processing Systems 31: Annual Conference on Neural Information Processing Systems 2018, NeurIPS 2018, 3–8 December 2018, Montréal, Canada. pp. 10338–10349 (2018). http://papers.nips.cc/paper/8233-learning-to-solve-smt-formulas

4. Biere, A., Fröhlich, A.: Evaluating CDCL variable scoring schemes. In: Heule, M., Weaver, S. (eds.) SAT 2015. LNCS, vol. 9340, pp. 405–422. Springer, Cham (2015). https://doi.org/10.1007/978-3-319-24318-4_29

5. Biere, A., Heule, M., van Maaren, H., Walsh, T. (eds.): Handbook of Satisfiability, Frontiers in Artificial Intelligence and Applications, vol. 185. IOS Press, Amsterdam (2009)

6. de Moura, L., Passmore, G.O.: The Strategy Challenge in SMT Solving. In: Bonacina, M.P., Stickel, M.E. (eds.) Automated Reasoning and Mathematics. LNCS (LNAI), vol. 7788, pp. 15–44. Springer, Heidelberg (2013). https://doi.org/10.1007/978-3-642-36675-8_2

7. Devlin, J., Uesato, J., Bhupatiraju, S., Singh, R., Mohamed, A.r., Kohli, P.: RobustFill: neural program learning under noisy I/O. In: Proceedings of the 34th International Conference on Machine Learning, vol. 70, pp. 990–998. JMLR. org (2017)

8. Heule, M., van Maaren, H.: Look-ahead based SAT solvers. In: Biere et al. [5], pp. 155–184. https://doi.org/10.3233/978-1-58603-929-5-155

9. Heule, M.J.: Schur number five. In: Thirty-Second AAAI Conference on Artificial Intelligence (2018)

10. Heule, M.J.H., Kullmann, O., Biere, A.: Cube-and-conquer for satisfiability. Handbook of Parallel Constraint Reasoning, pp. 31–59. Springer, Cham (2018). https://doi.org/10.1007/978-3-319-63516-3_2

11. Heule, M.J.H., Kullmann, O., Marek, V.W.: Solving and verifying the boolean pythagorean triples problem via cube-and-conquer. In: Creignou, N., Le Berre, D. (eds.) SAT 2016. LNCS, vol. 9710, pp. 228–245. Springer, Cham (2016). https://doi.org/10.1007/978-3-319-40970-2_15

12. Proceedings of sat competition 2018; solver and benchmark descriptions (2018). http://hdl.handle.net/10138/237063

13. Hinton, G., et al.: Deep neural networks for acoustic modeling in speech recognition. IEEE Sig. Process. Mag. 29 (2012)

14. Hochreiter, S., Schmidhuber, J.: Long short-term memory. Neural Comput. 9(8), 1735–1780 (1997)

15. Hoder, K., Reger, G., Suda, M., Voronkov, A.: Selecting the selection. In: Olivetti, N., Tiwari, A. (eds.) IJCAR 2016. LNCS (LNAI), vol. 9706, pp. 313–329. Springer, Cham (2016). https://doi.org/10.1007/978-3-319-40229-1_22

16. Huang, D., Dhariwal, P., Song, D., Sutskever, I.: GamePad: A learning environment for theorem proving. arXiv preprint arXiv:1806.00608 (2018)

17. Irving, G., Szegedy, C., Alemi, A.A., Een, N., Chollet, F., Urban, J.: DeepMath-deep sequence models for premise selection. In: Advances in Neural Information Processing Systems, pp. 2235–2243 (2016)

18. Kaliszyk, C., Chollet, F., Szegedy, C.: HolStep: A machine learning dataset for higher-order logic theorem proving. arXiv preprint arXiv:1703.00426 (2017)

19. Kingma, D.P., Ba, J.: Adam: A method for stochastic optimization. arXiv preprint arXiv:1412.6980 (2014)

20. Knuth, D.E.: Estimating the efficiency of backtrack programs. Math. Comput. 29(129), 122–136 (1975)

21. Knuth, D.E.: The Art of Computer Programming, vol. 4, Fascicle 6: Satisfiability (2015)
22. Krizhevsky, A., Sutskever, I., Hinton, G.E.: ImageNet classification with deep convolutional neural networks. In: Advances in neural information processing systems, pp. 1097–1105 (2012)
23. Kullback, S., Leibler, R.A.: On information and sufficiency. Ann. Math. Stat. **22**(1), 79–86 (1951)
24. Kullmann, O.: Fundaments of branching heuristics. In: Biere et al. [5], pp. 205–244. https://doi.org/10.3233/978-1-58603-929-5-205
25. Lawler, E.L., Wood, D.E.: Branch-and-bound methods: a survey. Oper. Res. **14**(4), 699–719 (1966)
26. Liang, J., K., H.G.V., Poupart, P., Czarnecki, K., Ganesh, V.: An empirical study of branching heuristics through the lens of global learning rate. In: Lang, J. (ed.) Proceedings of the Twenty-Seventh International Joint Conference on Artificial Intelligence, IJCAI 2018, 13–19 July 2018, Stockholm, Sweden, pp. 5319–5323. ijcai.org (2018). https://doi.org/10.24963/ijcai.2018/745, http://www.ijcai.org/proceedings/2018/
27. Loos, S.M., Irving, G., Szegedy, C., Kaliszyk, C.: Deep network guided proof search. In: Eiter, T., Sands, D. (eds.) LPAR-21, 21st International Conference on Logic for Programming, Artificial Intelligence and Reasoning, Maun, Botswana, May 7–12, 2017. EPiC Series in Computing, vol. 46, pp. 85–105. EasyChair (2017). http://www.easychair.org/publications/paper/340345
28. Marques-Silva, J.P., Sakallah, K.A.: GRASP: a search algorithm for propositional satisfiability. IEEE Transact. Comput. **48**(5), 506–521 (1999)
29. Mijnders, S., de Wilde, B., Heule, M.: Symbiosis of search and heuristics for random 3-sat. CoRR abs/1402.4455 (2010)
30. Moritz, P., et al.: Ray: a distributed framework for emerging {AI} applications. In: 13th USENIX Symposium on Operating Systems Design and Implementation OSDI 18), pp. 561–577 (2018)
31. Moskewicz, M.W., Madigan, C.F., Zhao, Y., Zhang, L., Malik, S.: Chaff: Engineering an efficient sat solver. In: Proceedings of the 38th annual Design Automation Conference. pp. 530–535. ACM (2001)
32. Oh, C.: Between SAT and UNSAT: The fundamental difference in CDCL SAT. In: Heule, M., Weaver, S. (eds.) SAT 2015. LNCS, vol. 9340, pp. 307–323. Springer, Cham (2015). https://doi.org/10.1007/978-3-319-24318-4_23
33. Parisotto, E., Mohamed, A.r., Singh, R., Li, L., Zhou, D., Kohli, P.: Neuro-symbolic program synthesis. arXiv preprint arXiv:1611.01855 (2016)
34. Ryan, L.: Efficient algorithms for clause-learning SAT solvers, Masters thesis (2004)
35. Schulz, S.: E-A brainiac theorem prover. Ai Communicat. **15**(2, 3), 111–126 (2002)
36. Schulz, S.: We know (nearly) nothing! but can we learn? In: Reger, G., Traytel, D. (eds.) ARCADE 2017, 1st International Workshop on Automated Reasoning: Challenges, Applications, Directions, Exemplary Achievements, Gothenburg, Sweden, 6th August 2017. EPiC Series in Computing, vol. 51, pp. 29–32. EasyChair (2017). http://www.easychair.org/publications/paper/6kgF
37. Selsam, D., Lamm, M., Bünz, B., Liang, P., de Moura, L., Dill, D.L.: Learning a SAT solver from single-bit supervision. In: International Conference on Learning Representations (2019). https://openreview.net/forum?id=HJMC_iA5tm
38. Silver, D., et al.: Mastering the game of go without human knowledge. Nature **550**(7676), 354–359 (2017)
39. Sutcliffe, G.: The CADE ATP system competition - CASC. AI Mag. **37**(2), 99–101 (2016)

40. Tammet, T.: Gandalf. J. Autom. Reason. **18**(2), 199–204 (1997). https://doi.org/10.1023/A:1005887414560
41. Urban, J., Sutcliffe, G., Pudlák, P., Vyskočil, J.: MaLARea SG1 - Machine learner for automated reasoning with semantic guidance. In: Armando, A., Baumgartner, P., Dowek, G. (eds.) IJCAR 2008. LNCS (LNAI), vol. 5195, pp. 441–456. Springer, Heidelberg (2008). https://doi.org/10.1007/978-3-540-71070-7_37
42. Wang, M., Tang, Y., Wang, J., Deng, J.: Premise selection for theorem proving by deep graph embedding. In: Advances in Neural Information Processing Systems, pp. 2786–2796 (2017)
43. Wetzler, N., Heule, M.J.H., Hunt, W.A.: DRAT-trim: Efficient checking and trimming using expressive clausal proofs. In: Sinz, C., Egly, U. (eds.) SAT 2014. LNCS, vol. 8561, pp. 422–429. Springer, Cham (2014). https://doi.org/10.1007/978-3-319-09284-3_31
44. Whalen, D.: Holophrasm: a neural automated theorem prover for higher-order logic. arXiv preprint arXiv:1608.02644 (2016)
45. Wu, Y., et al.: Google's neural machine translation system: Bridging the gap between human and machine translation. arXiv preprint arXiv:1609.08144 (2016)
46. Xu, L., Hutter, F., Hoos, H.H., Leyton-Brown, K.: Satzilla: Portfolio-based algorithm selection for SAT. J. Artif. Intell. Res. **32**, 565–606 (2008). https://doi.org/10.1613/jair.2490

Verifying Binarized Neural Networks by Angluin-Style Learning

Andy Shih[✉], Adnan Darwiche, and Arthur Choi

Computer Science Department, University of California,
Los Angeles, USA
{andyshih,darwiche,aychoi}@cs.ucla.edu

Abstract. We consider the problem of verifying the behavior of binarized neural networks on some input region. We propose an Angluin-style learning algorithm to compile a neural network on a given region into an Ordered Binary Decision Diagram (OBDD), using a SAT solver as an equivalence oracle. The OBDD allows us to efficiently answer a range of verification queries, including counting, computing the probability of counterexamples, and identifying common characteristics of counterexamples. We also present experimental results on verifying binarized neural networks that recognize images of handwritten digits.

Keywords: Verification · Neural networks · Decision Diagrams

1 Introduction

Neural networks are used for a wide array of tasks, including speech recognition, image classification, and language translation. They also power safety-critical applications, such as autonomous driving, where humans need to understand and formally verify the behavior of underlying neural networks. While recent advancements have improved the performance and scale of neural networks, there are not enough methods for providing formal guarantees about their behavior. In addition, the intricate structure of a neural network makes it impractical to reason about their behavior manually. This has sparked a recent line of research that aims to automatically verify neural network properties [6,20,26,28,35,36].

We propose in this paper an approach for verifying the properties of neural networks, which is based on knowledge compilation [5,11,12,29]. Our approach applies to the class of neural networks with discrete inputs and output, but we will highlight the special case of Binarized Neural Networks (BNNs) [14], which have binary weights and activations at runtime, leading to space and computational efficiencies. BNNs have been shown to achieve comparable performance on some standard datasets, compared to more traditional networks using floating-point precision [14].

One particular property of BNNs that has been studied is robustness [23]. Users of a BNN can pinpoint a particular input instance **x** and ask for guarantees on the behavior of the BNN for other inputs in the neighborhood of **x**, which

© Springer Nature Switzerland AG 2019
M. Janota and I. Lynce (Eds.): SAT 2019, LNCS 11628, pp. 354–370, 2019.
https://doi.org/10.1007/978-3-030-24258-9_25

we call an *input region*, denoted by $S_\mathbf{x}$. This has practical applications, e.g., for image classification, where users expect an image of, say, a dog to remain classified as a dog if only a few pixels are modified. Since the number of ways to tweak an image is exponential in the number of modified pixels, it is impractical to perform the verification by enumeration.

A method was recently proposed for detecting counterexamples in any input region $S_\mathbf{x}$ encode-able as a CNF [26]. Our proposed approach pushes this direction further by harnessing techniques from knowledge compilation, allowing one to also *reason* about counterexamples. For example, we can efficiently count the counterexamples in $S_\mathbf{x}$, compute their probability, enumerate a subset of them, and identify their common characteristics. Another useful query supported by our approach, the *prime-implicant query*, returns a subset of inputs that, if fixed, will guarantee that the neural network output will stick even if we vary the unfixed inputs [16,31].

Using the example of image classification, our new techniques allow us to perform reasoning on all images that are some pixels away from some target image, say, of a dog. Whereas previous methods only tell us that it is possible to classify another image in the neighborhood of the dog image as a cat, our new method can determine how many neighborhood images are classified as cats and identify key characteristics that are shared among all such images. Moreover, the prime-implicant query identifies a minimal set of pixels in the dog image that guarantees a correct classification even if we modify some of the unfixed pixels.

To reason about BNNs, we compile them into a tractable representation and then apply verification queries to the compiled representation. The compilation is done once per input region and, if successful, allows one to efficiently answer a range of queries that are otherwise NP-hard [30].

We now give an overview of our compilation algorithm. We compile BNNs into Ordered Binary Decision Diagrams (OBDDs), which are decision graphs that are tractable for many queries and transformations due to an enforced variable ordering [4,12,24,33]. Let B be a BNN, and let B_S represent the function of B on S, an input region of interest. To obtain B_S as an OBDD, we leverage an Angluin-style algorithm for learning the OBDD representation of B_S using standard membership and equivalence queries [1,25]. First, we construct a hypothesis OBDD and then iteratively call equivalence queries, adding OBDD nodes until its output agrees with B_S. To answer equivalence queries efficiently, we encode the BNN and the hypothesis OBDD into a CNF, and require that the region S can be encoded as a CNF as well. When the algorithm terminates, it returns an OBDD D such that $D(\mathbf{x}) = B(\mathbf{x}) : \forall \mathbf{x} \in S$, a notion related to the `Constrain` operator on OBDDs [24]. We then verify properties of BNN B by performing efficient verification queries on OBDD D.

Compared to the two main compilation paradigms of *bottom-up* and *top-down* compilation,[1] the Angluin-style learning algorithm is more similar to top-down

---

[1] Bottom-up compilation constructs constants and literals of a knowledge base and then composes them together using the *Apply* operation [11]. Top-down compilation recursively conditions the knowledge base and then combines the recursive compilations to obtain the final compilation [13,27].

approaches, in that it never creates unnecessary nodes [11]. The main feature that distinguishes the Angluin-style learning algorithm from top-down approaches is the support of incremental and anytime compilation. The Angluin-style learning algorithm can slowly increase the region of interest, so that the compiled OBDD of a smaller region can be used as the hypothesis OBDD for the compilation task of a larger region, without starting over. We can essentially save our progress, and build on it later if we decide the initial region is too small.

This paper is structured as follows. Section 2 provides an introduction to BNNs and OBDDs. Section 3 describes the encodings of BNNs and OBDDs into CNF. Section 4 goes over the Angluin-style learning algorithm, which is used by our compilation algorithm in Sect. 5. We report experiments on the efficiency of our compilation algorithm in Sect. 6, followed by a case study in Sect. 7. We finally discuss related work in Sect. 8 and conclude in Sect. 9.

2   Background

In this section, we describe Binarized Neural Networks and Ordered Binary Decision Diagrams in more detail.

2.1   Binarized Neural Networks

A Binarized Neural Network is a feed-forward neural network where the weights and activations are binarized using $\{-1, 1\}$. A BNN is composed of internal blocks and one output block. Internal blocks consist of three layers: a linear transformation (LIN), batch normalization (BN), and binarization (BIN).

- The LIN layer has parameters $\mathbf{a}$ (weights) and b (bias). Given an input $\mathbf{x}$, this layer returns $\langle \mathbf{a}, \mathbf{x} \rangle + b$.
- The BN layer has parameters μ (mean), σ (standard deviation), α (weight), and γ (bias). Given an input y, this layer returns $\alpha(\frac{y-\mu}{\sigma}) + \gamma$.
- The BIN layer returns the sign (1 or -1) of its input.

The output block consists of a LIN layer and an ARGMAX layer. The ARGMAX layer picks the output class with the highest activation. More details regarding these blocks and layers and their exact definitions is given by Narodytska et al. [26]. For convenience we consider a BNN with outputs 0 or 1.

2.2   Ordered Binary Decision Diagrams

An *Ordered Binary Decision Diagram* (OBDD) is a *tractable* representation of a Boolean function over variables $\mathbf{X} = X_1, \ldots, X_n$ [4,24,33]. An OBDD is a rooted, directed acyclic graph with two sinks called the 1-sink and 0-sink. Every node (except the sinks) in the OBDD is labeled with a variable X_i and has two labeled outgoing edges: the 1-edge and the 0-edge. The labeling of the OBDD nodes respects some global ordering of the variables $\mathbf{X}$: if there is an edge from

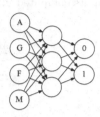

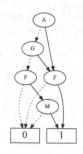

(a) BNN with two outputs $\{0,1\}$. The output with higher activation is the classification.

(b) OBDD with sinks $\{0,1\}$.

Fig. 1. A BNN and its corresponding OBDD on four inputs. The two representations compute the same function.

a node labeled X_i to a node labeled X_j, then X_i must come before X_j in the global ordering. To evaluate the OBDD on an instance $\mathbf{x}$, start at the root node of the OBDD. Let x_i be the value of variable X_i of the current node. Repeatedly follow the x_i-edge of the current node, until a sink node is reached. Reaching the 1-sink means $\mathbf{x}$ is evaluated to 1 and reaching the 0-sink means $\mathbf{x}$ is evaluated to 0 by the OBDD. Hence, an OBDD can be viewed as representing a function $f(\mathbf{X})$ that maps instances $\mathbf{x}$ into $\{0,1\}$.

Consider the BNN in Fig. 1a, which classifies a movie as a box-office success or not. It has four binary inputs: A (*Adapted Screenplay*), G (*Great Cinematography*), F (*Famous Cast*), and M (*Marketing*). The parameters of the BNN are not shown, but it computes the truth table as shown in Table 1. The OBDD in Fig. 1b also computes the truth table in Table 1, so we can verify properties of the BNN by performing verification on the OBDD. We can examine, for example, a movie that is an adapted screenplay, has great cinematography, a famous cast, heavy marketing, and is classified as being a box office success. This movie corresponds to input $\{A = 1, G = 1, F = 1, M = 1\}$ and a classification of 1. Using the OBDD in Fig. 1b we can deduce, in time linear in the size of the OBDD, that the movie could have had poor cinematography and low marketing, and would still be classified as being a box office success. In fact, the partial input $\{A = 1, F = 1\}$ completely determines that the movie will be classified as being successful, regardless of how the remaining input is set. This is an example of the many types of efficient verification queries that can be done on an OBDD [31].

3   CNF Encodings

We next provide the encoding of BNNs and OBDDs into CNF, which will serve an important role in our main compilation algorithm.

Table 1. The Boolean function on the 16 possible inputs computed by the BNN and OBDD in Fig. 1.

	A	G	F	M	$f(\mathbf{x})$		A	G	F	M	$f(\mathbf{x})$
1	-	-	-	-	-	9	+	-	-	-	-
2	-	-	-	+	-	10	+	-	-	+	-
3	-	-	+	-	-	11	+	-	+	-	+
4	-	-	+	+	+	12	+	-	+	+	+
5	-	+	-	-	-	13	+	+	-	-	-
6	-	+	-	+	-	14	+	+	-	+	-
7	-	+	+	-	+	15	+	+	+	-	+
8	-	+	+	+	+	16	+	+	+	+	+

3.1 BNN to CNF

We use the conversion given by Narodytska et al. [26]. An internal block of a BNN consists of three layers: a linear transformation (LIN), batch normalization (BN), and binarization (BIN). The LIN layer has parameters $\mathbf{a}$ (weights) and b (bias). The BN layer has parameters μ (mean), σ (std), α (weight), and γ (bias). Put together, the three layers of an internal block can be translated to the following output function $h(\mathbf{x})$ of a neuron on an input instance $\mathbf{x}$ [26].

$$h(\mathbf{x}) = 1 \iff \langle \mathbf{a}, \mathbf{x} \rangle \geq -\frac{\sigma}{\alpha}\gamma + \mu - b$$

Since the weights $\mathbf{a}$ and input $\mathbf{x}$ are binarized as $\{-1, 1\}$, the above computation reduces to a cardinality constraint of the form $\sum_{i=1}^{m} l_i \geq C$, where $l_i \in \{0, 1\}$ and $C \in \mathbb{R}$. This cardinality constraint can be encoded as a CNF.

The output block has a LIN layer followed by an ARGMAX layer, which can be encoded using a similar technique. First, we encode a cardinality constraint for all pairs of classes, which tells us the class that has a higher activation function in the pairing. Then, we use a final set of cardinality constraints to determine the class that was the winner in all of its pairings [26]. Since we focus on neural networks with binary output classes in this paper, a single CNF variable is enough to represent the output of the BNN.

The space complexity of this conversion is $O(NC^2)$, where N is the number of neurons in the BNN and C is the constant from the above cardinality constraint.

3.2 OBDD to CNF

We convert an OBDD into a CNF using the well-known Tseitin transformation [32], which converts a Boolean circuit into a CNF. Consider an OBDD node labelled by variable X. If the two children of this node compute Boolean functions C_0, C_1, then the OBDD node computes the Boolean function $R = (C_0 \wedge \neg X) \vee (C_1 \wedge X)$. We can then represent the Boolean function of this node by the following five clauses:

$$\neg R \vee C_0 \vee X$$
$$\neg R \vee C_1 \vee \neg X$$
$$\neg R \vee C_0 \vee C_1$$

$$R \vee \neg C_0 \vee X$$
$$R \vee \neg C_1 \vee \neg X$$

Applying this conversion to all OBDD nodes leads to a CNF representation of the Boolean function computed by the OBDD. The number of CNF clauses produced by this conversion is $5N$, where N is the number of OBDD nodes.

The above encodings allow us to convert a BNN into a CNF α and an OBDD into a CNF β. Let $\mathbf{X}$ be the CNF variables corresponding to the BNN inputs and O be the variable corresponding to its output. Then $\alpha \wedge \mathbf{x} \wedge O$ will be satisfiable iff the BNN outputs 1 under input $\mathbf{x}$. Similarly, $\alpha \wedge \mathbf{x} \wedge \neg O$ will be satisfiable iff the BNN outputs 0 under input $\mathbf{x}$. Now let $\mathbf{X}$ be the CNF variables corresponding to the OBDD variables and R be the variable we introduced for the OBDD root. Then $\beta \wedge \mathbf{x} \wedge R$ will be satisfiable iff the OBDD outputs 1 under input $\mathbf{x}$ and $\beta \wedge \mathbf{x} \wedge \neg R$ will be satisfiable iff the OBDD outputs 0 under input $\mathbf{x}$.

When the BNN and the OBDD share the same inputs $\mathbf{x}$, we can check for their *inequivalence* with the formula $\phi = \alpha \wedge \beta \wedge (O \vee R) \wedge (\neg O \vee \neg R)$ [26]. Then, ϕ is satisfiable iff there is some instantiation of $\mathbf{x}$ such that $(O \wedge \neg R) \vee (\neg O \wedge R)$ (i.e. BNN and OBDD disagree).

4   Angluin-Style Exact Learning of Finite Automaton

In this section we describe Angluin's algorithm for learning Deterministic Finite Automata (DFA) [1]. The DFA learning algorithm has an adaptation for learning OBDDs [25], which serves as the backbone for our neural network compilation algorithm. DFAs and OBDDs are initimately related: a Complete OBDD (an OBDD that does not skip variables [33]) is also a DFA (but a DFA is not necessarily an OBDD).

We roughly summarize the exposition on the topic of learning DFAs from the textbook by Kearns and Vazirani [21]. The learning algorithm falls under the category of *active* learning where the algorithm can learn through experimentation, as opposed to *passive* learning where the algorithm has no control over the sample of examples. To learn the DFA for a function f, the learning process requires access to oracles for two types of queries:

- Membership Queries: The learning process selects an instance $\mathbf{x}$ and the oracle returns the value of $f(\mathbf{x})$.
- Equivalence Queries: The learner submits a hypothesis automaton h. The oracle tells the learner if h computes the correct function (i.e. $h = f$), otherwise the oracle returns a counterexample $\mathbf{x}$ for which $h(\mathbf{x}) \neq f(\mathbf{x})$.

The main idea of the algorithm is as follows. Let S be the set of states of a minimal DFA we want to learn. Recall that each state represents a distinct equivalence class of input strings. At all times we keep a hypothesis DFA whose states $S^\star$ represent a partition of S. We iteratively refine the partition by splitting some partition element of $S^\star$ into two, so that $|S^\star|$ increases. When $|S^\star| = |S|$,

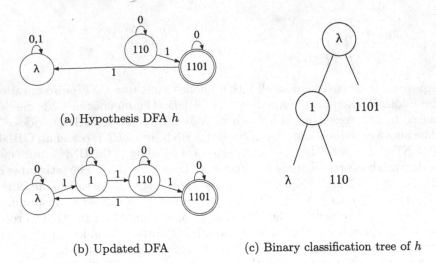

(a) Hypothesis DFA h

(b) Updated DFA   (c) Binary classification tree of h

Fig. 2. Learning the finite automaton for the 3 mod 4 counter. Using the counterexample 1101, we modify the hypothesis DFA into the updated DFA.

each element in the partition contains exactly one equivalence class from S, so our hypothesis DFA computes the target DFA.

Initially, we start with a one-node hypothesis DFA with just one state, which partitions all the states in S into one group. As long as our DFA is incorrect, we will receive counterexamples from the equivalence query. Given a counterexample e, we can simulate e on our hypothesis DFA and identify the first state s^* for which its following step in the simulation is provably incorrect. This can be done efficiently by maintaining a binary classification tree, the details of which we omit. We then refine the partition by splitting s^* into two nodes. This process repeats until we have learned all the states of S, at which point the equivalence query gives no more counterexamples and our algorithm terminates.

Suppose we wish to learn a DFA on binary inputs for the 3 mod 4 counter f, and we currently have the hypothesis DFA h in Fig. 2a and its binary classification tree in Fig. 2c. Since $h(1101) = 0 \neq f(1101)$, we get the string 1101 as a counterexample. Using the binary classification tree along with membership queries, the algorithm identifies the state λ in h as faulty, and splits it into two. This generates the updated DFA in Fig. 2b, which computes f correctly.

The automaton learning algorithm was adapted into an OBDD learning algorithm by Nakamura [25]. This variation requires n equivalence queries and $6n^2 + n \log(m)$ membership queries, where n is the number of nodes in the final OBDD and m is the number of variables in the OBDD.

Algorithm 1. CompileBNN($B, \mathbf{X}, S$)

input: A Binarized Neural Network B with input variables $\mathbf{X}$, and a CNF S encoding an input region

output: An OBDD D computing the function of B on S

main:

1: $\alpha, O \leftarrow$ BNNToCNF($B, \mathbf{X}$)
2: $D \leftarrow$ initial hypothesis OBDD
3: $\beta, R \leftarrow$ OBDDToCNF($D, \mathbf{X}$)
4: $\phi \leftarrow \alpha \wedge \beta \wedge (O \vee R) \wedge (\neg O \vee \neg R) \wedge S$
5: **while** ϕ has a satisfying assignment $\mathbf{s}$ **do**
6: $\quad \mathbf{x} \leftarrow$ projection of $\mathbf{s}$ on $\mathbf{X}$
7: $\quad D \leftarrow$ UpdateHypothesis($D, \mathbf{x}$)
8: $\quad \beta, R \leftarrow$ OBDDToCNF($D, \mathbf{X}$)
9: $\quad \phi \leftarrow \alpha \wedge \beta \wedge (O \vee R) \wedge (\neg O \vee \neg R) \wedge S$
10: **return** D

5 BNN Compilation Algorithm

We now describe our main contribution: a compilation algorithm from a BNN to an OBDD. Given a BNN B on n binary inputs and one binary output, we wish to obtain an OBDD D that computes the function of B on a region S (i.e. $D(\mathbf{x}) = B(\mathbf{x}) : \forall \mathbf{x} \in S$). We require region S to be encoded as a CNF.

Algorithm 1 implements our proposal. The subroutines BNNToCNF and OBDDToCNF perform the encodings described in Sect. 3. We encode the BNN B as a CNF α with output variable O. Then, we start the OBDD learning algorithm as described in Sect. 4 to learn the reduced OBDD representation of B. The learning algorithm creates a hypothesis OBDD D, which we encode as a CNF β with variable R representing the OBDD output. We set ϕ on Line 4 such that ϕ has a satisfying assignment iff the current hypothesis OBDD D does not compute the same function as BNN B on region S. While ϕ is satisfiable, we take the satisfying assignment and keep only the variables corresponding to the BNN/OBDD inputs as our counterexample $\mathbf{x}$. The subroutine UpdateHypothesis then edits our hypothesis OBDD using counterexample $\mathbf{x}$. Once we have an unsatisfiable ϕ, we return the OBDD D with the guarantee that it computes the same function as BNN B on S. Note that there are no guarantees on the output of OBDD D on instances outside S. The number of iterations of the **while** loop is N, where N is the number of nodes in the final output D.

In Algorithm 2 we propose the construction of an input region that captures all instances in the neighborhood of some instance $\mathbf{x}$ on n variables. More specifically, Algorithm 2 takes in an instance $\mathbf{x}$, a radius r, and outputs a CNF S on variables $X_1, \ldots, X_n$. An instance $\mathbf{x}^\star$ is a satisfying assignment for S iff the Hamming distance between $\mathbf{x}$ and $\mathbf{x}^\star$ is no greater than r. This becomes a cardinality constraint, which can be encoded in many ways [2]. For ease of exposition, we use an OBDD for the constraint and then convert it to CNF. In the algorithm, node $d_{i,j}$ stores the state with $n - i$ variables processed and a current Hamming

---

Algorithm 2. r-RadiusDomain($\mathbf{x}, r$)

---

input: An input $\mathbf{x} = x_1, \ldots, x_n$ and a radius $r \leq n$

output: A CNF that encodes all instances $\mathbf{x}^*$ such that $h(\mathbf{x}, \mathbf{x}^*) \leq r$, where h measures the Hamming distance

main:
1: $d \leftarrow$ a 2D array with dimensions $[0, n] \times [0, r]$
2: **for** $j \leftarrow 0$ to r **do**
3:     $d_{0,j} \leftarrow \top$
4: **for** $i \leftarrow 1$ to n **do**
5:     **for** $j \leftarrow 0$ to r **do**
6:         $h \leftarrow d_{i-1,j}$
7:         $l \leftarrow d_{i-1,j-1}$ **if** $j > 0$ **else** $\bot$
8:         $d_{i,j} \leftarrow$ OBDD node: label X_i, x_i-child h, $\neg x_i$-child l
9: **return** OBDDToCNF($d_{n,r}, \mathbf{X}$)

---

distance of $r - j$. On Line 8, the child edge of $d_{i,j}$ that agrees with x_i points to $d_{i-1,j}$. The other child edge points to $d_{i-1,j-1}$ if $j > 0$, otherwise it points to $\bot$. By using S as an input for Algorithm 1, we can compile an OBDD that exactly computes the function of a BNN for all instances close to some instance of interest, measured by the number of differing features. The time and space complexity of Algorithm 2 is $O(nr)$.

To extend our algorithm into an anytime compilation algorithm, we start with a small region of interest and increase its size over time. The compiled OBDD D will compute the same function as B on this small region. To compile the OBDD for a larger region, we can feed in D as the initial hypothesis OBDD in Algorithm 1 on Line 2, without the need to build D from scratch. Then, we can use the updated OBDD to verify the properties of B on the enlarged region. We can continue to enlarge this region until it becomes $\{0,1\}^n$, at which point $S = \top$ and the compiled OBDD computes the same function as B everywhere.

6 Experiments

In this section we present experiments on two types of neural networks:

- binarized neural networks (BNNs) [14], as described in Sect. 2. In particular, we assumed a fully-connected multi-layer feedforward architecture;
- convolutional neural networks (CNN), where we simply used step activations instead of the more commonly used ReLU activations [7].[2] In such a network, if the network inputs are binary, then the inputs and outputs of all neurons are binary (note that we do not use max-pooling in our experiments). Such a network corresponds to a Boolean circuit, although in general it will not be

---

[2] We first train the network using sigmoid activations, and then at test time we replace the sigmoid activations with step activations, while keeping the learned weights.

tractable. However, we can encode it as a CNF using the Tseitin transformation, and use the same algorithm described in Sect. 5 to learn its (tractable) OBDD.

We considered the USPS digits dataset, and binarized the inputs to get 16×16 black and white images [15]. We then trained our neural networks to distinguish between digit '0' images (false-class) and digit '8' images (true-class). We also tested on other pairs of digits, which gave similar results.

- We trained a BNN which achieved 94% accuracy using the training algorithm from [9]. We down-sampled the inputs to 8×8 images to get 64 input nodes. We further used 5 hidden nodes and 2 output nodes. The network was encoded into a CNF with $10,664$ variables and $41,553$ clauses. Using `riss-coprocessor` to pre-process auxiliary variables, we compressed the CNF to $3,438$ variables and $23,254$ clauses [19]. The original and compressed CNFs are equivalent after existentially quantifying out all variables except for the inputs/output, which is enough for the correctness of our algorithm.
- We trained a CNN which achieved 97% accuracy, using TensorFlow. The network used the original 16×16 images, and thus had 256 input nodes. We created two convolution layers, each with stride size 2. We first swept a 3×3 filter on the original 16×16 image (resulting in a 7×7 grid), followed by a second 2×2 filter (resulting in a 3×3 grid). These outputs were the inputs of a fully-connected layer with a single output. We encoded this network into a CNF with $10,547$ variables and $31,682$ clauses and using `riss-coprocessor`, we pre-processed the auxiliary variables to get a compressed CNF with $1,473$ variables and $11,638$ clauses [19].

Experiments were done using a single Intel Xeon CPU E5-2670 processor. We used a time limit of one hour for each compilation. In general, we find that the fully-connected architecture of the BNN was more challenging to compile (hence, the reason for down-sampling the input images). In fact, when we trained a CNN on the 8×8 inputs, we were able to compile the full network (i.e., over the space of all images, and not just for a fixed region around a given image).

For the BNN and CNN that we trained, we identified instances classified as digit '0' (Fig. 3a), and compiled the neighborhood around it using Algorithms 1 and 2. The variable order we used for the OBDD is the natural row-by-row left-to-right ordering of the pixels in the images. We used the `riss` SAT solver for our experiments [19]. Table 2 (BNN) and Table 3 (CNN) shows the compilation results for increasing values of r. We did the same for an instance that is classified as digit '8' (Fig. 3b). We also compiled around the neighborhood of an image that is neither a '0' nor an '8' (a smile, Fig. 3c). For experiments with small input spaces, we manually verified the correctness of the OBDD through enumeration.

We make a few observations. For both the BNN and CNN, compiling larger regions around the smile was more challenging than compiling the regions around a digit. This is perhaps because there is less structure around an image that the network was not trained with. Next, while we scaled to a larger radius r using

(a) A digit 0 that is classified as '0'.

(b) A digit 8 that is classified as '8'.

(c) A smile which is classified as '8' by the BNN and '0' by the CNN.

Fig. 3. Three 16×16 images: digit 0, digit 8, and a smile. For each image we compile around its r-neighborhood (the used 8×8 images are not shown).

the BNN, the space of images was still much larger for the smaller radius that we compiled with the CNN, since the input images were much bigger (16×16 for the CNN versus 8×8 for the BNN).

The bottleneck in our experiments is the average time for a SAT query, which is done once for each of the N equivalence queries, where N is the size, i.e., number of nodes, of the OBDD (sizes are given in Tables 2 and 3). As the OBDD grows, the membership queries become a bottleneck as well since the number of membership queries is quadratic on N.

7 Case Study

In this section we perform verification queries on the convolutional neural networks (CNNs) that we trained and compiled in Sect. 6. First, we counted the number of counterexamples. Second, we performed prime-implicant queries (PI queries for short), which give a subset of pixels that render the remaining pixels irrelevant for the classification [31], up to the region under consideration.[3]

Consider the instance visualized in Fig. 3a, classified as a '0' digit. For $r = 3$ in Table 3, the reduced OBDD is just the constant false ($\perp$). This means that there were no counterexamples in this region, and that flipping any $r = 3$ pixels in our image will still produce another image classified as digit '0' (the false class). Recall that an image has 256 pixels in our example, so this classification holds for all of the $2,796,417$ possible inputs within a radius of 3 around our image in Fig. 3a.

---

[3] Ignatiev et al. [16] also subsequently proposed to use (prime) implicants to explain the decisions made by neural networks. While they computed implicants directly, we learned the OBDD of a neural network. Having an OBDD not only facilitates the computation of prime implicants, but it also allows model counting to be performed efficiently [12], which provides more powerful tools for analysis, as we shall show.

Table 2. Compilation of a BNN on 64 variables around the r-neighborhood of an image of a digit 0, digit 8, and a smile.

r	Input space	OBDD size	Compile time (s)
Digit 0			
1	65	0 ($\bot$)	<1
2	2,081	0 ($\bot$)	<1
3	43,745	0 ($\bot$)	<1
4	679,121	0 ($\bot$)	<1
5	8,303,633	0 ($\bot$)	2
6	83,278,001	509	403
7	704,494,193	2,202	2,166
Digit 8			
1	65	0 ($\top$)	<1
2	2,081	0 ($\top$)	<1
3	43,745	0 ($\top$)	<1
4	679,121	0 ($\top$)	2
5	8,303,633	243	111
6	83,278,001	765	584
7	704,494,193	2,431	3,168
Smile			
1	65	0 ($\top$)	<1
2	2,081	258	31
3	43,745	1,437	420
4	679,121	6,048	3,336

For $r = 6$, we get a reduced OBDD of size $1,469$, indicating the existence of counterexamples. We first consider the number of assignments satisfying this OBDD (i.e., the number of counterexamples), which can be done in time linear in the size of the OBDD. In particular, we found that $20,413,779$ out of the $377,519,940,289$ images (0.005%) were classified incorrectly as the digit '8.' Hence, not only can we detect if a given instance is sensitive to perturbations (flips of the pixels), we can also *quantify* how robust it is by counting *how many* ways the instance can be flipped. This is in contrast to approaches to neural network verification based on solving NP-complete problems, such as those relying (just) on SAT-solvers, where counting is in general out of scope (counting is a #P-complete problem).

Next, using the PI query, we identified a minimal set of pixels that guaranteed a correct classification, regardless of how the other pixels are set, within a radius of 6 of Fig. 3a. The result is shown in Fig. 4a. This PI query tells us about the behavior of our CNN classifier, in the space of images around Fig. 3b. In particular, it suffices to have these particular white pixels near the border of the

Table 3. Compilation of a CNN on 256 variables around the r-neighborhood of an image of a digit 0, a digit 8, and a smile.

r	Input space	OBDD size	Compile time (s)
Digit 0			
1	257	0 ($\perp$)	<1
2	32,897	0 ($\perp$)	<1
3	2,796,417	0 ($\perp$)	<1
4	177,589,057	12	2
5	8,987,138,113	220	29
6	377,519,940,289	1,469	450
Digit 8			
1	257	0 ($\top$)	<1
2	32,897	0 ($\top$)	<1
3	2,796,417	0 ($\top$)	<1
4	177,589,057	64	18
5	8,987,138,113	573	250
6	377,519,940,289	3,345	3,486
Smile			
1	257	0 ($\perp$)	<1
2	32,897	8	<1
3	2,796,417	93	7
4	177,589,057	622	138
5	8,987,138,113	3,269	1,661

image, and these black pixels in the center of the image, for the classifier to fix its decision that the image is of a digit '0.'

We can ask the same queries for the instance visualized in Fig. 3b and classified as digit '8.' For $r = 3$ in Table 3 (middle), the OBDD is just the constant true ($\top$), which means that flipping any 3 pixels of our instance will still produce another image classified correctly as digit '8' (the true class). For $r = 6$, we get an OBDD of size 3,345. Using this OBDD, we found that 181,664,350 out of the 377,519,940,289 images (0.05%) are classified incorrectly as the digit '0.' The PI query identified the minimal set of pixels in Fig. 4b which guaranteed a correct classification regardless of how the remaining pixels are set (within a radius of 6 of Fig. 3b).

For the "smile" image in Fig. 3c, the compiled OBDD for the ($r = 5$)-neighborhood is larger than the corresponding OBDDs of the first two images (see each $r = 5$ row in Table 3). As well, for $r = 5$, the PI query for the "smile" requires 19 out of the 256 pixels to be fixed in order to guarantee a classification, while the PI queries for the digit '0' and digit '8' only require 4 and 12

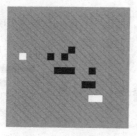   

(a) 12 out of 256 pixels   (b) 19 out of 256 pixels
fixed from Figure 3a   fixed from Figure 3b

Fig. 4. Prime implicant results for $r = 6$ for the images in Fig. 3a and b. The grey striped region represents 'don't care' pixels. If we fix the black/white pixels in Fig. 4a, any completing image within a radius of 6 from Fig. 3a must be classified as '0'. If we fix the black/white pixels in Fig. 4b, any completing image within a radius of 6 from Fig. 3b must be classified as '8'.

pixels respectively (Fig. 5). This suggests that the behavior of the BNN is less structured in the region around the image of the "smile", possibly because it is unclear how the image should be classified.

8 Related Work

The success of neural networks has led to the recent line of work on understanding and verifying their behaviors [6,20,26,28]. These works use, for example, solvers for NP-complete problems such as Mixed-Integer Linear Programming (MILP), satisfiability (SAT), or satisfiability modulo theory (SMT). These systems seek to verify a particular property of a neural network, or otherwise provide a counter-example. We push this line of work further by allowing one to reason about the distribution or the characteristics of counterexamples, which is enabled by learning the OBDD of a given neural network. These richer queries allow us to better understand the neural network behavior beyond detecting the presence of counterexamples.

Choi et al. [7] also consider the compilation of neural networks into a tractable representation, and in particular, into a Sentential Decision Diagram (SDD) [8, 10].[4] They focus on a different class of neural networks and take the approach of reducing a neural network to a Boolean circuit, and then compiling the circuit into a tractable one using classical knowledge compilation techniques. While this approach allows a larger set of verification queries, it does not allow for local or incremental compilation, so it may be less scalable, other things being equal.

Finally, there is also recent work on learning finite state automata from recurrent neural networks (RNNs) [22,34]. Weiss et al. [34] also use an Angluin-style approach for learning the finite state automaton of an RNN. More specifically,

---

[4] Note that SDDs are known to be exponentially more succinct than OBDDs [3].

(a) 4 out of 256 pixels
fixed from Figure 3a

(b) 12 out of 256 pixels
fixed from Figure 3b

(c) 19 out of 256 pixels
fixed from Figure 3c

Fig. 5. Prime implicant results for $r = 5$ for the images shown in Fig. 3. The grey striped region represents 'don't care' pixels. If we fix the black/white pixels in Fig. 5a, any completing image within a radius of 5 from Fig. 3a must be classified as '0'. If we fix the black/white pixels in Fig. 5b, any completing image within a radius of 5 from Fig. 3b must be classified as '8'. If we fix the black/white pixels in Fig. 5c, any completing image within a radius of 5 from Fig. 3b must be classified as '8'.

their approach is based on learning the finite state automaton of an iteratively-refined abstraction of an RNN's state space, and hence the final automaton learned is not necessarily equivalent to the original RNN. Koul et al. [22] trains an RNN and then quantizes the state space using an autoencoder. The result is a quantized network, whose corresponding state machine can be readily extracted. Angluin-style approaches, including ours, can be viewed as instances of *program synthesis*, where a program (a finite state automaton) is learned from a specification (a neural network). For more on formal synthesis, which lies at the increasingly important intersection of the fields of formal verification and machine learning, see, e.g., [17,18].

9   Conclusion

We presented new techniques for verifying the behavior of a binarized neural network on some input region. We outlined an algorithm for compiling a BNN into an OBDD on any input region that can be encoded efficiently as a CNF. Our algorithm combines existing methods for CNF encodings with an Angluin-style algorithm for learning OBDDs. The compiled OBDD gives us access to a range of efficient verification queries and allows us to reason about counterexamples, such as computing their probability and identifying their common characteristics. In domains such as image classification, our approach can let users pinpoint a specific input image I, and then reason about images that are some pixels away from I but classified differently from I. We showed some experiments on a digits classifier, performing verification queries and scaling to 256 inputs.

Acknowledgments. This work has been partially supported by NSF grant #IIS-1514253, ONR grant #N00014-18-1-2561 and DARPA XAI grant #N66001-17-2-4032.

References

1. Angluin, D.: Learning regular sets from queries and counterexamples. Inf. Comput. **75**(2), 87–106 (1987)
2. Bailleux, O., Boufkhad, Y.: Efficient CNF encoding of boolean cardinality constraints. In: Rossi, F. (ed.) CP 2003. LNCS, vol. 2833, pp. 108–122. Springer, Heidelberg (2003). https://doi.org/10.1007/978-3-540-45193-8_8
3. Bova, S.: SDDs are exponentially more succinct than OBDDs. In: Proceedings of the Thirtieth AAAI Conference on Artificial Intelligence, pp. 929–935 (2016)
4. Bryant, R.E.: Graph-based algorithms for boolean function manipulation. IEEE Trans. Comput. **C–35**, 677–691 (1986)
5. Cadoli, M., Donini, F.M.: A survey on knowledge compilation. AI Commun. **10**(3–4), 137–150 (1997)
6. Cheng, C.-H., Nührenberg, G., Huang, C.-H., Ruess, H.: Verification of binarized neural networks via inter-neuron factoring (Short Paper). In: Piskac, R., Rümmer, P. (eds.) VSTTE 2018. LNCS, vol. 11294, pp. 279–290. Springer, Cham (2018). https://doi.org/10.1007/978-3-030-03592-1_16
7. Choi, A., Shi, W., Shih, A., Darwiche, A.: Compiling neural networks into tractable Boolean circuits. In: AAAI Spring Symposium on Verification of Neural Networks (VNN19) (2019)
8. Choi, A., Xue, Y., Darwiche, A.: Same-decision probability: a confidence measure for threshold-based decisions. Int. J. Approximate Reasoning (IJAR) **53**(9), 1415–1428 (2012)
9. Courbariaux, M., Hubara, I., Soudry, D., El-Yaniv, R., Bengio, Y.: Binarized neural networks: training deep neural networks with weights and activations constrained to +1 or −1 (2016)
10. Darwiche, A.: SDD: a new canonical representation of propositional knowledge bases. In: Proceedings of the 22nd International Joint Conference on Artificial Intelligence (IJCAI), pp. 819–826 (2011)
11. Darwiche, A.: Tractable knowledge representation formalisms. In: Tractability: Practical Approaches to Hard Problems, pp. 141–172. Cambridge University Press (2014)
12. Darwiche, A., Marquis, P.: A knowledge compilation map. JAIR **17**, 229–264 (2002)
13. Huang, J., Darwiche, A.: The language of search. J. Artif. Intell. Res. **29**, 191–219 (2007)
14. Hubara, I., Courbariaux, M., Soudry, D., El-Yaniv, R., Bengio, Y.: Binarized neural networks. In: Advances in Neural Information Processing Systems (NIPS), pp. 4107–4115 (2016)
15. Hull, J.J.: A database for handwritten text recognition research. IEEE Trans. Pattern Anal. Mach. Intell. **16**(5), 550–554 (1994)
16. Ignatiev, A., Narodytska, N., Marques-Silva, J.: Abduction-based explanations for machine learning models. In: Proceedings of the Thirty-Third AAAI Conference on Artificial Intelligence (AAAI) (2019)
17. Jha, S., Raman, V., Pinto, A., Sahai, T., Francis, M.: On learning sparse boolean formulae for explaining AI decisions. In: Barrett, C., Davies, M., Kahsai, T. (eds.) NFM 2017. LNCS, vol. 10227, pp. 99–114. Springer, Cham (2017). https://doi.org/10.1007/978-3-319-57288-8_7
18. Jha, S., Seshia, S.A.: A theory of formal synthesis via inductive learning. Acta Informatica **54**(7), 693–726 (2017)

19. Kahlert, L., Krüger, F., Manthey, N., Stephan, A.: Riss solver framework v5. 05 (2015)
20. Katz, G., Barrett, C., Dill, D.L., Julian, K., Kochenderfer, M.J.: Reluplex: an efficient SMT solver for verifying deep neural networks. In: Majumdar, R., Kunvcak, V. (eds.) CAV 2017. LNCS, vol. 10426, pp. 97–117. Springer, Cham (2017). https://doi.org/10.1007/978-3-319-63387-9_5
21. Kearns, M., Vazirani, U.V.: An Introduction to Computational Learning Theory. MIT Press, Cambridge (1994)
22. Koul, A., Fern, A., Greydanus, S.: Learning finite state representations of recurrent policy networks. In: Proceedings of the Seventh International Conference on Learning Representations (ICLR) (2019)
23. Leofante, F., Narodytska, N., Pulina, L., Tacchella, A.: Automated verification of neural networks: advances, challenges and perspectives. CoRR abs/1805.09938 (2018). http://arxiv.org/abs/1805.09938
24. Meinel, C., Theobald, T.: Algorithms and Data Structures in VLSI Design: OBDD - Foundations and Applications. Springer, Heidelberg (1998)
25. Nakamura, A.: An efficient query learning algorithm for ordered binary decision diagrams. Inf. Comput. **201**(2), 178–198 (2005)
26. Narodytska, N., Kasiviswanathan, S.P., Ryzhyk, L., Sagiv, M., Walsh, T.: Verifying properties of binarized deep neural networks. In: Proceedings of the Thirty-Second AAAI Conference on Artificial Intelligence (AAAI) (2018)
27. Oztok, U., Darwiche, A.: A top-down compiler for sentential decision diagrams. In: Proceedings of the 24th International Joint Conference on Artificial Intelligence (IJCAI), pp. 3141–3148 (2015)
28. Pulina, L., Tacchella, A.: An abstraction-refinement approach to verification of artificial neural networks. In: Touili, T., Cook, B., Jackson, P. (eds.) CAV 2010. LNCS, vol. 6174, pp. 243–257. Springer, Heidelberg (2010). https://doi.org/10.1007/978-3-642-14295-6_24
29. Selman, B., Kautz, H.A.: Knowledge compilation and theory approximation. J. ACM **43**(2), 193–224 (1996)
30. Shih, A., Choi, A., Darwiche, A.: Formal verification of Bayesian network classifiers. In: Proceedings of the 9th International Conference on Probabilistic Graphical Models (PGM) (2018)
31. Shih, A., Choi, A., Darwiche, A.: A symbolic approach to explaining Bayesian network classifiers. In: Proceedings of the 27th International Joint Conference on Artificial Intelligence (IJCAI) (2018)
32. Tseitin, G.: On the complexity of derivation in propositional calculus. In: Studies in Constructive Mathematics and Mathematical Logic, pp. 115–125 (1968)
33. Wegener, I.: Branching Programs and Binary Decision Diagrams. SIAM, Philadelphia (2000)
34. Weiss, G., Goldberg, Y., Yahav, E.: Extracting automata from recurrent neural networks using queries and counterexamples. In: Proceedings of the 35th International Conference on Machine Learning (ICML), pp. 5244–5253 (2018)
35. Weng, T.W., et al.: Towards fast computation of certified robustness for ReLU networks. In: Proceedings of the Thirty-Fifth International Conference on Machine Learning (ICML) (2018)
36. Zhang, H., Zhang, P., Hsieh, C.J.: RecurJac: an efficient recursive algorithm for bounding jacobian matrix of general neural networks and its applications. In: Proceedings of the Thirty-Third AAAI Conference on Artificial Intelligence (AAAI) (2019)

CrystalBall: Gazing in the Black Box of SAT Solving

Mate Soos[1]([✉]), Raghav Kulkarni[2], and Kuldeep S. Meel[1]

[1] School of Computing, National University of Singapore, Singapore, Singapore
[2] Chennai Mathematical Institute, Chennai, India

Abstract. Boolean satisfiability is a fundamental problem in computer science with a wide range of applications including planning, configuration management, design and verification of software/hardware systems. The annual SAT competition continues to witness impressive improvements in the performance of the winning SAT solvers largely thanks to the development of new heuristics arising out of intensive collaborative research in the SAT community. Modern SAT solvers achieve scalability and robustness with sophisticated heuristics that are challenging to understand and explain. Consequently, the development of new algorithmic insights has been primarily restricted to *expert intuitions* and evaluation of the new insights have been restricted to performance measurement in terms of the runtime of solvers or a proxy for the runtime of solvers. In this context, one may ask: *whether it is possible to develop a framework to provide white-box access to the execution of SAT solver that can aid both SAT solver developers and users to synthesize algorithmic heuristics for modern SAT solvers?*

The primary focus of our project is precisely such a framework, which we call CrystalBall. More precisely, we propose to view modern *conflict-driven clause learning* (CDCL) solvers as a composition of classifiers and regressors for different tasks such as *branching, clause memory management, and restarting*. The primary objective of this paper is to introduce a framework to peek inside the SAT solvers – CrystalBall– to the AI and SAT community. The current version of CrystalBall focuses on deriving a classifier to keep or throw away a learned clause. In a departure from recent machine learning based techniques, CrystalBall employs supervised learning and uses extensive, multi-gigabyte data extracted from runs of a single SAT solver to perform predictive analytics.

1 Introduction

Boolean satisfiability is a fundamental problem in computer science with a wide range of applications including planning, configuration management, design and verification of software/hardware systems. While the mention of SAT can be traced to the early 19th century, efforts to develop practically successful SAT

The detailed technical report along with experimental results and source code is available at https://meelgroup.github.io/crystalball/.

© Springer Nature Switzerland AG 2019
M. Janota and I. Lynce (Eds.): SAT 2019, LNCS 11628, pp. 371–387, 2019.
https://doi.org/10.1007/978-3-030-24258-9_26

solvers go back to 1960s [5]. The annual SAT competition continues to witness impressive improvements in the performance of the winning SAT solvers largely thanks to the development of new heuristics arising out of intensive collaborative research in SAT community [1]. While the presence of scores of heuristics has contributed to the robustness and scalability of the SAT solvers, it has come at the cost of lack of explainability and understanding the behavior of SAT solvers. Consequently, the development of new algorithmic insights has been primarily restricted to *expert intuitions* and evaluation of the new insights has been restricted to performance measurement in terms of the runtime of solvers or a proxy for the runtime of solvers.

One of the most critical, but not well-understood heuristic in SAT solvers is learned clause database management, even though it plays a crucial role in the performance of CDCL SAT solvers. For effective database management, a SAT solver needs to decide which learned clauses to keep in the memory as this will affect both memory usage, and, more importantly, runtime - thanks to the potentially useless clause being checked for potential propagation or conflict. While there are several different heuristics used by different modern SAT solvers, the poor understanding of when learned clauses are used during the proof generation makes it hard for the SAT community to develop and improve the heuristics further.

A promising direction in better understanding heuristics in modern SAT solvers was pursued recently by Liang et al. [13,15,16]. It focused on using machine learning for the development of a new variable branching heuristic. Their project, MapleSAT, achieved a significant milestone by winning two gold medals in the 2016 SAT competition and two silver medals in the 2017 SAT competition. While the success of MapleSAT shows the potential of machine learning techniques in SAT solving, the framework for designing heuristics is still primarily restricted to a black box view of SAT solving by focusing on runtime or a proxy of runtime, similarly to [12] which focuses on learning efficient heuristics for QBF formulas. In this context, one may ask: *whether it is possible to develop a framework to provide white-box access to the execution of SAT solver, which can aid the SAT solver developer to synthesize algorithmic heuristics for modern SAT solvers?*

The purpose of our project, called CrystalBall, is to answer the above question affirmatively. We view modern CDCL solvers as a composition of classifiers and regressors for different tasks such as *branching* (which variable to branch on), *clause memory management* (which learned clauses to keep in the memory and which ones to throw), *restarts* (when to terminate a branch and restart), and the like. To gain a deeper understanding of the underlying classifiers, as a first step, we have built a framework to provide white box access to SAT solvers during the solving phase. We envision that such a framework to allow the end user to gain an in-depth data-driven understanding of the performance of their heuristics and aid them in designing better heuristics. We do not aim to replace *expert intuition* but propose an *expert-in-the-loop* approach where the expert is aided with statistically sound explainable classifiers by CrystalBall.

The current version of CrystalBall focuses on inferring two classifiers to predict whether a learnt clause should be kept or thrown away. We take a supervised learning approach that required the design of a sophisticated architecture for data-collection from the execution trace of a SAT solver. We then use supervised learning to infer a set of interpretable classifiers, which are translated to C++ code[1], compiled into CryptoMiniSat and executed along with the solver.

It is worth mentioning that the classifiers were inferred using only a small set of UNSAT instances. The ability of the learned classifier to handle SAT instances almost as well as UNSAT instances, along with being able to out-perform the state-of-the-art solver of 2017, provides strong evidence in support for the choices of different components in our framework. CrystalBall is released as an open-source framework and we believe CrystalBall could serve as a back-bone for designing algorithmic ideas for modern CDCL solvers via a data-driven understanding of their heuristics.

The rest of the paper is organized as follows. We first introduce notation and preliminaries in Sect. 2. We describe in detail the feature engineering, large-scale extraction of data from SAT solvers, labeling of data and classifier in Sect. 3 and present preliminary results in Sect. 4. We finally conclude in Sect. 5 with an outlook for future work.

2   Notations and Preliminaries

We borrow the preliminaries and terminology from [8]. Let X be the set of n Boolean variables. A literal a is either a variable x or its negation $\bar{x}$. A clause $C = a_1 \vee \ldots \vee a_k$ is a disjunction of literals. A clause of size 1 is called a unit clause. A formula F over X is in Conjunctive Normal Form (CNF) if it is a conjunction over clauses.

A resolution derivation of C from a formula F is a sequence of clauses $(C_1, C_2, \ldots C_\tau)$ such that $C_\tau = C$ and every C_i is either a clause in F (an axiom) or is derived from clauses C_j, C_k with $j, k < i$, by the propositional res-olution rule, $A \vee a \diamond B \vee \bar{a} \rightarrow A \vee B$ We refer to this resolution step $\diamond$ as "$A \vee a$ and $B \vee \bar{a}$ are resolved over a". A unit propagation is a special propositional rule when $B = \emptyset$.

Given an input formula F, the run of a CDCL solver consists following sequence of actions:

1. Choose a variable to branch on and assign a value (0 or 1) to the chosen variable (if x_i is assigned 0 then it is equivalent to adding $\bar{x}_i$ to the set of clauses).
2. Use unit propagation rules until some clause gets falsified. This is known as *conflict*.
3. Derive a *learned clause* from the conflict, add the learned clause to the database $\mathcal{D}$ of clauses, and *backtrack*.
4. *Restart* the search after some time and start branching again from the top

---

[1] Translation of the predictor happens by recursively walking the decision tree(s) and emitting human-readable C++ code.

When we start branching, i.e., choose a variable and assign a value to it then the literals implied by this assignment are said to belong to the first *decision level*. Subsequently if we make another decision to branch on another variable then the new literals implied at this step are said to belong to the second decision level. A literal can belong to at most one decision level. We define *LBD* (Literal Block Distance) score of a clause to be the number of distinct decision levels to which the literals of the clause belong at the time of creation of the clause [2].

One can view modern CDCL solver as composition of classifiers and regressors to perform the following actions such as branching, learned clause cleaning, and restarting. It is worth noting that much of the prior work in the SAT community has implicitly focused on designing better classifiers for each of the above components even though viewing the different components as classifiers has not always been explicit [17]. The classifiers employed in state of the art SAT solvers have significantly improved over the years in their empirical performance, but there has been lack of rigorous analysis or theoretical understanding of the reasons behind the performance of these models. In this work, we focus on classifiers for *learned clause cleaning*, and we review two of the most prominent classifiers, employed in MiniSat and Maple_LCM_Dist. It is worth noting that MiniSat has been one of the most prolific SAT solvers, significantly faster than any other solver at the time of its release and Maple_LCM_Dist won the 2017 SAT Competition Main track.

The Classifier of MiniSat. MiniSat maintains a limit on the maximum number of clauses in the memory, which is geometrically increased every time clause cleaning is performed. To this end, MiniSat keeps track of the *activity* of learned clauses. For a clause C, its *activity* is incremented every time C participates in the *1st Unique Implication Point (1st UIP)* conflict [4]. During clause cleaning, learned clauses are sorted in decreasing order by their *activity* and the bottom half of the learned clauses are thrown away.

The Classifier of Maple_LCM_Dist [14]. This 2017 SAT Competition winning solver has a 3-tier system for keeping learned clauses: *Tier 0*, *Tier 1*, and *Tier 2*. *Tier 0* is never cleaned, i.e., clauses in *Tier 0* are never removed, while *Tier 1* is cleaned every 25K conflicts and *Tier 2* is cleaned every 10 K conflicts. The different tiers' classifiers and the movement between the tiers is relatively complex, and we refer the interested reader to the technical report for a detailed description of them.

2.1   Related Work

We assume that the reader is familiar with the SAT problem and for lack of space, we refer the reader to [4] for an extensive survey of related literature. While CrystalBall, to the best of our knowledge, is the first framework to provide white-box access to the execution of SAT solver; our work, nonetheless, makes use of several ideas from the extensive research pursued by the SAT community.

Xu et al. [25] proposed one of the earliest approaches to using machine learning for SAT solving. Their approach, SATZilla, focused on predicting the best

SAT solver from a given portfolio of different solvers. SATZilla employed a supervised machine learning training process and focused on runtime as the metric. Recently, the SAT community has focused efforts to understand the performance of SAT solvers from different angles such as empirical studies focused on runtime [3,10,11], through the lens of proof complexity [7], and the like. In a series of papers, Liang et al. [13,15,16] have proposed usage of metrics other than runtime such as learning rate, global learning rate and the like [17]. Similarly, NeuroSAT [22] showcases the potential for ML in SAT, but does so using non-explainable deep learning with single-bit supervision.

3   CrystalBall: An Overview of the Framework

We now present the primary technical contribution of our work, CrystalBall focusing on designing the algorithmic ideas for keeping and throwing away learned clauses. Ideally, one would want to record the entire trace of the execution of the SAT solver for a given instance and perform classification on the collective traces on several instances. Given the complexity of modern SAT solvers, recording the entire trace is time and space consuming and cannot realistically scale beyond small SAT instances [23]. Since a learned clause can be viewed as derived by the application of propositional resolution, our insight is to employ DRAT to reconstruct a close approximation to the significant aspects of the trace of the SAT solver. Since we are using DRAT, our focus is limited to the execution of SAT solver on UNSAT instances. We designed the framework of CrystalBall to allow integration of most modern CDCL-based SAT solvers. For our work, we have integrated CrystalBall with CryptoMiniSat, a modern competitive SAT solver that was placed 3rd in the recently held SAT'18 competition.

CrystalBall consists of four phases: (i) feature engineering, (ii) data collection, (iii) data labeling, and (iv) classifier creation. *Feature engineering* focuses on the design of features that can be used by the classifier. *Data collection* focuses on the modifications to the SAT solver required for an efficient collection of reliable data and computation of labels corresponding to each learned clause. *Labeling* focuses on labeling the learned clauses whether to keep them or throw them away, based on total knowledge, i.e., past and future, of the learned clause *Classifier creation* focuses on employing state of the art supervised machine learning techniques to predict the label based on the past performance (i.e., data available while solving) of the learned clause.

The objective of CrystalBall is not to replace *expert intuition* but to allow for an *expert in the loop* paradigm where a significant amount of relevant data is made available to the expert to allow both for validation of ideas as well as to inspire new ones.

3.1   The Base Solver

As discussed in Sect. 2, the state of the art SAT solvers are comprised of multiple interdependent components, e.g., the learned clauses kept in memory influence clause learning, which influences the activity of variables and therefore branching and in turn, affecting the restart and clause deletion. The interdependence of the various components is both a challenge and an opportunity in the collection of data. The complexity of interactions is a challenge, as it influences the data we gather in ways that are sometimes hard to understand. However, the interdependence of these components is what makes a SAT solver useful in solving real-world problems, and hence collecting this (sometimes messy) data makes the data useful. Collecting "clean" data would make little, if any, sense as the data would be useful neither to make inferences about what modern SAT solvers do nor to train a system to delete learned clauses in a modern SAT solver.

Our base solver to collect data exhibits the following set of dynamic behaviors, all of which are part of standard CryptoMiniSat, except for not explicitly deleting learned clauses:

1. We employ standard VSIDS variable branching heuristic as introduced in Chaff [19] as well as a learning rate based heuristic [16] along with polarity caching [21].
2. We use a mix of geometric [6] and Luby sequence [18] based static restart heuristics as well as a LBD score-based dynamic restart heuristic.
3. We perform inprocessing as standard for CryptoMiniSat. CryptoMiniSat does not perform preprocessing.
4. We strive to keep all learned clauses in memory since we want to know when every learned clause is useful in the unsatisfiability proof. Note that inprocessing may delete learned clauses in some cases.

3.2 Feature Engineering

The accuracy of typical supervised models is often correlated with presence of large number of features and large training data. This comes at the cost of training time and additional complexity of training process. Since training is an offline process and needs to be performed only once, we focus on design of a large number of features corresponding to learned clause. Our features can be categorized into four classes: (i) global features, (ii) contextual features, (iii) restart features, and (iv) performance features. We describe below different features in more detail with an intuitive rational for their inclusion.

Global Features. The global features of a learned clause are the property of the CNF formula at the time of clause creation. For example distribution statistics of horn clauses, irredundant clauses, number of variables, and the like. In particular, we use the features employed by SATZilla in the development of *portfolio solvers*. Since the underlying CNF formula undergoes substantial modifications during the solving, our solver recomputes these features regularly (in particular, at every 100K conflicts) unlike SATZilla that focuses on features only at startup. For every learned clause, we use the latest generated set of global features. Intuitively, inclusion of these features allows the classifier to avoid overfitting to particular types of CNF instances.

Contextual Features. To capture the context in which a clause is learned, we store features corresponding to the context of generation of a clause. In particular, contextual features are computed at the time of generation of the clause and relate to the generated clause. Contextual features include the number of literals in the clause or its LBD score.

Restart Features. Restarts constitute a core component of the modern SAT solvers and one can view every restart corresponding to a phase in the execution of a SAT solver. Intuitively, one expects restart to capture the state of the solver and the progress achievable from that state. We focus on features that capture the execution of the SAT solver in the current and preceding restarts. In particular, the restart features correspond to statistics (average and variance) on the size and LBD score of clauses, branch depth, trail depth during the current and previous restart.

Performance Features. The modern CDCL-based SAT solver maintain several performance parameters about learned clauses, which influence cleaning of the learned clauses. For example, as stated above, the classifiers in Maple_LCM_Dist employ *touched* and *activity* for ordering of clauses in Tier 1 and Tier 2 respectively. Activity is the clause activity as measured by MiniSat [6] and *touched* is the last conflict at which the system played part in a 1st UIP conflict clause generation. Consequently, we compute and maintain several performance features such as the number of times the solver played part of a 1st UIP conflict clause generation, the number of times it caused a conflict and the number of times it caused a propagation.

Normalization. Ideally we want our features to be independent of the problem so that we can compare the values of a feature across problems. Our original features are the property of the particular run of the SAT solver. For different problems the absolute value of same feature can differ drastically. Therefore we can not directly compare the feature values across different problems. Instead, we have to rescale the feature values so that they become somewhat comparable across different problems. In order to achieve this we *relativize* the feature values by taking average feature values in the history as a guideline and measuring the ratio of the actual feature value and this average instead. This normalization is not perfect. However it does help in reducing the difference of scales of the the same feature across different problems. For instance the absolute learned clause size can vary drastically (10 vs 100) across different problems. However, the *relative learned clause size* feature becomes comparable across different problems

3.3   Data Collection

The data collection consists of two passes: a forward pass and a backward pass.

- **Forward pass:** The SAT solver is run on each of the benchmark formulae. During execution, we keep track of a set fraction of randomly chosen learned clauses, which we call *marked* learned clauses. For each *marked* learned clause C, we track and calculate its features and characteristics and continuously write them to a database file for later analysis. A DRAT proof is produced while running.
- **Backward pass:** A modified DRAT-trim [24] is used to parse the DRAT proof. This modified proof checker writes data into the same database file about each and every use of all *marked* learned clause in the unsatisfiability proof.

It is worth noting that while we keep track of learned clauses during the forward pass, we do not track how they were learned, i.e., which resolution rules were used to learn them. Tracking the resolution proof during the forward pass has been long thought to be computationally intractable for all except toy benchmarks. Recent works in progress have made some interesting headway; sustained development in this area may lead to versions of CrystalBall with the ability to keep track of proof in the forward trace.[2]

Our approach of using DRAT-trim has a number of advantages and disadvantages. The primary disadvantage arises from the fact that we are referring to two different proof trees during the forward and backward passes, i.e., proof tree generated by solver (which we are not tracking) might be different from the proof generated by DRAT-trim during the backward pass. However, this division of responsibility between the solver and the proof checker saves significantly on computational, and implementation efforts, and, most importantly allows the system to be used universally with minimal modification, as all competitive SAT solvers, and even many older ones, contain DRAT-trim support thanks to it being mandatory to participate in modern SAT Competitions.

Tracking and Sampling

We attach an ID, C_{ID}, to each clause C so that C can be correctly and fully tracked. We only track some randomly selected set of clauses due to size and timing constraints. All non-tracked clauses' C_{ID} is set to 0. We modified both the SAT solver to output and DRAT-trim to read and store, the 64-bit clause ID for each clause in the binary DRAT-trim format. We randomly set 96% of clauses' C_{ID} to 0 to have sufficient data without severe adverse effect on performance, thus being able to handle larger instances. Neither the forward nor the backward passes then wrote any data about clauses with C_{ID} of 0.

Forward Pass Data Gathering with SQLite

Dumping gigabytes of data for later analysis is a non-trivial task because it has to achieve simultaneously the convenience of data access, speed, and consistency

---

[2] Private communication: J. Nordström.

of data collected across different runs. The data thus collected can significantly affect the quality of classifiers built on top. To dump the data, we chose SQLite because (a) it is self-contained, requiring no separate SQL server process, and (b) the data created by SQLite is a single file that can be efficiently collated with other files and copied from cloud and cluster systems, which is not the case for most of the other SQL servers' data files. Furthermore, SQLite is a mature, well-performing database with a complex query language, subqueries, indexes and the like.

By default, SQLite is synchronous. However, in our case, if execution fails, DRAT-trim also fails to generate data, and the run would be unusable in any case. The default SQLite choice of synchronicity comes at the expense of significantly increased runtime, with no benefits in our case. Therefore we set `PRAGMA synchronous = OFF` and `PRAGMA journal_mode = MEMORY` to increase the speed of writing data to disk substantially. Whenever possible, we also `INSERT` multiple data in one transaction, using SQLite's `BEGIN/END TRANSACTION` methods to lower synchronization overhead. All data dumping is done using *prepared queries*, creating precompiled queries for all data-dumping operations that later use raw C/C++ operations to write data. Hence, query strings are interpreted only once at the start, and the data is copied only once from the SAT solver's memory into SQLite library memory space, to be written to disk later.

To minimize overhead, we do not create indexes for any tables at table creation – instead, we create all needed indexes before querying. This eliminates the overhead of index maintenance during data collection. Indexes are also dropped and re-created when needed during data analysis for the same reason. Significant performance improvement can come from *not* having certain indexes, or more properly, having the right indexes only.

These usage details turn SQLite into a structured, fast, raw data-dumping solution that can later be used as a full-fledged SQL query system. This was key to a viable solution.

Backward Pass Data Gathering with DRAT-trim

Given a propositional formula φ and a clausal proof, DRAT-trim validates that the proof is a certificate of unsatisfiability of the formula φ. To this end, DRAT-trim first forward-searches the CNF, and the SAT-solver generated a proof file for the empty clause. Once it finds the empty clause, it runs through the proof file in a reverse fashion, recursively marking all clauses that contributed to the empty clause. To use DRAT-trim, we modified both the solver and DRAT-trim to write and, respectively, read, the conflict number ConflNo at which each clause is generated. DRAT-trim then knows for each learned clause what conflict number it is generated at. When verifying the proof, DRAT-trim uses this information to infer what conflict numbers each clause is used at. During DRAT-trim's backward pass, for all clauses $C_{ID} > 0$, the data pair of C_{ID}, ConflNo is dumped into a database drat − data.

Recall, DRAT-trim does not have the information about the participation of the clauses in conflict generation during the forward pass of the solver.

DRAT-trim can only infer that given the clauses in the database, the conflict could have taken place. It is possible that there are two sets of clauses, A and B, $A \neq B$, both (potentially overlapping) sets are in the clause database of the solver and given either set, the conflict could have taken place. DRAT-trim employs a greedy algorithm to pick one of the two sets.

In general, it is possible to construct examples where an exponential number of clause sets could have caused a conflict and finding a minimum set may be neither trivial nor necessarily useful in building a classifier. It is, however, an exciting avenue of research to optimize the proof by making DRAT-trim take the smallest set at every point, for a particular post-processing overhead. This could allow training a classifier that could optimize for proof size. Exploring such extensions is beyond the scope of this work, and we leave it to future work.

3.4 Data Labeling

To infer a classifier via supervised learning, the inference engine requires the data corresponding to each clause be labeled a clause to be kept or thrown away. A naive strategy would be to label a clause *useful* if the clauses is used at all in the UNSAT proof generated by DRAT-trim. Analysis of data gathered from the backward pass indicates that proofs generated by CryptoMiniSat use close to 50% of the kept learned clauses while simultaneously throwing away approximately 99% of its learned clauses. This apparent contradiction is due to two factors: (1) as evidenced from the data gathered, most clauses are used in a hot spot close to where they are learned and not (or rarely) used later (2) SAT solver developers understand that there is a cost that is paid (both in memory usage and, more importantly, CPU clock cycles) for keeping a clause. Hence, if a clause is mostly useful in the near future, but may have, say, a single use far in the future, it may be beneficial to throw it away after a short amount of time, having served most of its purpose, and hoping that the solver will find a way to the proof anyway.

Given this analysis, it is clear that there are two corresponding guiding factors that govern whether we should label a clause to be kept: (1) whether the clause is useful at all later and (2) whether the distribution of future uses merits the solver to keep the clause, i.e. whether the cost should be paid to keep the clause until the future point(s) when it's useful. Satisfying requirement (1) is trivial when labeling a data point for keep/throw, given that we know when a clause will be useful and hence we know its last-use point. However, there are no existing cost models to satisfy (2). While a detailed study of construction of cost models is beyond the scope of this work and deferred to future work, we define the usefulness of a clause and the desired classifiers as follows:

1. A clause c is labeled to be kept for an interval of t conflicts if the number of times it was used in the final unsatisfiability proof (as computed by DRAT-trim) is greater than the average of the number of times all the clauses in databases are useful over the interval.
2. We employ two labelings (and associated classifiers) to handle the short-term and long-term usefulness of a clauses. The classifier *keep-short* (resp. *keep-long*) is short-sighted (resp. long-sighted) and attempts to predict whether the

clause is to be kept for the next 10K (resp. 100K) conflicts and is trained by using the data labeled by setting $t = 10,000$ (resp. $t = 100,000$) as explained above.

3.5 Inference of a Classifier via Supervised Learning

Given the labeled data, we considered several choices for our classifier including SVM, decision trees, random forests, and logistic regression. The desired learning algorithm was chosen based on the following primary constraints: (1) Our 218 features, comprised of four different categories, are mixed and heterogeneous, (2) there is no straightforward way to normalize all of our features, (3) the model must be easily convertible to C++ code as the decision inside the SAT solver has to very fast, and (4) the model must provide meaningfully good prediction accuracy.

Although we have created good classifiers using both SVM and logistic regression, we found that satisfying requirement (3) is relatively complicated for these and tuning them in our chosen framework, scikit-learn [20], is harder than decision trees and random forests. Decision trees satisfy all requirements and allow for easy visualization but give relatively worse prediction accuracy than random forests. We therefore chose random forests as the classifier for our classifier when running in the solver, and decisions trees when visualizing and debugging the decision logic during training.

For demonstration purposes, both a *keep-short* and a *keep-long* trained decision tree is visualized in the technical report. As expected, the actual random forests used are too big to be visualized, containing approx. 320 decision nodes for each classifier's tree, where 10 trees make up a decision forest. The prediction accuracy of the decision trees visualized are only slightly lower than the final decision forests, and reviewing them can lead to interesting insights.

3.6 Feature Ranking

As discussed in Sect. 2, state of the art solvers over the past decade have relied on finding and exploiting strong features to estimate learned clause quality. These relatively few features form the core of their heuristic for learned clauses database management. The identification of these heuristics has largely been driven by expert intuition and on runtime measurements.

In contrast, CrystalBall employs a data-driven heuristic along with an expert-driven cost model, for identifying the distinguishing features, which we can chose to be a relatively large set, given that the classifier will be inferred automatically and complicated relationships between them will be handled by the classifier. We use feature importance method of the Random Forest to rank the features by their importance. A Random Forest consists of several decision trees. The root of a decision tree corresponds to the most important decision in the decision tree as the root affects the classification for a large fraction of inputs. Furthermore, the decision nodes closer to the root are more important while the ones farther away from the root are less important. Quantitatively CART 4.5 uses the *average decrease in Gini impurity* as the measure of feature importance. Gini impurity is a measure of how often a randomly chosen element from the set would be incorrectly labeled if it was randomly labeled according to the distribution of the labels in the subset. The importance of a feature in a Random Forest is the average importance of the features in each decision tree. The importance of a feature in a decision tree is the decrease in Gini impurity caused by decisions involving that feature in the decision tree [9].

4   Results

As stated in the introduction, the mission of CrystalBall is to develop a framework to provide white-box access to the execution of SAT solver. The current version of CrystalBall aims to understand the algorithmic ideas for keeping and throwing away learned clauses.

To conduct experiments, we used a high-performance computer cluster, where each node has an E5-2690 v3 CPU with 24 cores and 96 GB of RAM. We used all the 934 unique CNFs from SAT Competitions' 2014,'16 and '17 both to obtain training data then to evaluate the speed of the final solver executable—however, following standard practice, we split the data into 70% for training and 30% for testing and also did not (and could not) use satisfiable instances for training, which constituted about 45% of all instances. The experimental results presented in this paper required over 250,000 CPU hours (equivalent to 28 CPU years).

Training Phase. When collecting data, we used a 12 h timeout for both the solver and DRAT-trim and generated over 37 GB of SQLite data. Since the number of clauses learned for different problems varied widely, we sampled a fixed set of data points from each benchmark to ensure fair representation and discarded problems that were solved too fast to be meaningful. After sampling, we have ≈429 K data points for the *keep-short* classifier and ≈85 K data points for the *keep-long* classifier. Each data point contained the 200+ features plus the label to keep or throw away the clause.

Testing Phase. While solving, we used a 5000 s timeout and 4 GB memory limit, which is in line with general SAT Competition rules. Recall, we used two classifiers *keep-short* and *keep-long*. If either of the two classifiers triggers, the clause is kept. However, if the *long-keep* triggers, the clause will not be deleted for the next 100K conflicts.

Table 1. Confusion matrix for the trained classifiers

		Prediction Throw Keep				Prediction Throw Keep	
Ground truth	Throw	0.64	0.36	Ground truth	Throw	0.63	0.37
	Keep	0.11	0.89		Keep	0.09	0.91
		(a) *keep-short*				(b) *keep-long*	

4.1   Accuracy of the Classifiers

As described in Sect. 3.4, we train two binary classifiers whether to keep or throw away a clause at every $N = 10,000$ conflict ticks. The confusion matrix for the classifiers *keep-short* and *keep-long* for the test data are shown in Table 1.

A careful design of the objective function is mandated by the solver's widely different actions corresponding to the predicted label. In case one incorrectly predicts that a clause should be thrown away at tick k and then would correctly predict to keep it at tick $k + 1$, it is already thrown away and the system has already failed. Hence, it is important to try to err on the side of caution and sway our classifiers towards keeping clauses. Hence, we gave twice the sample weight to predicting to keep a clause relative to throwing it away. This focuses on minimizing the error cases where the classifier incorrectly predicts that the clause should be thrown away, the false negative scenario. Therefore the false positive and false negative error rates are not symmetric for our case, ≈ 0.35 vs. ≈ 0.10. All in all, the high accuracy of cells in the confusion matrix shows the potential of the data-driven approach for classification of whether to keep a clause or not.

4.2   Insights from Feature Ranking

Using the feature ranking method through a random forest classifier (as describe above), we obtain the feature rankings and importance scores present in Table 2. Note that since we train two classifiers, with two different labelings, we obtain two rankings and two set of importance scores. In fact, the two differ in ways that are quite interesting. The *keep-short* classifier identifies the well-known "last used in a 1st UIP conflict" as the most important feature. On the other hand, the *keep-long* classifier identifies the total number of times a learned clause participated in a 1st UIP conflict as the most important feature. A generally interesting observation is that all features identified in the top 10 for both classifiers are dynamic features i.e., the said features are not computed at clause creation such LDB (Literal Block Distance), used by most solvers. A possible explanation would be that features such as LBD may be most useful for a classifier that would predict to keep the clause forever (what one could call *keep-forever*), instead of re-examining the clause every N conflicts. We defer the design of such a classifier to future work.

4.3   Solving SAT Competition CNFs

Given the insights from the feature ranking, it is crucial to perform a performance analysis of the learned model for validity and more in-depth insight. The straightforward method is to augment the base solver, CryptoMiniSat v5.6.8, with the model inferred by CrystalBall. To this end, we have performed a preliminary study by using 22 features selected using the guidance given by the top feature list, as marked in Table 2. Implementation and testing of classifiers based on a more extensive set of features is beyond the scope of this study and left

Table 2. Table of feature rankings. We refer the reader to technical report for interpretation of each of the features. Features marked with a * were used in both classifiers, except for the ones marked with ** that were left out of the *keep-short* classifier. The only feature not present in these rankings but used is dump_no, the number of times a learned clause has been up for deletion.

Feature	Relative impor-tance	Feature	Relative impor-tance
rdb0.used_for_uip_creation*	.1121	rdb0.sum_uip1_used*	.1304
rdb0.last_touched_diff*	.1052	rdb1.sum_uip1_used*	.0983
rdb0.activity_rel	.0813	rdb0.used_for_uip_creation*	.0774
rdb0.sum_uip1_used**	.0635	rdb0.act_ranking	.0740
rdb1.sum_uip1_used**	.0631	rdb0.act_ranking_top_10*	.0511
rdb1.activity_rel	.0521	rdb0.last_touched_diff*	.0489
rdb1.last_touched_diff*	.0486	rdb1.act_ranking	.0481
rdb1.act_ranking_top_10*	.0457	rdb0.sum_delta_confl_uip1_used*	.0435
rdb0.act_ranking	.0442	rdb1.used_for_uip_creation	.0435
rdb0.act_ranking_top_10*	.0416	rdb0.activity_rel	.0416
rdb1.act_ranking	.0403	rdb1.act_ranking_top_10*	.0351
rdb0.sum_delta_confl_uip1_used**	.0304	rdb1.last_touched_diff*	.0346
rdb1.used_for_uip_creation	.0296	rdb1.sum_delta_confl_uip1_used	.0293
cl.antecedents_lbd_long_reds_var*	.0189	cl.lbd_rel*	.0217
rdb1.sum_delta_confl_uip1_used	.0183	cl.lbd_rel_queue*	.0176
cl.lbd_rel*	.0172	cl.size_rel*	.0152
rdb.rel_last_touched_diff*	.0162	cl.size*	.0148
cl.lbd_rel_queue*	.0138	cl.antecedents_lbd_long_reds_max	.0146
rdb.rel_used_for_uip_creation*	.0135	rdb1.activity_rel	.0139
cl.lbd_rel_long*	.0126	cl.lbd_rel_long*	.0135

(a) Best features for *keep-short*	(b) Best features for *keep-long*

for future work. We compare the augmented solver vis-a-vis Maple_LCM_Dist, whose tiered set of classifiers serves as inspiration for our classifiers *short-keep* and *long-keep*. We performed the comparison over all the unique 934 instances from SAT Competitions 2014-17 with a timeout of 5000 s.

In summary, CryptoMiniSat augmented with our classifiers, referred to as PredCryptoMiniSat, could solve 612 formulas, obtaining a PAR-2 score[3] of 3761077 while Maple_LCM_Dist could only solve 591 obtaining a PAR-2 score of 4039152. It is worth recalling that our classifiers are learned only using UNSAT instances and therefore, we had data corresponding to only 236 out of 945 formulas. In particular, PredCryptoMiniSat solved 271 satisfiable instances and 341 unsatisfiable instances, which is in line with the distribution of known SAT and

---

[3] PAR-2 score is defined as the sum of all runtimes for solved instances + 2*timeout for unsolved instances, lowest score wins. This scoring mechanism has been used in most recent SAT Competitions.

UNSAT instances among the problems. The ability of the learned classifier to handle SAT instances almost as well as UNSAT instances, along with being able to outperform the state-of-the-art solver of 2017, provides strong evidence in support for the component design choices of CrystalBall.

It is essential to analyze the above results through an appropriate lens. First, the latest hand tuned model in CryptoMiniSat allows it to solve 637 formulas with a PAR-2 score of 3506488. Although this is significantly better than PredCryptoMiniSat, it has been tuned over many years. Secondly, our model is not optimized for memory consumption which leads to significantly increased cache misses, hence increased runtime. To be able to use all 22 features, PredCryptoMiniSat keeps an additional 68 bytes of data for each clause. Furthermore, a fine-tuned version with fewer features and perhaps more than two classifiers is likely to result in improved runtime. Therefore, overall we believe that our learned model not only highlights surprising power of several features but could also be a starting point to design state of the art solvers by using auto-generated data-driven yet interpretable models.

5   Conclusion

In this paper, we introduced to the SAT community our framework, CrystalBall, to analyze and generate classifiers using significant amounts of behavioral data collected from the run of a SAT solver. CrystalBall combines data collection of its forward pass with proof data using DRAT-trim in its backward pass to allow studying more than 260 UNSAT instances from SAT Competitions. Our preliminary results demonstrate the potential of data-driven approach to accurately predict whether to keep or throw away learned clauses and to rank features that are useful and in this prediction. Our experiments were able to not only derive interesting set of features but also demonstrate the strength of our solver on competition benchmarks.

As a next step, we plan to extend CrystalBall to allow easier integration with other state of the art SAT solvers. In the long term, we believe CrystalBall will both enable to better understand SAT solvers and lower the barrier to designing heuristics for high-performance SAT solvers. Finally, given the requirement of tight integration of learned models into state of the art SAT solvers, CrystalBall presents exciting opportunities for the design of interpretable machine learning models.

Acknowledgements. The authors are grateful for Marijn Heule's help with an early version of DRAT-Trim with Clause IDs. This research is supported in part by the National Research Foundation Singapore under its AI Singapore Programme AISG-RP-2018-005 and NUS ODPRT Grant R-252-000-685-133. The computational work for this article was performed on resources of the National Supercomputing Centre, Singapore https://www.nscc.sg.

References

1. Proceedings of SAT Competition 2018; solver and benchmark descriptions (2017). https://helda.helsinki.fi/handle/10138/224324
2. Audemard, G., Simon, L.: Predicting learnt clauses quality in modern SAT solvers. In: Boutilier, C. (ed.) IJCAI, pp. 399–404 (2009)
3. Biere, A., Fröhlich, A.: Evaluating CDCL variable scoring schemes. In: Heule, M., Weaver, S. (eds.) SAT 2015. LNCS, vol. 9340, pp. 405–422. Springer, Cham (2015). https://doi.org/10.1007/978-3-319-24318-4_29
4. Biere, A., Heule, M., van Maaren, H., Walsh, T.: Handbook of Satisfiability. IOS Press, Amsterdam (2009)
5. Davis, M., Putnam, H., Robinson, J.: The decision problem for exponential diophantine equations. Ann. Math. **74**, 425–436 (1961)
6. Eén, N., Sörensson, N.: An extensible SAT-solver. In: Giunchiglia, E., Tacchella, A. (eds.) SAT 2003. LNCS, vol. 2919, pp. 502–518. Springer, Heidelberg (2004). https://doi.org/10.1007/978-3-540-24605-3_37
7. Elffers, J., Giráldez-Cru, J., Gocht, S., Nordström, J., Simon, L.: Seeking practical CDCL insights from theoretical SAT benchmarks. In: Proceedings of IJCAI, pp. 1300–1308 (2018)
8. Elffers, J., Johannsen, J., Lauria, M., Magnard, T., Nordström, J., Vinyals, M.: Trade-offs between time and memory in a tighter model of CDCL SAT solvers. In: Creignou, N., Le Berre, D. (eds.) SAT 2016. LNCS, vol. 9710, pp. 160–176. Springer, Cham (2016). https://doi.org/10.1007/978-3-319-40970-2_11
9. Friedman, J., Hastie, T., Tibshirani, R.: The Elements of Statistical Learning, vol. 1. Springer Series in Statistics, New York (2001)
10. Jamali, S., Mitchell, D.: Centrality-based improvements to CDCL heuristics. In: Beyersdorff, O., Wintersteiger, C.M. (eds.) SAT 2018. LNCS, vol. 10929, pp. 122–131. Springer, Cham (2018). https://doi.org/10.1007/978-3-319-94144-8_8
11. Katebi, H., Sakallah, K.A., Marques-Silva, J.P.: Empirical study of the anatomy of modern SAT solvers. In: Sakallah, K.A., Simon, L. (eds.) SAT 2011. LNCS, vol. 6695, pp. 343–356. Springer, Heidelberg (2011). https://doi.org/10.1007/978-3-642-21581-0_27
12. Lederman, G., Rabe, M.N., Seshia, S.A.: Learning heuristics for automated reasoning through deep reinforcement learning. In: Proceedings of ICLR (2019)
13. Liang, J.H., Ganesh, V., Poupart, P., Czarnecki, K.: Exponential recency weighted average branching heuristic for SAT solvers. In: Proceedings of AAAI, pp. 3434–3440 (2016)
14. Liang, J.H., Ganesh, V., Poupart, P., Czarnecki, K.: Learning rate based branching heuristic for SAT solvers. In: Creignou, N., Le Berre, D. (eds.) SAT 2016. LNCS, vol. 9710, pp. 123–140. Springer, Cham (2016). https://doi.org/10.1007/978-3-319-40970-2_9
15. Liang, J.H., Oh, C., Mathew, M., Thomas, C., Li, C., Ganesh, V.: Machine learning-based restart policy for CDCL SAT solvers. In: Beyersdorff, O., Wintersteiger, C.M. (eds.) SAT 2018. LNCS, vol. 10929, pp. 94–110. Springer, Cham (2018). https://doi.org/10.1007/978-3-319-94144-8_6
16. Liang, J.H., Hari Govind, V.K., Poupart, P., Czarnecki, K., Ganesh, V.: An empirical study of branching heuristics through the lens of global learning rate. In: Gaspers, S., Walsh, T. (eds.) SAT 2017. LNCS, vol. 10491, pp. 119–135. Springer, Cham (2017). https://doi.org/10.1007/978-3-319-66263-3_8

17. Liang, J.: Machine learning for SAT solvers. Ph.D. thesis, Technical report, University of Waterloo (2018)
18. Luby, M., Sinclair, A., Zuckerman, D.: Optimal speedup of Las Vegas algorithms. Inf. Process. Lett. **47**(4), 173–180 (1993)
19. Moskewicz, M.W., Madigan, C.F., Zhao, Y., Zhang, L., Malik, S.: Chaff: engineering an efficient SAT solver. In: Proceedings of DAC, pp. 530–535. ACM (2001)
20. Pedregosa, F., et al.: Scikit-learn: machine learning in Python. J. Mach. Learn. Res. **12**, 2825–2830 (2011)
21. Pipatsrisawat, K., Darwiche, A.: A lightweight component caching scheme for satisfiability solvers. In: Marques-Silva, J., Sakallah, K.A. (eds.) SAT 2007. LNCS, vol. 4501, pp. 294–299. Springer, Heidelberg (2007). https://doi.org/10.1007/978-3-540-72788-0_28
22. Selsam, D., Lamm, M., Bünz, B., Liang, P., de Moura, L., Dill, D.L.: Learning a SAT solver from single-bit supervision. In: Proceedings of ICLR (2019)
23. Shacham, O., Yorav, K.: On-the-fly resolve trace minimization. In: Proceedings of DAC, pp. 594–599. ACM (2007)
24. Wetzler, N., Heule, M.J.H., Hunt, W.A.: DRAT-trim: efficient checking and trimming using expressive clausal proofs. In: Sinz, C., Egly, U. (eds.) SAT 2014. LNCS, vol. 8561, pp. 422–429. Springer, Cham (2014). https://doi.org/10.1007/978-3-319-09284-3_31
25. Xu, L., Hutter, F., Hoos, H., Leyton-Brown, K.: SATzilla: portfolio-based algorithm selection for SAT. J. Artif. Intell. Res. **32**, 565–606 (2008)

Clausal Abstraction for DQBF

Leander Tentrup[1]([✉])[iD] and Markus N. Rabe[2]

[1] Reactive Systems Group, Saarland University, Saarbrücken, Germany
tentrup@react.uni-saarland.de
[2] Google Research, Mountain View, CA, USA

Abstract. Dependency quantified Boolean formulas (DQBF) is a logic admitting existential quantification over Boolean functions, which allows us to elegantly state synthesis problems in verification such as the search for invariants, programs, or winning regions of games. In this paper, we lift the clausal abstraction algorithm for quantified Boolean formulas (QBF) to DQBF. Clausal abstraction for QBF is an abstraction refinement algorithm that operates on a sequence of abstractions that represent the different quantifier levels. For DQBF we need to generalize this principle to partial orders of abstractions. The two challenges to overcome are: (1) Clauses may contain literals with incomparable dependencies, which we address by the recently proposed proof rule called Fork Extension, and (2) existential variables may have spurious dependencies, which we prevent by tracking consistency requirements during the execution. Our implementation DCAQE solves significantly more formulas than the existing DQBF algorithms.

1 Introduction

The search for functions given declarative specifications is often called the synthesis problem and it is considered to be an extremely hard algorithmic problem. The synthesis of invariants, programs, or winning regions of (finite) games can all be expressed as the existence of a function $f\colon \mathbb{B}^m \to \mathbb{B}^n$ such that for all tuples of inputs $x_1, \ldots, x_k \in \mathbb{B}^m$ some relation $\varphi(x_1, f(x_1), \ldots, x_k, f(x_k))$ over function applications of f is satisfied. While it is possible to specify these problems in SMT or in first-order logic, existing algorithms struggle to solve even simple instances of synthesis queries.

In order to develop a new algorithmic approach for synthesis problems, we focus on the simplest logic admitting the existential quantification over Boolean functions, dependency quantified Boolean formulas (DQBF). However, existing algorithms for DQBF perform poorly, in particular on synthesis problems [3]. This is not surprising: Typical synthesis queries contain two or more function applications, i.e. are of the form $\exists f. \forall x_1, x_2. \varphi(x_1, f(x_1), x_2, f(x_2))$, and involve bit-vector variables, e.g. $x_1, x_2 \in \mathbb{B}^n$. The so far best performing algorithm for DQBF needs to expand either x_1 or x_2 in order to reach a linear quantifier prefix,

M. N. Rabe—Work partially done while at University of California, Berkeley.

which can then be converted to a QBF [11]. This means that they often reduce to QBF formulas that are exponential in n.

Abstraction refinement algorithms have been very successful for QBF, winning the recent editions of the annual QBF competition [15, 16, 19, 21]. Inspired by this success story, we lift the abstraction refinement algorithm called *clausal abstraction* [21] to DQBF. The idea of clausal abstraction for QBF is to split the given quantified problem into a sequence of propositional problems, one for each quantifier in the quantifier prefix, and instantiate a SAT solver for each of them. The SAT solvers solve the quantified problem by communicating assignments (representing examples and counter-examples) to their neighbors.

Lifting clausal abstraction to DQBF comes with two major challenges: First, clausal abstraction is based on Q-resolution [23] and Q-resolution is sound but incomplete for DQBF [1]. In particular, clauses may contain variables from incomparable quantifiers, so called information forks [20] which characterizes the reason for incompleteness. We address this problem using the *Fork Extension* [20] proof rule, which allows us to split clauses with information fork into a set of clauses without information fork by introducing new variables. Second, clausal abstraction relies on the linear quantifier order of QBFs in prenex normal form. For DQBF, however, quantifiers can form an arbitrary partial order. When building a linear order by over-approximating the dependencies of existential variables and applying clausal abstraction naively, those variables may have *spurious dependencies*, i.e. they may only be able to satisfy all the constraints, if they depend on variables that are not allowed by the Henkin quantifiers. We show how to record consistency requirements, i.e., partial Skolem functions, that guarantee that existential variables solely depend on their stated dependencies.

In this paper, we present the first abstraction based solving approach for DQBF. The algorithm successfully applies recent insight in solving quantified Boolean formulas: It is based on the versatile and award-winning clausal abstraction framework [12, 16, 21–23] and leverages progress in DQBF proof systems [20]. Their integration in this work is non-trivial. To handle the non-linear dependencies, we use an over-approximation of the dependencies together with consistency requirements. Further, we turn clausal abstraction into an incremental algorithm that can accept new clauses and variables during solving. Our experiments show that our approach consistently outperforms first-order reasoning [8] on the DQBF benchmarks and it is especially well-suited for the synthesis benchmark set [3] where expansion-based solvers fall short.

2 Preliminaries

Let $\mathcal{V}$ be a finite set of propositional variables. We use the convention to denote universally quantified variables (short also *universals*) by x and existentially quantified variables (or *existentials*) by y. The set of all universals is denoted $\mathcal{X}$, and the set of all existentials is denoted $\mathcal{Y}$. For sets of universals and existentials we use X and Y, respectively. We consider DQBF of the form $\forall x_1 \ldots \forall x_n. \exists y_1(H_1) \ldots \exists y_m(H_m). \varphi$, that is, DQBF begin with universal quantifiers followed by *Henkin quantifiers* and the quantifier-free part φ. A Henkin quantifier $\exists y(H)$ introduces a new variable y, like a normal quantifier, but also specifies a set $H \subseteq \mathcal{X}$ of *dependencies*. A *literal* l is either a variable $v \in \mathcal{V}$ or its negation $\overline{v}$. We call the disjunction $C = (l_1 \vee l_2 \cdots \vee l_n)$ over literals a *clause*, and assume w.l.o.g. that the propositional part of DQBFs are given as a conjunction of clauses, i.e., in conjunctive normal form (CNF). We call the propositional part φ of a DQBF in CNF the *matrix* and we use C_i to refer to the ith clause of φ where unambiguous. For convenience, we treat clauses also as a set of literals and we treat matrices as a set of clauses and use the usual set operations for their manipulation. We denote by $var(l)$ the variable v corresponding to literal l. For literals l of existential variables with dependency set H we define $dep(l) = H$. For literals of universal variables we define $dep(l) = \{var(l)\}$. We lift the operator dep to clauses by defining $dep(C) = \bigcup_{l \in C} dep(l)$. We define $C|_V$ for some clause C and set of variables V as the clause $\{l \in C \mid var(l) \in V\}$.

Given a set of variables $V \subseteq \mathcal{V}$, an *assignment* of V is a function $\alpha : V \to \mathbb{B}$ that maps each variable $v \in V$ to either true ($\top$) or false ($\bot$). A *partial assignment* is a partial function from V to $\mathbb{B}$, i.e. it may be *undefined* on some inputs. To improve readability, we represent (partial) assignments also as a conjunction of literals (i.e., a cube), e.g., we write $x_1\overline{x_2}$ to denote the assignment $\{x_1 \mapsto \top, x_2 \mapsto \bot\}$. We use $\alpha \sqcup \alpha'$ as the update of partial assignment α with α', formally defined as $(\alpha \sqcup \alpha')(v) = \begin{cases} \alpha'(v) & \text{if } v \in dom(\alpha') , \\ \alpha(v) & \text{otherwise} . \end{cases}$

We write $\alpha \sqsubseteq \alpha'$ if $\alpha(v) = \alpha'(v)$ for every $v \in dom(\alpha)$. To restrict the domain of an assignment α to a set of variables V, we write $\alpha|_V$. We denote *the set of assignments* of a set of variables V by $\mathcal{A}(V)$. A *Skolem function* $f_y : \mathcal{A}(dep(y)) \to \mathbb{B}$ maps an assignment of the dependencies of y to an assignment of y. The truth of a DQBF Φ with matrix φ is equivalent to the existence of a Skolem function f_y for every variable y of the existentially quantified variables $\mathcal{Y}$, such that substituting all existentials y in φ by their Skolem function f_y results in a valid formula. We use $\Phi[\alpha]$ to denote the replacement of variables bound by α in Φ with the corresponding value.

Relation to QBF in Prenex Form. In QBF the dependencies of a variable are implicitly determined by the universal variables that occur before the quantifier in the quantifier prefix. This gives rise to the notion that QBF have a *linear* quantifier prefix, whereas DQBF allows for partially ordered quantifiers.

3   Lifting Clausal Abstraction

In this section, we lift clausal abstraction to DQBF. We begin with a high level explanation of the algorithm for QBF and a discussion of the invariants that hold for QBF but are no longer valid for DQBF. For each of those we identify the underlying problem and show how we need to modify clausal abstraction. In the following subsections we then explain those extensions in detail. For the remainder of this section, we assume w.l.o.g. that we are given a DQBF Φ with matrix φ, that φ does not contain clauses with information forks, and that every clause is universally reduced. If a formula contains information forks initially, they can be removed as described in Sect. 3.4.

The clausal abstraction algorithm assigns existential and universal variables, where the order of assignments is determined by the quantifier prefix, until all clauses in the matrix are satisfied or there is a conflict, i.e., a set of clauses that cannot be satisfied simultaneously. Those variable assignments are generated by propositional formulas, one for every quantifier, which we call *abstractions*. In case of a conflict, the reason for this conflict is excluded by refining the abstraction at an outer quantifier.

The assignment order is based on the quantifier prefix. Thus, for QBF it holds that an existential variable is only assigned if its dependencies are assigned. In DQBF, Henkin quantifiers allow us to introduce incomparable dependency sets, and hence, in general, there is no linear order of assignments. We thus weaken this invariant by requiring that for every existential variable y, all of its dependencies have to be assigned before assigning y. We ensure this by creating a graph-based data structure, the dependency lattice, described in Sect. 3.1. As an immediate consequence, and in contrast to QBF, an existential variable may be assigned different values depending on assignments to non-dependencies, and we call this phenomenon a *spurious dependency*. To eliminate those spurious dependencies, we enhance the certification approach of clausal abstraction [21] to build, incrementally, a constraint system that enforces that an existential variable only depends on its dependencies. These *consistency requirements* represent partial Skolem functions. Section 3.5 describes how the consistency requirements are derived, how they are integrated in the algorithm, and when they are invalidated.

We build an abstraction for every existential quantifier $\exists Y$, splitting every clause C of the matrix into three parts, based on whether a literal $l \in C$ is (1) a dependency, (2) a literal of a variable in Y, or (3) neither of the two. Section 3.2 gives a formal description of the abstraction. As mentioned, all dependencies of Y must be assigned before we query the abstraction of the quantifier $\exists Y$ for a candidate assignment of variables Y. From the perspective of this abstraction, assignments to non-Y variables are equivalent when they satisfy the same set of clauses. Vice versa, the only information that matters for other abstractions is the set of clauses satisfied by variables Y or their dependencies. The abstraction for Y therefore defines a set of interface variables consisting of *satisfaction variables* and *assumption variables*, one for every clause C, where the satisfaction variable indicates whether the clause is satisfied by a dependency of Y and the assumption variable indicates whether C must still be satisfied by variables outside of Y. Conflicts are represented by a set of assumption variables that turned out to be not satisfiable only by variables outside of Y. Refinements are clauses over those assumption variables, requiring that at least one of those contained clauses is satisfied by an assignment to Y.

Those refinements correspond to conflict clauses in search-based algorithms and can be formalized as derived clauses in the Q-resolution calculus [23]. Since Q-resolution is incomplete for DQBF and the incompleteness can be character-

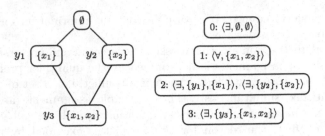

Fig. 1. Dependency lattice (left) and quantifier levels (right) for the DQBF given in Example 1.

ized by clauses with information fork, we check if a conflict clause derived by the algorithm contains such a fork. If this is the case, we *split* this clause into a set of clauses that are fork-free. As a byproduct, new existential variables are created. We show in Sect. 3.4 how clauses with information fork are split and how the clausal abstraction algorithm is extended to *incrementally* accept new clauses and variables.

Example 1. We will use the following formula with the dependency sets $\{x_1\}$, $\{x_2\}$, and $\{x_1, x_2\}$ as a running example.

$$\forall x_1, x_2. \, \exists y_1(x_1). \, \exists y_2(x_2). \, \exists y_3(x_1, x_2).$$
$$\underbrace{(x_1 \vee \overline{x_2} \vee \overline{y_2} \vee y_3)}_{C_1} \underbrace{(\overline{x_1} \vee y_2 \vee y_3)}_{C_2} \underbrace{(\overline{y_1} \vee x_2 \vee \overline{y_3})}_{C_3} \underbrace{(y_1 \vee \overline{y_3})}_{C_4} \underbrace{(x_1 \vee y_1)}_{C_5}$$

3.1   Dependency Lattice and Quantifier Levels

To lift clausal abstraction to DQBF, we need to deal with *partially* ordered dependency sets. Given a DQBF Φ, the algorithm starts with closing the dependency sets under intersection, which can also be described as building the meet-semilattice $\langle \mathcal{H}, \subseteq \rangle$. That is, $\mathcal{H}$ contains all dependency sets of variables in Φ and we add $H \cap H'$ to $\mathcal{H}$ for every $H, H' \in \mathcal{H}$ until a fixed point is reached. We call this meet-semilattice the *dependency lattice*. For our running example, we have to add the empty dependency set, resulting in the dependency lattice depicted on the left of Fig. 1. In addition to the dependency sets $\mathcal{H}$ and the edge relation $\subseteq$, we depict the existential variables next to their dependency sets.

Quantifier Levels and Nodes. We continue with building the data structure on which the algorithm operates. A *node* binds a variable of the DQBF. A universal node $\langle \forall, X \rangle$ binds universal variables X and an existential node $\langle \exists, Y, H \rangle$ binds existential variables Y with dependency set H. Nodes are grouped together in *quantifier levels*, where each universal level contains exactly one universal node and existential levels may contain multiple existential nodes. We index levels by natural numbers i, starting with 0. On the right of Fig. 1 is an example for the data structure obtained from the dependency lattice on its left. Before describing the construction of quantifier levels, we state their invariants. For some node

N, let $bound_\forall(N)$ be the set of variables bound at N, i.e., the union of all X where $\langle \forall, X \rangle$ is in a level with smaller index than node N. Let $bound_\exists(N)$ be the analogously defined set of bound existential variables. The set of bound variables is $bound(N) := bound_\exists(N) \,\dot\cup\, bound_\forall(N)$.

Proposition 1. *The quantifier levels data structure has the following properties.*

1. *Every variable is bound exactly once, i.e., for every variable v in Φ, there is exactly one node $\langle \forall, X \rangle$ or $\langle \exists, Y, H \rangle$ such that $v \in X$ or $v \in Y$.*
2. *Every pair of nodes $\langle \exists, Y, H \rangle$ and $\langle \exists, Y', H' \rangle$ with $Y \neq Y'$ contained in an existential level have incomparable dependencies, i.e., $H \not\subseteq H'$ and $H \not\supseteq H'$.*
3. *For every pair of nodes $\langle \exists, Y_i, H_i \rangle$ and $\langle \exists, Y_j, H_j \rangle$ contained in existential levels i and j with $i < j$, it holds that either $H_i \subset H_j$ or H_i and H_j are incomparable.*
4. *For every existential node $\langle \exists, Y, H \rangle$ it holds that $H \subseteq bound_\forall(\langle \exists, Y, H \rangle)$.*
5. *There is a unique maximal $\langle \exists, Y, H \rangle$ with $H \supset H'$ for every other $\langle \exists, Y', H' \rangle$.*

In the following, we describe the construction of quantifier levels from a dependency lattice. Every element of the dependency lattice $H \in \mathcal{H}$ makes one existential node, $\langle \exists, Y, H \rangle$, where Y is the set of existential variables with dependency set H, i.e. $dep(y) = H$ for all $y \in Y$. Some existential nodes (like the root node in our example) may thus be initially empty. The existential levels are obtained by an antichain decomposition of the dependency lattice (satisfying Proposition 1.2 and 1.3). If the dependency lattice does not contain a unique maximal element, we add an empty existential node $\langle \exists, \mathcal{X}, \emptyset \rangle$ (Proposition 1.5).

Universal variables are placed in the universal node just before the existential level they first appear in as a dependency. This is achieved by a top-down pass through the existential quantifier levels, adding a universal level with node $N = \langle \forall, X \rangle$ before existential level with nodes $\langle \exists, Y_1, H_1 \rangle, \ldots, \langle \exists, Y_k, H_k \rangle$ such that $X = \left(\bigcup_{1 \leq i \leq k} H_i \right) \setminus bound_\forall(N)$ (Proposition 1.4). Empty universal levels $\langle \forall, \emptyset \rangle$ are omitted. Level numbers follow the inverse order of the dependency sets, such that the "outer" quantifiers have smaller level numbers than the "inner" quantifiers; see Fig. 1.

If the formula is a QBF, it holds that $bound_\forall(\langle \exists, Y, H \rangle) = H$. For QBF, this construction yields a strict alternation between universal and existential levels, but for DQBF existential levels can succeed each other, as shown in Fig. 1.

Algorithmic Overview. The overall approach of the algorithm is to construct a propositional formula θ for every node, that represents which clauses it can satisfy (for existential nodes) or falsify (for universal nodes). We describe their initialization in detail below. In every iteration of the loop in algorithm SOLVE (Fig. 2) the variable assignment α_V is extended (case CandidateFound), which we assume to be globally accessible, or node abstractions are refined by adding an additional clauses (case Conflict).

The nodes are responsible for determining candidate assignments to the variables bound at that node, or to give a reason why there is no such assignment. If a node is able to provide a candidate assignment, we proceed to the successor

```
1: procedure SOLVE(DQBF Φ)
2:     levels ← build quantifier levels
3:     initialize every node in levels, i.e., build abstraction θ, set entries ← []
4:     αᵥ ← {}, lvl ← 0
5:     loop
6:         match SOLVELEVEL(lvl)
7:             CandidateFound ⇒ lvl ← lvl + 1
8:             Conflict(jmpBackToLvl) ⇒ lvl ← jmpBackToLvl
9:             Result(res) ⇒ return res
```

Fig. 2. Main solving algorithm that iterates over the quantifier levels.

level (Fig. 2, line 7). A conflict occurs when the algorithm determines that the current assignment α_V definitely violates the formula (unsat conflict) or satisfies it (sat conflict). When conflicts are inspected (explained in Sect. 3.3), they indicate a level that tells the main loop how far we have to jump back (Fig. 2, line 8). The last alternative in the main loop is that we have found a result, which allows us to terminate (line 9).

3.2 Initialization of the Abstractions θ

The formula θ for each node represents how the node's variables interact with the assignments on other levels. The algorithm guarantees that whenever we generate a candidate assignment for a node, all variables on outer (=smaller) levels have a fixed assignment, and thus some set of clauses is satisfied already. Existential nodes then try to satisfy more clauses with their assignment, while universal nodes try to find an assignment that makes it harder to satisfy all clauses. An existential variable y may not only depend on assignments of its dependencies, but also on assignments of existential variables with strict smaller dependency as they are in a strictly smaller level (see Sect. 3.1) and thus are assigned before y. We call this the extended dependency set, written $exdep(y)$, and it is defined as $dep(y) \,\dot\cup\, \{y' \in \mathcal{Y} \mid dep(y') \subset dep(y)\}$. For a set $Y \subseteq \mathcal{Y}$, we define $exdep(Y) = \bigcup_{y \in Y} exdep(y)$.

The interaction of abstractions is established by a common set of *clause satisfaction* variables S, one variable $s_i \in S$ for every clause $C_i \in \varphi$. Given some existential node $\langle \exists, Y, H \rangle$ with extended dependency set $D = exdep(Y)$ and assignment α_V of outer variables V (w.r.t. $\exists Y$, i.e., $V = bound(\langle \exists, Y, H \rangle)$). For every clause $C_i \in \varphi$ it holds that if s_i is assigned to true, one of its dependencies has satisfied the clause, that is, $\alpha_V \vDash C_i|_D$. Thus, an assignment of the satisfaction variables α_S is an abstraction of the concrete variable assignment α_V as multiple assignments could led to the same satisfied clauses.

For universal quantifiers, this abstraction is sufficient as the universal player tries to satisfy as few clauses as possible. For existential quantifiers, however, the existential player can choose to either satisfy the clause directly or *assume* that the clause will be satisfied by an inner quantifier. Thus, we add an additional type of variables A, called *assumption variables*, with the intended semantics

```
1: procedure SOLVELEVEL(lvl)
2:     if lvl is universal then return SOLVE∀(levels[lvl])
3:     for each node n in levels[lvl] do
4:         if SOLVE∃(n) = Conflict(jmpBackToLvl) then
5:             return Conflict(jmpBackToLvl)
6:     return CandidateFound
```

Fig. 3. Algorithm for solving a quantifier level by iterating over the contained nodes.

that a_i is set to false at some existential quantifier $\exists Y$ implies that the clause C_i is satisfied at this quantifier (either by an assignment α_Y to variables Y of the current node or an assignment of dependencies represented by an assignment α_S to the satisfaction variables S), formally, $\alpha_V \dot{\sqcup} \alpha_Y \models C_i|_{D \dot{\cup} Y}$ if a_i is false.

We continue by defining the abstraction that implements this intuition. Formally, for every node $\langle \exists, Y, H \rangle$ and every clause C_i, we define $C_i^< := \{l \in C_i \mid var(l) \in exdep(Y)\}$ as the set of literals on which the current node may depend, $C_i^= := \{l \in C_i \mid var(l) \in Y\}$ as the set of literals which the current node binds, and $C_i^> := \{l \in C_i \mid var(l) \notin exdep(Y) \cup Y\}$ as the set of literals on which the current node may not depend. By definition, it holds that $C = C_i^< \dot{\cup} C_i^= \dot{\cup} C_i^>$. The clausal abstraction θ_Y for this node is defined as $\bigwedge_{C_i \in \varphi} (a_i \vee s_i \vee C_i^=)$. Note, that s_i and a_i are omitted if $C_i^< = \emptyset$ and $C_i^> = \emptyset$, respectively.

Over time, the algorithm calls each node potentially many times for candidate assignments, and it adds new clauses learnt from refinements. The new clauses for existential nodes will only contain literals from assumption variables $L \subseteq A$, representing sets of clauses that together cannot be satisfied by the inner levels. The refinement $\bigvee_{a_i \in L} \overline{a_i}$ ensures that some clause C_i with $a_i \in L$ is satisfied at this node.

Universal nodes $\langle \forall, X \rangle$ have the objective to falsify clause. We define the abstraction θ_X for this node as $\bigwedge_{C_i \in \varphi} (s_i \vee \neg C_i^=) = \bigwedge_{C_i \in \varphi} \left(s_i \vee \bigwedge_{l \in C_i^=} \overline{l} \right)$. Observe that universal nodes do not have separate sets of variables A and S, but just one copy S. This is just a minor simplification, exploiting the formula structure of universal nodes. Note that s_i set to false implies that α_X falsifies the literals in the clause, that is, $\alpha_X \models \neg C_i^=$. Refinements are represented as clauses $\bigvee_{s_i \in L} \overline{s_i}$ over literals in S.

In our running example, clauses 3–5 $(\overline{y_1} \vee x_2 \vee \overline{y_3})(y_1 \vee \overline{y_3})(x_1 \vee y_1)$ are represented at node $\langle \exists, \{y_1\}, \{x_1\} \rangle$ by clauses $(a_3 \vee \overline{y_1})(a_4 \vee y_1)(y_1)$. Note especially, that $x_2 \notin exdep(y_1) = \{x_1\}$, thus there is no s variable in the first encoded clause, despite x_2 being assigned earlier in the algorithm (Fig. 1).

3.3 Solving Levels and Nodes

SOLVELEVEL in Fig. 3 directly calls SOLVE∀ or SOLVE∃ on all the nodes in the level. For existential levels, if any node returns a conflict, the level returns that conflict (Fig. 3, line 5).

We assume a SAT solver interface $\text{SAT}(\psi, \alpha)$ for matrices ψ and assumptions (represented by an assignment) α. It returns either $\text{Sat}(\alpha')$, which means the

```
 1: procedure SOLVE∃(node ≡ ⟨∃, Y, H⟩)        11: procedure SOLVE∀(node ≡ ⟨∀, X⟩)
 2:    αY ← CHECKCONSISTENCY(αV)                12:    αS ← prj∀(X, αV)
 3:    αS ← prj∃(Y, αV)                         13:    match SAT(θX, αS)
 4:    match SAT(θY, αY ⊔ αS)                   14:      Unsat(α) ⇒
 5:      Unsat(α) ⇒                             15:        return REFINE(sat, α|S, node)
 6:        return REFINE(unsat, α|S, node)      16:      Sat(α) ⇒ update αV with α|X
 7:      Sat(α) ⇒ update αV with α|Y           17:      return CandidateFound
 8:        if node is maximal element then
 9:          return REFINE(sat, αS, node)
10:        return CandidateFound
```

Fig. 4. Process existential and universal nodes.

formula is satisfiable with assignment $\alpha' \sqsupseteq \alpha$, or $\mathsf{Unsat}(\alpha')$, which means the formula is unsatisfiable and $\alpha' \sqsubseteq \alpha$ are the failed assumptions (i.e. the unsat core), that is, $\mathrm{SAT}(\psi, \alpha')$ is unsatisfiable as well.

We process universal and existential nodes with the two procedures shown in Fig. 4. The SAT solvers generate a candidate assignment to the variables (lines 4 and 13) of that node, which is then used to extend the (global) assignment α_V (lines 7 and 16). In case the SAT solver returns Unsat, the unsat core represents a set of clauses that cannot be satisfied (for existential nodes) or falsified (for universal nodes). The unsat core is then used to refine an outer node (lines 6 and 15) and we proceed with the level returned by REFINE.

Solving Existential Nodes. There are some differences in the handling of existential and universal nodes that we look into now. The linear ordering of the levels in our data structure means that there may be a variable assigned that an existential node must not depend on. We therefore need to *project* the assignment α_V to those variables in the node's dependency set. We define a function $prj_\exists : 2^Y \times \mathcal{A}(V) \to \mathcal{A}(S)$ that maps variable assignments α_V to assignments of satisfaction variables S such that s_i is set to true if, and only if, some literal $l \in C_i^<$ is assigned positively by α_V. Thus, the projection function only considers actual dependencies of $\langle \exists, Y, H \rangle$:

$$prj_\exists(Y, \alpha_V)(s_i) = \begin{cases} \top & \text{if } \alpha_V \vDash C_i^< \\ \bot & \text{otherwise} \end{cases}$$

For our running example, at node $\langle \exists, \{y_2\}, \{x_2\} \rangle$, the projection for the first clause $C_1 = (x_1 \vee \overline{x_2} \vee \overline{y_2} \vee y_3)$ is $prj_\exists(\{y_2\}, \overline{x_1}\boldsymbol{x_2})(s_1) = prj_\exists(\{y_2\}, x_1\boldsymbol{x_2})(s_1) = \bot$ and $prj_\exists(\{y_2\}, \overline{x_1}\overline{\boldsymbol{x_2}})(s_1) = prj_\exists(\{y_2\}, x_1\overline{\boldsymbol{x_2}})(s_1) = \top$ because $C_1^< = (\overline{x_2})$.

If the SAT solver returns a candidate assignment at the maximal existential node (i.e., the node on innermost level), we know that all clauses have been satisfied, and we have therefore refuted the candidate assignment of some universal node. This is handled by calling REFINE in line 9. For existential nodes we additionally have to check for consistency, which we discuss in Sect. 3.5 (called in line 2).

```
1: procedure REFINE(res, αS, node)
2:     if res = unsat then
3:         Cconflict = ⋃Ci∈φ,αS(si)=⊥ Ci|bound(node)          ▷ Cconflict is universally reduced
4:         if Cconflict contains information fork then
5:             fork elimination ⇒ add clauses and variables, update abstractions θ
6:             RESETCONSISTENCY for all nodes
7:             return Conflict(lvl = 0)
8:     if next ← DETERMINEREFINEMENTNODE(res, αS, node.level) then
9:         return Conflict(next.level)
10:     else                                                   ▷ conflict at outermost ∃/∀ node
11:         return Result(res)
```

Fig. 5. Refinement algorithm that applies Fork Extension in case of information forks.

Solving Universal Nodes. Similar to the projection for existential nodes, we need an (almost symmetric) projection for universal nodes (line 12). It has to differ slightly from $prj_\exists$, because we use just one set of variables S for universal nodes. A universal quantifier cannot falsify the clause if it is already satisfied.

$$prj_\forall(X, \alpha_V)(s_i) = \begin{cases} \top & \text{if } \alpha_V \vDash C_i|_{bound\langle\forall,X\rangle} \\ undef & \text{otherwise} \end{cases}$$

3.4 Refinement

Algorithm REFINE in Fig. 5 is called whenever there is a conflict, i.e. whenever it is clear that α_V satisfies the formula (*sat* conflict) or violates it (*unsat* conflict). In case there is an unsat conflict at an existential node, we build the (universally reduced) conflict clause from α_S [23] in line 3. If the clause is fork-free, we can apply the standard refinement for clausal abstraction [21] with the exception that we need to find the unique refinement node first (line 8). This backward search over the quantifier levels is shown in Fig. 6. For an *unsat* conflict, we traverse the levels backwards until we find an existential node that binds a variable contained in the conflict clause. Because the conflict clause is fork-free, the target node of the traversal is unique. For a *sat* conflict, we do the same for universal nodes but the uniqueness comes from the fact that universal levels are singletons. We then add the refinement clause to the SAT solver at the corresponding node (lines 6 and 10) and proceed. For *sat* conflicts, we have to additionally learn consistency requirements at existential nodes (line 13) that make sure that the node produces the same result if the assignment (restricted to the dependencies of that node) repeats. In case the conflict propagated beyond the root node, we terminate with the given result.

```
1: procedure DETERMINEREFINEMENTNODE(res, αS, lvl)
2:     while lvl ≥ 0 do
3:         if res = unsat and lvl is existential then
4:             for node ⟨∃, Y, H⟩ in levels[lvl] do    ▷ check if Y is contained in conflict clause
5:                 if Ci|Y ≠ ∅ for some Ci ∈ φ with αS(si) = ⊥ then
6:                     θY ← θY ∧ ⋁Ci∈φ,αS(si)=⊥ āi              ▷ refine abstraction
7:                     return ⟨∃, Y, H⟩
8:             else if res = sat and lvl is universal with node ⟨∀, X⟩ then
9:                 if Ci|X ≠ ∅ for some Ci ∈ φ with αS(si) = ⊤ then
10:                     θX ← θX ∧ ⋁Ci∈φ,αS(si)=⊤ s̄i              ▷ refine abstraction
11:                     return ⟨∀, X⟩
12:             else if res = sat and lvl is existential then        ▷ add consistency requirements
13:                 LEARNENTRY(N) for each N = ⟨∃, Y, H⟩ in levels[lvl] with H ⊂ bound∀(N)
14:         lvl ← lvl − 1
15:     return res
```

Fig. 6. Backward search algorithm to determine refinement node.

Fork Extension. In case that the conflict clause contains a fork, we apply Fork Extension [20][1]. *Fork Extension* allows us to split a clause $C_1 \vee C_2$ by introducing a fresh variable y. The dependency set of y is defined as the intersection $dep(C_1) \cap dep(C_2)$ and represents that the question whether C_1 or C_2 satisfies the original clause needs to be resolved based on the information that is available to both of them. Fork Extension is usually only applied when C_1 and C_2 have incomparable dependencies ($dep(C_1) \not\subseteq dep(C_2)$ and $dep(C_1) \not\supseteq dep(C_2)$), as only then the dependency set of y is smaller than those of C_1 and of C_2. The formal definition of the rule is

$$\frac{C_1 \cup C_2 \qquad y \text{ is fresh}}{\exists y(dep(C_1) \cap dep(C_2)). \ C_1 \cup \{y\} \ \wedge \ C_2 \cup \{\overline{y}\}}\text{FEx}$$

Example 2. As an example of applying Fork Extension, consider the quantifier prefix $\forall x_1 x_2. \exists y_1(x_1). \exists y_2(x_2)$ and clause $(\overline{x_1} \vee y_1 \vee y_2)$. Applying **FEx** with the decomposition $C_1 = \{\overline{x_1}, y_1\}$ and $C_2 = \{y_2\}$ results in the clauses $(\overline{x_1} \vee y_1 \vee \mathbf{y_3})(\overline{\mathbf{y_3}} \vee y_2)$ where y_3 is a fresh existential variable with dependency set $dep(y_3) = \emptyset$ ($dep(C_1) = \{x_1\}$ and $dep(C_2) = \{x_2\}$).

After applying Fork Extension, we encode the newly created clauses and variables within their respective nodes. We update the abstractions with those fresh variables and clauses as for the initial abstraction discussed in Sect. 3.2. Additionally, we reset learned Skolem functions as they may be invalidated by the refinement (Fig. 5, line 6).

---

[1] Fork Extension as introduced in [20] is incomplete for general DQBF. However, it is complete for a normal form of DQBF. We refer to the full version [24] for details.

3.5   Consistency Requirements

The algorithm described so far produces correct refutations in case the DQBF is false. For positive results, the *consistency* of Skolem functions of *incomparable* existential variables may be violated. Consider for example the formula $\forall x_1 \forall x_2. \exists y_1(x_1). \exists y_2(x_2). \exists y_3(x_1, x_2). \varphi$ and assume that for the assignment $\overline{x_1}\overline{x_2}$, there is a corresponding satisfying assignment $\overline{y_1}\overline{y_2}\overline{y_3}$. If the next assignment is $\overline{x_1}x_2$, then the assignment to y_1 has to be the same as before ($y_1 \to \perp$) as the value of its sole dependency x_1 is unchanged.

We enhance the certification capabilities of clausal abstraction [21] to build consistency requirements that represent partial Skolem functions in our algorithm during solving. We incrementally build a list of *entries*, where the first component in an entry is a propositional formula over the dependencies and the second component is the corresponding assignment α_Y. Before generating a candidate assignment in SOLVE$_\exists$, we call CHECKCONSISTENCY (Fig. 7) to check if the assignment α_Y for the given assignment α_V of dependencies is already determined, by iterating through the learned entries (Fig. 7, lines 2–3). If it is the case, we get an assignment α_Y that is then assumed for the candidate generation. Note that in this case, the SAT call in line 4 of SOLVE$_\exists$ is guaranteed to return Sat (we already verified this assignment, otherwise it would not have been learned). Further, consistency requirements are only needed for existential nodes $\langle \exists, Y, H \rangle$ with $H \subset bound_\forall(\langle \exists, Y, H \rangle)$, i.e., that observe an over-approximation of their dependency set. For those nodes, the consistency requirements enforce that whenever two assignments of the dependencies are equal, the assignment of α_Y returns the same value as well. We call RESETCONSISTENCY (Fig. 7) to reset the consistency requirements in case we applied Fork Extension (Fig. 5, line 6) as the new clauses may affect already learned parts of the function. We, further, have to reset the clauses learned at universal nodes (Fig. 7, line 7).

We *learn* a new consistency requirement by calling LEARNENTRY (Fig. 7) on the backward search on sat conflicts, that is in line 13 in Fig. 6. When we determine the refinement node for sat conflicts, we call LEARNENTRY in every existential node $\langle \exists, Y, H \rangle$ with $H \subset bound_\forall(\langle \exists, Y, H \rangle)$ on the path to that node. In our example, when the base case of $\langle \exists, \{y_3\}, \{x_1, x_2\} \rangle$ returns (all clauses are satisfied, line 9 in Fig. 4), we add consistency requirements at nodes $\langle \exists, \{y_2\}, \{x_2\} \rangle$ and $\langle \exists, \{y_1\}, \{x_1\} \rangle$ before refining at $\langle \forall, \{x_1, x_2\} \rangle$.

3.6   Example

We consider a possible execution of the presented algorithm on our running example. For the sake of readability, we combine unimportant steps and focus on the interesting cases. Assume the following initial assignment $\alpha_1 = x_1\overline{x_2}y_1\overline{y_2}$ before node $N_{max} \equiv \langle \exists, \{y_3\}, \{x_1, x_2\} \rangle$. The result of projecting function $prj_\exists(\{y_3\}, \alpha_1)$ is $s_1\overline{s_2}s_3\overline{s_4}s_5$ and the SAT solver (Fig. 4, line 4) returns Unsat(α_1') with core $\alpha_1' = \overline{s_2}\overline{s_4}$ as there is no way to satisfy both clauses $(s_2 \vee y_3)$ and $(s_4 \vee \overline{y_3})$ of the abstraction. The refinement algorithm (Fig. 5) builds the conflict clause $C_{conflict} = C_2|_{bound(N_3)} \cup C_4|_{bound(N_3)} = (\overline{x_1} \vee y_2 \vee y_1)$ at line 3 which contains

```
1: procedure CHECKCONSISTENCY(α_V)
2:     for (cond, α_Y) in entries do
3:         if SAT(cond, α_V) is Sat then return α_Y
4:     return empty assignment
5: procedure RESETCONSISTENCY
6:     entries ← []
7:     reset learned clauses at universal nodes
8: procedure LEARNENTRY(node ≡ ⟨∃, Y, H⟩)
9:     let α_S and α_Y be from line 7 of Fig. 4.
10:        entries.push((⋀_{C_i | α_S(s_i)=⊤} C_i^<, α_Y))
```

Fig. 7. Algorithms for handling consistency requirements.

an information fork between y_1 and y_2. We have already seen in Example 2 that the fork can be eliminated resulting in fresh variable y_4 with $dep(y_4) = \emptyset$ and the clauses 6 and 7 $(\overline{x_1} \vee y_1 \vee y_4)(\overline{y_4} \vee y_2)$.

Now, the root node contains variable y_4, for which we assume assignment $\{y_4 \mapsto \top\}$. For the same universal assignment as before $(x_1\overline{x_2})$, the assignment of y_2 has to change to $\{y_2 \mapsto \top\}$ due to the newly added clause 7, leading to $\alpha_2 = x_1\overline{x_2}\overline{y_1}y_2y_4$ before node N_{max}. The only unsatisfied clause is C_4 which can be satisfied using $\{y_3 \mapsto \bot\}$, leading to the base case (Fig. 4, line 9). During refinement, we learn Skolem function entries $(x_1 \wedge y_4, \overline{y_1})$ and $(\overline{x_2} \wedge y_4, y_2)$ at nodes $\langle \exists, \{y_1\}, \{x_1\}\rangle$ and $\langle \exists, \{y_2\}, \{x_2\}\rangle$ as $prj_\exists(\{y_1\}, x_1\overline{x_2})$ and $prj_\exists(\{y_2\}, x_1\overline{x_2})$ assign s_1, s_5, s_6 and s_1, s_6 positively, respectively.

For the following universal assignment $\overline{x_1}\overline{x_2}$, the value of y_2 is already determined by the consistency requirements (Fig. 4, line 2) to be positive. There is a continuation of the algorithm without further unsat conflict, determining that the instance is true.

3.7   Correctness

We sketch the correctness argument for the algorithm, which relies on formal arguments regarding correctness and certification of the clausal abstraction algorithm [21] and the subsequent analysis of the underlying proof system [23].[2] For soundness, the algorithm has to guarantee that existential variables are assigned consistently, that is for an existential variable y with dependency $dep(y)$ it holds that $f_y(\alpha) = f_y(\alpha')$ if $\alpha|_{dep(y)} = \alpha'|_{dep(y)}$ for every α and α'. Our algorithm maintains this property at every point during the execution by a combination of over-approximation and consistency requirements. Completeness relies on the fact that the underlying proof system is refutationally complete for DQBF. Progress is guaranteed as there are only finitely many different conflict clauses and, thus, only finitely many Skolem function resets.

---

[2] A formal correctness proof is given in the full version [24].

4   Evaluation

We compare our prototype implementation, called DCAQE[3], against the publicly available DQBF solvers, IDQ [9], HQS [11], and IPROVER [17]. We ran the experiments on machines with a 3.6 GHz quad-core Xeon processor with timeout and memout set to 10 min and 8 GB, respectively. We used the DQBF preprocessor HQSPRE [26] for every solver except HQS. We evaluate our solver on the DQBF case studies regarding reactive synthesis [3] and the partial equivalence checking problem (PEC) [6,11].

Table 1. Number of instances solved within 10 min. For every solver, we give the number of solved instances overall (#) and broken down by satisfiable (⊤), unsatisfiable (⊥), and uniquely solved instances (∗).

Benchmark	#	DCAQE				IDQ				HQS				IPROVER			
		#	⊤	⊥	∗	#	⊤	⊥	∗	#	⊤	⊥	∗	#	⊤	⊥	∗
PEC1 [6]	1000	**839**	7	**832**	**224**	37	0	37	0	636	10	626	32	71	0	71	0
PEC2 [11]	720	342	71	271	12	214	45	169	0	**401**	**104**	**297**	**60**	288	60	228	0
BoSy [3]	1216	**1006**	**389**	**617**	**66**	924	335	589	2	735	231	504	0	946	370	576	20
	2936	**2187**				1175				1772				1305			

The first two benchmark sets consider the partial equivalence checking (PEC) problem [10], that is, the problem whether a circuit containing not-implemented (combinatorial) parts, so called "black boxes", can be completed such that it is equivalent to a reference circuit. The inputs to the circuit are modeled as universally quantified variables and the outputs of the black boxes as existentially quantified variables. Since the output of a black box should only depend on the inputs that are actually visible to the black box, we need to restrict the dependencies of the existentially quantified variables to subsets of the universally quantified variables. The benchmark sets PEC1 and PEC2 refers to [6] and [11], respectively. The second case study (BoSy) considers the problem of synthesizing sequential circuits from specifications given in linear-time temporal logic (LTL) [3]. The benchmarks were created using the tool BoSy [4] and the LTL benchmarks from the Reactive Synthesis Competition [13,14]. Each formula encodes the existence of a sequential circuit that satisfies the LTL specification.

The results are presented in Table 1. The PEC instances contain over-proportionally many unsatisfiable instances and we conjecture that the differences between DCAQE and IDQ/IPROVER can be explained by the effectiveness of the resolution-based refutations that DCAQE is based on. HQS performs well on those benchmarks as well, which could be due to the fact that it implements the fast refutation technique [6] that was introduced alongside the benchmark set PEC1. The reactive synthesis benchmark set is were DCAQE excels. The benchmark set contains many easily solvable benchmarks, indicated by the high

---

[3] Available at https://github.com/ltentrup/caqe.

number of instances that are commonly solved by all solvers. However, there
are also a fair amount of hard instances and DCAQE solves significantly more
of those than any other solver. Further, we can see the effect mentioned in the
introduction of infeasibility of expansion-based methods as shown by the result of
HQS. The cactus plot given in Fig. 8 shows that DCAQE makes more progress,
especially with a larger runtime where the other solvers solve very few instances
after 100 s. These results give rise to the hope that the scalability of more expressive synthesis approaches [2,5,7] can be improved by employing DQBF solving.

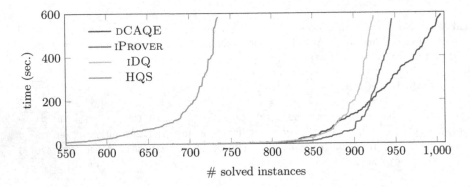

Fig. 8. Cactus plot for the BoSy benchmark.

5 Related Work

The satisfiability problem for DQBF was shown to be NExpTime-complete [18].
Fröhlich et al. [8] proposed a first detailed solving algorithm for DQBF based
on DPLL. They already encountered many challenges of lifting QBF algorithms
to DQBF, like Skolem function consistency, replay of Skolem functions, forks in
conflict clauses, but solved them differently. Their algorithm, called DQDPLL,
has some similarities to our algorithm (in the same way that clausal abstraction
and QDPLL share the same underlying proof system [23]), but performs significantly worse [8]. We highlight a few differences which we believe to be crucial:
(1) Our algorithm tries to maintain as much order as possible. Placing universal nodes at the latest possible allows us to apply the cheaper QBF refinement
method more often. (2) We learn consistency requirements only if they have
been verified to satisfy the formula, while DQDPLL learns them on decisions.
Consequently, in DQDPLL, learned Skolem functions become part of the clauses,
thus, making conflict analysis more complicated and less effective as they may be
undone during solving. We keep the consistency requirements distinct from the
clauses, all learned clauses at existential nodes are thus valid during solving. (3)
Skolem functions in DQDPLL are represented as clauses representing truth-table
entries, thus, become quickly infeasible. In contrast, we use a separate certification mechanism as in QBF solvers [21]. iDQ [9] uses an instantiation-based

algorithm which is based on the Inst-Gen calculus, a state-of-the art decision procedure for the effectively propositional fragment of first-order logic (EPR), which is also NExpTime-complete. HQS [11] is an expansion based solver that expands universal variables until the resulting instance has a linear prefix and applies QBF solving afterwards. Bounded unsatisfiability [6] asserts the existence of a partial (bounded) expansion tree that guarantees that no Skolem function exists. QBF preprocessing techniques have been lifted to DQBF [25,26]. Our solving technique is based on clausal abstraction [21] (also called clause selection [16]) for QBF, which can provide certificates [21]. Later, it was shown that refutation in clausal abstraction can be simulated by Q-resolution [23].

6   Conclusions

We lifted the clausal abstraction algorithm to DQBF. This algorithm is the first to exploit the new Fork Resolution proof system and it significantly increases performance of DQBF solving on synthesis benchmarks. In particular, in the light of the past attempts to define search algorithms [8] (which are closely related to clausal abstraction) for DQBF this is a surprising success. It appears that the Fork Extension proof rule was the missing piece in the puzzle to build search/abstraction algorithms for DQBF.

Acknowledgments. We thank Bernd Finkbeiner for his valuable feedback on earlier versions of this paper. This work was partially supported by the German Research Foundation (DFG) as part of the Collaborative Research Center "Foundations of Perspicuous Software Systems" (TRR 248, 389792660) and by the European Research Council (ERC) Grant OSARES (No. 683300).

References

1. Balabanov, V., Chiang, H.K., Jiang, J.R.: Henkin quantifiers and boolean formulae: a certification perspective of DQBF. Theor. Comput. Sci. **523**, 86–100 (2014). https://doi.org/10.1016/j.tcs.2013.12.020
2. Coenen, N., Finkbeiner, B., Sánchez, C., Tentrup, L.: Verifying hyperliveness. In: Proceedings of CAV (2019, to appear)
3. Faymonville, P., Finkbeiner, B., Rabe, M.N., Tentrup, L.: Encodings of bounded synthesis. In: Legay, A., Margaria, T. (eds.) TACAS 2017. LNCS, vol. 10205, pp. 354–370. Springer, Heidelberg (2017). https://doi.org/10.1007/978-3-662-54577-5_20
4. Faymonville, P., Finkbeiner, B., Tentrup, L.: BoSy: an experimentation framework for bounded synthesis. In: Majumdar, R., Kunčak, V. (eds.) CAV 2017. LNCS, vol. 10427, pp. 325–332. Springer, Cham (2017). https://doi.org/10.1007/978-3-319-63390-9_17
5. Finkbeiner, B., Hahn, C., Lukert, P., Stenger, M., Tentrup, L.: Synthesizing reactive systems from hyperproperties. In: Chockler, H., Weissenbacher, G. (eds.) CAV 2018. LNCS, vol. 10981, pp. 289–306. Springer, Cham (2018). https://doi.org/10.1007/978-3-319-96145-3_16

6. Finkbeiner, B., Tentrup, L.: Fast DQBF refutation. In: Sinz, C., Egly, U. (eds.) SAT 2014. LNCS, vol. 8561, pp. 243–251. Springer, Cham (2014). https://doi.org/10.1007/978-3-319-09284-3_19

7. Finkbeiner, B., Tentrup, L.: Detecting unrealizability of distributed fault-tolerant systems. Log. Methods Comput. Sci. **11**(3) (2015). https://doi.org/10.2168/LMCS-11(3:12)2015

8. Fröhlich, A., Kovásznai, G., Biere, A.: A DPLL algorithm for solving DQBF. In: Proceedings of POS@SAT (2012)

9. Fröhlich, A., Kovásznai, G., Biere, A., Veith, H.: iDQ: instantiation-based DQBF solving. In: Proceedings of SAT. EPiC Series in Computing, vol. 27, pp. 103–116. EasyChair (2014)

10. Gitina, K., Reimer, S., Sauer, M., Wimmer, R., Scholl, C., Becker, B.: Equivalence checking of partial designs using dependency quantified boolean formulae. In: Proceedings of ICCD, pp. 396–403. IEEE Computer Society (2013). https://doi.org/10.1109/ICCD.2013.6657071

11. Gitina, K., Wimmer, R., Reimer, S., Sauer, M., Scholl, C., Becker, B.: Solving DQBF through quantifier elimination. In: Proceedings of DATE, pp. 1617–1622. ACM (2015)

12. Hecking-Harbusch, J., Tentrup, L.: Solving QBF by abstraction. In: Proceedings of GandALF. EPTCS, vol. 277, pp. 88–102 (2018). https://doi.org/10.4204/EPTCS.277.7

13. Jacobs, S., et al.: The 4th reactive synthesis competition (SYNTCOMP 2017): benchmarks, participants & results. In: Proceedings of SYNT@CAV. EPTCS, vol. 260, pp. 116–143 (2017). https://doi.org/10.4204/EPTCS.260.10

14. Jacobs, S., et al.: The 3rd reactive synthesis competition (SYNTCOMP 2016): benchmarks, participants & results. In: Proceedings of SYNT@CAV. EPTCS, vol. 229, pp. 149–177 (2016). https://doi.org/10.4204/EPTCS.229.12

15. Janota, M., Klieber, W., Marques-Silva, J., Clarke, E.M.: Solving QBF with counterexample guided refinement. Artif. Intell. **234**, 1–25 (2016). https://doi.org/10.1016/j.artint.2016.01.004

16. Janota, M., Marques-Silva, J.: Solving QBF by clause selection. In: Proceedings of IJCAI, pp. 325–331. AAAI Press (2015)

17. Korovin, K.: iProver – an instantiation-based theorem prover for first-order logic (system description). In: Armando, A., Baumgartner, P., Dowek, G. (eds.) IJCAR 2008. LNCS (LNAI), vol. 5195, pp. 292–298. Springer, Heidelberg (2008). https://doi.org/10.1007/978-3-540-71070-7_24

18. Peterson, G., Reif, J., Azhar, S.: Lower bounds for multiplayer non-cooperative games of incomplete information. Comput. Math. Appl. **41**, 957–992 (2001)

19. Pulina, L., Seidl, M.: The 2016 and 2017 QBF solvers evaluations (QBFEVAL'16 and QBFEVAL'17). Artif. Intell. **274**, 224–248 (2019). https://doi.org/10.1016/j.artint.2019.04.002

20. Rabe, M.N.: A resolution-style proof system for DQBF. In: Gaspers, S., Walsh, T. (eds.) SAT 2017. LNCS, vol. 10491, pp. 314–325. Springer, Cham (2017). https://doi.org/10.1007/978-3-319-66263-3_20

21. Rabe, M.N., Tentrup, L.: CAQE: a certifying QBF solver. In: Proceedings of FMCAD, pp. 136–143. IEEE (2015)

22. Tentrup, L.: Non-prenex QBF solving using abstraction. In: Creignou, N., Le Berre, D. (eds.) SAT 2016. LNCS, vol. 9710, pp. 393–401. Springer, Cham (2016). https://doi.org/10.1007/978-3-319-40970-2_24

23. Tentrup, L.: On expansion and resolution in CEGAR based QBF solving. In: Majumdar, R., Kunčak, V. (eds.) CAV 2017. LNCS, vol. 10427, pp. 475–494. Springer, Cham (2017). https://doi.org/10.1007/978-3-319-63390-9_25
24. Tentrup, L., Rabe, M.N.: Clausal abstraction for DQBF (full version). CoRR abs/1808.08759 (2019). http://arxiv.org/abs/1808.08759
25. Wimmer, R., Gitina, K., Nist, J., Scholl, C., Becker, B.: Preprocessing for DQBF. In: Heule, M., Weaver, S. (eds.) SAT 2015. LNCS, vol. 9340, pp. 173–190. Springer, Cham (2015). https://doi.org/10.1007/978-3-319-24318-4_13
26. Wimmer, R., Reimer, S., Marin, P., Becker, B.: HQSpre – an effective preprocessor for QBF and DQBF. In: Legay, A., Margaria, T. (eds.) TACAS 2017. LNCS, vol. 10205, pp. 373–390. Springer, Heidelberg (2017). https://doi.org/10.1007/978-3-662-54577-5_21

On Super Strong ETH

Nikhil Vyas$^{(\boxtimes)}$ and Ryan Williams

MIT, Cambridge, MA, USA
{nikhilv,rrw}@mit.edu

Abstract. Multiple known algorithmic paradigms (backtracking, local search and the polynomial method) only yield a $2^{n(1-1/O(k))}$ time algorithm for k-SAT in the worst case. For this reason, it has been hypothesized that the worst-case k-SAT problem cannot be solved in $2^{n(1-f(k)/k)}$ time for any unbounded function f. This hypothesis has been called the "Super-Strong ETH", modeled after the ETH and the Strong ETH. We give two results on the Super-Strong ETH:

1. It has also been hypothesized that k-SAT is hard to solve for randomly chosen instances near the "critical threshold", where the clause-to-variable ratio is $2^k \ln 2 - \Theta(1)$. We give a randomized algorithm which refutes the Super-Strong ETH for the case of random k-SAT and planted k-SAT for any clause-to-variable ratio. For example, given any random k-SAT instance F with n variables and m clauses, our algorithm decides satisfiability for F in $2^{n(1-\Omega(\log k)/k)}$ time, with high probability (over the choice of the formula and the randomness of the algorithm). It turns out that a well-known algorithm from the literature on SAT algorithms does the job: the PPZ algorithm of Paturi, Pudlák and Zane [17].

2. The Unique k-SAT problem is the special case where there is at most one satisfying assignment. Improving prior reductions, we show that the Super-Strong ETHs for Unique k-SAT and k-SAT are equivalent. More precisely, we show the time complexities of Unique k-SAT and k-SAT are very tightly correlated: if Unique k-SAT is in $2^{n(1-f(k)/k)}$ time for an unbounded f, then k-SAT is in $2^{n(1-f(k)(1-\varepsilon)/k)}$ time for every $\varepsilon > 0$.

1   Introduction

The k-SAT problem is the canonical NP-complete problem for $k \geq 3$. Tremendous effort has been devoted to finding faster worst-case algorithms for k-SAT. Because it is widely believed that $\mathsf{P} \neq \mathsf{NP}$, the search has been confined to superpolynomial-time algorithms. Despite much effort, there are no known algorithms for k-SAT which run in $(2-\epsilon)^n$ time for a universal constant $\epsilon > 0$, independent of k. This inability to find algorithms has led researchers to the following two popular hypotheses, which strengthen $\mathsf{P} \neq \mathsf{NP}$:

N. Vyas—Supported by NSF Grant CCF-1741615.
R. Williams—Supported by NSF CCF-1741615.

M. Janota and I. Lynce (Eds.): SAT 2019, LNCS 11628, pp. 406–423, 2019.
https://doi.org/10.1007/978-3-030-24258-9_28

- **Exponential Time Hypothesis (ETH)** [14] There is an $\alpha > 0$ such that no 3-SAT algorithm runs in $2^{\alpha n}$ time.
- **Strong Exponential Time Hypothesis (SETH)** [5] There does not exist a constant $\epsilon > 0$ such that for all k, k-SAT can be solved in $(2 - \epsilon)^n$ time.

The present situation for worst-case k-SAT algorithms looks even worse than hypothesized. The current best known algorithms for k-SAT all have running time bounds of the form $2^{n\left(1-\Omega\left(\frac{1}{k}\right)\right)}$, i.e., $2^{n\left(1-\frac{c}{k}\right)}$ for some constant $c > 0$. It is a very interesting phenomenon that the same running time upper bound is achieved by radically different algorithmic paradigms, such as randomized backtracking [16,17], local search [19], the polynomial method [6] and linear programming based methods [3]. Even for simpler variants such as unique-k-SAT, no significantly faster algorithms are known (with a better dependence on k in the exponent). Hence it is possible that the runtime behavior of $2^{n\left(1-\Omega\left(\frac{1}{k}\right)\right)}$ is actually optimal for k-SAT algorithms. This was termed the Super-Strong ETH in a 2015 talk by the second author [24]. We state the Super-SETH as follows:

Super-SETH: Super Strong Exponential Time Hypothesis
For every unbounded function $f : \mathbb{N} \to \mathbb{N}$, there is no (randomized) algorithm for k-SAT running in $2^{n\left(1-\frac{f(k)}{k}\right)}$ time.

Intuitively, Super-SETH says that the $\Omega(1/k)$ "savings" in the exponent is optimal: not even an $f(k)/k$ savings can be achieved, for any unbounded f. In this paper, we study Super-SETH in two natural restricted scenarios:

- **Random/Planted k-SAT.** We consider two general cases: (a) finding solutions to random k-SAT instances where each clause is drawn uniformly and independently from all possible k-width clauses, and (b) finding solutions to planted k-SAT instances, where a random (hidden) solution σ is sampled, then each clause is drawn uniformly and independently from all possible k-width clauses that satisfy σ.
  Random k-SAT has a well-known threshold behaviour in which, for $\alpha_{sat} = 2^k \ln 2 - \Theta(1)$ and for all constant $\epsilon > 0$, random k-SAT instances are SAT w.h.p. (with high probability) for $m < (\alpha_{sat} - \epsilon)n$ and UNSAT w.h.p. for $m > (\alpha_{sat} + \epsilon)n$. Note that, as far as decidability is concerned, for instances below (respectively, above) the threshold we may simply output "SAT" (respectively, "UNSAT") and we will be correct whp. It has been conjectured [10,20] that random instances at the threshold $m = \alpha_{sat}n$ are the hardest random instances, and it is difficult to determine their satisfiability. We are motivated by the following strengthening of this conjecture: **Are random instances near the threshold as hard as the worst-case instances of k-SAT?**
- **Unique k-SAT.** This is the special case of finding a SAT assignment to a k-CNF, when one is promised that there is at most one satisfying assignment. It is well-known to be NP-complete under randomized reductions [22]. As

mentioned earlier, the best known algorithms for Unique-k-SAT have the same running time behaviour of $2^{n\left(1-O\left(\frac{1}{k}\right)\right)}$ as k-SAT. In fact some of the best-known k-SAT algorithms [16,17] have an easier analysis when restricted to the case of Unique-k-SAT. PPSZ [16], the current best known algorithm for k-SAT (when $k \geq 5$) has only been derandomized for Unique-k-SAT. **Could worst-case algorithms for Unique k-SAT be marginally faster than those for k-SAT?**

In principle, in this "ultra fine-grained" setting we are studying (where the exponential dependence on k matters), both above special cases could potentially be just as hard as k-SAT, or both of them could be easier. In this paper, we prove that Super-SETH is false for Random k-SAT, and the Super-SETH for Unique k-SAT is equivalent to the general Super-SETH: the dependence on k in the exponent is the *same* for the two problems.

1.1 Prior Work

As mentioned earlier, many algorithmic paradigms have been introduced for solving k-SAT in the worst case, but none are known to run in $2^{n(1-\omega_k(1/k))}$ time. There also has been substantial work on polynomial-time algorithms for random k-SAT that return solutions for m below the threshold. Note that even though we know that these instances are satisfiable whp, that does not immediately give a way to *find* a solution. Chao and Franco [7] first proved that the unit clause heuristic (the same key component of the PPZ algorithm) finds solutions with high probability for random k-SAT when $m \leq c2^k n/k$ for some constant $c > 0$. The current best known polynomial-time algorithm in this regime is by Coja-Oghlan [8] and it can find a solution whp for random k-SAT when $m \leq c2^k n \log(k)/k$ for some constant $c > 0$. Interestingly, we also know of polynomial time algorithms for large m. Specifically, it is known that for a certain constant $C_0 = C(k)$ and $m > C_0 \cdot n$ there are polynomial-time algorithms finding solutions to planted k-SAT instances by Krivelevich and Vilenchik [15] and random k-SAT (conditioned on satisfiability) by Coja-Oghlan, Krivelevich and Vilenchik [9]. However, both of these results require that $m \geq 4^k n/k$ [23]. To our knowledge, no improvements over worst-case k-SAT algorithms have yet been reported for random k-SAT very close to the threshold.

Valiant and Vazirani [22] gave poly-time randomized reductions from SAT instances F on n variables to Unique-SAT instances F' on n variables such that, if F is SAT then F' a unique satisfying assignment with probability at least $\Omega(1/n)$, and if F is UNSAT then F' is UNSAT. This reduction is not applicable to convert k-SAT instances to Unique-k-SAT instances, as they do not preserve the clause width (and when we perform a reduction to reduce the clause width, the number of variables blows up too much for exponential-time algorithms). To address this, Calabro, Impagliazzo, Kabanets and Paturi [4] gave a randomized polynomial-time reduction with one-sided error from k-SAT to Unique-k-SAT which works with probability $2^{-O(n \log^2(k)/k)}$. The probability bound was further improved by Traxler [21] to $2^{-O(n \log(k)/k)}$. Both of these reductions imply that k-SAT and Unique k-SAT either both have $2^{\delta n}$ time algorithms for some universal $\delta < 1$, or neither of them do (i.e., the SETH and the SETH for Unique-k-SAT are equivalent). However these results are not sufficient for an equivalence w.r.t. Super-SETH: for example, from these results it is still possible that k-SAT has no $2^{n(1-\omega_k(1/k))}$ time algorithms, while Unique-k-SAT has a $2^{n(1-\Omega(\log k/k))}$ time algorithm.

1.2   Our Results

Average-Case k-SAT Algorithms. First we present an algorithm which breaks the Super-Strong ETH for random k-SAT. In particular, we give a $2^{n\left(1-\Omega\left(\frac{\log k}{k}\right)\right)}$-time algorithm which finds a solution whp for random-k-SAT (conditioned on satisfiability) for all values of m. In fact, our algorithm is an old one from the SAT algorithms literature: the PPZ algorithm of Paturi, Pudlak and Zane [17].

In order to show that PPZ breaks Super-Strong ETH in the random case, we first show that PPZ yields a faster algorithm for random *planted* k-SAT for large enough m.

Theorem 1. *There is a randomized algorithm that, given a* **planted** *k-SAT instance F sampled from $P(n, k, m)^1$ with $m > 2^{k-1} \ln(2)$, outputs a satisfying assignment to F in $2^{n\left(1-\Omega\left(\frac{\log k}{k}\right)\right)}$ time with $1 - 2^{-\Omega\left(n\left(\frac{\log k}{k}\right)\right)}$ probability (over the planted k-SAT distribution and the randomness of the algorithm).*

Next, we give a reduction from random k-SAT (conditioned on satisfiability, we denote this distribution by R^+) to planted k-SAT. Similar reductions/equivalences have been observed before in [1,2].

Theorem 2. *Suppose there is an algorithm A for planted k-SAT $P(n, k, m)$, for all $m \geq 2^k \ln 2(1 - f(k)/2)n$, which finds a solution in time $2^{n(1-f(k))}$ and with probability $1 - 2^{-nf(k)}$, where $1/k < f(k) = o_k(1)$. Then for any m', given a random k-SAT instance sampled from $R^+(n, k, m')$, a satisfying assignment can be found in $2^{n(1-\Omega(f(k)))}$ time whp.*

Combining Theorems 1 and 2 yields:

Theorem 3. *Given a random k-SAT instance F sampled from $R^+(n, k, m)$, we can find a solution in $2^{n\left(1-\Omega\left(\frac{\log k}{k}\right)\right)}$ time whp.*

Remark 1. We obtain a randomized algorithm for random k-SAT which always reports UNSAT on unsatisfiable instances, and finds a SAT assignment whp on satisfiable instances. Feige's Hypothesis for k-SAT [12] conjectures that there are no efficient *refutations* for random k-SAT near the threshold, i.e., there are no

---

[1] See "Three k-SAT Distributions" in Sect. 2 for formal definitions of different k-SAT distributions.

efficient algorithms which always report SAT on satisfiable instances, and report UNSAT on unsatisfiable instances with probability at least 1/2. Refuting Feige's hypothesis in our setting remains an intriguing open problem.

Theorems 1 and 3 imply that at least one of the following are true:

- Either the random instances of k-SAT at the threshold are *not* the hardest instances of k-SAT, or
- Super-Strong ETH is false.

For the PPZ algorithm (randomized branching with unit propagation), time **lower bounds** of the form $2^{n(1-O(\frac{1}{k}))}$ are in fact known [18]. Thus we can say that, with respect to the PPZ algorithm, random k-SAT instances are *provably* more tractable than worst-case k-SAT instances. On the other hand, for the PPSZ algorithm (randomized branching with limited resolution on small sets of clauses) which gives the current best known running time for k-SAT (when $k \geq 4$) we only know $2^{n(1-O(\frac{\log k}{k}))}$ lower bounds [18], matching our upper bounds for the random case. Hence it is possible that PPSZ actually runs in $2^{n(1-\Omega(\frac{\log k}{k}))}$ time for worst-case k-SAT.

We observe that our techniques can be used to get algorithms running faster than $2^{n(1-\Omega(\frac{\log k}{k}))}$ for planted k-SAT and random k-SAT (conditioned on satisfiability), depending on how large m/n is compared to the threshold density. These results appear in the full version.

Unique k-SAT Equivalence. In Sect. 5 we give a "low exponential" time reduction from k-SAT to Unique-k-SAT, which proves that the two problems are equivalent w.r.t. Strong-SETH: i.e., there is a $2^{n(1-\omega_k(1/k))}$ time algorithm for Unique-k-SAT if and only if there is a $2^{n(1-\omega_k(1/k))}$ time algorithm for k-SAT. In fact, our reduction has the following stronger property:

Theorem 4. *If Unique k-SAT is solvable in $2^{(1-f(k)/k)n}$ time for some unbounded function f, then k-SAT is solvable in $2^{(1-f(k)/k+O(\log(f(k))/k))n}$ time.*

As mentioned earlier, the current best algorithm for k-SAT PPSZ [16] has a much easier analysis for Unique k-SAT, and in fact it was an open question to show that its running time on general instances of k-SAT matches the running time for Unique k-SAT; this was eventually resolved by Hertli [13]. Theorem 4 implies that, in order to obtain faster algorithms for k-SAT which break Super-Strong ETH, it would be sufficient to restrict ourselves to Unique k-SAT, which might simplify the analysis as in the case of PPSZ.

2   Preliminaries

Notation. In this paper, we generally assume $k \geq 3$ is an arbitrarily large integer. Throughout the paper, we compare time bounds that have the form

$2^{n(1-\Omega(\log k)/k)}$ with $2^{n(1-O(1/k))}$ time, where the big-Ω and the big-O hide multiplicative constants; such notation only makes sense when k can grow unboundedly.

We use the terms "solution", "SAT assignment", and "satisfying assignment" interchangeably. For an n-variable assignment $s \in \{0,1\}^n$ and an index set $I \subseteq [n]$, we use $s_{|I}$ to denote the length-$|I|$ substring of s projected on the index set I. We use the notation $x \in_r \chi$ to denote that x is randomly sampled from the distribution χ. By $poly(n)$, we mean some function $f(n)$ which satisfies $f(n) = O(n^c)$ for a universal constant $c \geq 1$. Letting n be the number of variables in a k-CNF, a random event about k-CNF holds whp (with high probability) if it holds with probability $1 - f(n)$, where $f(n) \to 0$ as $n \to \infty$. We use log and ln to denote the logarithm base-2 and base-e respectively, and $H(p) = -p\log(p) - (1-p)\log(1-p)$ denotes the binary entropy function, and $\tilde{O}(f(n))$ denotes $O(f(n)\log(f(n)))$.

Three k-SAT Distributions. We consider three distributions for random k-SAT:

- $R(n,k,m)$ is the distribution over formulas F of m clauses, where each clause is drawn i.i.d. from the set of all k-width clauses. This is the standard k-SAT distribution.
- $R^+(n,k,m)$ is the distribution over formulas F of m clauses where each clause is drawn i.i.d. from the set of all k-width clauses and we condition F on being satisfiable i.e. $R(n,k,m)$ conditioned on satisfiability.
- $P(n,k,m,\sigma)$ is the distribution over formulas F of m clauses where each clause is drawn i.i.d. from the set of all k-width clauses which satisfy σ. $P(n,k,m)$ is the distribution over formulas F formed by sampling $\sigma \in \{0,1\}^n$ uniformly and then sampling F from $P(n,k,m,\sigma)$.

Note that an algorithm solving the search problem (finding SAT assignments) for instances sampled from R^+ is stronger than deciding satisfiability for instances sampled from R: given an algorithm for the search problem on R^+, we may run it on a random instance from R, returning SAT if and only if the algorithm returns a valid satisfying assignment.

2.1 Structural Properties of Planted and Random k-SAT

A few structural results about planted and random k-SAT will be useful in analyzing our algorithms. In particular, we consider bounds on the expected number of solutions of planted k-SAT instances and random k-SAT instances (conditioned on satisfiability).

A well-known conjecture is that the satisfiability of random k-SAT displays a threshold behaviour for all k. The following lemma which states that the conjecture holds for all k (larger than a fixed constant) was proven by Ding, Sly and Sun [11].

Lemma 1 ([11]). *There is a constant k_0 such that for all $k > k_0$, there exists an $\alpha_{sat} = 2^k \ln 2 - \Theta(1)$ and for all constant $\epsilon > 0$, we have that:*

$$\text{For } m < (1 - \epsilon)\alpha_{sat}n, \lim_{n \to \infty} \Pr_{F \in_r R(n,k,m)} [F \text{ is satisfiable}] = 1$$

$$\text{For } m > (1 + \epsilon)\alpha_{sat}n, \lim_{n \to \infty} \Pr_{F \in_r R(n,k,m)} [F \text{ is satisfiable}] = 0$$

We will also need the fact that, whp, the ratio of the number of solutions and its expected value is not too small, as long as m is not too large.

Lemma 2 (Achlioptas, Lemma 22 of [1]). *For $F \in_r R(n, k, m)$, let $\mathcal{S}$ be the set of solutions of F. Then $E[|\mathcal{S}|] = 2^n(1 - \frac{1}{2^k})^m$. Furthermore, for $\alpha_d = 2^k \ln 2 - k$ and $m < \alpha_d n$ we have*

$$\lim_{n \to \infty} \Pr[|\mathcal{S}| < E[|\mathcal{S}|]/2^{O(nk/2^k)}] = 0.$$

Together, the above two results have the following useful consequence. Intuitively, the below lemma states that if our random k-SAT instance is slightly below the threshold, then (conditioned on being satisfiable) we can fairly tightly bound the expected number of SAT assignments.

Lemma 3. *For $F \in_r R^+(n, k, m)$ let Z denote the number of solutions of F. For all constant $\delta > 0$, if $m < (1 - \epsilon)\alpha_{sat}$ for some constant $\epsilon > 0$, then $2^n(1 - \frac{1}{2^k})^m \leq E[Z] \leq (1 + \delta)2^n(1 - \frac{1}{2^k})^m$. Furthermore, for $\alpha_d = 2^k \ln 2 - k$ and $m < \alpha_d n$,*

$$\lim_{n \to \infty} \Pr[Z < E[Z]/2^{O(nk/2^k)}] = 0.$$

Proof. Let $F' \in_r R(n, k, m)$ and let Z' denote the number of solutions of F'. Letting p_n denote the probability that F' is unsatisfiable, we have $E[Z'] = (1 - p_n)E[Z]$. By Lemma 1, $\lim_{n \to \infty} p_n \to 0$, hence $2^n(1 - \frac{1}{2^k})^m \leq E[Z] \leq (1 + \delta)2^n(1 - \frac{1}{2^k})^m$.

Observe that $\Pr[Z < E[Z]/2^{O(nk/2^k)}] \leq \Pr[Z' < E[Z]/2^{O(nk/2^k)}]$, as Z is just Z' conditioned on being positive. Furthermore $\Pr[Z' < E[Z]/2^{O(nk/2^k)}] \leq \Pr[Z' < E[Z']/2^{O(nk/2^k)}]$ as $E[Z] \leq 2E[Z']$. By Lemma 2, $\Pr[Z' < E[Z']/2^{O(nk/2^k)}]$ tends to 0. $\qquad\square$

We will use a planted k-SAT algorithm to solve random k-SAT instances conditioned on their satisfiability. The basic idea was introduced in an unpublished manuscript by Ben-Sasson, Bilu, and Gutfreund [2]. We will use the following lemma therein.

Lemma 4 (Lemma 3.3 of [2]). *For a given F in $R^+(n, k, m)$ with Z solutions, it is sampled from $P(n, k, m)$ with probability Zp, where p only depends on n, k, and m.*

Corollary 1. *For $F \in_r R^+(n, k, m)$ and $F' \in_r P(n, k, m)$ let Z and Z' denote their number of solutions respectively. Then for $\alpha_d = 2^k \ln 2 - k$ and for $m < \alpha_d n$, $\lim_{n \to \infty} \Pr[Z' < E[Z]/2^{O(nk/2^k)}] = 0.$*

Proof. We have $\lim_{n\to\infty} \Pr[Z < E[Z]/2^{O(nk/2^k)}] = 0$ by Lemma 3. Lemma 4 shows that the planted k-SAT distribution $P(n, k, m)$ is biased toward satisfiable formulas with more solutions. The distribution $R^+(n, k, m)$ instead chooses all satisfiable formulas with equal probability. Hence $\lim_{n\to\infty} \Pr[Z' < E[Z]/2^{O(nk/2^k)}] = 0$. □

So far, our lemmas have only handled the case where $m > \alpha_{sat}n$. Next we prove a lemma bounding the number of expected solutions when $m > \alpha_{sat}n$.

Lemma 5. *For $m \geq (\alpha_{sat} - 1)n$, the expected number of solutions of $F \in_r R^+(n, k, m)$ and $F' \in_r P(n, k, m)$ is at most $2^{O(n/2^k)}$ in both cases.*

Proof. Lemma 4 shows that the planted k-SAT distribution $P(n, k, m)$ is biased toward satisfiable formulas with more solutions. In particular, the expected number of solutions of $F' \in_r P(n, k, m)$ upper bounds the expected number for $F \in_r R^+(n, k, m)$. So it suffices to upper bound the expected number of solutions of $F' \in_r P(n, k, m)$.

Let Z be the random variable denoting the number of solutions of F'. Let σ denote the planted solution in F, and let x be some assignment which has hamming distance i from σ. For a clause C satisfied by σ but not by x, all of C's satisfied literals must come from the i bits where σ and x differ, and all its unsatisfied literals must come from the remaining $n - i$ bits. Letting j denote the number of satisfying literals in C, the probability that a randomly sampled clause C is satisfied by σ but not by x is $\sum_{j=1}^{k} \frac{\binom{k}{j}}{2^k-1}(\frac{i}{n})^j(1-\frac{i}{n})^{k-j} = \frac{1-(1-\frac{i}{n})^k}{2^k-1}$. We will now upper bound $E[Z]$.

$$E[Z] = \sum_{y \in \{0,1\}^n} \Pr[y \text{ satisfies } F']$$

$$= \sum_{i=1}^{n} \binom{n}{i} \Pr[\text{Assignment } x \text{ that differs from } \sigma \text{ in } i \text{ bits satisfies } F']$$

$$= \sum_{i=1}^{n} \binom{n}{i} \Pr[\text{A random clause satisfying } \sigma \text{ satisfies } x]^m$$

$$= \sum_{i=1}^{n} \binom{n}{i} (1 - \Pr[\text{A random clause satisfying } \sigma \text{ does not satisfy } x])^m$$

$$= \sum_{i=1}^{n} \binom{n}{i} \left(1 - \frac{1-(1-i/n)^k}{2^k-1}\right)^m \quad [\text{As shown above}]$$

$$\leq \sum_{i=1}^{n} \binom{n}{i} e^{-m\left(\frac{1-(1-i/n)^k}{2^k-1}\right)} \quad [\text{As } 1-x \leq e^{-x}]$$

$$\leq \sum_{i=1}^{n} \binom{n}{i} e^{-(\alpha_{sat}-1)n\left(\frac{1-(1-i/n)^k}{2^k-1}\right)}$$

$$\leq 2^{O(n/2^k)} \sum_{i=1}^{n} \binom{n}{i} e^{-((2^k-1)\ln 2)n\left(\frac{1-(1-i/n)^k}{2^k-1}\right)} \text{ [As } m \geq (2^k \ln 2 - O(1))n]$$

$$\leq 2^{O(n/2^k)} \sum_{i=1}^{n} \binom{n}{i} 2^{-n\left(1-(1-i/n)^k\right)}$$

$$\leq 2^{O(n/2^k)} \sum_{i=1}^{n} 2^{n\left(H(i/n)-1+(1-i/n)^k\right)} \leq 2^{O(n/2^k)} \max_{0 \leq p \leq 1} 2^{n\left(H(p)-1+(1-p)^k\right)}.$$

Let $f(p) = H(p) - 1 + (1-p)^k$. Then $f'(p) = -\log\left(\frac{p}{1-p}\right) - k(1-p)^{k-1}$ and $f''(p) = \frac{-1}{p(1-p)} + k(k-1)(1-p)^{k-2}$. Thus $f''(p) = 0 \iff p(1-p)^{k-1} = \frac{1}{k(k-1)}$. Note that $f''(p)$ has only two roots in $[0,1]$, hence $f'(p)$ has at most 3 roots in $[0,1]$. It can be verified that for sufficiently large k, $f'(p)$ indeed has three roots at $p = \Theta(1/2^k)$, $\Theta(\log k/k)$, and $1/2 - \Theta(k/2^k)$. At all these three values of p, $f(p) = O(1/2^k)$. Hence $E[Z] \leq 2^{O(n/2^k)}$.   □

3   Planted k-SAT and the PPZ Algorithm

In this section, we establish that the PPZ algorithm solves random planted k-SAT instances faster than $2^{n-n/O(k)}$ time.

Reminder of Theorem 1. *There is a randomized algorithm that given a planted k-SAT instance F sampled from $P(n,k,m)$ with $m > 2^{k-1}\ln(2)$, outputs a satisfying assignment to F in $2^{n\left(1-\Omega\left(\frac{\log k}{k}\right)\right)}$ time with $1 - 2^{-\Omega\left(n\left(\frac{\log k}{k}\right)\right)}$ probability (over the planted k-SAT distribution and the randomness of the algorithm).*

We will actually prove a stronger claim:

For any σ and F sampled from $P(n,k,m,\sigma)$, we can find a set S of $2^{n\left(1-\Omega\left(\frac{\log k}{k}\right)\right)}$ variable assignments in $2^{n\left(1-\Omega\left(\frac{\log k}{k}\right)\right)}$ time, such that $\sigma \in S$ with probability $1 - 2^{-\Omega\left(n\left(\frac{\log k}{k}\right)\right)}$ (the probability is over the planted k-SAT distribution and the randomness of the algorithm).

Theorem 1 yields an algorithm that (always) finds a solution for k-SAT instance F sampled from $P(n,k,m)$, and runs in *expected* time $2^{n\left(1-\Omega\left(\frac{\log k}{k}\right)\right)}$. In fact, the algorithm of Theorem 1 is a simplification of the PPZ algorithm [17], a well-known worst case algorithm for k-SAT. PPZ runs in polynomial time, and outputs a SAT assignment (on any satisfiable k-CNF) with probability $p \geq 2^{-n+n/O(k)}$. It can be repeatedly run for $O(n/p)$ times to obtain a worst-case algorithm that is correct whp. We consider a simplified version which is sufficient for analyzing planted k-SAT:

Algorithm 1. Algorithm for planted k-SAT

1: **procedure** SIMPLE-PPZ(F)
2:     **while** $i \leq n$ **do**
3:         **if** there is a unit clause C in the formula **then**
4:             Assign the variable in C so that C is true
5:         **else if** x_i is unassigned **then**
6:             Assign x_i randomly. Set $i \leftarrow i + 1$
7:         **else**
8:             Set $i \leftarrow i + 1$
9:     Output the assignment if it satisfies F.

Our Simple-PPZ algorithm (Algorithm 1) only differs from PPZ in that PPZ also performs an initial random permutation of variables. For us, a random permutation is unnecessary: a random permutation of the variables in the planted k-SAT distribution yields the same distribution of instances. That is, the original PPZ algorithm would have the same behavior as Simple-PPZ on random instances.

We will start with a few useful definitions.

Definition 1 ([17]). *A clause C is* critical *with respect to variable x_i and SAT assignment σ if x_i is the only variable in C whose corresponding literal is satisfied by σ.*

Definition 2. *A variable x_i in F is* good *for an assignment σ if there is a clause C in F which is critical with respect to x_i and σ, and i is the largest index among all variables in C. We say that x_i is* good *with respect to C in such a case. A variable which is not good is called* bad.

Observe that for every good variable x_i, if all variables x_j for $j < i$ are assigned correctly with respect to σ, then Simple-PPZ sets x_i correctly, due to the unit clause rule. As such, given a formula F with z good variables for σ, the probability that Simple-PPZ finds σ is at least $2^{-(n-z)}$: if all $n - z$ bad variables are correctly assigned, the algorithm is forced to set all good variables correctly as well. Next, we prove a high-probability lower bound on the number of good variables in a random planted k-SAT instance.

Lemma 6. *For $m > n2^{k-1}\ln 2$, a planted k-SAT instance sampled from $P(n, k, m, \sigma)$ has $\Omega(n \log k / k)$ good variables with respect to σ, with probability $1 - 2^{-\Omega\left(\frac{n \log k}{k}\right)}$.*

Proof. Let $F \in_r P(n, k, m, \sigma)$ and let L be the last (when sorted by index) $n \ln k/(2k)$ variables. Let L_g, L_b be the good and bad variables respectively, with respect to σ, among the variables in L. Let E be the event that $|L_g| \leq n \ln k/(500k)$. Our goal is to prove a strong upper bound on the probability that E occurs. For all $x_i \in L$, we have that $i \geq n(1 - \ln k/(2k))$. Observe that if a clause C is such that $x_i \in L_b$ is good with respect to C, then C does not occur in F. We will lower bound the probability of such a clause occurring in F, with respect to a fixed variable $x_i \in L$. Recall that in planted k-SAT, each clause is drawn uniformly at random from the set of clauses satisfied by σ. Fixing σ and a variable x_i and sampling one clause C, we get that

$$\Pr_{C \text{ which satisfies } \sigma}[x_i \in L \text{ is good with respect to } C]$$

$$= \frac{\text{number of clauses for which } x_i \in L \text{ is good}}{\text{total number of clauses satisfying } \sigma} = \frac{\binom{i-1}{k-1}}{\binom{n}{k}(2^k - 1)}$$

$$\geq \frac{1}{2} \left(\frac{i}{n}\right)^{k-1} \frac{k}{2^k n} \, [\text{As } i \geq n(1 - \ln k/(2k))]$$

$$\geq \frac{1}{2} \left(\frac{i}{n}\right)^{k} \frac{k}{2^k n} \geq \frac{1}{2} \left(1 - \frac{\ln k}{2k}\right)^{k} \frac{k}{2^k n} \, [\text{As } i \geq n(1 - \ln k/(2k))]$$

$$\geq \frac{1}{2} \left(e^{-\ln k/k}\right)^{k} \frac{k}{2^k n} \, [\text{As } k \text{ is large, and } e^{-w} \leq 1 - w/2 \text{ for small enough } w > 0]$$

$$\geq \frac{1}{2^{k+1} n}$$

If the event E is true, then $|L_b| > n \ln k/(4k)$. Therefore, *every time we sample a clause C*, the probability that C makes some variable $x_i \in L_b$ good is at least $\frac{\ln k}{k 2^{k+3}}$, as the sets of clauses which make different variables good are disjoint sets. Now we upper bound the probability of E occurring:

$$\Pr[E] \leq \sum_{i=1}^{n \ln k/(500k)} \Pr[\text{exactly } i \text{ vars among the last } n \ln k/(2k) \text{ vars are good}]$$

$$\leq \sum_{i=1}^{n \ln k/(500k)} \binom{n \ln k/(2k)}{i} \left(1 - \frac{\ln k}{k 2^{k+3}}\right)^{m}$$

$$\leq n \binom{n \ln k/(2k)}{n \ln k/(500k)} \left(1 - \frac{\ln k}{k 2^{k+3}}\right)^{n 2^{k-1} \ln 2} \quad .[\text{As } m > n 2^{k-1} \ln 2]$$

Applying the inequality $1 - x \leq e^{-x}$ for $x > 0$, the above is at most

$$n \binom{n \ln k/(2k)}{n \ln k/(500k)} \left(e^{-\frac{\ln k}{k 2^{k+3}}}\right)^{n 2^{k-1} \ln 2} \leq n \binom{n \ln k/(2k)}{n \ln k/(500k)} \left(2^{-\frac{n \ln k}{16k}}\right) \leq 2^{-\delta \frac{n \ln k}{k}}$$

for appropriately small but constant $\delta > 0$, which proves the lemma statement. $\square$

We are now ready to prove Theorem 1.

Proof of Theorem 1. By Lemma 6, we know that with probability at least $1 - p$ for $p = 2^{-\Omega(n(\frac{\log k}{k}))}$, the number of good variables with respect to a hidden planted solution σ in F is at least $\gamma n \log k/k$ for a constant $\gamma > 0$. For such instances, a single run of PPZ outputs σ with probability at least $2^{-n(1-\gamma \log k/k)}$. Repeating

PPZ for $\mathrm{poly}(n)2^{n(1-\gamma\log k/k)}$ times implies a success probability at least $1-1/2^n$. Hence the overall error probability is at most $p + 1/2^n \leq 2^{-\Omega(n(\frac{\log k}{k}))}$. $\square$

We proved that PPZ runs in time $2^{n(1-\Omega(\frac{\log k}{k}))}$ when m is "large enough", i.e., $m > n2^{k-1}\ln 2$. When $m \leq n2^{k-1}\ln 2$, we observe that the much simpler approach of merely randomly sampling assignments already works, whp! This is because by Corollary 1 (in the Preliminaries), the number of solutions of $F \in_r P(n,k,m)$ for $m \leq n2^{k-1}\ln 2$ is at least $2^{n/2}2^{-O(nk/2^k)}$ whp. When this event happens, randomly sampling $\mathrm{poly(n)}2^{n/2}2^{O(nk/2^k)}$ assignments will uncover a solution whp.

4 Reducing from Random k-SAT to Planted Random k-SAT

In this section we observe a reduction from random k-SAT to planted k-SAT, which yields the desired algorithm for random k-SAT (see Theorem 3). The following lemma is similar to results in Achlioptas [1], and we present it here for completeness.

Lemma 7 ([1]). *Suppose there exists an algorithm A for planted k-SAT $P(n,k,m)$, for some $m \geq 2^k\ln 2(1 - f(k)/2)n$, which finds a solution in time $2^{n(1-f(k))}$ and with probability $1 - 2^{-nf(k)}$, where $1/k < f(k) = o_k(1)^2$. Then given a random k-SAT instance sampled from $R^+(n,k,m)$, we can find a satisfiable solution in $2^{n(1-\Omega(f(k)))}$ time with $1 - 2^{-n\Omega(f(k))}$ probability.*

Proof. Let F be sampled from $R^+(n,k,m)$, and let Z denote its number of solutions with s its expected value. As $f(k) > 1/k$ and $m \geq 2^k\ln 2(1-f(k)/2)n$, Lemmas 3 and 5 together imply that $s \leq 2 \cdot 2^{nf(k)/2}$.

We will now run Algorithm A. Note that if Algorithm A gives a solution it is correct hence we can only have error when the formula is satisfiable but algorithm A does not return a solution. We will now upper bound the probability of A making an error.

$$\Pr_{F\in R^+(n,k,m),A}[A \text{ returns no solution}]$$

$$\leq \sum_{\sigma\in\{0,1\}^n} \Pr_{F\in R^+(n,k,m),A}[\sigma \text{ satisfies } F \text{ but } A \text{ returns no solution}]$$

$$\leq \sum_{\sigma\in\{0,1\}^n} \Pr_{F\in R^+(n,k,m),A}[A \text{ returns no sol} \mid \sigma \text{ satisfies F}] \Pr_{F\in R^+(n,k,m)}[\sigma \text{ satisfies F}]$$

$$\leq \sum_{\sigma\in\{0,1\}^n} \Pr_{F\in P(n,k,m,\sigma),A}[A \text{ returns no solution}] \Pr_{F\in R^+(n,k,m)}[\sigma \text{ satisfies F}]$$

where the last inequality used the fact that $R^+(n,k,m)$ conditioned on having σ as a solution is the distribution $P(n,k,m,\sigma)$. Now note that

---

[2] Note we can assume wlog that $f(k) > 1/k$, as we already have a $2^{n(1-1/k)}$ algorithm for worst-case k-SAT.

$\Pr_{F \in R^+(n,k,m)}[\sigma \text{ satisfies } F] = s/2^n$, and $P(n,k,m) = P(n,k,m,\sigma)$, where σ is sampled uniformly from $\{0,1\}^n$. Hence the expression simplifies to

$$\frac{s}{2^n}\left(2^n \Pr_{F \in P(n,k,m),A}[A \text{ does not return a solution}]\right)$$
$$= s \Pr_{F \in P(n,k,m),A}[A \text{ does not return a solution}].$$

Since $s \leq 2 \cdot 2^{nf(k)/2}$, the error probability is $\leq 2 \cdot 2^{nf(k)/2} 2^{-nf(k)} \leq 2 \cdot 2^{-nf(k)/2} = 2^{-\Omega(nf(k))}$. $\qquad\square$

Next, we give another reduction from random k-SAT to planted k-SAT. This theorem is different from Lemma 7, in that, given a planted k-SAT algorithm that works in a certain regime of m, it implies a random k-SAT algorithm for *all* values of m.

Reminder of Theorem 2. *Suppose there is an algorithm A for planted k-SAT $P(n,k,m)$, for all $m \geq 2^k \ln 2(1 - f(k)/2)n$, which finds a solution in time $2^{n(1-f(k))}$ and with probability $1 - 2^{-nf(k)}$, where $1/k < f(k) = o_k(1)$. Then for any m', given a random k-SAT instance sampled from $R^+(n,k,m')$, a satisfying assignment can be found in $2^{n(1-\Omega(f(k)))}$ time whp.*

Proof. Let F be sampled from $R^+(n,k,m)$, and let Z denote its number of solutions with s its expected value. The expected number of solutions of F' sampled from $R(n,k,m')$ serves as a lower bound for s. Hence if $m' \leq 2^k \ln 2(1 - f(k)/2)n \leq \alpha_d n$, then $s > 2^{nf(k)/2}$ and furthermore, as we have $f(k) > 1/k$, Lemma 3 implies that, $\lim_{n\to\infty}\Pr[Z < s/2^{O(nk/2^k)}] = 0$. So if we randomly sample $O(2^n 2^{O(nk/2^k)}/s) \leq 2^{n(1-\Omega(f(k)))}$ assignments, one of them will satisfy F whp. Otherwise if $m' \geq 2^k \ln 2(1 - f(k)/2)n$ then we can use Lemma 7 to solve it in required time. $\qquad\square$

Finally, we combine Algorithm 1 for planted k-SAT and the reduction in Theorem 2 to obtain an algorithm for finding solutions of random k-SAT (conditioned on satisfiability). This disproves Super-SETH for random k-SAT.

Reminder of Theorem 3. *Given a random k-SAT instance F sampled from $R^+(n,k,m)$ we can find a solution in $2^{n(1-\Omega(\frac{\log k}{k}))}$ time whp.*

Proof. By Theorem 1 we have an algorithm for planted k-SAT running in $2^{n(1-\Omega(\frac{\log k}{k}))}$ time with $1 - 2^{-\Omega(n(\frac{\log k}{k}))}$ probability for all $m > (2^{k-1} \ln 2)n$. This algorithm satisfies the required conditions in Theorem 2 with $f(k) = \Omega(\log k/k)$ for large enough k. The implication in Theorem 2 proves the required statement. $\qquad\square$

Just as in the case of planted k-SAT, when $m < n(2^k \ln 2 - k)$ we can find solutions of $R^+(n,k,m)$ whp, by merely random sampling assignments. The correctness of random sampling follows from Lemma 3.

5   k-SAT and Unique k-SAT

In this section we give a randomized reduction from k-SAT to Unique k-SAT which implies their equivalence for Super Strong ETH:

Reminder of Theorem 4. *If Unique k-SAT is solvable in $2^{(1-f(k)/k)n}$ time for some unbounded $f(k)$, then k-SAT is solvable in $2^{(1-f(k)/k+O((\log f(k))/k))n}$ time.*

We start with a slight modification of the Valiant-Vazirani lemma.

Lemma 8 (Weighted-Valiant-Vazirani). *Let $S \subseteq \{0,1\}^k = R$ be a set of assignments on variables $x_1, x_2, \ldots x_k$, with $2^{j-1} \leq |S| < 2^j$. Suppose that for each $s \in S$ there exists a weight $w_s \in \mathbb{Z}^+$, and let $\bar{w}$ denote the average weight over all $s \in S$. There is a randomized polytime algorithm **Weighted-Valiant-Vazirani** that on input (R, j) outputs a matrix $A \in \mathbb{F}_2^{j \times n}$ and a vector $b \in \mathbb{F}_2^j$ such that*

$$\Pr_{A,b} [|\{x \mid Ax = b \wedge x \in S\}| = 1, w_s \leq 2\bar{w}] > \frac{1}{16}.$$

*If the condition in the probability expression is satisfied, we say **Weighted-Valiant-Vazirani** on (R, j) has succeeded.*

Proof. The original Valiant-Vazirani Lemma [22] gives a randomized polytime algorithm to generate A, b such that for all $s \in S$, $\Pr_{A,b}[\{s\} = \{x \mid Ax = b \wedge x \in S\}] > \frac{1}{8|S|}$. Moreover, by Markov's inequality, we have $\Pr_{s \in S}[w_s \leq 2\bar{w}] \geq 1/2$. Hence the set of $s \in S$ with $w_s \leq 2\bar{w}$ has size at least $|S|/2$. This implies $\Pr_{A,b}[\exists s, \{s\} = \{x \mid Ax = b \wedge x \in S\}, w_s \leq 2\bar{w}] > \left(\frac{1}{8|S|}\right)\left(\frac{|S|}{2}\right) = \frac{1}{16}$. $\quad\square$

Proof of Theorem 4. Let A be an algorithm for Unique k-SAT which runs in time $2^{(1-f(k)/k)n}$.

Let S be the set of SAT assignments to F. Suppose $|S| \geq 2^{nf(k)/k}n$. Then the probability that the random search in lines 1 to 4 never finds a solution is

$$(1 - n2^{nf(k)/k}/2^n)^{2^{n(1-f(k)/k)}} \leq e^{-n}.$$

Thus if $|S| \geq 2^{nf(k)/k}n$ then the algorithm finds a solution whp. From now on, we assume $|S| < 2^{nf(k)/k}n$.

In line 6, we define a sequence of probabilities $p_1, p_2, \ldots, p_k$. Note that

$$\sum_{i=1}^{k} p_i = \sum_{i=1}^{f(k)} p_i + \sum_{i=f(k)+1}^{k} p_i \leq 1/2 + 1/(2f(k)) \sum_{j=1}^{\infty} (1/2f(k))^{j/f(k)}$$

$$\leq \frac{1}{2} + \frac{1}{f(k)(1 - (1/2f(k))^{1/f(k)})} \leq 1,$$

as $f(k)$ is unbounded, and $\lim_{x \to \infty} x(1 - (1/2x)^{1/x}) = \infty$.

Algorithm 2. Algorithm for k-SAT.

Input: k-SAT formula F

We assume that there is an algorithm A for Unique k-SAT running in time $2^{n(1-f(k)/k)}$.

1: **for** $i = 0$ to $2^{n(1-f(k)/k)}$ **do**
2:    sample random solution s
3:    **if** s satisfies F **then**
4:       Return s
5: Divide n variables into n/k equal parts $R_1, R_2 \ldots R_{n/k}$ and let x^i denote the variables in set R_i
6: Define $p = p_1 = p_2 \ldots = p_{f(k)} = 1/(2f(k))$ and $p_j = p^{j/f(k)}$ for $f(k) \leq j \leq k$
7: $F_0 = F$
8: **for** $u = 1$ to $2^{cn \log(f(k))/k}$ **do**
9:    **for** $i = 1$ to n/k **do**
10:       Sample z_i from $[k]$ choosing $z_i = j$ with probability p_j
11:       $(A_i, b_i) = $ **Weighted-Valiant-Vazirani**(R_i, z_i)
12:       $F_i = F_{i-1} \wedge (A_i x^i = b_i)$
13:    $s = A(F_{n/k})$
14:    Return s if it satisfies F
15: Return unsatisfiable

We will now analyze the ith run of the loop from line 9 to line 14. Let $S_0 = S$, and let S_i be the set of solutions to the formula F_i defined in line 12.

Let E_i be the event that:

1. $2^{z_i-1} \leq |\{s_{|R_i} \mid s \in S_{i-1}\}| < 2^{z_i}$.   [As defined in line 10]
2. for all $s \in S_i$, the restriction on R_i is the same, i.e., $|\{s_{|R_i} \mid s \in S_i\}| = 1$.
3. $|S_{i-1}|/|S_i| \geq 2^{z_i-2}$, $|S_i| \neq 0$.

In Line 11 we apply **Weighted-Valiant-Vazirani** to (R_i, z_i) with the set of assignments being $\{s_{|R_i} \mid s \in S_{i-1}\}$ where an assignment v has weight $w_v = |\{v = s_{|R_i} \mid s \in S_{i-1}\}|$. For **Weighted-Valiant-Vazirani** to apply, we require that z_i indeed represents an estimate of number of possible assignments to variables of R_i in a satisfying assignment i.e. $2^{z_i-1} \leq |\{s_{|R_i} \mid s \in S_{i-1}\}| < 2^{z_i}$ which is exactly the condition 1 in E_i. If the call to **Weighted-Valiant-Vazirani** succeeds, then we have that only a unique assignment to R_i remains, i.e, $|\{s_{|R_i} \mid s \in S_i\}| = 1$ which is the condition 2 of E_i. Similarly condition 3 is also implied by the success of **Weighted-Valiant-Vazirani**.

Let y_i satisfy $2^{y_i-1} \leq |\{s_{|R_i} \mid s \in S_{i-1}\}| < 2^{y_i}$. Then for E_i to be true we need that the sample z_i is equal to y_i, and **Weighted-Valiant-Vazirani** on (R_i, z_i) succeeds.

Let $E = \bigcap_i E_i$. If event E occurs, then the restrictions of all solutions on each R_i's are the same, and there is a solution as $|S_{n/k}| \neq 0$, hence there is a unique satisfying assignment. We wish to lower bound the probability of E occurring.

$$\Pr[E] = \prod_i \Pr[E_i \mid \bigwedge_{j<i} E_j]$$

$$\geq \prod_i \Pr[z_i = y_i \mid \bigwedge_{j<i} E_j] \cdot \prod_i \Pr[\mathbf{WVV}(R_i, z_i) \mid \forall j < i, E_j]$$

$$\geq \prod_i p_{y_i} \prod_i \left(\frac{1}{16}\right) \qquad \text{[By Lemma 8]} \tag{1}$$

$$\geq 16^{-n/k} \prod_i p_{y_i}$$

When E holds, $|S| = |S_0| = \prod_i |S_{i-1}|/|S_i|$, as $|S_{n/k}| = 1$, Furthermore $\prod_i |S_{i-1}|/|S_i| \geq \prod_i 2^{y_i - 2}$, by condition 3. Since $|S| \leq 2^{nf(k)/k}n$, we have $\prod_i 2^{y_i-2} \leq 2^{nf(k)/k}n$. Taking logarithms, $\sum_i y_i \leq O(n/k) + nf(k)/k \leq O(nf(k)/k)$. Therefore

$$\Pr[E] \geq 16^{-n/k} \prod_i p_{y_i} \text{[Restating equation (1)]}$$

$$\geq 16^{-n/k} \prod_{y_i \leq f(k)} p_{y_i} \prod_{y_i > f(k)} p_{y_i}$$

$$\geq 16^{-n/k} \cdot (1/2f(k))^{n/k} \cdot \prod_{y_i > f(k)} (1/2f(k))^{(y_i/f(k))} \tag{2}$$

$$\geq 16^{-n/k} \cdot (1/2f(k))^{n/k} \cdot (1/2f(k))^{\sum_{y_i > f(k)}(y_i/f(k))}$$

$$\geq 16^{-n/k} \cdot (1/2f(k))^{n/k} \cdot (1/2f(k))^{O(n/k)}$$

$$\geq 16^{-n/k} \cdot 2^{-O(n \log f(k)/k)} \geq 2^{-O(n \log f(k)/k)}.$$

As mentioned earlier, if E occurs, then there is a unique SAT assignment and it is found by our Unique k-SAT algorithm A. The probability E does not happen over all $2^{cn(\log f(k))/k}$ runs of the loop on line 8 is at most $1 - 2^{-O(n(\log f(k))/k)})2^{cn(\log f(k))/k} \ll 2^{-n}$, for sufficiently large c. The total running time is $2^{n(1-f(k)/k)} + 2^{cn(\log f(k))/k} \cdot 2^{(1-f(k)/k)n} \leq 2^{(1-f(k)/k+O((\log f(k))/k))n}$. $\square$

Theorem 4 immediately implies an "ultra fine-grained" equivalence between k-SAT and Unique-k-SAT:

Corollary 2. *Unique k-SAT is in $2^{(1-\omega_k(1/k))n}$ time $\Leftrightarrow$ k-SAT is in $2^{(1-\omega_k(1/k))n}$ time.*

Acknowledgement. We thank Erik Demaine for organizing an open problems session for 6.S078 at MIT, during which some results of this paper were proved.

References

1. Achlioptas, D., Coja-Oghlan, A.: Algorithmic barriers from phase transitions. In: IEEE 49th Annual IEEE Symposium on Foundations of Computer Science, FOCS 2008, pp. 793–802. IEEE (2008)

2. Ben-Sasson, E., Bilu, Y., Gutfreund, D.: Finding a randomly planted assignment in a random 3CNF (2002, manuscript)
3. Brakensiek, J., Guruswami, V.: Bridging between 0/1 and linear programming via random walks. CoRR abs/1904.04860 (2019). http://arxiv.org/abs/1904.04860
4. Calabro, C., Impagliazzo, R., Kabanets, V., Paturi, R.: The complexity of unique k-SAT: an isolation lemma for k-CNFs. J. Comput. Syst. Sci. **74**(3), 386–393 (2008)
5. Calabro, C., Impagliazzo, R., Paturi, R.: The complexity of satisfiability of small depth circuits. In: Chen, J., Fomin, F.V. (eds.) IWPEC 2009. LNCS, vol. 5917, pp. 75–85. Springer, Heidelberg (2009). https://doi.org/10.1007/978-3-642-11269-0_6
6. Chan, T.M., Williams, R.: Deterministic apsp, orthogonal vectors, and more: quickly derandomizing razborov-smolensky. In: Proceedings of the Twenty-Seventh Annual ACM-SIAM Symposium on Discrete Algorithms, SODA 2016, Arlington, VA, USA, January 10–12 2016, pp. 1246–1255 (2016)
7. Chao, M.T., Franco, J.: Probabilistic analysis of a generalization of the unit-clause literal selection heuristics for the k satisfiability problem. Inf. Sci.: Int. J. **51**(3), 289–314 (1990)
8. Coja-Oghlan, A.: A better algorithm for random k-SAT. SIAM J. Comput. **39**(7), 2823–2864 (2010)
9. Coja-Oghlan, A., Krivelevich, M., Vilenchik, D.: Why almost all k-CNF formulas are easy (2007, manuscript)
10. Cook, S.A., Mitchell, D.G.: Finding hard instances of the satisfiability problem: a survey. In: Satisfiability Problem: Theory and Applications, Proceedings of a DIMACS Workshop, Piscataway, New Jersey, USA, 11–13 March 1996, pp. 1–18 (1996)
11. Ding, J., Sly, A., Sun, N.: Proof of the satisfiability conjecture for large k. In: Proceedings of the Forty-Seventh Annual ACM on Symposium on Theory of Computing, STOC 2015, Portland, OR, USA, 14–17 June 2015, pp. 59–68 (2015)
12. Feige, U.: Relations between average case complexity and approximation complexity. In: Proceedings on 34th Annual ACM Symposium on Theory of Computing, 19–21 May 2002, Montréal, Québec, Canada, pp. 534–543 (2002)
13. Hertli, T.: 3-SAT faster and simpler - unique-SAT bounds for PPSZ hold ingeneral. SIAM J. Comput. **43**(2), 718–729 (2014). https://doi.org/10.1137/120868177
14. Impagliazzo, R., Paturi, R.: On the complexity of k-SAT. J. Comput. Syst. Sci. **62**(2), 367–375 (2001). https://doi.org/10.1006/jcss.2000.1727
15. Krivelevich, M., Vilenchik, D.: Solving random satisfiable 3CNF formulas in expected polynomial time. In: Proceedings of the Seventeenth Annual ACM-SIAM Symposium on Discrete Algorithms, SODA 2006, Miami, Florida, USA, 22–26 January 2006, pp. 454–463 (2006)
16. Paturi, R., Pudlák, P., Saks, M.E., Zane, F.: An improved exponential-time algorithm for k-SAT. J. ACM **52**(3), 337–364 (2005). https://doi.org/10.1145/1066100.1066101
17. Paturi, R., Pudlák, P., Zane, F.: Satisfiability coding lemma. Chicago J. Theor. Comput. Sci. 1999 (1999). http://cjtcs.cs.uchicago.edu/articles/1999/11/contents.html
18. Pudlák, P., Scheder, D., Talebanfard, N.: Tighter hard instances for PPSZ. In: 44th International Colloquium on Automata, Languages, and Programming, ICALP 2017, 10–14 July 2017, Warsaw, Poland, pp. 85:1–85:13 (2017)
19. Schöning, U.: A probabilistic algorithm for k-SAT and constraint satisfaction problems. In: 40th Annual Symposium on Foundations of Computer Science, FOCS 1999, 17–18 October 1999, New York, NY, USA, pp. 410–414 (1999)

20. Selman, B., Mitchell, D.G., Levesque, H.J.: Generating hard satisfiability problems. Artif. intell. **81**(1–2), 17–29 (1996)
21. Traxler, P.: The time complexity of constraint satisfaction. In: Grohe, M., Niedermeier, R. (eds.) IWPEC 2008. LNCS, vol. 5018, pp. 190–201. Springer, Heidelberg (2008). https://doi.org/10.1007/978-3-540-79723-4_18
22. Valiant, L.G., Vazirani, V.V.: NP is as easy as detecting unique solutions. Theor. Comput. Sci. **47**(3), 85–93 (1986). https://doi.org/10.1016/0304-3975(86)90135-0
23. Vilenchik, D.: Personal communication
24. Williams, R.: Circuit analysis algorithms. Talk at Simons Institute for Theory of Computing, August 2015. https://youtu.be/adJvi7tL-qM?t=925

Author Index

Printed in the United States
By Bookmasters

Printed in the United States
By Bookmasters